Contemporary Clinical Neuroscience

Series Editor

Mario Manto, Division of Neurosciences, Department of Neurology
CHU-Charleroi, Charleroi, Belgium, University of Mons
Mons, Belgium

Contemporary Clinical Neurosciences bridges the gap between bench research in the neurosciences and clinical neurology work by offering translational research on all aspects of the human brain and behavior with a special emphasis on the understanding, treatment, and eradication of diseases of the human nervous system. These novel, state-of-the-art research volumes present a wide array of preclinical and clinical research programs to a wide spectrum of readers representing the diversity of neuroscience as a discipline. The book series considers proposals from leading scientists and clinicians. The main audiences are basic neuroscientists (neurobiologists, neurochemists, geneticians, experts in behavioral studies, neurophysiologists, neuroanatomists), clinicians (including neurologists, psychiatrists and specialists in neuroimaging) and trainees, graduate students, and PhD students.

Volumes in the series provide in-depth books that focus on neuroimaging, ADHD (attention deficit hyperactivity disorder and other neuropsychiatric disorders, neurodegenerative diseases, G protein receptors, sleep disorders, addiction issues, cerebellar disorders, and neuroimmune diseases. The series aims to expand the topics at the frontiers between basic research and clinical applications. Each volume is available in both print and electronic form.

Maria Gabriela Valdovinos
Editor

Intellectual and Developmental Disabilities

A Dynamic Systems Approach

Springer

Editor
Maria Gabriela Valdovinos
Department of Psychology and Neuroscience
Drake University
Des Moines, IA, USA

ISSN 2627-535X ISSN 2627-5341 (electronic)
Contemporary Clinical Neuroscience
ISBN 978-3-031-66931-6 ISBN 978-3-031-66932-3 (eBook)
https://doi.org/10.1007/978-3-031-66932-3

This Springer imprint is published by the registered company Springer Nature Switzerland AG
The registered company address is: Gewerbestrasse 11, 6330 Cham, Switzerland

To my family for their unwavering support, my mentor Steven R. Schroeder who fostered an appreciation for the intersection of biology and behavior, and for the individuals and families who have taught me so much over the years.

Preface

This publication was born from the need to find a text for my Psychology of Developmental Disabilities course which often provided an introduction to the field of developmental disabilities to upper-class, undergraduate psychology students. With this book, students (upper graduate, graduate, and professional students) are introduced to an integrated perspective of developmental disabilities across fields of research and clinical practice.

The book title includes the term "intellectual disability," to encompass the range of diagnostic terms used. For example, the Diagnostic and Statistical Manual of Mental Disorders, Fifth Edition, Text Revision (DSM-5-TR), the term used is "Intellectual Developmental Disorder (Intellectual Disability)," and within the International Classification of Mortality and Morbidity Statistics (ICD-11), the term used is "Disorders of Intellectual Development." Additionally, the title includes the term "developmental disability" in reference to the impact several factors may have on physical, cognitive, language, and behavioral development. Many of the conditions discussed in this text relate to neurodevelopmental disorders (conditions impacting brain and neurological development).

The initial chapters of this book introduce intellectual and developmental disabilities with Tenorio and Arango describing the definition and assessment process for a diagnosis of intellectual disability. Wehmeyer then provides an overview of the history of the field of intellectual and developmental disability tracing social movements and policy changes. Magaña, Yu, and Hussein Al-Mamari discuss current issues facing individuals with intellectual and developmental disabilities, their family members, and caregivers.

The chapters that follow discuss the biological contributors to intellectual and developmental disabilities, beginning with an overview of genetic and epigenetic factors by Kliethermes and Campbell, respectively. Additionally, neurobiological development is described by Thompson. Schrier Vergano then describes inborn errors of metabolism and the impact these conditions have on development. Concerning prenatal factors that influence development, Bick, Spann, and colleagues explore maternal conditions, teratogenic exposure, and environmental conditions impact on development. Gigliotti-Manessis, Rutter, and colleagues discuss non-genetic, perinatal factors that influence the trajectory of development.

Continuing with the examination of the lived experiences of individuals with intellectual and developmental disabilities, Hangen, Weyman, and Valdovinos discuss the common occurrence of challenging behavior (e.g., aggression, self-injury), the factors that contribute to their presence, and behavioral approaches to evaluation and treatment. Halvorsen, Kildahl, and Helverschou describe mental health disorders in intellectual and developmental disabilities, addressing challenges in assessment and treatment. Next discussed, although often overlooked, is the sexual and reproductive health of women with intellectual and developmental disabilities by Brown and Tarasoff.

Three disorders were selected for closer focus. The first condition described is Down syndrome, and Esbensen, Schworer, and Hartley provide an overview of the lifespan of individuals with Down syndrome. The second condition described by Arnold, Roberts, and colleagues is fragile X syndrome, the most commonly inherited contributor to intellectual disability. Finally, Gordon and Dimitropoulos provide an overview of autism spectrum disorder, an increasingly common neurodevelopmental disorder.

The final two chapters of the book discuss issues related to research on intellectual and developmental disabilities. Lee, O'Brien, and Grosman provide a review of neuroimaging technologies and what can be learned from these approaches. And, Yang, Merrill, and Conde discuss research methods often employed to study factors related to intellectual and developmental disabilities.

Des Moines, IA, USA Maria Gabriela Valdovinos

Acknowledgments

This work was supported by the College of Arts and Sciences at Drake University.

Contents

Contributors

Paulina Sofía Arango U. Department of Psychology, Universidad de los Andes, Santiago, Chile

Millenium Institute for Care Research (MICARE), Santiago, Chile

Holley P. Arnold Department of Psychology, Institute for Mind and Brain, University of South Carolina, Columbia, SC, USA

Johanna Bick Department of Psychology, University of Houston, Houston, TX, USA

Conner J. Black Department of Psychology, Institute for Mind and Brain, University of South Carolina, Columbia, SC, USA

Hilary K. Brown Department of Health & Society, University of Toronto Scarborough, Toronto, ON, Canada

Dalla Lana School of Public Health, University of Toronto, Toronto, ON, Canada

Maria Catalina Camacho Department of Psychiatry, Washington University in St. Louis, St. Louis, MO, USA

Daniel B. Campbell Department of Pediatrics and Human Development, Michigan State University, Grand Rapids, MI, USA

Sonia Conde Department of Psychology, Montclair State University, Montclair, NJ, USA

Kevin Cook Developing Brain Institute, Children's National Hospital, Washington, DC, USA

Dana DeMaster Univeristy of Texas Health Houston, Houston, TX, USA

Anastasia Dimitropoulos Department of Psychological Sciences, Case Western Reserve University, Cleveland, OH, USA

Anna J. Esbensen Division of Developmental and Behavioral Pediatrics, Cincinnati Children's Hospital Medical Center, Cincinnati, OH, USA

Department of Pediatrics, University of Cincinnati College of Medicine, Cincinnati, OH, USA

Fetal, Infant, and Toddler Neuroimaging Group (FIT'NG)

Angela Gigliotti-Manessis Department of Health Studies and Applied Educational Psychology, Teachers College, Columbia University, New York, NY, USA

Rachel A. Gordon Department of Psychological Sciences, Case Western Reserve University, Cleveland, OH, USA

Hannah E. Grosman Department of Psychological and Brain Sciences, Drexel University, Philadelphia, PA, USA

Marianne Berg Halvorsen Department of Pediatric Rehabilitation, University Hospital of North Norway, Tromsø, Norway

Marcella Hangen Department of Behavior Analysis, School of Sciences and Health Professions, Simmons University, Boston, MA, USA

Rachel M. Hantman Department of Psychology, Institute for Mind and Brain, University of South Carolina, Columbia, SC, USA

Sigan L. Hartley Waisman Center, University of Wisconsin-Madison, Madison, WI, USA

Department of Human Development and Family Studies, University of Wisconsin-Madison, Madison, WI, USA

Sissel Berge Helverschou NevSom Norwegian Centre of Expertise for Neurodevelopmental Disorders and Hypersomnias, Oslo University Hospital, Oslo, Norway

Cristin M. Holland Department of Psychiatry, Vagelos College of Physicians and Surgeons, Columbia University, New York, NY, USA

Erin E. Hunt Department of Psychology, Institute for Mind and Brain, University of South Carolina, Columbia, SC, USA

Qusay S. Hussein Al-Mamari Steve Hicks School of Social Work, University of Texas at Austin, Austin, TX, USA

Sanjana Inala Department of Psychiatry, Vagelos College of Physicians and Surgeons, Columbia University, New York, NY, USA

Arvid Nikolai Kildahl NevSom Norwegian Centre of Expertise for Neurodevelopmental Disorders and Hypersomnias, Oslo University Hospital, Oslo, Norway

Regional Section Mental Health, Intellectual Disabilities/Autism, Oslo University Hospital, Oslo, Norway

Christopher L. Kliethermes Department of Psychology and Neuroscience, Drake University, Des Moines, IA, USA

Chandler E. Knott Department of Psychology, Institute for Mind and Brain, University of South Carolina, Columbia, SC, USA

Marta Korom Department of Psychological and Brain Science, University of Delaware, Newark, DE, USA

Nancy Raitano Lee Department of Psychological and Brain Sciences, Drexel University, Philadelphia, PA, USA

Danielle R. Lynch Department of Psychology, Institute for Mind and Brain, University of South Carolina, Columbia, SC, USA

Sandy Magaña Steve Hicks School of Social Work, University of Texas at Austin, Austin, TX, USA

Venus Mahmoodi Department of Psychiatry, Vagelos College of Physicians and Surgeons, Columbia University, New York, NY, USA

Edward C. Merrill Department of Psychology, The University of Alabama, Tuscaloosa, AL, USA

Livia Merrill Department of Psychology, University of Houston, Houston, TX, USA

Isabelle Mueller Department of Psychiatry, Vagelos College of Physicians and Surgeons, Columbia University, New York, NY, USA

Meghan O'Brien Department of Psychological and Brain Sciences, Drexel University, Philadelphia, PA, USA

Andrea Ortiz-Jimenez Department of Psychology, University of Houston, Houston, TX, USA

Jane E. Roberts Department of Psychology, Institute for Mind and Brain, University of South Carolina, Columbia, SC, USA

Natural Sciences, College of Arts and Sciences, University of South Carolina, Columbia, SC, USA

Tara M. Rutter Department of Psychiatry, Oregon Health & Science University, Portland, OR, USA

Samantha A. Schrier Vergano Division of Genetics, Seattle Children's Hospital, Edmonds, WA, USA

University of Washington, Seattle, WA, USA

Emily K. Schworer Waisman Center, University of Wisconsin-Madison, Madison, WI, USA

Thiviya Selvanathan Department of Pediatrics, BC Children's Hospital and University of British Columbia, Vancouver, BC, Canada

Department of Pediatrics, The Hospital for Sick Children, Toronto, ON, Canada

Marisa Spann Department of Psychiatry, Vagelos College of Physicians and Surgeons, Columbia University, New York, NY, USA

Lesley Tarasoff Department of Health & Society, University of Toronto Scarborough, Toronto, ON, Canada

Marcela Tenorio D. Department of Psychology, Universidad de los Andes, Santiago, Chile

Millenium Institute for Care Research (MICARE), Santiago, Chile

Barbara Thompson Department of Pediatrics and Human Development, College of Human Medicine, Michigan State University, Grand Rapids, MI, USA

Maria G. Valdovinos Department of Psychology and Neuroscience, Drake University, Des Moines, IA, USA

Michael L. Wehmeyer Department of Special Education and The Beach Center on Disability, University of Kansas, Lawrence, KS, USA

Jennifer R. Weyman Department of Special Education and Counseling, California State University, Los Angeles, Los Angeles, CA, USA

Yingying Yang Department of Psychology, Montclair State University, Montclair, NJ, USA

Amy Pei-Lung Yu Steve Hicks School of Social Work, University of Texas at Austin, Austin, TX, USA

About the Editor

Maria Gabriela Valdovinos, PhD, is Ellis and Nelle Levitt Distinguished Professor of Psychology, Drake University, Des Moines, Iowa. She received her doctorate in Developmental and Child Psychology from the University of Kansas and completed post-doctoral work at Vanderbilt University's Kennedy Center for Research on Human Development. Prior to beginning her studies in Kansas, Dr. Valdovinos worked in residential and day treatment settings with adults diagnosed with intellectual and developmental disabilities. Her experiences in these settings lead to her research interest in evaluating the pharmacological treatment of challenging behavior and the intersection between biological and behavioral events. Dr. Valdovinos is a Fellow of the American Association on Intellectual and Developmental Disabilities and of the American Psychological Association, Division 33, Intellectual and Developmental Disabilities/Autism Spectrum Disorder.

Assessment and Identification Process in Intellectual Disability

Marcela Tenorio D. and Paulina Sofía Arango U.

Abstract Intellectual disability is one of the most prevalent neurodevelopmental conditions, characterized by difficulties in reasoning and adaptive behaviors apparent during the developmental period. The identification process for intellectual disability involves using culturally adapted and normed standardized tests, alongside a complex clinical assessment conducted by professionals with sufficient knowledge and training. This enables them to accurately interpret assessment results and determine individualized support needs for each person. In this chapter, we commence with a definition of intellectual disability. We then provide an overview of the current standards for the identification process, encompassing the necessary assessment areas, available evaluation instruments, and clinical considerations essential for ensuring quality. Finally, we conclude the chapter with practical recommendations for conducting identification processes in developing or low-income countries.

Keywords Intellectual disability · Identification · Diagnosis · IQ · Adaptive behavior

Intellectual disability is one of the most prevalent neurodevelopmental conditions worldwide. According to available data, there is an estimated prevalence rate of 10.36 per 1000 people, with variations based on the country's level of development, increasing prevalence in middle- and low-income per capita countries (Maulik et al., 2011). Considering these data, the urgency to enhance the knowledge and competence of personnel responsible for the identification process becomes evident. This should be carried out with a full understanding of the international standard as a way of acknowledging the rights of everyone.

M. Tenorio D. (✉) · P. S. Arango U.
Department of Psychology, Universidad de los Andes, Santiago, Chile

Millenium Institute for Care Research (MICARE), Santiago, Chile
e-mail: mtenorio@uandes.cl

M. G. Valdovinos (ed.), *Intellectual and Developmental Disabilities*, Contemporary Clinical Neuroscience,
https://doi.org/10.1007/978-3-031-66932-3_1

This chapter provides a detailed review of the current identification process for intellectual disability as suggested by the international standard. The first part includes a definition of intellectual disability, followed by an examination of the complexities faced by a professional in the identification process. About these complexities, three sections are considered, including a discussion on IQ, the approach to adaptive behaviors as a central axis of the identification process, and the identification of the onset period as a verifier that it is a neurodevelopmental condition. The chapter concludes with two crucial sections: understanding the importance of determining the frequency and intensity of supports as a pivotal part of the identification process and a set of practical recommendations for professionals conducting these processes in developing or low-income countries.

Following an identification process based on a broad literacy of intellectual disability, in addition to being respectful of international standards, can be considered a minimum requirement in the practical process of fighting the stigma against persons with intellectual disability (Scior, 2016; Scior et al., 2013; Zeilinger et al., 2020) and ensuring the rights of people with disabilities as outlined in the Convention on the Rights of Persons with Disabilities (United Nations, 2006). This convention is a fundamental piece from the approach based on the rights of people with disabilities. It was led by the United Nations, presented in 2006, and has been ratified by 147 nations. In its content, it offers an approach from the social model of disability, announcing the explicit rights that must be guaranteed to this group.

Definition of Intellectual Disability

Intellectual disability is part of the neurodevelopmental conditions and is broadly addressed by the general construct of disability (Salvador-Carulla & Saxena, 2009). Disability is currently understood as the presence of long-term challenges at the motor, sensory, intellectual, developmental, or mental levels in an individual, which interact with environmental barriers, hindering their full participation on equal terms (Shakespeare, 2018; World Health Organization, 2019).

Intellectual disability was formerly referred to as mental retardation, but this term was completely replaced in international diagnostic manuals (both the American Psychiatric Association and the World Health Organization in the fifth and eleven versions, respectively). The replacement responds to the request of groups of individuals with intellectual disability and their stakeholders, who point out the stigmatization represented by the term (Ferguson et al., 2013).

According to the current international standard, the term intellectual disability encompasses a range of neurodevelopmental conditions characterized by challenges at the level of cognitive processing and/or reasoning, as well as a compromise in adaptive behaviors (Schalock et al., 2021). An important shift occurred some decades ago in the understanding of intellectual disability under a biopsychosocial framework. Instead of classifying the "severity" of the condition (as something stable, permanent, and defined by one's IQ), what is graded is the support required by the individual in particular contexts and at specific moments in life. The level of support needed is determined based on the evaluation of adaptive behaviors and can change throughout development, in different contexts, and according to the changes associated with interventions and supports provided (American Educational Research Association et al., 2014; Schalock et al., 2021).

Both the DSM-5-TR (American Psychiatric Association, 2013) and the ICD-11 (World Health Organization, 2019) list intellectual disability under the umbrella of neurodevelopmental conditions. Both manuals share similar classification, listing the following conditions: disorders of intellectual development, developmental speech or language disorders, autism spectrum disorder, developmental learning disorder, developmental motor coordination disorder, attention deficit hyperactivity disorder, stereotyped movement disorder, Primary tics or tic disorders, secondary neurodevelopmental syndrome, other specified neurodevelopmental disorders, and neurodevelopmental disorders, unspecified. The names of the conditions have minimal variations between the two manuals but refer to the same conditions. In this chapter, the names listed in the ICD-11 have been incorporated.

Identification of Intellectual Disability: A Comprehensive Approach

In this text, we decided to use the term "identification process" instead of "diagnostic" to align the stance with the biopsychosocial perspective of disability, allowing room to perceive intellectual disabilities as neurodevelopmental conditions rather than illnesses or disorders. When discussing identification processes, we typically think of procedures aimed at identifying the criteria, characteristics, or symptoms associated with a specific condition (Townsend, 2014). However, it is essential to broaden this conception to incorporate the central objectives of these processes, which include identifying the supports, adjustments, accommodations, and interventions that the person may need. The focus should extend beyond determining the presence or absence of a developmental disability to include the quality of life, well-being, and mental health of the individual, along with the necessary procedures and adjustments to promote and maintain them (Buntinx & Schalock, 2010; Simões et al., 2016).

For the identification processes of intellectual disability, a formal assessment is necessary to establish if an individual, whether a child or an adult, meets the three central criteria established by international standards (Schalock et al., 2011; Tassé et al., 2016). These criteria, outlined in diagnostic manuals such as the DSM-5-TR (Diagnostic and Statistical Manual of Mental Disorders; American Psychiatric Association, 2013), ICD-11 (International Classification of Diseases; World Health Organization, 2019), and AAIDD (American Association on Intellectual and Developmental Disabilities; Schalock et al., 2021), encompass (1) significant limitations in intellectual functioning, (2) significant limitations in adaptive behavior, and (3) onset in the developmental period. Beyond merely identifying these criteria, a robust assessment should delve deeper, aiming to answer critical questions regarding the need for supports and the individual's everyday functioning (Thompson et al., 2009; Zaagsma et al., 2023).

Intellectual disability, as stated earlier in this chapter, is a complex condition that requires a multidimensional approach to assessment. Recognizing and understanding the multifaceted nature of this disability is crucial for accurate identification and subsequent intervention planning.

Like cognitive and psychological assessments with any other person, the testing environment can impact the performance of individuals with an intellectual disability, and the relationship between the examiner and the individual being assessed is crucial (Davies et al., 2017). This is especially important when assessing persons with neurodevelopmental conditions who may have difficulty focusing and sustaining attention (Spaniol & Danielsson, 2022), expressing themselves (Næss et al., 2011), and feeling comfortable with someone they are not familiar with (Manzanero et al., 2012). Creating a comfortable and supportive environment is essential. Taking the time to get to know the person and establish rapport can mitigate anxiety and encourage optimal performance (DiStefano et al., 2020).

Another important consideration is the provision of supports and adjustments according to the person's needs and characteristics. The evaluator needs to be able to adapt to the individual's communication style, using visual aids, simple language, and allowing sufficient time for responses. Additionally, clinicians should be mindful of potential distractions in the room where the assessment takes place, as these may affect test results. The evaluator must ensure that all conditions are provided so that the true abilities of the person are accurately assessed (Thompson et al., 2018).

IQ Assessment in Intellectual Disability: Navigating Complexity

The first criterion for identifying intellectual disability, significant limitations in intellectual functioning, pertains to an individual's cognitive abilities. This involves evaluating the person's IQ through standardized tests that measure cognitive skills like reasoning, problem-solving, and learning (Fletcher & Miciak, 2024). IQ tests are designed to assess various aspects of intelligence, covering a range of cognitive skills (Kaufman et al., 2006). Typically, these tests include subtests for verbal comprehension, perceptual reasoning, working memory, and processing speed. The overall IQ score provides an estimate of an individual's general cognitive ability, usually expressed as a standardized score with a mean of 100 and a standard deviation of 15. To identify intellectual disability, the individual must obtain an IQ score at least two standard deviations below the mean, that is, below 70 points (Orsini et al., 2015; Rosas et al., 2014).

It is crucial to note that while IQ tests offer valuable insights into cognitive functioning, they are not exhaustive measures of intelligence and may not capture the full spectrum of an individual's abilities or potential (Loring & Bauer, 2010). In navigating the complexities of IQ assessment for identifying intellectual disability, clinicians play a pivotal role in ensuring the reliability and validity of the process. This involves not only selecting appropriate tests with psychometric evidence for the cultural context but also adapting assessment strategies based on the unique needs and characteristics of the individual being evaluated (Man et al., 2018).

Most IQ tests consider potential use for assessing individuals with intellectual disability and provide adaptations, such as changing the starting items on subtests to avoid frustration (e.g., WISC and WAIS). However, these adaptations may sometimes be insufficient, necessitating additional support to ensure that results reflect the individual's abilities (Tenorio et al., 2014). For instance, individuals with an intellectual disability may face language or communication challenges that heavily impact their performance on verbally reliant tests. Clinicians should choose tests considering non-verbal communication and providing alternative means of assessment, preventing language barriers from unduly influencing results.

Probably, the most commonly used tests for assessing intellectual functioning are the Wechsler Intelligence Scales: WISC (6:00–16:11; Wechsler, 2014) for children and adolescents and WAIS for adults (16:00–89:00 years of age; Wechsler, 2008). The WAIS and WISC are widely recognized for their evidence of reliability and validity and are one of the few gold standards for cognitive assessment with adaptation for Latin American countries, making them valuable instruments in clinical settings (Vinet et al., 2023). Other frequently employed measures for IQ are the Stanford-Binet Intelligence Scales (Peterson et al., 2020), the Kaufman Assessment Battery for Children (KABC; Kaufman & Kaufman, 2004), and The Woodcock-Johnson Tests of Cognitive Abilities (Shrank et al., 2014). These tests collectively contribute to the comprehensive evaluation of intellectual abilities, helping in the understanding of an individual's cognitive strengths and areas for potential support and intervention.

Cultural sensitivity is crucial in the IQ assessment, because cultural or linguistic factors may influence the performance in the tests. A culturally fair evaluation ensures that the assessment accurately reflects an individual's cognitive abilities without bias from external factors. IQ tests may contain cultural biases affecting performance, potentially leading to misinterpretations of intellectual functioning (Shuttleworth-Edwards, 2016). Test items may be sensitive to cultural particularities, especially in tasks assessing factual knowledge or vocabulary (Webb & Sasao, 2013). Selecting and administering tests with cultural sensitivity involves using standardized tests adapted to the individual's cultural background with evidence of validity across diverse populations.

Embracing cultural competence, considering language and communication nuances, and adopting an approach beyond IQ scores contribute to a more accurate and nuanced understanding of intellectual functioning in identifying intellectual disability. Ongoing professional development is crucial for clinicians to stay informed about the latest advancements in assessment tools and methodologies. Also, integrating multiple data sources and collaborating with interdisciplinary teams further enhances the precision and reliability of the assessment process (Merkley et al., 2022).

Assessing Adaptive Behavior: Navigating the Challenges of Daily Living

The second criterion, which refers to significant limitations in adaptive behavior, delves into an individual's ability to effectively navigate the challenges of everyday life (Schalock et al., 2021). Adaptive behavior encompasses a broad range of skills required for effective daily functioning, including communication, self-care, socialization, and daily living activities (Daunhauer, 2011; Schalock, 1999).

To assess adaptive behavior for the identification of intellectual disability, standardized assessments must be used, and the person must obtain a score two standard deviations below the mean. However, relevant qualitative information can be obtained also from direct observations, and information gathered from interviews with individuals, family members, and carers (Schalock et al., 2021). These sources provide valuable insights into an individual's strengths and challenges across various domains of adaptive behavior that are relevant when planning for support and interventions. It is essential to recognize that adaptive behavior is context-dependent, and evaluations should consider the individual's cultural background and environmental factors that may influence their adaptive functioning (Arango et al., 2018; Simões et al., 2016; Woolf et al., 2010).

The instruments for the assessment of adaptive behavior are questionnaires usually including versions to be completed by a third person who knows well the individual (usually a family member or a carer) and to be answered by the person being assessed.

Three scales are widely used for the assessment of adaptive behavior. Adaptive Behavior Assessment System (ABAS; Oakland & Harrison, 2008) can be used for the assessment of persons between 0 and 89 years of age and provides a profile of an individual's strengths and weaknesses in the three indexes of adaptive behavior: conceptual, social, and daily-living abilities. Each one of these indexes is composed of areas, which helps get a more comprehensive idea of the individual's profile of strengths and difficulties. The Vineland Adaptive Behavior Scales (VABS; Sparrow et al., 2016) is another of the most widely used assessments for adaptive behavior that can be used with individuals aged 0–90 years and covers three major domains of adaptive functioning: communication, daily living skills, and socialization. Finally, the Diagnostic Adaptive Behavior Scale (DABS; Tassé et al., 2017) was developed for the assessment of conceptual, social, and practical skills and can be used with individuals from 4 to 21 years of age.

As with the assessment of IQ, the assessment of adaptive behaviors requires that clinicians serve as orchestrators of a comprehensive and nuanced evaluation (Schalock & Luckasson, 2014; Saulnier & Klaiman, 2018). This involves not only selecting appropriate assessment tools but also considering the cultural relevance of these tools and the individual's unique context. Additionally, clinicians must recognize the dynamic nature of adaptive behavior and its interaction with intellectual functioning. By incorporating information from multiple sources, utilizing culturally sensitive assessments, and adopting a functional and developmental perspective, clinicians contribute to a more accurate and comprehensive understanding of adaptive behavior. The collaborative engagement with carers, educators, and the individuals themselves further enriches the assessment process, ensuring that adaptive behavior is evaluated in its multifaceted reality.

The information provided by the scales for assessing adaptive behavior is necessary to determine if this criterion for identification is met, but it also helps us determine the level of support required by the individual being evaluated.

According to international standards (Schalock et al., 2021), the level of support should be identified for each domain of adaptive behavior, considering that scores between 50 and 55 and 70 and 75 indicate mild support requirements, between 40 and 45 and 50–55 indicate moderate support requirements, between 25 and 30 and 40 and 45 indicate severe support requirements, and scores lower than 20–25 indicate profound and systematic support needs.

This classification categories consider a 95% confidence interval in the scores obtained in the assessment. One important consideration is that, beyond the test scores, it is the clinician who should determine the required level of supports. It is also important to note that available instruments often provide standard scores starting at 40, so it will be the clinician's task to determine when this case arises. Finally, as mentioned at the beginning of the chapter, the required level of support is dynamic, depending on many factors, so it is important to have periodic reassessments.

Temporal Considerations: Onset in the Developmental Period

The third criterion, onset in the developmental period, underscores that intellectual disability is a neurodevelopmental condition. Evaluating the onset of intellectual disability necessitates a retrospective analysis of developmental milestones, medical history, and family reports. The clinical interview stands as a cornerstone in the identification process for intellectual disability, particularly when determining the onset of the developmental period. While standardized tests and behavioral assessments offer valuable data, the clinical interview provides a unique opportunity for clinicians to delve into an individual's developmental history, gaining crucial insights into the timing and trajectory of cognitive and adaptive challenges (McConaughy & Whitcomb, 2022).

In conducting the clinical interview to assess the onset of the developmental period, clinicians play a pivotal role in creating a safe and collaborative environment. The clinician's skills in active listening, empathy, and cultural competence are crucial for eliciting accurate information. Additionally, clinicians should adapt their approach based on the individual's communication abilities, ensuring that the interview process is accessible and inclusive (Sherman et al., 2022).

The clinical interview allows clinicians not only to inquire about the signs and characteristics of neurodevelopmental conditions but also to understand the individual and their family. It provides context to comprehend their traits, including strengths and weaknesses and gives us an idea of the impact the condition may be having on the person's quality of life and well-being. A good clinical interview should include questions about the patient's developmental history, risk and protective factors, and other relevant elements, not only for the identification process but also for planning interventions and supports (Kildahl et al., 2023; Sinclair et al., 2023).

While guidelines and formats exist for clinical interviews, they are insufficient for a thorough identification process. As mentioned before, the interview goes beyond mere data collection for completing a clinical record; it constitutes a complex clinical process guided by case formulation and hypotheses that the clinician constructs and contrasts throughout the entire process.

The first step for a clinical interview useful in the identification process is to formulate hypotheses and then contrast them with guided questions that delve into elements that catch attention or may be relevant to determining the presence or absence of the criteria and characteristics of the hypothesized condition, in this case, intellectual disability. This requires the clinician to have profound knowledge of the neurodevelopmental condition, its identification criteria, and possible presentations (DiStefano et al., 2020).

The interview is the central tool for assessing, in the hypothesis of intellectual disability, whether the developmental criterion is met or not. Through questions about the developmental trajectory of the child, adolescent, or adult undergoing the evaluation process, we can get an idea of when the characteristics started to manifest if there is any event associated with their onset, and how their presentation has evolved.

When assessing a child, adolescent, or adult for the interview about the developmental history, it is useful to include the parents or someone who can provide information about the person's early development. They offer valuable perspectives on milestones, potential delays, and any observed challenges. However, when the identification process is carried out for an adult, and depending on their age, it is often not possible to have information from someone who has known them since childhood. In these cases, information that may help determine the onset of difficulties can be found in documents such as school records, reports from professionals who have attended to them at some point during their development, or even their medical history. Also, in the case of assessments with adolescents or adults, it is advisable to include a session with them, allowing the understanding of their current situation, experiences, and needs, as well as what they can recall about their difficulties and challenges growing up (Thompson et al., 2018).

Some areas that should be explored during the clinical interview include a thorough exploration of medical history, including prenatal and perinatal factors, contributing to understanding potential risk factors or complications that could impact early development and may be related to the etiology. This historical context helps pinpoint the onset of cognitive and adaptive challenges. Also, understanding educational history provides insights into academic progress, learning difficulties, and the recognition of intellectual and adaptive challenges by educators. This information helps establish a timeline for the onset of developmental difficulties. Social interactions and behavioral patterns should also be considered, as understanding an individual's social experiences sheds light on adaptive behavior challenges.

It is useful to complement the information collected in the interview with data from other people who know the person assessed in various contexts and from documents such as school or medical records, if available. By synthesizing information from diverse sources, clinicians can achieve a comprehensive understanding of the developmental history. This nuanced understanding aids in establishing the onset of cognitive and adaptive challenges, guiding the diagnostic process for intellectual disability (McConaughy & Whitcomb, 2022). The clinical interview, when conducted thoughtfully and inclusively, serves as a powerful tool for unraveling the complexities of developmental trajectories and informing targeted interventions for individuals with intellectual disability (Schalock et al., 2021).

Beyond the Identification: Assessing the Need for Supports

A comprehensive assessment of intellectual disability goes beyond confirming the presence of the condition; it also focuses on understanding the individual's unique strengths, challenges, and support needs. This involves evaluating the intensity and types of support required for the person to effectively participate in various life domains, such as education, employment, and community involvement (Buntinx & Schalock, 2010). Several instruments aid in assessing the supports needed by an individual, contributing to a person-centered approach in evaluating and addressing the unique needs of individuals with intellectual and developmental disabilities.

One of the most used instruments for assessing support needs is the Supports Intensity Scale (SIS; Kuppens et al., 2010; Thompson et al., 2014; Wehmeyer et al., 2009). Developed by the American Association on Intellectual and Developmental Disabilities (AAIDD), the SIS is specifically designed to assess the intensity and nature of support needed by individuals with intellectual and developmental disabilities. The SIS evaluates support requirements across three major life areas: personal, social, and community participation. It provides a comprehensive profile of an individual's support needs, assisting in the development of person-centered plans that aim to maximize independence and community participation. The individual's support needs are measured on a scale from intermittent to extensive, considering factors such as the complexity, duration, and frequency of support required in various life domains. The assessment involves gathering information through interviews with the individual, their caregivers, and professionals familiar with them.

As mentioned previously in this chapter, the need for support for a person is graded on a scale ranging from mild to profound and systematic support needs (Schalock et al., 2021). Individuals who require mild support can perform most daily tasks independently but may need occasional support, guidance, or reminders with complex tasks or situations. Those requiring moderate support may need assistance with some daily activities such as household chores, managing finances, transportation, or medications, as well as in some social situations or when making complex decisions.

People with intellectual disability who have severe support needs may require assistance with a wide range of daily activities, including personal care, communication, managing emotions, and understanding social cues. These persons present significant challenges in multiple areas of functioning: they may need constant assistance with daily living tasks, personal care, and communication, including feeding, mobility, and medical needs (Beadle-Brown et al., 2016).

Another crucial aspect that may require careful assessment beyond the identification process is behaviors that challenge. People with intellectual disability often exhibit behaviors that challenge and can significantly impact their participation in various contexts (Bowring et al., 2019; Hastings et al., 2013; Janssen et al., 2002). Usually, the term behaviors that challenge is associated with aggressive behavior; however, it can refer to both externalizing and internalizing behaviors. Externalizing behaviors are related to difficulty in regulation and are expressed as fighting, aggression, hyperactivity, risky sexual behavior, substance use, non-compliance, and delinquency (Memmott-Elison et al., 2020). Internalizing behaviors are cognitive or affective difficulties associated with over-regulation, for example, withdrawal, low self-esteem, self-injuring, attentional problems, and anxiety and depression (Emerson & Einfeld, 2011; Memmott-Elison et al., 2020).

Understanding the causes of these behaviors and planning support accordingly is an integral part of the assessment. In such cases, it is important to have special considerations to ensure that behaviors do not interfere with the assessment, as well as to conduct a specific evaluation of behaviors that allows for consideration of their impact and intervention needs. On one hand, it will be crucial, before assessment sessions, to understand the characteristics of behaviors that challenge and to prepare to promote the individual's motivation and participation while reducing the likelihood of their occurrence. On the other hand, it is relevant to conduct a Functional Behavior Analysis by a specialist in the field. Functional Behavior Analysis is a systematic and structured approach that helps understand the purpose or function of a behavior, with the goal of identifying the underlying factors contributing to behaviors that challenge and developing effective interventions to address them (Ali et al., 2014; Fynn et al., 2023; Prior et al., 2023).

Special Considerations for Developing Countries

As shown in this chapter, the formal identification process of intellectual disability involves a complex interdisciplinary approach that relies on the use of measurement tools enabling the assessment of cognitive development, represented by the intelligence quotient, and the measurement of adaptive behaviors. In both cases, it is necessary to have instruments aligned with the measurement recommendations for psychology and education, as this is the only way to analyze the robustness of the reliability and validity evidence of a particular instrument in each socio-cultural context (American Educational Research Association et al., 2014). Additionally, identification must be carried out by an interdisciplinary team with specific training in the field of intellectual disability.

For developing countries and countries with low per capita income, meeting these requirements can be a challenge. Some studies have documented the lack of measurement instruments that meet the minimum expected requirements for use in identification processes (Arango-Lasprilla, 2020; Mascialino et al., 2022; Morlett Paredes et al., 2021; Vinet et al., 2023). Furthermore, there is a very low number of specialists in psychology and psychiatry, which does not allow for the necessary coverage. Naturally, a consequence of this low supply of professionals compared to high demands for mental health is reflected in long waiting lists for consultations, a lack of specialists outside urban areas, and prices so high that they are inaccessible to most of the population (Chaskel et al., 2015; Chuaqui & Linn, 2021; Minoletti et al., 2012; Thompson et al., 2018).

While it is expected that all countries that signed and ratified the Convention on the Rights of Persons with Disabilities (United Nations, 2006) are implementing strategies to improve identification processes, generate statistics supporting decision-making, and increase the effective participation rates of people with disabilities in all decisions directly affecting them, the scarce available data, mostly from international cooperation agencies' reports (both official and those submitted by each country to the United Nations), make the lack of actions evident.

Given this scenario, for professionals working in developing countries or those with low per capita income, it is necessary to recommend caution in identification processes, making them aware of the challenges. We find it relevant to point out that highly skilled professionals who carry out identification processes in developing countries should be aware of the limitations but not remain indifferent to the deficiencies. It is essential that they actively participate in the review and formulation of public policies and official protocols so that existing gaps can be closed.

A non-exhaustive but as comprehensive a list as possible regarding the precautions professionals should take when working with this population in these territories include:

1. It is important that professionals involved in the identification processes of intellectual disability, whether carried out at the health or education services level, are familiar with the definition and ethical procedures agreed upon for identification. This should not affect the real possibility of meeting the standard or not.
2. Every professional involved in identification processes needs to be familiar with the Convention on the Rights of Persons with Disabilities, as well as the state of local regulations governing this collective in their country. When there are no specific regulations regarding intellectual disability, general disability norms, and non-discrimination policies will apply.
3. In all cases where an identification is carried out, the professional should strive to use the best available instrument for evaluating the defining criteria discussed in this chapter. In cases where assessment tools are not available in the country,

it is recommended to use the most updated version of gold standard instruments that are closest, either geographically or culturally.

4. In all cases where the presence of some form of intellectual disability is suspected, an individual measurement by standardized tests should be sought, conducted by a professional with specific training in the use of assessment instruments.
5. Whenever the presence of intellectual disability is suspected, the professional should seek the application of assessment instruments, preferring direct data collection on the person being evaluated. If, after implementing all possible clinical efforts, direct evaluation cannot be achieved, hetero-information should be preferred. This means that the professional should not pre-judge a person, discarding the possibility of someone participating in an assessment based solely on an interview.
6. It is highly recommended, as a best practice, that hetero-information-based assessment integrates different sources of information, with the sources chosen based on the individual's life cycle stage. For example, with a child, it would be ideal to have information from parents and teachers, while for adults, information from a professional mediating in job inclusion and a friend or partner of the person being evaluated might be more relevant.
7. In any identification process initiated, it will be necessary to have the person's permission, considering their will. For minors, the signing of informed consent by parents and informed assent by minors is required. In the case of adults, the evaluated individual should express their informed consent. For countries where the interdiction figure is still maintained, it is a recommended practice that the legal representative of the person with a disability signs the informed consent, but this should be ratified by the person with a disability.
8. It is crucial that, in cases where the presence of intellectual disability is identified, this information is communicated clearly and directly to the person. For minors and interdicted individuals, the information must be received by legal guardians, and they will decide what information is disclosed to the person, although it is always advisable to fully inform the directly involved individual. In the case of adults, the results of the identification process should be communicated in clear language, and ideally, a report in Easy Read should be provided.

Conclusion

The process of identification in intellectual disability is a complex area that requires knowledge of the condition, advanced training in the use of assessment tools, and high competence in result interpretation and consolidation. The availability of an international standard derived from the agreement of an international working group undoubtedly represents a very positive step forward. Recognizing this progress, it is also necessary to acknowledge the challenges faced by professionals responsible for these processes in developing countries, challenges that manifest in all areas.

In the future, it will be relevant for the generation of standards to provide specific recommendations for work in diverse cultural environments, always considering that the identification of the condition and the identification of support are fundamental areas for the inclusion of people with intellectual disability in all aspects of their lives. A good identification process, free from biases and stigma, is undoubtedly the gateway to the full guarantee of rights.FundingThe authors were funded by Chilean Government vía Fondecyt Regular 1221400 and Fondecyt Regular 1221349, and also by Millenium Program vía Millenium Institute for Care Research ICS2019_024.

References

Ali, A., Blickwedel, J., & Hassiotis, A. (2014). Interventions for challenging behaviour in intellectual disability. *Advances in Psychiatric Treatment, 20*(3), 184–192. https://doi.org/10.1192/APT.BP.113.011577

American Educational Research Association, American Psychological Association, & National Council on Measurement in Education. (2014). *The standards for educational and psychological testing*. AERA Publications Sales.

American Psychiatric Association. (2013). *Diagnostic and statistical manual of mental disorders: DSM-5*. American Psychiatric Association.

Arango-Lasprilla, J. C. (2020). Spanish-speaking pediatric neuropsychology: Recent research and new frontiers. *Developmental Neuropsychology, 45*(4), 167–168. https://doi.org/10.1080/87565641.2020.1775233

Arango, P., Aparicio, A., & Tenorio, M. (2018). Developmental trajectories of children with Down syndrome by socio-economic status: The case of Latin America. *Journal of Intellectual Disability Research, 62*(9), 759–774. https://doi.org/10.1111/jir.12516

Beadle-Brown, J., Leigh, J., Whelton, B., Richardson, L., Beecham, J., Baumker, T., & Bradshaw, J. (2016). Quality of life and quality of support for people with severe intellectual disability and complex needs. *Journal of Applied Research in Intellectual Disabilities, 29*(5), 409–421. https://doi.org/10.1111/JAR.12200

Bowring, D. L., Painter, J., & Hastings, R. P. (2019). Prevalence of challenging behaviour in adults with intellectual disabilities, correlates, and association with mental health. *Current Developmental Disorders Reports, 6*(4), 173–181. https://doi.org/10.1007/s40474-019-00175-9

Buntinx, W. H. E., & Schalock, R. L. (2010). Models of disability, quality of life, and individualized supports: Implications for professional practice in intellectual disability. *Journal of Policy and Practice in Intellectual Disabilities, 7*(4), 283–294. https://doi.org/10.1111/j.1741-1130.2010.00278.x

Chaskel, R., Gaviria, S. L., Espinel, Z., Taborda, E., Vanegas, R., Shultz, J. M., Psychiatry, A., & Santa, F. (2015). Mental health in Colombia. *BJPsych International, 12*(4), 95–97. https://doi.org/10.1192/S2056474000000660

Chuaqui, J., & Linn, J. G. (2021). Attempts to achieve parity and coverage in Chile in mental health. In *Innovations in global mental health* (Vol. 1–2(1), pp. 437–442). Springer. https://doi.org/10.1007/978-3-030-57296-9_23/COVER

Daunhauer, L. A. (2011). The early development of adaptive behavior and functional performance in young children with Down syndrome: Current knowledge and future directions. *International Review of Research in Developmental Disabilities, 40*(1), 109–137. https://doi.org/10.1016/B978-0-12-374478-4.00005-8

Davies, D. K., Stock, S. E., King, L., Wehmeyer, M. L., & Shogren, K. A. (2017). An accessible testing, learning and assessment system for people with intellectual disability. *International*

Journal of Developmental Disabilities, 63(4), 204–210. https://doi.org/10.1080/20473869.2017.1294313

DiStefano, C., Sadhwani, A., & Wheeler, A. C. (2020). Comprehensive assessment of individuals with significant levels of intellectual disability: Challenges, strategies, and future directions. *American Journal on Intellectual and Developmental Disabilities, 125*(6), 434–448. https://doi.org/10.1352/1944-7558-125.6.434

Emerson, E., & Einfeld, S. L. (2011). *Challenging behaviour*. Cambridge University Press.

Ferguson, D., Ferguson, P., & Wehmeyer, M. L. (2013). The self-advocacy movement: Late modern times (1980 CE to Present). In *The story of intellectual disability: An evolution of meaning, understanding, and public perception* (pp. 233–278). Paul Brookes Publishing.

Fletcher, J. M., & Miciak, J. (2024). Assessment of specific learning disabilities and intellectual disabilities. *Assessment, 31*(1), 53–74. https://doi.org/10.1177/10731911231194992

Fynn, G., Porter, M., Borchard, T., Kazzi, C., Zhong, Q., & Campbell, L. (2023). The effectiveness of cognitive behavioural therapy for individuals with an intellectual disability and anxiety: A systematic review. *Journal of Intellectual Disability Research, 67*(9), 816–841. https://doi.org/10.1111/JIR.13046

Hastings, R. P., Baker, P., & Hughes, J. C. (2013). A conceptual framework for understanding why challenging behaviours occur in people with developmental disabilities. *International Journal of Positive Behavioural Support, 3*(2), 5–13. https://www.researchgate.net/publication/263527512

Janssen, C. G. C., Schuengel, C., & Stolk, J. (2002). Understanding challenging behaviour in people with severe and profound intellectual disability: A stress-attachment model. *Journal of Intellectual Disability Research, 46*(Pt 6), 445–453. https://doi.org/10.1046/J.1365-2788.2002.00430.X

Kaufman, A. S., & Kaufman, N. L. (2004). *Kaufman test of educational achievement* (2nd ed.). American Guidance Service.

Kaufman, A. S., Flanagan, D. P., Alfonso, V. C., & Mascolo, J. T. (2006). Test review: Wechsler Intelligence Scale for Children, Fourth Edition (WISC-IV). *Journal of Psychoeducational Assessment, 24*(3), 278–295. https://doi.org/10.1177/0734282906288389/ASSET/0734282906288389.FP.PNG_V03

Kildahl, A. N., Oddli, H. W., & Helverschou, S. B. (2023). Bias in assessment of co-occurring mental disorder in individuals with intellectual disabilities: Theoretical perspectives and implications for clinical practice. *Journal of Intellectual Disabilities*. https://doi.org/10.1177/17446295231154119

Kuppens, S., Bossaert, G., Buntinx, W., Molleman, C., Van Den Abbeele, A., & Maes, B. (2010). Factorial validity of the Supports Intensity Scale (SIS). *American Journal on Intellectual and Developmental Disabilities, 115*(4), 327–339. https://doi.org/10.1352/1944-7558-115.4.327

Loring, D. W., & Bauer, R. M. (2010). Testing the limits: Cautions and concerns regarding the new Wechsler IQ and Memory scales. *Neurology, 74*(8), 685–690. https://doi.org/10.1212/WNL.0B013E3181D0CD12

Man, J., Kangas, M., Trollor, J., & Sweller, N. (2018). Clinical practices and barriers to best practice implementation of psychologists working with adults with intellectual disability and comorbid mental ill health. *Journal of Policy and Practice in Intellectual Disabilities, 15*(3), 256–266. https://doi.org/10.1111/JPPI.12256

Manzanero, A. L., Contreras, M. J., Recio, M., Alemany, A., & Martorell, A. (2012). Effects of presentation format and instructions on the ability of people with intellectual disability to identify faces. *Research in Developmental Disabilities, 33*(2), 391–397. https://doi.org/10.1016/J.RIDD.2011.09.015

Mascialino, G., Adana-Diaz, L., Rodriguez-Lorenzana, A., Rivera, D., & Arango-Lasprilla, J. C. (2022). Practice of neuropsychology in Ecuador. *Revista Ecuatoriana de Neurología, 31*(1), 49–58. https://doi.org/10.46997/REVECUATNEUROL31100049

Maulik, P. K., Mascarenhas, M. N., Mathers, C. D., Dua, T., & Saxena, S. (2011). Prevalence of intellectual disability: A meta-analysis of population-based studies. *Research in Developmental Disabilities, 32*(2), 419–436. https://doi.org/10.1016/j.ridd.2010.12.018

McConaughy, S. H., & Whitcomb, S. A. (2022). *Clinical interviews for children and adolescents: Assessment to intervention* (3rd ed.). The Guilford Press.

Memmott-Elison, M. K., Holgren, H. G., Padilla-Walker, L. M., & Hawkins, A. J. (2020). Associations between prosocial behavior, externalizing behaviors, and internalizing symptoms during adolescence: A meta-analysis. *Journal of Adolescence, 80*, 98–114.

Merkley, T. L., Esopenko, C., Zizak, V. S., Bilder, R. M., Strutt, A. M., Tate, D. F., & Irimia, A. (2022). Challenges and opportunities for harmonization of cross-cultural neuropsychological data. *Neuropsychology, 37*(3), 237–246. https://doi.org/10.1037/NEU0000818

Minoletti, A., Galea, S., & Susser, E. (2012). Community mental health services in Latin America for people with severe mental disorders. *Public Health Reviews, 34*(2), 1–23. https://doi.org/10.1007/BF03391681/METRICS

Morlett Paredes, A., Gooding, A., Artiola I Fortuny, L., Rivera Mindt, M., Suárez, P., Scott, T. M., Heaton, A., Heaton, R. K., Cherner, M., & Marquine, M. J. (2021). The state of neuropsychological test norms for Spanish-speaking adults in the United States. *The Clinical Neuropsychologist, 35*(2), 236–252. https://doi.org/10.1080/13854046.2020.1729866

Næss, K. A. B., Lyster, S. A. H., Hulme, C., & Melby-Lervåg, M. (2011). Language and verbal short-term memory skills in children with Down syndrome: A meta-analytic review. *Research in Developmental Disabilities, 32*(6), 2225–2234. https://doi.org/10.1016/J.RIDD.2011.05.014

Oakland, T., & Harrison, P. L. (2008). *Adaptive behavior assessment system-II: Clinical use and interpretation*. Elsevier.

Orsini, A., Pezzuti, L., & Hulbert, S. (2015). Beyond the floor effect on the Wechsler Intelligence Scale for Children--4th Ed. (WISC-IV): Calculating IQ and Indexes of subjects presenting a floored pattern of results. *Journal of Intellectual Disability Research, 59*(5), 468–473. https://doi.org/10.1111/jir.12150

Peterson, K. C., Peterson, M., & Carducci, B. J. (2020). Stanford-Binet intelligence scale. In *The Wiley encyclopedia of personality and individual differences* (pp. 451–455). https://doi.org/10.1002/9781118970843.CH145

Prior, D., Win, S., Hassiotis, A., Hall, I., Martiello, M. A., & Ali, A. K. (2023). Behavioural and cognitive-behavioural interventions for outwardly directed aggressive behaviour in people with intellectual disabilities. *Cochrane Database of Systematic Reviews, 2023*(2). https://doi.org/10.1002/14651858.CD003406.PUB5/INFORMATION/EN

Rosas, R., Tenorio, M., Pizarro, M., Cumsille, P., Bosch, A., Arancibia, S., Carmona-Halty, M., Pérez-Salas, C. P., Pino, E., & Vizcarra, B. (2014). Estandarización de la Escala Wechsler de Inteligencia Para Adultos-Cuarta Edición en Chile. *Psykhe (Santiago), 23*(1), 1–18. https://doi.org/10.7764/psykhe.23.1.529

Salvador-Carulla, L., & Saxena, S. (2009). Intellectual disability: Between disability and clinical nosology. *The Lancet, 374*(9704), 1798–1799. https://doi.org/10.1016/S0140-6736(09)62034-1

Saulnier, C. A., & Klaiman, C. (2018). *Essentials of adaptive behavior assessment of neurodevelopmental disorders*. Wiley.

Schalock, R. L. (1999). The merging of adaptive behavior and intelligence: Implications for the field of mental retardation. In S. A. Borthwick-Duffy (Ed.), *Adaptive behavior and its measurement: Implications for the field of mental retardation*. American Association on Mental Retardation.

Schalock, R. L., & Luckasson, R. A. (2014). *Clinical Judgment* (Second ed). American Association on Intellectual and Developmental Disabilities.

Schalock, R. L., Verdugo, M. A., & Gomez, L. E. (2011). Evidence-based practices in the field of intellectual and developmental disabilities: An international consensus approach. *Evaluation and Program Planning, 34*, 273–282. https://doi.org/10.1016/j.evalprogplan.2010.10.004

Schalock, R. L., Luckasson, R., & Tassé, M. J. (2021). *AAIDD manual: Intellectual disability: Definition, diagnosis, classification, and systems of supports* (12th ed.). AAIDD.

Scior, K. (2016). Toward understanding intellectual disability stigma: Introduction. In K. Scior & S. Werner (Eds.), *Intellectual disability and stigma* (pp. 3–13). Palgrave Macmillan.

Scior, K., Potts, H. W., & Furnham, A. F. (2013). Awareness of schizophrenia and intellectual disability and stigma across ethnic groups in the UK. *Psychiatry Research, 208*(2), 125–130. https://doi.org/10.1016/J.PSYCHRES.2012.09.059

Shakespeare, T. (2018). *Disability: The basics*. Routledge.

Sherman, E. M. S., Ee Tan, J., & Hrabok, M. (2022). *A compendium of neuropsychological tests: Fundamentals of neuropsychological assessment and test reviews for clinical practice*. Oxford University Press.

Shrank, F., McGrew, K., & Mather, N. (2014). *The Woodcock-Johnson IV test of cognitive abilities (WJ IV COG)*. Houghton Miffling Harcourt-Riverside.

Shuttleworth-Edwards, A. B. (2016). Generally representative is representative of none: Commentary on the pitfalls of IQ test standardization in multicultural settings. *The Clinical Neuropsychologist, 30*(7), 975–998. https://doi.org/10.1080/13854046.2016.1204011

Simões, C., Santos, S., Biscaia, R., & Thompson, J. R. (2016). Understanding the relationship between quality of life, adaptive behavior and support needs. *Journal of Developmental and Physical Disabilities, 28*(6), 849–870. https://doi.org/10.1007/s10882-016-9514-0

Sinclair, J., Hollway, J. A., Sanford, C., Cameto, R., Benson, B., Arnold, L. E., & Seeley, J. R. (2023). Developing the diagnostic interview for adolescents and adults with mild/moderate intellectual aisabilities: An interview schedule of mental disorders (DIAAID). *Journal of Mental Health Research in Intellectual Disabilities*. https://doi.org/10.1080/19315864.2023.2214089

Spaniol, M., & Danielsson, H. (2022). A meta-analysis of the executive function components inhibition, shifting, and attention in intellectual disabilities. *Journal of Intellectual Disability Research, 66*(1–2), 9–31. https://doi.org/10.1111/JIR.12878

Sparrow, S., Cicchetti, D. V., & Saulnier, C. A. (2016). *Vineland adaptive behavior scales, Third Edition (Vineland-3)*. Pearson Clinical.

Tassé, M. J., Luckasson, R., & Schalock, R. L. (2016). The relation between intellectual functioning and adaptive behavior in the diagnosis of intellectual disability. *Intellectual and Developmental Disabilities, 54*(6), 381–390. https://doi.org/10.1352/1934-9556-54.6.381

Tassé, M. J., Schalock, R. L., Baldoni, G., Bersani, H., Borthwick-Duffy, S. A., Spreat, S., Thissen, D., Widaman, K. F., & Zhang, D. (2017). *Diagnostic Adaptive Behavior Scale (DABS) user's manual*. AAIDD.

Tenorio, M., Campos, R., & Karmiloff-Smith, A. (2014). What standardized tests ignore when assessing individuals with neurodevelopmental disorders. *Estudios de Psicología, 35*(2), 426–437. https://doi.org/10.1080/02109395.2014.922264

Thompson, J. R., Bradley, V. J., Buntinx, W. H. E., Schalock, R. L., Shogren, K. A., Snell, M. E., Wehmeyer, M. L., Borthwick-Duffy, S., Coulter, D. L., Craig, E. M., Gomez, S. C., Lachapelle, Y., Luckasson, R. A., Reeve, A., Spreat, S., Tassé, M. J., Verdugo, M. A., & Yeager, M. H. (2009). Conceptualizing supports and the support needs of people with intellectual disability. *Intellectual and Developmental Disabilities, 47*(2), 135–146. https://doi.org/10.1352/1934-9556-47.2.135

Thompson, J. R., Wehmeyer, M. L., Hughes, C., Shogren, K. A., Palmer, S. B., & Seo, H. (2014). The supports intensity scale–children's version: Preliminary reliability and validity. *Inclusion, 2*(2), 140–149. https://doi.org/10.1352/2326-6988-2.2.140

Thompson, T., Coleman, J. M., Riley, K., Snider, L. A., Howard, L. J., Sansone, S. M., & Hessl, D. (2018). Standardized assessment accommodations for individuals with intellectual disability. *Contemporary School Psychology, 22*(4), 443–457. https://doi.org/10.1007/S40688-018-0171-4

Townsend, M. C. (2014). *Essentials of psychiatric mental health nursing: Concepts of care in evidence-based practices*. F.A. Davis Company.

United Nations. (2006). *Convention on the rights of persons with disabilities – articles | United Nations Enable*. https://www.un.org/development/desa/disabilities/convention-on-the-rights-of-persons-with-disabilities/convention-on-the-rights-of-persons-with-disabilities-2.html

Vinet, E. V., Rodríguez-Cancino, M., Domínguez, A. S., Rojas-Mora, P., & Saiz, J. L. (2023). The use of test by chilean psychologists: A disturbing panorama. *Psykhe, 32*(1), 1–19. https://doi.org/10.7764/PSYKHE.2020.25293

Webb, S. A., & Sasao, Y. (2013). New directions in vocabulary testing. *RELC Journal, 44*(3), 263–277. https://doi.org/10.1177/0033688213500582

Wechsler, D. (2008). *Wechsler adult intelligence scale-fourth edition*. Pearson. https://doi.org/10.1037/T15169-000

Wechsler, D. (2014). *WISC-V*. NCS Pearson, Incorporated.

Wehmeyer, M., Chapman, T. E., Little, T. D., Thompson, J. R., Schalock, R., & Tassé, M. J. (2009). Efficacy of the Supports Intensity Scale (SIS) to predict extraordinary support needs. *American Journal on Intellectual and Developmental Disabilities, 114*(1), 3–14. https://doi.org/10.1352/2009.114:3-14

Woolf, S., Woolf, C. M., & Oakland, T. (2010). Adaptive behavior among adults with intellectual disabilities and its relationship to community independence. *Intellectual and Developmental Disabilities, 48*(3), 209–215. https://doi.org/10.1352/1944-7558-48.3.209

World Health Organization. (2019). *International classification of diseases 11th revision*. World Health Organization.

Zaagsma, M., van de Velde, D., Koning, M. H. M., Volkers, K. M., Schippers, A. P., & van Hove, G. (2023). 'When I need them, I call them and they will be there for me'. Experiences of independently living people with intellectual disabilities with 24/7 available online support. *Disability & Society, 38*(4), 659–682. https://doi.org/10.1080/09687599.2021.1932756

Zeilinger, E. L., Stiehl, K. A. M., Bagnall, H., & Scior, K. (2020). Intellectual disability literacy and its connection to stigma: A multinational comparison study in three European countries. *PLoS One, 15*(10), e0239936. https://doi.org/10.1371/journal.pone.0239936

A History of Intellectual and Developmental Disabilities

Michael L. Wehmeyer

Abstract This chapter traces the history of the field of intellectual and developmental disabilities from the beginning of the nineteenth century to the early part of the twenty-first century. The earliest attempts at habilitation for people with intellectual and developmental disabilities began in France in the early 1800s and spread to the United States in the mid-1800s. The last half of the nineteenth century saw a rapid growth of state-run institutions for people with intellectual and developmental disabilities in the United States, along with changes in the purpose of such institutions, from habilitative to ensuring segregation. Over the first half of the twentieth century, institutions grew, and conditions worsened such that in the mid-1900s, a parent movement emerged that created a community-based system. In the last few decades of the twentieth century, a self-advocacy movement was established, providing opportunities for people with intellectual and developmental disabilities to have a voice in their lives. The chapter concludes with suggestions as to what the history of the field teaches us for today and the future.

Keywords History · Institutions · Eugenics · Community · Supports

This chapter, and to a significant degree the field of intellectual and developmental disabilities, begins in the region of Aveyron in South Central France in early January 1800 where a young boy was found alone, seemingly a "wild child," outside the woods near the village of Saint-Sernin-sur-Rance (Lane, 1976). This child, thought to be about 12 years of age and subsequently called Victor, had been spotted intermittently in wooded areas from 1797 onward. Although it may seem extraordinary today, narratives about feral children were not unusual in the late eighteenth and early nineteenth centuries, so much so that by the late nineteenth century, Rudyard

M. L. Wehmeyer (✉)
Department of Special Education and The Beach Center on Disability,
University of Kansas, Lawrence, KS, USA
e-mail: wehmeyer@ku.edu

M. G. Valdovinos (ed.), *Intellectual and Developmental Disabilities*,
Contemporary Clinical Neuroscience,
https://doi.org/10.1007/978-3-031-66932-3_2

Kipling would include, as the central character in a series of stories titled *The Jungle Book*, Mowgli, a boy who was raised by wolves in the jungle (Newton, 2002).

If narratives of feral children were not unusual, what happened to Victor was. He was eventually taken into the household of a physician at the National Institute for Deaf-Mutes in Paris, Jean-Marc Gaspard Itard, who chose to disregard the professional opinions of the time, including those of his mentor, Philippe Pinel, to try to teach Victor. Pinel, who was widely respected and is considered the founder of the field of psychiatry, had examined Victor and declared that he was incurably defective. It was only around that time that a distinction began to be made between so-called "idiocy" and "insanity."[1] Pinel's beliefs, and the beliefs of virtually all psychiatrists at the time, was that people who were deemed to be "idiots" were ineducable.

The 26-year-old Itard argued that Victor's developmental impairments were due to his isolation in the woods and not, as Pinel believed, some incurable "defect." Itard approached Victor as an instance of developmental "deficiency": a so-called "noble savage" whose development had been impeded by his deprived circumstances growing up. And so, for several years, Itard sought to teach Victor to speak, to perform basic daily living tasks, and so forth. Ultimately, Itard considered his experiment with Victor a failure. As he explained in his summary report, Victor's development was "the slow and laborious result of a very active education in which the most powerful methods are used to obtain most insignificant results" (Itard, 1806/1962, p. 100). But it should be noted that it was not that Victor could not learn at all, it was that he did not learn enough. The progress that Victor showed under Itard's tutelage was deemed to be insufficient (Ferguson & Ferguson, 1986).

There was, certainly, a history of the lives of people with intellectual and developmental disabilities that preceded Victor and Itard, reaching all the way back to pre-history (Wehmeyer, 2013). However, a modern history of intellectual and developmental disabilities that includes habilitative and educative efforts begins with Itard and his efforts to educate Victor. Itard's legacy continued in his pupil, Edouard Séguin. Seguin picked up on his mentor's work, initiating efforts to educate children and adults who had been deemed as ineducable and, eventually, developing a system that he described in detail in his book, *Traitment Moral, Hygiene, et Education des Idiots et des Autres Enfants Arrieres* (Seguin, 1846), later translated into English as *Idiocy and Its Treatment by the Physiological Method* (Seguin, 1866). To Itard goes the credit for being the first to try to educate someone with intellectual and developmental disabilities, but it was Seguin who first created a systematic process by which to do so.

[1] It is difficult, if not impossible, to relate the history of intellectual and developmental disabilities without using terms that, today, are pejorative and offensive. I try to minimize such usage and, when necessary, place these terms in quotes.

The Institutional Era

Coming to America

Seguin's *Physiological Method* spread through France and Europe as a model treatment for people deemed to be "idiots." In 1842, Samuel Gridley Howe, who was the superintendent of the Perkins Institute for the Blind in Boston, and Horace Mann, considered the father of the public school movement in America, visited Europe and met with Seguin. When they returned to the United States, they publicized Seguin's work in talks and in newspaper articles and articles in medical journals. Howe went a step further and pressured the Massachusetts legislature to establish a school for "idiots" and in April of 1846, he was appointed chair of a committee to determine what could be done to further their cause (Smith & Wehmeyer, 2022).

The report from Howe's committee was issued in March 1847 and a full version was published in 1848 titled *On the Causes of Idiocy* (Howe, 1848). In October 1948, with funding from the legislature, Howe converted one wing of the Perkins Institute for the Blind into an experimental school for four boys classified as "idiots."

Although the first public school for people with intellectual disability in America, Howe's school was not the first school for this population in the United States. Hervey Wilbur, a physician who had read about Seguin's work in Europe, possibly from writings by Howe and Mann, started a private school in his home in Barre, Massachusetts, that opened 6 months before Howe opened his school (Katovitch, 2016). Both Hervey's school, called the Elm Hill School and Home for the Education of Feebleminded Youth, and Howe's public school implemented Seguin's method.

Seguin based his intervention on the rudimentary knowledge of the neural system that existed at that time emphasizing sensory stimulation via auditory and visual input, and on his understanding of typical child development, particularly motor and cognitive development. *Physiological Treatment* was composed of three primary areas. The first was physical training, using various exercise and motor development apparatus to stimulate sensory and motor development. For example, students climbed onto swingboards, suspended by a rope from the ceiling, and were pushed so that their feet touched a wall near the swing, thus giving them the sensation of the pressure on their soles. The final two components of Seguin's method involved academic and speech instruction to promote cognitive and language development, as well as moral training and socialization.

It is worth noting that at this point, the purpose of schools such as Elm Hill was not to prepare students for future employment or advanced learning, but simply to improve their cognitive, language, and motor development so that they might become better citizens in society.

Growth of Institutions

Seguin's *Physiological Method*, which was the basis for pioneering work by Maria Montessori in early childhood education, became the standard for the growing number of facilities opened by states in the United States to educate and habilitate people deemed as "feebleminded." In 1848, Edouard Seguin moved to America and began consulting with superintendents of these schools. In 1852, Howe asked Seguin to take charge of the Massachusetts school as it moved from the Perkins School for the Blind to its own facility. By then, Hervey Wilbur had become the superintendent of a new institution in Syracuse, New York.

The number of state-run institutions rapidly increased through the late 1800s and early 1900s. While the intent of early institutions was to educate and habilitate people who, to that point, had not had access to such efforts, by the late 1800s the purposes of these institutions had changed. Institutions were intended not to habilitate, but to protect their inmates from the rigors and potential abuses of everyday society. In the early 1900s, this mission changed again: with people with intellectual disability viewed as menaces to society, institutions became places to segregate them for the benefit of society and to impose social control over the "unfit" to the benefit of the "fit."

The changing purposes of state-run institutions mirrored changes in society, including early twentieth century beliefs about eugenics. Eugenics is a term coined by Charles Darwin's half-cousin, Sir Francis Galton, and refers to the so-called science of the improvement of the human race by better breeding (Galton, 1904). The British eugenics movement focused primarily on efforts to improve the race by having *successful* people (that is, people like Galton) breed at greater rates. Some focus was turned to reducing the rate at which "the unfit" or less-desirable people bred in England, but in America, this negative form of eugenics (e.g., focusing on keeping some people deemed as unfit from reproducing) became the primary focus. The "unfit" included immigrants, people with intellectual or other disabilities, people of color in general, and often, just poor people (Smith & Wehmeyer, 2022).

Social historian Wolf Wolfensberger (1975) identified three trends in institutionalization—isolation, enlargement, and economization—that by 1910 dictated many aspects of the daily lives of the people who lived there. *Isolation* refers to the idea, mentioned previously, that people with intellectual disabilities need to be isolated from the rest of society. *Enlargement* emphasized the fact that the number of people being served by these institutions continued to grow through the first half of the twentieth century, from 30 in 1900 to 90 in 1950. The number of people residing in these state-run institutions increased even more dramatically, from about 60,000 in 1930 to almost 140,000 in 1950 (Wolfensberger, 1975). Institutions that were built to hold 1000 residents now held three or four times that number. Although institution populations grew, state funding for them did not. Thus, *economization* refers, in part, to the practices in institutions of using inmates to perform work that might otherwise be performed by paid staff, from caring for other inmates, to cleaning, to cooking and serving food, and so on. This system of peonage was among the reasons nobody ever left these institutions for the community. More capable inmates were needed to run the institution (Beckwith, 2016).

In 1927, the U.S. Supreme Court verdict in Buck v. Bell declared involuntary sterilization of some people, including people with intellectual disability, as constitutional. American institutions became the focal point for so-called asexualization operations, which were intended as eugenic measures. Women who were deemed "feebleminded" were targeted for forced, involuntary sterilization, in part so they could be released from the increasingly over-burdened and crowded institutions. By the 1970s, it was estimated that 50,000 Americans with disabilities were forcibly sterilized.

As detailed in Michael D'Antonio's book *The State Boys Rebellion*, from 1946 to 1953 young men who were residents of the Walter E. Fernald State School, named after an early superintendent who was an active part of the eugenics movement, were subjects of experiments jointly run by researchers at Harvard and MIT in which they were fed trace amounts of radioactive iron in their oatmeal as part of a study of the absorption of calcium and iron, without their knowledge or consent or the knowledge or consent of their parents or guardian (D'Antonio, 2004).

During the first half of the twentieth century, professionals shaped how people with intellectual disability were viewed and how disability itself was understood. Disability was conceptualized as a pathology internal to the person and attributable to disease or disorder; it was a characteristic of the person and people with disabilities were viewed by the public and professionals alike as broken, diseased, pathological, atypical, or aberrant, as different from the rest of "us" (Wehmeyer, 2013). Inevitably, disability was associated with negative stereotypes; people with intellectual disability were viewed as menaces to society, and were segregated, sterilized, and shunned.

The Community Era

The Parent Movement

In the 1950s parents began to reject the idea that their sons or daughters had to be institutionalized. The burgeoning parent movement sought to create a system in which their children might live at home and go to school like other children. This movement gained a powerful ally when, in 1960, John F. Kennedy was elected President of the United States. President Kennedy was the sibling of a sister with intellectual disability and in 1961 he established the "President's Panel on Mental Retardation" (Braddock, 2010). The panel issued 97 recommendations to create a community-based system, and only weeks before his assassination and urged on by his sister Eunice Kennedy Shriver, he signed into law the *Maternal and Child Health and Mental Retardation Planning Amendment to the Social Security Act*, taking this nation's first steps toward a community-based system of supports (Wehmeyer, 2013).

In 1965, President Kennedy's brother, Senator Robert Kennedy, visited the Willowbrook State School on Staten Island, in which almost 7000 people with intellectual disability were warehoused, almost twice the number of people the institution was designed to hold (Goode et al., 2013). As he was interviewed by reporters outside the institution, Kennedy was deeply disturbed by what he saw. He told the reporters:

> I think that particularly at Willowbrook, we have a situation that borders on a snake pit, and that the children live in filth, that many of our fellow citizens are suffering tremendously because lack of attention, lack of imagination, lack of adequate manpower. There is very little future for these children, for those who are in these institutions. .... (Minnesota Developmental Disabilities Council, n.d.)

That same year, disability scholar Burton Blatt, who was professor and chair of the special education department at Boston University, toured four institutions in the northeast and brought with him a photographer who surreptitiously snapped images of the horrific conditions in the facilities. The resulting photo essay, titled *Christmas in Purgatory* (Blatt & Kaplan, 1965), juxtaposed stark black and white photographs of frequently naked, often undernourished, severely impaired residents huddled on benches or sitting on the concrete floors, or showing rows of iron beds with young children confined to them with poetry verses and essays selected by Blatt. *Christmas in Purgatory* began with these words:

> There is a hell on earth and in America there is a special inferno. We were visitors there during Christmas, 1965 (Blatt & Kaplan, 1965, p. v.).

In 1972, a physician at Willowbrook let Geraldo Rivera and a film crew into Willowbrook through a back entrance. His report on a local television station became national news. Writing about this later, Rivera (1972) said:

> When Dr. Wilkins slid back the heavy metal door of B Ward, building No. 6, the horrible smell of the place staggered me. It was so wretched that my first thought was that the air was poisonous and would kill me ... [there] was a wailing sound that I still hear and that I will never forget. I said out loud, but to nobody in particular, "My God, they're children."
>
> Wilkins looked at me and said, 'Welcome to Willowbrook.' (p. 3)

Normalization

Leaders in the parent movement like Elizabeth Boggs and Gunnar Dybwad and Eunice Kennedy Shriver, along with some professionals like Burton Blatt, began to advocate for alternatives to institutions for people with intellectual and developmental disabilities. A framework for a community-based system was forwarded by Swedish philosopher Bengt Nirje (1969) who proposed the "Normalization Principle" as a guide for creating services that let people with intellectual disability "obtain an existence as close to the normal as possible" (Nirje, p. 363). Nirje stated:

> As I see it, the normalization principle means making available to the mentally retarded patterns and conditions of everyday life which are as close as possible to the norms and patterns of the mainstream of society. (p. 363)

Nirje (1969) identified eight "facets" and implications of the normalization principle. Normalization means that people with intellectual disability experience a normal rhythm of the day, the year, and life; had the opportunity to have typical developmentally appropriate experiences, to make choices, have intimate relationships, have opportunities to earn a living, and live, learn, work, and play in their community.

Scheerenberger (1987) noted that "at this stage in its development, the normalization principle basically reflected a lifestyle, one diametrically opposed to many prevailing institutional practices" (p. 117). The parent movement, the Kennedy administration's establishment of community-based services, and the wide adoption of the normalization principle had dramatic impacts on the use of institutions in the United States The 1954 institution census in the United States was almost 175,000 people, a figure that would rise to nearly 200,000 people by the 1970s. But, in September of 1987, the Willowbrook State School shut its doors, with every person who had lived there moving into the community, and by 1988 the institution census in America dropped below 100,000 for the first time in decades. As the institution census declined, the community options exploded: in 1977, slightly more than 20,000 people with intellectual and developmental disabilities lived in state-funded or private community-based residences. By the end of the 1980s, that figure had risen by almost 100,000 people, and by the end of the twentieth century, that number topped 500,000 (Wehmeyer, 2015). These increases occurred despite some opposition to community-based residences from neighborhood associations, challenges to zoning laws, and other forms of opposition (Rizzo et al., 1992).

The parent movement made significant progress in their other main objective, access to education. Two landmark court decisions, Pennsylvania Association for Retarded Citizens v. Commonwealth (1971) and Mills v. Board of Education of the District of Columbia (1972) established the rights of children with intellectual disability to receive a public education under the 14th amendment to the Constitution and in 1975, the U.S. Congress passed and President Gerald Ford signed Public Law 94-142, the Education for All Handicapped Children Act, which we know today as the Individuals with Disabilities Education Act, into law.

People First

While the parent movement changed the landscape from segregation to community inclusion, another movement was growing, the self-advocacy movement. In some ways, the self-advocacy movement, or the People First movement as it is often referred to with regard to people with intellectual and developmental disabilities, began with the normalization principle and Bengt Nirje. In a chapter in the book that introduced the normalization principle to a North American Audience, Nirje wrote:

> One major facet of the normalization principle is to create conditions through which a handicapped person experiences the normal respect to which any human is entitled. Thus, the choices, wishes, desires and aspirations of a handicapped person have to be taken into consideration as much as possible in actions affecting him. To assert oneself with one's family, friends, neighbors, co-workers, other people, or vis-a-vis an agency is difficult for many people. It is especially difficult for someone who has a disability or is otherwise perceived as devalued. Thus, the road to self-determination is indeed both difficult and all-important for a person who is impaired. (Nirje, 1972, p. 177)

The People First movement began with people with intellectual disability around the world. In the United States in the late 1960s, a small group of self-advocates in Salem, Oregon, came together and adopted the phrase "We are people first" (Edwards, 1982). They joined other self-advocates in a conference in British Columbia in 1973 then organized the first American conference, from which the idea of People First language emerged (Dybwad, 1996). Slowly, people with intellectual disability began to have a voice in decisions that impacted their lives and to influence policy and practice. Nancy Ward, who is a self-advocacy leader originally from Nebraska, wrote about those accomplishments in her state in the early years of the movement:

> One of the things that People First of Nebraska is real proud of is, and this really blows my mind, because you would think that since Nebraska is one of the leading states for community services for people that have disabilities, we would have done this a long time ago. It took People First of Nebraska to notice that we had a lot of obsolete language in our state statutes, like "moron," "idiot," "imbecile," and to do something about it. We got the legislature to get rid of the language. It was a compromise, because we had to replace it with "'people who have mental retardation,' but at least it's better than nothing." (Ward & Schultz, 1996, p. 209)

The Supports Era

Over the past few decades, the field has adopted a new paradigm of supports. Previous eras focused on "places" (e.g., institutions, communities). The supports era focuses not on a place but on what is needed to enable people with disabilities to live, learn, work, and play wherever they choose (Thompson et al., 2017). Supports are defined as resources or strategies that are designed for the person to support him/her in everyday life, and that aim to promote the development, education, interests, functioning, and individual's personal well-being (Thompson et al., 2009).

A key element in mobilizing the support era has been a focus on promoting and enhancing the self-determination of people with intellectual and developmental disabilities. That began with Nirje's 1972 call for the right to self-determination of people with intellectual disability quoted previously. Being self-determined implies that people themselves cause things to happen in their own lives, instead of other people (Mithaug, 1998). People who are self-determined act volitionally; they make choices based on their preferences (Wehmeyer, 1992). In our own work, we have defined being self-determined as acting as the causal agent in one's own life, where causal agent refers to self-caused action (Shogren et al., 2015; Wehmeyer, 1992)

What Should We Learn from the History of Intellectual Disability?

Much has changed since Jean Marc Gaspard Itard decided it was worth trying to teach Victor how to talk, care for himself, and so forth. From Seguin's initial attempts at education to today, the everyday lives of people with intellectual disability have changed dramatically. And yet, there is still far to go. In Washington DC, two statues flank the entrance to the National Archives. One statue depicts an elderly man sitting with a scroll clutched in his right hand and a closed book on his lap. In statuary symbolism, scrolls and books represent knowledge and wisdom. Because the book is closed in this statue, it refers to the knowledge and wisdom of past generations. The inscription on the base admonishes the viewer to "study the past." The second statue depicts a young woman who lifts her eyes from the pages of an open book. The open book is knowledge of the future, to which the young woman looks. The inscription in the base, from Shakespeare's play The Tempest, says: "the past is prologue."

We think of the phrase "what is past is prologue" to mean that history is a preview of what is to come and if we are not careful, we might repeat that history. That is one reason to study the past and learn the history of intellectual disability. But the phrase has a more nuanced meaning. The term appeared in The Tempest in scene I of Act II. In that context, it meant that all that happened before, in Act I, was prologue, but that there is more to come; that this is, in essence, a new beginning. So, what do we need to have learned from the past so that we can better prepare for the future? I would suggest five things.

1. First, it is inevitable that *if we conceptualize and understand people as different, it will lead to discrimination and maltreatment*. And this is not only pertinent to events from many decades ago, like the peonage and indentured servitude of the institutional era. In May 2013, a jury in Davenport, Iowa awarded $240 million dollars to 32 men with intellectual disability who had been forced by their employers at an Iowa turkey processing plant to engage in difficult work over many decades eviscerating turkeys for 41 cents per hour. If you read New York Times reporter Dan Barry's book about this horrific situation, *The Boys in the Bunkhouse: Servitude and Salvation in the Heartland* (Barry, 2016), you will understand that you cannot treat people the way these men were treated and, simultaneously, believe they are fully human.

2. Second, it is a simple lesson that we should have learned by now. *Separate is never equal.* When you enter the Brown versus Board of Education National Historic Site in Topeka, Kansas, the first thing you see is a plaque stating: "We conclude that in the field of public education the doctrine of 'separate but equal' has no place. Separate educational facilities are inherently unequal." The Olmstead Act of 1999 stated fundamentally the same thing about segregating people with intellectual disability in institutions, with the U.S. Supreme Court ruling that unjustified segregation of persons with disabilities constitutes discrimination in violation of Title II of the Americans with Disabilities Act. When we separate and segregate people, we are proclaiming their differentness and their peculiarity, and we head down the path we, as a nation, have traveled far too often of discrimination and maltreatment.
3. Third, I return to Robert Kennedy's statement after touring Willowbrook. The injustices and tragedies of places like Willowbrook reflect, primarily, on our lack of imagination; our inability or unwillingness to imagine how we can provide supports that enable people with intellectual and developmental disabilities to live richer, fuller lives in their communities. We do, in fact, know how to do better, we just choose, as a society, not to do so.
4. Fourth, if you look at the expectations held about people with intellectual disability from the 1950s onward, one trend holds true decade after decade, whether it be in employment, independent living, or education; people with intellectual disability have always achieved more than what the expectations from the previous decade or generation suggested they would. We have been wrong about our expectations in the past, and we're probably wrong now... people with intellectual disability can achieve more than we expect.
5. Fifth, one cannot learn about this history and not be struck by a fundamental truth, embodied, really, in each of the points. That is, when we conceptualize people as different and peculiar, when we separate and segregate them, and when we believe them to be incompetent and incapable and hold them to low expectations, we fundamentally deny them the dignity and value due to all human beings. The General Principles of the United Nations Convention on the Rights of Persons with Disabilities (CRPD) begin with the principle of "respect for inherent dignity, individual autonomy including the freedom to make one's own choices, and the independence of persons" (Committee on the Rights of Persons with Disabilities, 2016).

The past is prologue, but there is more to come. We can begin the second act with a focus on achieving the purposes of the CRPD "to promote, protect and ensure the full and equal enjoyment of all human rights and fundamental freedoms by all persons with disabilities, and to promote respect for their inherent dignity" (Committee on the Rights of Persons with Disabilities, 2016).

References

Barry, D. (2016). *The boys in the bunkhouse: Servitude and salvation in the heartland*. Harper.

Beckwith, R. M. (2016). *Disability servitude: From peonage to poverty*. Springer.

Blatt, B., & Kaplan, F. (1965). *Christmas in purgatory: A photographic essay on mental retardation*. Human Policy Press.

Braddock, D. (2010). Honoring Eunice Kennedy Shriver's legacy in intellectual disability. *Intellectual and Developmental Disabilities, 48*(1), 63–72.

Committee on the Rights of Persons with Disabilities. (2016). *General comment No. 4*. Accessed online January 10, 2024 at https://www.ohchr.org/Documents/HRBodies/CRPD/GC/RighttoEducation/CRPD-C-GC-4.doc

D'Antonio, M. (2004). *The state boys rebellion*. Simon and Schuster.

Dybwad, G. (1996). Setting the stage historically. In G. Dybwad & H. Bersani Jr. (Eds.), *New voices: Self-advocacy by people with disabilities* (pp. 1–17). Brookline Books.

Edwards, J. (1982). *We are people first: Our handicaps are secondary*. Ednick.

Ferguson, D. L., & Ferguson, P. M. (1986). The new Victors: A progressive policy analysis of work reform for people with very severe handicaps. *Mental Retardation, 24*, 331–338.

Galton, F. (1904). Eugenics: Its definition, scope and aims. *Nature, 70*, 82.

Goode, D., Hill, D., Reiss, J., & Bronston, W. (2013). *A history and sociology of the Willowbrook State School*. American Association on Intellectual and Developmental Disabilities.

Howe, S. G. (1848). *On the causes of idiocy: Being a supplement to a report by Dr S.G. Howe and the other commissioners appointed by the Governor of Massachusetts to inquire into the condition of the idiots of the Commonwealth, dated February 26, 1848 with An Appendix*. MacLachlan and Steward.

Itard, J-M. G. (1962). *The wild boy of Aveyron* (G. Humphrey & M. Humphrey, Trans.). Appleton-Century-Crofts. (Original work published 1806).

Katovitch, D. M. (2016). *Beautiful children: The story of the Elm Hill School and Home for feeble-minded children and youth*. Museum of Disability History.

Lane, H. (1976). *The wild boy of Aveyron*. Harvard University Press.

Minnesota Governor's Council on Developmental Disabilities. (n.d.). *Parallels in time: A six hour history of disability*. Accessed online at https://mn.gov/mnddc/parallels/five/5b/bobby-kennedy-snakepits.html on January 10, 2024.

Mithaug, D. (1998). Your right, my obligation? *Journal of the Association for Persons with Severe Handicaps, 23*, 41–43.

Newton, M. (2002). *Savage girls and wild boys: A history of feral children*. St. Martin's Press.

Nirje, B. (1969). The normalization principle and its human management implications. In R. B. Kugel & W. Wolfensberger (Eds.), *Changing residential patterns for the mentally retarded* (pp. 227–254). President's Committee on Mental Retardation.

Nirje, B. (1972). The right to self-determination. In W. Wolfensberger (Ed.), *Normalization: The principle of normalization* (pp. 176–200). National Institute on Mental Retardation.

Rivera, G. (1972). *Willowbrook: A report on how it is and why it doesn't have to be that way*. Random House.

Rizzo, A. M., Zipple, A. M., Pisciotta, J., & Bycoff, S. (1992). Strategies for responding to community opposition in an existing group home. *Psychosocial Rehabilitation Journal, 15*(3), 85–95.

Scheerenberger, R. C. (1987). *A history of mental retardation: A quarter century of promise*. Paul H. Brookes.

Seguin, E. (1846). *Traitment moral, hygiene, et education des idiots et des autres enfants arrieres*. J. B. Bailliere.

Seguin, E. (1866). *Idiocy and its treatment by the physiological method*. William Wood.

Shogren, K. A., Wehmeyer, M. L., Palmer, S. B., Forber-Pratt, A. J., Little, T. J., & Lopez, S. (2015). Causal agency theory: Reconceptualizing a functional model of self-determination. *Education and Training in Autism and Developmental Disabilities, 50*(3), 251.

Smith, J. D., & Wehmeyer, M. L. (2022). *Good blood, bad blood: Science, nature and the myth of the Kallikaks* (2nd ed.). American Association on Intellectual and Developmental Disabilities.
Thompson, J. R., Buntinx, W., Schalock, R. L., Shogren, K. A., Snell, M. E., Wehmeyer, M. L., et al. (2009). Conceptualizing supports and the support needs of people with intellectual disability. *Intellectual and Developmental Disabilities, 47*, 135–146.
Thompson, J. R., Wehmeyer, M. L., Shogren, K. A., & Seo, H. J. (2017). The supports paradigm and intellectual and developmental disabilities. In K. A. Shogren, M. L. Wehmeyer, & N. N. Singh (Eds.), *Handbook of positive psychology in intellectual and developmental disabilities: Translating research into practice* (pp. 23–35). Springer.
Ward, N., & Schoultz, B. (1996). People first of Nebraska. In G. Dybwad & H. Bersani Jr. (Eds.), *New voices: Self-advocacy by people with disabilities* (pp. 203–215). Brookline Books.
Wehmeyer, M. L. (1992). Self-determination and the education of students with mental retardation. *Education and Training in Mental Retardation, 27*, 302–314.
Wehmeyer, M. L. (2013). *The story of intellectual disability: An evolution of meaning, understanding, and public perception*. Paul H. Brookes.
Wehmeyer, M. L. (2015). Changing the place and perception: From institutions to community inclusion. In F. Spooner, M. Agran, & K. Gee (Eds.), *TASH: 40 years of progressive leadership* (pp. 96–105). TASH.
Wolfensberger, W. (1975). *Changing patterns in residential services for the mentally retarded.* President's Committee on Mental Retardation.

Current Issues in Intellectual and Developmental Disabilities: Families, Self-Advocates, and Intersectionality

Sandy Magaña, Amy Pei-Lung Yu, and Qusay S. Hussein Al-Mamari

Abstract In this chapter, the authors discuss three interrelated issues related to individuals with intellectual and developmental disabilities (IDD). The authors begin with the issue of family caregiving, including conceptual frameworks for understanding caregiving in IDD and caregivers' experiences across the life course of the individual with IDD. The next issue discussed is that of self-advocacy among individuals with IDD including a discussion of the history and current roles of the self-advocacy movement among individuals with IDD and autism. Finally, the authors discuss the issue of intersectional identities of individuals with IDD, particularly on the topics of race and ethnicity, gender, sexual orientation and gender identity, and immigrants and refugees with IDD. It is important to highlight these issues related to those with lived experience for people who support individuals with IDD and their families to understand their effects on the quality of life of individuals with IDD. In conclusion, the case is made for individuals with IDD, family members, and service providers (including teachers, social workers, physicians, nurses, therapists, and policymakers) to strive to address the differentiating needs of individuals with IDD and their families, and tailor their services and policies accordingly.

Keywords Families · Self-advocates · Intersectionality · Race · Ethnicity · Immigrants

While there are many current issues with respect to intellectual and developmental disabilities (IDD), in this chapter, we focused on three important ones that relate to those with lived experience: families of individuals with IDD, the engagement of

---

S. Magaña (✉) · A. P.-L. Yu · Q. S. Hussein Al-Mamari
Steve Hicks School of Social Work, University of Texas at Austin, Austin, TX, USA
e-mail: smagana@austin.utexas.edu

M. G. Valdovinos (ed.), *Intellectual and Developmental Disabilities*,
Contemporary Clinical Neuroscience,
https://doi.org/10.1007/978-3-031-66932-3_3

self-advocates with IDD in research and practice, and intersectional identities of individuals with IDD. We use both *person first* and *identity first* languages, which we discuss in more detail in the section on self-advocacy. When referring to individuals with IDD, we use *person first*, and when referring to autistic individuals, we use *identity first* language.

Families of Individuals with IDD

In the United States, approximately 72% of 7.39 million individuals with IDD live at home with family caregivers (Residential Information Systems Project [RISP], 2023). From early childhood to older adulthood, families play a critical role in the quality of life of individuals with IDD. Often, parents of children with IDD provide the most support for their child, ranging from supporting activities of daily living, personal development, full participation, meaningful choices, and ensuring the physical and financial well-being of their child with IDD (Hewitt & Nye-Lengerman, 2019; Ryff & Seltzer, 1996). Furthermore, parents of children with IDD have historically been strong advocates in disability legislation, particularly policies related to special education (Burke et al., 2021). Depending on the need, parents facilitate interactions between their child with IDD and their extended family members, friends, community members, school professionals, healthcare professionals, and direct support professionals. Often, multiple family members, including siblings, grandparents, and other extended family members, are collectively involved in supporting their relatives with IDD, especially if they live in the same household or in proximity (Tanis et al., 2022). As such, caregiving responsibilities often shift around different family members, depending on the caregivers' time, ability, and outside demands. This is particularly evident during moments of transition, such as from childhood to adolescence or adulthood to older adulthood, when life circumstances require family caregivers' flexibility (Grossman & Magaña, 2016; Meadan et al., 2015).

Caregiving and Support Across the Life Course

Ecological and life course theories are essential for understanding and interpreting the experiences of the families of individuals with IDD. Bronfenbrenner's ecological systems theory states that an individual is embedded within multiple levels of influence and that the interactions at each level concurrently affect one another (Bronfenbrenner, 1979; Sallis & Owen, 2002; Ravindran & Myers, 2012). These multiple levels of influence are conceptualized as the individual's microsystem, mesosystem, exosystem, and macrosystem, and individual characteristics (i.e., the disability) would naturally affect the systems' interactions and caregiving dynamics. Understanding the complex interplay of psychological, social, economic, political, and physical forces in these multiple systems is helpful in explaining the care and support dynamics at home (Gitterman et al., 2013; Glidden et al., 2021). The life course perspective is an interdisciplinary theoretical framework that explores how social and cultural contexts holistically influence a person over time (Elder et al., 2015). Employing the life-course perspective as we examine the literature on family caregiving helps us understand the diversity and fluidity of paths across different life stages (Grossman & Magaña, 2016).

The life course perspective is an interdisciplinary theoretical framework that explores how social and cultural contexts holistically influence a person over time (Elder et al., 2015). According to this theory, age-differentiated life stages shape life events, transitions, and trajectories (Elder, 1998), and the trajectories of one stage of life are inherently influenced by the trajectories in the previous stage of life (Johnson et al., 2011). For example, the trajectories of adolescents are inherently influenced by their childhood. Furthermore, various contextual changes, developmental stages, transitioning from childhood to adolescence, young adulthood to middle adulthood, and into and through old age could affect the family system's interactions and caregiving experience (Avieli, 2020; Macmillian & Copher, 2005).

It is important to distinguish between lifespan and life-course approaches, as the literature often uses them interchangeably. While the lifespan approach refers to the progression of the "typical life stages" of a person, the life course approach recognizes that there are multiple variations and pathways across individuals and families (Elder & Giele, 2009). The life course perspective is, therefore, suitable for individuals with IDD, as it allows for flexibility in reaching these life stages (Grossman & Magaña, 2016).

Families of Children and Adolescents with IDD

Family caregivers, particularly parents, play an important role in the early development of children with IDD. These include neurological, cognitive, motor, linguistic, socioemotional, and behavioral development (Perrin et al., 2016; Provenzi et al., 2018). Family caregivers often work alongside educators and professionals to support full participation, expectations, and meaningful choices early in life (Hewitt & Nye-Lengerman, 2019). Meta-analytic studies have shown that good parenting and positive parent–child interactions predict positive developmental outcomes in neurotypical children (Prime et al., 2023) and children with IDD (Dyches et al., 2012).

However, parents of children with IDD often exhibit higher levels of caregiving burden than do parents of neurotypical children (Raina et al., 2004). This could lead to higher risks of psychological stress, poorer social functioning, and other health-related issues than in parents of neurotypical children (Masefield et al., 2020; Trute et al., 2007). Notably, parents of individuals with autism tend to have elevated stress and poorer quality of life than parents of neurotypical children or other types of disabilities. Manifestations of autism, including specialized or focused behaviors, are often more pronounced and complex to manage than those of different types of IDD (Dabrowska & Pisula, 2010; Hayes & Watson, 2013; Vasilopoulou & Nisbet, 2016).

While most research has illuminated negative impacts and trajectories in raising children and adolescents with IDD, some have highlighted positive aspects of the experience. Notably, a systematic review screened 3181 articles and examined 22 studies on positive aspects of caring for a child with IDD. The review found eight prominent themes across the studies. These were (1) gained personal strength, (2) increased personal development, (3) gained a new perspective on life, (4) the child as a source of happiness and fulfillment, (5) improved/expanded relationships, (6) the positive effect the child has on others, (7) increased spirituality/religiosity, and (8) caring role (Beighton & Wills, 2019).

Families of Adults and Older Adults with IDD

Many adults with IDD continue to live at home, often due to personal choice or a lack of alternative housing options (Tanis et al., 2022). This percentage is usually higher for those who come from racially and ethnically marginalized groups, as living at home is often the most viable option (Bradley et al., 2021; Gary et al., 2009; Scharlach et al., 2006; Thai et al., 2015). Depending on the level of support that the individual with IDD needs, family caregivers may provide a range of support, including maintaining the health and well-being of the relative with IDD, exercising choice, and facilitating social activities and participation in the community (Schalock et al., 2007). For example, families may support individuals in managing finances, making important life decisions, medical care, daily living skills, work, and navigating romantic relationships (Hewitt & Nye-Lengerman, 2019).

With advancements in medicine and technology, individuals with IDD now have longer life expectancy (Landes et al., 2020). With most parental caregivers of IDD, either aging or middle-aged, many parents will likely delegate caregiving responsibilities to their other children, especially when they are physically unable to provide support (Heller & Arnold, 2010; Tanis et al., 2022). Therefore, it is crucial to include siblings in family-centered services (King et al., 2010), service coordination (Coyle et al., 2014), and transition planning (Chambers et al., 2004; Lee et al., 2021), as they typically have the longest familial and caregiving relationships with their siblings with IDD.

Culturally Diverse Families of Individuals with IDD

It is important to highlight protective factors and strengths among racial and ethnic minoritized communities. Black, Latino, and Asian parents of children with IDD report the importance of relying on their families for social and instrumental support (Rivera-Figueroa et al., 2022; Xu et al., 2022), which may help them to buffer the challenges and stressors they experience. Black and Latino parents also demonstrate a more positive attitude toward their child with IDD compared to White parents, such as reporting greater satisfaction with and positive contributions of their child (Ekas et al., 2019; Magana & Smith, 2006). Latino parents report a greater sense of mastery, family cohesion, personal growth, and less pessimism about their child's future than White parents (Lopez et al., 2018; Magaña & Smith, 2006). Black parents focus on promoting activities of daily living for their children and preventing harm from discrimination and racism (Burkett et al., 2017). Another area of strength for many Latino and Black families is faith and spirituality (Rivera-Figueroa et al., 2022). Black and Latino communities often report that their child is a blessing from God and rely on God and prayer to guide them (Ekas et al., 2019; Salkas et al., 2016). Faith can help these families cope and advocate for their children (Burke et al., 2019).

Self-Advocacy and IDD

While families are important for children and adults with IDD, there is a growing acknowledgment that self-determination can be overlooked for individuals with IDD as they move into adulthood. The self-advocacy movements of individuals with IDD and autism aimed to address this neglect of allowing people with IDD to have their own agency in their own lives. Here, we provide a very brief history of the self-advocacy movements and then discuss how self-advocacy has developed in recent times to emphasize including individuals who identify with autism and other IDDs in decisions about research. The self-advocacy movement of individuals with IDD came to fruition in the 1980s and 1990s and emphasized the importance of individuals with IDD speaking up for themselves with respect to programs, services, and societal issues that pertain to them (McDonald et al., 2022). However, its roots began in the 1960s and 1970s as the Disability Rights Movement began to emerge (Dybwad & Bersani Jr., 1996). There were many advocacy organizations that represented individuals with IDD, however, individuals with IDD themselves were rarely allowed to participate and share their voices as it was mostly family members and professionals who spoke and led the organizations (Dybwad & Bersani Jr., 1996). This began to change with self-advocates pushing for more involvement in decisions about their own lives and formed organizations like *People First* (http://peoplefirstwv.org/) and *Self-Advocates Becoming Empowered* (*SABE*; https://publications.ici.umn.edu/frontline-initiative/3-2/self-advocates-becoming-empowered%2D%2Dsabe). *People First* began in 1974 in Salem, Oregon, as an organization charged with organizing a conference in which individuals with IDD could speak for themselves (Dybwad & Bersani Jr., 1996). *People First* became an international organization. In the 1990s, a new generation of self-advocates in the United States saw the need for a national organization and formed, *SABE* with approval from *People First*. This movement has been critical in shaping programs and services for individuals with IDD by advocating for the elimination of draconian practices such as institutionalization and the mistreatment of individuals with IDD. Furthermore, it pushed for the basic rights of individuals with IDD, including self-determination and being accepted and valued as a person (Dybwad & Bersani Jr., 1996). The disability rights movement, family members, and individuals with IDD contributed to the push for legislation starting in 1975 that resulted in the Developmental Disabilities Assistance and Bill of Rights Act of 2000, also known as the DD Act (Public Law 106-402—Oct. 30, 2000). This legislation aims to "assure that individuals with developmental disabilities and their families participate in the design of and have access to needed community services, individualized supports, and other forms of assistance that promote self-determination, independence, productivity, and integration and inclusion in all facets of community life through culturally competent programs authorized under this title..." (DD Act, 17b). This Act authorizes programs such as State Councils on Developmental Disabilities (DD Councils), University Centers for Excellence in Developmental Disabilities (UCEDDs), and Protection and Advocacy Centers in each State. These programs are required to maintain advisory councils or committees that include individuals with IDD as advisors.

The autistic self-advocacy and neurodiversity movements began to arise in the 1990s and 2000s, with the emergence of the internet as it provided a more accessible form of communication for many autistic individuals (Leadbitter et al., 2021). The neurodiversity movement builds on aspects of the disability rights movement and disability studies frameworks, such as the social model of disability. Key premises of this movement are that there is a range of human variation in neurological functioning therefore, differences should not be considered pathological, and that ableism and the environment contribute to disabling experiences of autistic individuals as it does with individuals with other disabilities (Leadbitter et al., 2021). Here, ableism is defined as a set of beliefs, ideals, and attributes about oneself, one's body, and one's relationship with others, that are based on the abilities that one exhibits or values (Wolbring, 2008). Ableism is the consensus of a particular social group that values and promotes specific abilities over others (Wolbring, 2008). Thus, people with disabilities are often devalued and discriminated against like those from other minoritized groups. Although the definition and understanding of disability vary across groups, most people unconsciously interpret disability based on normative definitions of functioning (Friedman & Owen, 2017).

People with mental disabilities experience multiple layers of marginalization and encounter unique inequities. There is notably limited research directly exploring how ableism and/or racism affect people's overall quality of life, but in the research that does exist, it is shown that "ableism causes society to perceive individuals with disabilities differently" (Friedman & Owen, 2017). Expressing prejudiced feelings toward disabled people may be considered taboo, however, the beliefs and actions that result from ableist thoughts remain powerful forces, creating barriers and struggles for those with IDD.

One of the first autistic-run organizations, the Autism Network International (ANI) began in the early nineties and rejected the term *low functioning* as many of the original members were labeled with that term yet went on to earn university degrees (Silberman, 2016). ANI is credited with coming up with the term *neurotypical,* often shortened to NT, which refers to individuals with the limitations of not having an autistic brain. While the People First movement valued what is referred to as *people first language*, for example, person with Down syndrome, or person with autism, the autistic self-advocacy movement preferred *identity first language*, referring to themselves as autistic. *Identity first language* aligns with what was going on in the disability rights movement and disability studies in which people were embracing their disabilities as part of their identities and taking back their labels that may have been considered derogatory in the past (Berger, 2013). Several autistic runs organizations have emerged since ANI such as Wrong Planet in the United Kingdom (Silberman, 2016) and the Autistic Self Advocate Network (ASAN) in the United States (https://autisticadvocacy.org/about-asan/who-we-are/). ASAN was incorporated in 2006 and has developed into a robust autistic-run nonprofit that provides advocacy, leadership development, technical assistance, and other important functions.

There has been a growing body of research on autism, with more than $400,000,000 in federal dollars going toward it as of 2020 (Interagency Autism Coordinating Committee [IACC], 2023). Autistic self-advocate groups have been critical of the research being conducted about autism and whether it is asking the right questions or even useful to their own lives (Nicolaidis et al., 2011). The Interagency Autism Coordinating Committee (IACC) is a federal advisory committee that advises federal agencies funded by the Autism Cares Act (Public Law 116-60) that noted a shift in the autism community toward allowing autistic self-advocates to weigh in on priorities in autism research (IACC, 2023), and has been intentional in including self-advocates in the IACC. There is a growing movement for individuals with IDD and autism to be included in participatory research committees that advise research on IDD and autism. Community-based participatory research (CBPR) has been long recognized as a research approach to study health equity issues and emphasizes the inclusion of the community being studied in all aspects of the research study (Minkler & Wallerstein, 2008). More researchers and disability communities are recognizing this approach by including individuals with IDD on research advisory and decision-making boards (Nicolaidis et al., 2011; Stack & McDonald, 2018). One example is an academic community partnership called The Academic Autistic Spectrum Partnership in Research and Education (AASPIRE) which was established as an ongoing partnership that included autistic adults to advise and conduct several studies aimed at improving the lives of autistic individuals (Nicolaidis et al., 2011). Another example is a partnering project that included individuals with IDD on a steering committee and community advisory boards for a two-site study (Stack & McDonald, 2018). Other researchers have recognized the importance of including family members of individuals with IDD in a similar participatory approach, particularly parents of young children, and those from minoritized backgrounds (Magaña et al., 2017).

Intersectionality of Individuals with IDD

Intersectionality is a concept that is of interest to and/or is experienced by IDD self-advocates, individuals with IDD, and their families. Intersectionality was introduced in the late 1980s as a heuristic approach that rejects single-axis thinking and focuses on understanding the dynamics of differences and the fluidity of identities (Cho et al., 2013; Collins, 2015). The discourse originally began in the context of anti-discrimination and social movement politics and has since been facilitating discussions around race, class, gender, sexuality, ethnicity, nation, ability, and age (Cho et al., 2013). Over the past decades, intersectionality has been used as a conceptual framework for examining differences and sameness in multiple disciplines (Cho et al., 2013). Intersectionality was originally a legal theory coined by Kimberlé Crenshaw that argues for the consideration of identities within minoritized groups that may be discriminated against differently (Crenshaw, 1989). For example, we know that individuals with IDD experience discrimination and marginalization. What about Black individuals compared to White individuals with IDD, or an individual with IDD who identifies as transgender compared to an individual with IDD who is cisgender? How are their experiences unique? While the discourse around intersectionality is growing, the discussions of interactions between different social identities and demographic characteristics of individuals with IDD are often overlooked. In this section, we will explore intersectional identities that individuals with IDD may experience.

Race/Ethnicity and IDD

For young children with IDD and their families, access to early diagnosis and treatment of IDDs is experienced differently by those from racial and ethnic minoritized backgrounds, often leading to over or under detection of certain IDDs (Magaña & Vanegas, 2021). For example, Latino children have been historically underdiagnosed with autism, and Black children have been overrepresented in intellectual disability (Magaña & Vanegas, 2021). These differences in detection and diagnosis are likely caused by differences in social determinants of health, such as socioeconomic status, discrimination, language barriers, and limited educational and diagnostic resources available to people from low-resource and rural communities. The rates of diagnosis of specific IDDs are continually fluctuating. A report from the Centers for Disease Control and Prevention (CDC) autism surveillance project based on data between 2002 and 2016 found that in 2016, Latino and Black children had a higher prevalence of autism than White children in contrast to previous years when prevalence was lower especially for Latino children (Shaw et al., 2022). These authors also found that Latino and Black autistic children were more likely to have intellectual disability compared to White autistic children (Shaw et al., 2022). A study of data from the National Survey of Children's Health using years from 2016 to 2021 reported similar results (Goldblum et al., 2023). These findings may reflect that the systems are improving in diagnosing children from minoritized backgrounds, however there is still a significant disparity in the prevalence of Black and Latino children who have autism but do not have intellectual disability (Shaw et al., 2022). It is likely that many of these children are undetected by our diagnostic and service systems. Once diagnosed, racially and ethnically diverse children are less likely to receive services compared to their White peers (Bishop-Fitzpatrick & Kind, 2017; Magaña et al., 2013; Smith et al., 2020). Furthermore, individuals with IDD from racial and ethnic minoritized backgrounds are often misdiagnosed by school psychologists and special education professionals. Instead of ID or IDD, they may be classified with oppositional defiant disorder or conduct disorder, two diagnoses that do not qualify for Individuals with Disabilities Education Act (IDEA) protections (National Council on Disability, 2015).

In addition, individuals with IDD are more likely to have unmet healthcare needs than those without IDD (Mulcahy et al., 2022). These needs are exacerbated for those who come from racially and ethnically minoritized groups as they often experience many contextual barriers and, therefore are susceptible to greater racial, economic, and health disparities (Magaña et al., 2016; Miller et al., 2023). Individuals with IDD who identify as a racial minority and/or lesbian, gay, bisexual, transgender, and queer (LGBTQ+) often have even more challenging experiences in accessing and receiving healthcare than those who do not share both minoritized identities (Abbott & Burns, 2007; Wilson et al., 2018). The effects of these disparities are often intertwined and can be detrimental as individuals with IDD seek healthcare across their life course (Miller et al., 2023). Moreover, these health disparities are often unaddressed and continue into adulthood.

Adults with IDD are also prone to higher rates of obesity and other chronic illnesses compared to adults without IDD (García-Domínguez et al., 2020; Hsieh et al., 2014). Within this community, Black and Latino individuals with IDD have higher rates of chronic diseases such as diabetes, asthma, and hypertension (Xu et al., 2023). They may also experience higher rates of mental health problems. For example, one study found that Black autistic youth had significantly higher rates of depressive symptoms than White youth (Williams et al., 2022). Another study found that Latino and non-Latino Black adults with IDD are more likely to exhibit fair or poor physical and mental health than non-Latino White counterparts (Magaña et al., 2016). The risk and expression of co-occurring conditions are heightened for individuals with IDD from diverse backgrounds as they often face numerous contextual stressors such as language barriers, financial burden, discrimination, and social isolation (Updegraff & Umaña-Taylor, 2015).

Youth and adults with IDD may experience more negative encounters with law enforcement than those without IDD. This is often due to law enforcement officers misinterpreting behaviors that individuals with IDD may exhibit such as communication challenges, reaction to loud noises, and difficulty with social cues (Anthony & Bills, 2021). In particular, individuals of IDD from racial and ethnic minoritized communities are at an increased risk for law enforcement interactions as many of them experience contextual stressors and social and communication challenges resulting from the disability simultaneously (Anthony & Bills, 2021; Rava et al., 2017). It is well known that Black and Latino youth and adults experience more negative encounters, and these encounters can be exacerbated by law enforcement's perception and limited knowledge about disability. One study found that Black individuals with disabilities are three times more likely to be killed by police than White individuals (Abram, 2020).

Gender, Sexual Orientation, and Gender Identity

Children who were assigned as boys at birth were twice as likely to have been diagnosed with IDD between 2019 and 2021 (Zablotsky et al., 2023). This rate may be driven by autism diagnoses, as boys are almost four times as likely to be diagnosed with autism when compared with girls and this has been consistently reported in research (Goldblum et al., 2023). However, there is emerging research that finds that autistic children assigned as girls at birth may go undetected because autism may present differently in girls than in boys (Allely, 2019). In a systematic review, several studies have reported that girls may be better at social imitation, which camouflages symptoms (Allely, 2019). Others argue that how autism is characterized is based on males, including diagnostic instruments that may not be sensitive to other genders (Chester, 2019). Furthermore, assumptions by professionals that autism is mainly manifested in boys may lead to girls being overlooked (Bargiela et al., 2016; Chester, 2019). This scholarship suggests that the prevalence rate for girls may be underestimated.

Women and girls with intellectual disabilities are especially vulnerable to gender-based violence, including physical, sexual, psychological, and economic violence (The World Bank Group et al., 2019). In addition, women and girls with IDD face sterilization, coerced abortion, and removal of children (The World Bank Group et al., 2019). Women and girls with IDD are often unwilling to disclose gender-based violence due to stigmas, and when they do, disclosure of any kind is often more complicated due to communication barriers and comprehension challenges with others (Van Der Heijden et al., 2019). In many situations, women and girls with IDD are susceptible to stereotypes and vulnerabilities, which often make them unreliable reporters of violence (United Nations Population Fund, 2018).

Sexuality is often downplayed for individuals with IDD as professionals and parents often presume individuals with IDD as asexual (McCann et al., 2016). Subsequently, there is very little discussion on gender diversity and sexual preferences among individuals with IDD, their families, and the professionals who work with them. Sexual minorities with IDD have unique intersectional experiences and needs that are often neglected (McCann et al., 2016). However, despite the increased understanding that individuals with IDD experience gender diversity and diverse sexual preferences at least as much as the general population (McCann et al., 2016), they often encounter numerous personal and socio-environmental barriers in their sexual lives (Matin et al., 2021). Individuals with IDD and in particular, women and girls with IDD require sex education that is tailored to their needs (Matin et al., 2021). It is important as sexual knowledge (e.g., knowledge of contraceptive methods, sexual behaviors, abuse, and the process of intercourse) is key to healthy sexual relationships (Matin et al., 2021). However, many women with ID reported the lack of support, and sexual education in schools, in part due to cultural prohibitions and stigmas (Martin et al., 2021). As a result, women with IDD are often more vulnerable to sexual exploitation compared to their peers without IDD (Martin et al., 2021).

Individuals with IDD, especially those from Sexual and Gender Minority (SGM) communities, experience significantly more sexual harassment and other forms of mistreatment by healthcare providers, including being denied or receiving a lower quality of care, than those without disabilities (Lamba et al., 2023). SGM individuals with disabilities, who are often younger, belong to minoritized racial and ethnic groups, are more likely to be unemployed and uninsured, and have increased odds of delayed care or accessing care compared to those without disabilities (Lamba et al., 2023).

Emerging studies have found increased rates of gender diversity among autistic children and adults compared to those in the general population (Warrier et al., 2020). For example, a large study of 641,850 individuals found that compared to cisgender individuals (gender aligns with their sex assigned at birth), transgender and gender diverse individuals (gender does not always align with sex assigned at birth) were five times as likely to be diagnosed with autism (Warrier et al., 2020). This study also found higher rates of neurodevelopmental disabilities, such as attention deficit hyperactivity disorder and learning disorders among gender-diverse individuals compared to cisgender individuals (Warrier et al., 2020). It is not yet understood why autism and neurodevelopmental disabilities are overrepresented in gender-diverse populations.

Immigrants and Refugees with IDD

Refugees and migrants with IDD are often overlooked by settlement services (Luce, 2018). Such exclusion greatly impacts individuals with IDD and their families who are already facing barriers as refugees and migrants, such as increased challenges in accessing resources and community participation (Luce, 2018). Furthermore, refugees and migrants with IDD often experience heightened isolation due to a lack of awareness of the IDD migrant communities and inclusive practices (Luce, 2018).

Refugees and migrants with IDD often face additional challenges not only due to their multiple minoritized identities but also stigma and discrimination in the host country (Hultman et al., 2023). Essential resources such as accessible housing and knowledge about societal resources are often lacking, impacting their ability to lead a fulfilling life (Hultman et al., 2023). It is important that service providers recognize the multiple intersectional identities (i.e., race/ethnicity, sexual orientation, gender preferences, spirituality) that refugees and migrants may have so that they may provide appropriate services accordingly (Hultman et al., 2023).

Like racially and ethnically minoritized individuals in the United States, refugees and migrants with disabilities also find comfort in their faith and spirituality (Bridi et al., 2023; Rayes et al., 2021). Studies in Berlin, Germany, have explored faith-based coping strategies among Arabic-speaking refugee adults seeking mental healthcare services (Rayes et al., 2021). These coping mechanisms extend beyond IDD to encompass post-traumatic stress disorder, anxiety, or depression (Rayes et al., 2021). Many refugees and migrants believe that events are predetermined by God, fate, or destiny, influencing their perceptions of mental and cognitive health (Bridi et al., 2023). These spiritual beliefs, in the absence of resources, serve as a source of hope and support for refugees and migrants dealing with IDD or other disabilities (Bridi et al., 2023).

Conclusion

This chapter highlighted issues in research and practice on topics of family members of individuals with IDD, self-advocates with IDD, and diversity and intersectionality among individuals with IDD and their families. In response to these issues, professionals in direct support and service settings, such as social workers, teachers, physicians, nurses, and therapists, should strive to include individuals with IDD and their families in all conversations about them. This includes but is not limited to screening and diagnosis, informational meetings, individual service/education plan meetings, transition meetings, and future planning. It is important that service providers recognize the unique needs of each individual and their families and provide accommodations and flexibility accordingly.

With the increase in racially and ethnically diverse groups in the United States, service providers should place more emphasis on providing services that are culturally and linguistically responsive to the needs of individuals with IDD and their families. Part of this may be understanding coping mechanisms such as spirituality among minoritized populations, including immigrants and refugees. Given that the individual's identities and contextual factors greatly affect how they experience IDD, it is imperative to understand their intersectional identities socially and culturally so that professionals and family caregivers can better support the individual with IDD. An important aspect of intersectionality is gender and sexual orientation. Individuals with IDD need support from family members and professionals who care for them to understand and embrace their sexual identities (Hughes, 2022). To address challenges faced by migrants and refugees with IDD, it is recommended that authorities and organizations that work with migrants and individuals with disabilities should collaborate in developing a comprehensive training program to support them (Luce, 2018). In addition, all operational and informational documents provided to migrants should have easy-read versions, plain language, and options for other languages.

To promote health for individuals with IDD, it is important for healthcare providers to focus on building a community and messaging that connects to both IDD and their racial and ethnic identities (Biswas, 2016). When people from minoritized populations with IDD go to clinics or hospitals, they face multiple stigmas, therefore, healthcare providers should communicate directly with their patients, even those who have communication or cognitive differences and those with caregivers (Kripke, 2018). Communication might include sign language, speech, writing, letter boards, voice output devices, facial expressions, and gestures.

Individuals with IDD should be included in decision-making boards of organizations that provide services to them or that work on IDD policy. It is important to have individuals with IDD and, where relevant, family members in research advisory committees and as researchers and peer-interventionists. Those with lived experience can advise on all aspects of research, from the research questions to the interpretation of findings. Overall, to better serve individuals with IDD and their families, including them in all aspects of policy, practice, and research is warranted by making the phrase, *Nothing About Us Without Us* (Charlton, 2000) a reality.

References

Abbott, D., & Burns, J. (2007). What's love got to do with it?: Experiences of lesbian, gay, and bisexual people with intellectual disabilities in the United Kingdom and views of the staff who support them. *Sexuality Research & Social Policy: A Journal of the NSRC, 4*(1), 27–39. https://doi-org.ezproxy.lib.utexas.edu/10.1525/srsp.2007.4.1.27

Abrams, A. (2020, June 25). Black, disable–d and at risk: The overlooked problem of police violence against Americans with disabilities. *Time*. https://time.com/5857438/police-violence-black-disabled/

Allely, C. S. (2019). Understanding and recognising the female phenotype of autism spectrum disorder and the "camouflage" hypothesis: A systematic PRISMA review. *Advances in Autism, 5*(1), 14–37. https://doi.org/10.1108/AIA-09-2018-0036

Anthony, N., & Bills, K. (2021). Black autistic, and interactions with law enforcement. *Academia Letters,* Article 395.

Avieli, H. (2020). How middle-aged siblings of adults with intellectual disability experience their roles: A qualitative analysis. *Journal of Developmental and Physical Disabilities, 32*(4), 633–651. https://doi.org/10.1007/s10882-019-09710-3

Bargiela, S., Steward, R., & Mandy, W. (2016). The experiences of late-diagnosed women with autism spectrum conditions: An investigation of the female autism phenotype. *Journal of Autism and Developmental Disorders, 46*(10), 3281–3294. https://doi.org/10.1007/s10803-016-2872-8

Beighton, C., & Wills, J. (2019). How parents describe the positive aspects of parenting their child who has intellectual disabilities: A systematic review and narrative synthesis. *Journal of Applied Research in Intellectual Disabilities, 32*(5), 1255–1279. https://doi.org/10.1111/jar.12617

Berger, R. (2013). *Introducing disability studies*. Lynne Rienner publishers.

Bishop-Fitzpatrick, L., & Kind, A. J. H. (2017). A scoping review of health disparities in autism spectrum disorder. *Journal of Autism and Developmental Disorders, 47*(11), 3380–3391. https://doi.org/10.1007/s10803-017-3251-9

Biswas, A. (2016). *Obesity in people with intellectual disabilities*. University of Hertfordshire: Intellectual Disability and Health. https://www.intellectualdisability.info/physical-health/obesity-in-people-with-intellectual-disabilities

Bradley, V. J., Hiersteiner, D., Goode, T. D., Bonardi, A., Giordano, S., Bourne, M. L., & Vegas, L. (2021). *What do NCI data tell us about significant racial and ethnic disparities across qual-*

ity of life and health domains? National Core Indicators Data Brief. Retrieved from: https://legacy.nationalcoreindicators.org/upload/core-indicators/NCI_DB_RacialEquity_final.pdf

Bridi, L., Kaki, D. A., Albahsahli, B., Abu Baker, D., Khan, X., Aljenabi, R., Bencheikh, N., Schiaffino, M. K., Moore, A. A., & Al-Rousan, T. (2023). The influences of faith on illness representations and coping procedures of mental and cognitive health among aging Arab refugees: A qualitative study. *Frontiers in Psychiatry, 14*, 1083367. https://doi.org/10.3389/fpsyt.2023.1083367

Bronfenbrenner, U. (1979). *The ecology of human development: Experiments by nature and design*. Harvard University Press.

Burke, M. M., Rios, K., Garcia, M., Sandman, L., Lopez, B., & Magaña, S. (2019). Examining the perspectives of Latino families of children with autism spectrum disorder towards advocacy. *Exceptionality, 27*(3), 201–214. https://doi.org/10.1080/09362835.2018.1433041

Burke, M. M., Rossetti, Z., Aleman-Tovar, J., & Rios, K. (2021). Examining the relation between empowerment and civic engagement among parents of individuals with intellectual and developmental disabilities. *Journal of applied research in intellectual disabilities: JARID, 34*(6), 1569–1581. https://doi-org.ezproxy.lib.utexas.edu/10.1111/jar.12901

Burkett, K., Morris, E., Anthony, J., Shambley-Ebron, D., & Manning-Courtney, P. (2017). Parenting African American children with autism: The influence of respect and faith in mother, father, single-, and two-parent care. *Journal of Transcultural Nursing, 28*(5), 496–504. https://doi.org/10.1177/1043659616662316

Chambers, C. R., Hughes, C., & Carter, E. W. (2004). Parent and sibling perspectives on the transition to adulthood. *Education and Training in Developmental Disabilities, 39*, 79–94.

Charlton, J. (2000). *Nothing about us without us: Disability oppression and empowerment*. University of California Press.

Chester, V. (2019). Autistic women and girls: Under-recognised, under-researched and under-served. *Advances in Autism, 5*(1), 1–1. https://doi.org/10.1108/AIA-01-2019-049

Cho, S., Crenshaw, K. W., & McCall, L. (2013). Toward a field of intersectionality studies: Theory, applications, and praxis. *Signs: Journal of Women in Culture and Society, 38*(4), 785–810. https://doi.org/10.1086/669608

Collins, P. H. (2015). Intersectionality's definitional dilemmas. *Annual Review of Sociology, 41*(1), 1–20. https://doi.org/10.1146/annurev-soc-073014-112142

Coyle, C. E., Kramer, J., & Mutchler, J. E. (2014). Aging together: Sibling carers of adults with intellectual and developmental disabilities. *Journal of Policy & Practice in Intellectual Disabilities, 11*(4), 302–312. https://doi.org/10.1111/jppi.12094

Crenshaw, K. (1989). Demarginalizing the intersection of race and sex: A Black feminist critique of antidiscrimination doctrine, feminist theory and antiracist politics. *University of Chicago Legal Forum, 1989*(1), 139–167.

Dabrowska, A., & Pisula, E. (2010). Parenting stress and coping styles in mothers and fathers of pre-school children with autism and Down syndrome. *Journal of Intellectual Disability Research, 54*(3), 266–280. https://doi.org/10.1111/j.1365-2788.2010.01258.x

Dybwad, G., & Bersani, H. A. (1996). In G. Dybwad & H. Bersani Jr. (Eds.), *New voices: Self-advocacy by people with disabilities*. Brookline Books.

Dyches, T. T., Smith, T. B., Korth, B., Roper, S. O., & Mandleco, B. (2012). Positive parenting of children with developmental disabilities: A meta-analysis. *Research in Developmental Disabilities, 33*, 2213–2220.

Ekas, N. V., Tidman, L., & Timmons, L. (2019). Religiosity/spirituality and mental health outcomes in mothers of children with autism spectrum disorder: The mediating role of positive thinking. *Journal of Autism and Developmental Disorders, 49*(11), 4547–4558. https://doi.org/10.1007/s10803-019-04165-z

Elder, G. H. (1998). The life course as a developmental theory. *Child Development, 69*(1), 1–12. https://doi.org/10.1111/j.1467-8624.1998.tb06128.x

Elder, G. H., Jr., & Giele, J. Z. (2009). Life course studies: An evolving field. In G. H. Elder Jr. & J. Z. Giele (Eds.), *The craft of life course research* (pp. 1–24). The Guilford Press.

Elder, G. H., Shanahan, M. J., & Jennings, J. A. (2015). Human development in time and place. In M. H. Bornstein & T. Leventhal (Eds.), *Handbook of child psychology and developmental science* (pp. 1–49). Wiley.

Friedman, C., & Owen, A. L. (2017). Defining disability: Understandings of and attitudes towards ableism and disability. *Disability Studies Quarterly, 37*(1). https://doi.org/10.18061/dsq.v37i1.5061

García-Domínguez, L., Navas, P., Verdugo, M. Á., & Arias, V. B. (2020). Chronic health conditions in aging individuals with intellectual disabilities. *International Journal of Environmental Research and Public Health, 17*(9), 3126. https://doi.org/10.3390/ijerph17093126

Gary, K. W., Arango-Lasprilla, J. C., & Stevens, L. F. (2009). Do racial/ethnic differences exist in post-injury outcomes after TBI? A comprehensive review of the literature. *Brain Injury, 23*(10), 775–789. https://doi.org/10.1080/02699050903200563

Gitterman, A., Germain, C. B., & Knight, C. (2013). *Ecological framework. Encyclopedia of social work*. NASW Press and Oxford University Press.

Glidden, L. M., Floyd, F. J., Hastings, R. P., & Mailick, M. R. (2021). Family impact and adjustment across the lifespan: Parents of children with intellectual and developmental disabilities. In L. M. Glidden et al. (Eds.), *APA handbook of intellectual and developmental disabilities: Clinical and educational implications* (pp. 213–246). American Psychological Association. https://doi.org/10.1037/0000194-000

Goldblum, J. E., McFayden, T. C., Bristol, S., Putnam, O. C., Wylie, A., & Harrop, C. (2023). Autism prevalence and the intersectionality of assigned sex at birth, race, and ethnicity on age of diagnosis. *Journal of Autism and Developmental Disorders*. https://doi.org/10.1007/s10803-023-06104-5

Grossman, B. R., & Magaña, S. (2016). Introduction to the special issue: Family support of persons with disabilities across the life course. *Journal of Family Social Work, 19*(4), 237–251. https://doi.org/10.1080/10522158.2016.1234272

Hayes, S. A., & Watson, S. L. (2013). The impact of parenting stress: A meta-analysis of studies comparing the experience of parenting stress in parents of children with and without autism spectrum disorder. *Journal of Autism and Developmental Disorders, 43*(3), 629–642. https://doi.org/10.1007/s10803-012-1604-y

Heller, T., & Arnold, C. K. (2010). Siblings of adults with developmental disabilities: Psychosocial outcomes, relationships, and future planning. *Journal of Policy and Practice in Intellectual Disabilities, 7*, 16–25. https://doi.org/10.1111/j.1741-1130.2010.00243.x

Hewitt, A., & Nye-Lengerman, K. M. (2019). *Community living and participation for people with intellectual and developmental disabilities*. American Association on Intellectual and Developmental Disabilities.

Hsieh, K., Rimmer, J. H., & Heller, T. (2014). Obesity and associated factors in adults with intellectual disability. *Journal of Intellectual Disability Research: JIDR, 58*(9), 851–863. https://doi-org.ezproxy.lib.utexas.edu/10.1111/jir.12100

Hughes, O. E. (2022). Supporting gender self-determination for people with intellectual and developmental disabilities. *Inclusive Practices, 1*(2), 70–74. https://doi.org/10.1177/27324745211044452

Hultman, L., Asaba, E., Riedel, D., Abdu, S., Afe, H., Atafnu, R., Ejigu, L., Bolling, J., Negussie, M., Ntobua, J., & Mondaca, M. (2023). Migration and disability narratives from an intersectional perspective: A photovoice study. *Disability & Society*, 1–26. https://doi.org/10.1080/09687599.2023.2271157

Interagency Autism Coordinating Committee, IACC. (2023). *2021–2023 IACC Strategic plan for Autism Research, Services and Policy*. Retrieved from the U.S. Department of Health and Human Services Interagency Autism Coordinating Committee website. https://iacc.hhs.gov/publications/strategic-plan/2023/

Johnson, M., Crosnoe, R., & Elder, G. (2011). Insights on adolescence from a life course perspective. *Journal of Research on Adolescence, 21*, 273–280. https://doi.org/10.1111/j.1532-7795.2010.00728.x

King, M., Ruggles, S., Alexander, J. T., Flood, S., Genadek, K., Schroeder, M. B., Trampe, B., & Vick, R. (2010). *Integrated public use microdata series, current population survey: Version 3.0. [machine-readable database]* (p. 20). University of Minnesota.

Kripke, C. (2018). Adults with developmental disabilities: A comprehensive approach to medical care. *American Family Physician, 97*(10), 649–656.

Lamba, S., Obedin-Maliver, J., Mayo, J., Flentje, A., Lubensky, M. E., Dastur, Z., & Lunn, M. R. (2023). Self-reported barriers to care among sexual and gender minority people with disabilities: Findings from The PRIDE Study, 2019-2020. *American Journal of Public Health (1971), 113*(9), 1009–1018. https://doi.org/10.2105/AJPH.2023.307333

Landes, S. D., McDonald, K. E., Wilmoth, J. M., & Carter Grosso, E. (2020). Evidence of continued reduction in the age-at-death disparity between adults with and without intellectual and/or developmental disabilities. *Journal of Applied Research in Intellectual Disabilities, 34*(3), 916–920. https://doi.org/10.1111/jar.12840

Leadbitter, K., Buckle, K. L., Ellis, C., & Dekker, M. (2021). Autistic self-advocacy and the neurodiversity movement: Implications for autism early intervention research and practice. *Frontiers in Psychology, 12*. https://doi.org/10.3389/fpsyg.2021.635690

Lee, C., Burke, M. M., & Arnold, K. (2021). Sibling participation in service planning meetings for their brothers and sisters with intellectual and developmental disabilities in the United States. *Journal of Policy and Practice in Intellectual Disabilities, 18*(2), 104–112. https://doi.org/10.1111/jppi.12359

Lopez, K., Xu, Y., Magaña, S., & Guzman, J. (2018). Parents' reaction to autism diagnosis: A qualitative analysis comparing Latino and white parents. *Journal of Rehabilitation, 84*, 41–50.

Luce, A. (2018). *Asylum seekers and refugees with intellectual disabilities in Europe*. Samuel Centre for Social Connectedness.

Macmillian, R., & Copher, R. (2005). Families in the life course: Interdependency of roles, role configurations, and pathways: Theoretical and methodological issues in studying families. *Journal of Marriage and Family, 67*(4), 858–879.

Magaña, S., & Smith, M. J. (2006). Psychological distress and well-being of Latina and nonLatina White mothers of youth and adults with an autism spectrum disorder: Cultural attitudes towards coresidence status. *American Journal of Orthopsychiatry, 76*(3), 346–357. https://doi.org/10.1037/0002-9432.76.3.346

Magaña, S., & Vanegas, S. B. (2021). Culture, race, and ethnicity and intellectual and developmental disabilities. In L. M. Glidden, L. Abbeduto, L. L. McIntyre, & M. J. Tassé (Eds.), *APA handbook of intellectual and developmental disabilities: Foundations* (pp. 355–382). American Psychological Association. https://doi-org.ezproxy.lib.utexas.edu/10.1037/0000194-014

Magaña, S., Lopez, K., Aguinaga, A., & Morton, H. (2013). Access to diagnosis and intervention services among Latino children with autism spectrum disorders. *Intellectual and Developmental Disabilities, 51*(3), 141–153. https://doi.org/10.1352/1934-9556-51.3.141

Magaña, S., Parish, S., Morales, M. A., Li, H., & Fujiura, G. (2016). Racial and ethnic Health disparities among people with intellectual and developmental disabilities. *Intellectual and Developmental Disabilities, 54*(3), 161–172. https://doi.org/10.1352/1934-9556-54.3.161

Magaña, S., Lopez, K., & Machalicek, W. (2017). Parents taking action: A psycho-educational intervention for Latino parents of children with autism spectrum disorders. *Family Process, 56*, 50–74. https://doi.org/10.1111/famp.12169

Masefield, S. C., Prady, S. L., Sheldon, T. A., Small, N., Jarvis, S., & Pickett, K. E. (2020). The caregiver health effects of caring for young children with developmental disabilities: A meta-analysis. *Maternal & Child Health Journal, 24*(5), 561–574. https://doi-org.ezproxy.lib.utexas.edu/10.1007/s10995-020-02896-5

Matin, B. K., Ballan, M., Darabi, F., Karyani, A. K., Soofi, M., & Soltani, S. (2021). Sexual health concerns in women with intellectual disabilities: A systematic review in qualitative studies. *BMC Public Health, 21*(1), 1965. https://doi.org/10.1186/s12889-021-12027-6

McCann, E., Lee, R., & Brown, M. (2016). The experiences and support needs of people with intellectual disabilities who identify as LGBT: A review of the literature. *Research in Developmental Disabilities, 57*, 39–53. https://doi-org.ezproxy.lib.utexas.edu/10.1016/j.ridd.2016.06.013

McDonald, K. E., Gibbons, C., Conroy, N., & Olick, R. S. (2022). Facilitating the inclusion of adults with intellectual disability as direct respondents in research: Strategies for fostering trust, respect, accessibility and engagement. *Journal of Applied Research in Intellectual Disabilities: JARID, 35*(1), 170–178. https://doi.org/10.1111/jar.12936

Meadan, H., Stoner, J. B., & Angell, M. E. (2015). Fathers of children with autism: Perceived roles, responsibilities, and support needs. *Early Child Development & Care, 185*(10), 1678–1694.

Miller, H. L., Thomi, M., Patterson, R. M., & Nandy, K. (2023). Effects of intersectionality along the pathway to diagnosis for autistic children with and without co-occurring attention deficit hyperactivity disorder in a nationally-representative sample. *Journal of Autism and Developmental Disorders, 53*(9), 3542–3557. https://doi-org.ezproxy.lib.utexas.edu/10.1007/s10803-022-05604-0

Minkler, M., & Wallerstein, N. (Eds.). (2008). *Community-based participatory research for health: From processes to outcomes*. Jossey-Bass.

Mulcahy, A., Streed, C. G., Wallisch, A. M., Batza, K., Kurth, N., Hall, J. P., & McMaughan, D. J. (2022). Gender identity, disability, and unmet healthcare needs among disabled people living in the community in the United States. *International Journal of Environmental Research and Public Health, 19*(5), 2588. https://doi.org/10.3390/ijerph19052588

National Council on Disability. (2015). *Breaking the school-to-prison pipeline for students with disabilities*. https://www.ncd.gov/report/breaking-the-school-to-prison-pipeline-for-students-with-disabilities/

Nicolaidis, C., Raymaker, D., McDonald, K., Dern, S., Ashkenazy, E., Boisclair, C., Roberson, S., & Baggs, A. (2011). Collaboration strategies in nontraditional community-based participatory research partnerships: Lessons from an academic-community partnership with autistic self-advocates. *Progress in Community Health Partnerships: Research, Education, & Action, 5*(2), 143–150. https://doi.org/10.1353/cpr.2011.0022

Perrin, E. C., Leslie, L. K., & Boat, T. (2016). Parenting as primary prevention. *JAMA Pediatrics, 170*(7), 637–638. https://doi.org/10.1001/jamapediatrics.2016.0225

Prime, H., Andrews, K., Markwell, A., Gonzalez, A., Janus, M., Tricco, A. C., Bennett, T., & Atkinson, L. (2023). Positive parenting and early childhood cognition: A systematic review and meta-analysis of randomized controlled trials. *Clinical Child and Family Psychology Review, 26*(2), 362–400. https://doi.org/10.1007/s10567-022-00423-2

Provenzi, L., Scotto di Minico, G., Giusti, L., Guida, E., & Müller, M. (2018). Disentangling the dyadic dance: Theoretical, methodological and outcomes systematic review of mother-infant dyadic processes. *Frontiers in Psychology, 9*(348). https://doi.org/10.3389/fpsyg.2018.00348

Raina, P., O'Donnell, M., Schwellnus, H., Rosenbaum, P., King, G., Brehaut, J., et al. (2004). Caregiving process and caregiver burden: Conceptual models to guide research and practice. *BMC Pediatrics, 4*, 1. https://doi.org/10.1186/1471-2431-4-1

Rava, J., Shattuck, P., Rast, J., & Roux, A. (2017). The prevalence and correlates of involvement in the criminal justice system among youth on the autism spectrum. *Journal of Autism and Developmental Disorders, 47*(2), 340–346. https://doi.org/10.1007/s10803-016-2958-3

Ravindran, N., & Myers, B. J. (2012). Cultural influences on perceptions of health, illness, and disability: A review and focus on autism. *Journal of Child and Family Studies, 21*(2), 311–319. https://doi.org/10.1007/s10826-011-9477-9

Rayes, D., Karnouk, C., Churbaji, D., Walther, L., & Bajbouj, M. (2021). Faith-based coping among Arabic-speaking refugees seeking mental health services in Berlin, Germany: An exploratory qualitative study. *Frontiers in Psychiatry, 12*. https://doi.org/10.3389/fpsyt.2021.595979

Residential Information Systems Project. (2023). *People with IDD in the United States*. University of Minnesota, RISP, Research and Training Center on Community Living, Institute on Community Integration. Retrieved from: https://risp.umn.edu

Rivera-Figueroa, K., Marfo, N., & Eigsti, I. (2022). Parental perceptions of Autism spectrum disorder in Latinx and Black sociocultural contexts: A systematic review. *American Journal on Intellectual and Developmental Disabilities, 127*(1), 42–63. https://doi.org/10.1352/1944-7558-127.1.42

Ryff, C. D., & Seltzer, M. M. (1996). The uncharted years of midlife parenting. In C. D. Ryff & M. M. Seltzer (Eds.), *The parental experience in midlife*. University of Chicago Press.

Salkas, K., Magaña, S., Marques, I., & Mirza, M. (2016). Spirituality in Latino families of children with autism spectrum disorder. *Journal of Family Social Work, 19*(1), 1–18. https://doi.org/10.1080/10522158.2015.1114060

Sallis, J. F., & Owen, N. (2002). Ecological models of health behavior. In K. Glanz, B. K. Rimer, & F. M. Lewis (Eds.), *Health behavior and health education: Theory, research, and practice* (3rd ed., pp. 462–484). Jossey-Bass.

Schalock, R. L., Gardner, J. F., & Bradley, V. J. (2007). *Quality of life for people with intellectual and other developmental disabilities. Applications across individuals, organizations, communities, and systems*. American Association on Intellectual and Developmental Disabilities.

Scharlach, A. E., Kellam, R., Ong, N., Baskin, A., Goldstein, C., & Fox, P. J. (2006). Cultural attitudes and caregiver service use: Lessons from focus groups with racially and ethnically diverse family caregivers. *Journal of Gerontological Social Work, 47*(1-2), 133–156. https://doi.org/10.1300/J083v47n01_09

Shaw, K., McArthur, D., Hughes, M., Bakian, A., Lee, L., Pettygrove, S., & Maenner, M. (2022). Progress and disparities in early identification of autism spectrum disorder: Autism and developmental disabilities monitoring network, 2006-2016. *Journal of the American Academy of Child and Adolescent Psychiatry, 61*(7), 905–914.

Silberman, S. (2016). *Neuro tribes: The legacy of autism and the future of neurodiversity.* Avery; Reprint edition.

Smith, K. A., Gehricke, J. G., Iadarola, S., Wolfe, A., & Kuhlthau, K. A. (2020). Disparities in service use among children with autism: A systematic review. *Pediatrics, 145*(Suppl. 1), S35–S46. https://doi.org/10.1542/peds.2019-1895G

Stack, E., & McDonald, K. (2018). We are "Both in charge, the academics and self-advocates": Empowerment in community-based participatory research. *Journal of Policy and Practice in Intellectual Disabilities, 15*(1), 80–89. https://doi.org/10.1111/jppi.12236

Tanis, E. S., Lulinski, A., Wu, A., Braddock, D., & Hemp, R. (2022). *The state of the states in intellectual and developmental disabilities*. Kansas University Center on Developmental Disabilities, The University of Kansas. Retrieved from: https://www.StateoftheStates.org

Thai, J. N., Barnhart, C. E., Cagle, J., & Smith, A. K. (2015). "It just consumes your life": Quality of life for informal caregivers of diverse older adults with late-life disability. *American Journal of Hospice and Palliative Medicine*, 1–7. https://doi.org/10.1177/1049909115583044

The World Bank Group, The Global Women's Institute at George Washington University, Inter American Development Bank, & International Center for Research on Women. (2019). *Brief on violence against women and girls with disabilities* [Brief]. The World Bank Group. https://documents1.worldbank.org/curated/en/864511600841231218/pdf/Brief-on-Violence-Against-Women-and-Girls-with-Disabilities.pdf

Trute, B., Hiebert-Murphy, D., & Levine, K. (2007). Parental appraisal of the family impact of childhood developmental disability: Times of sadness and times of joy. *Journal of Intellectual & Developmental Disability, 32*(1), 1–9. https://doi.org/10.1080/13668250601146753

United Nations Population Fund. (2018). *Five things you didn't know about disability and sexual violence*. UNFPA ESARO. https://esaro.unfpa.org/en/news/five-things-you-didnt-know-about-disability-and-sexual-violence

Updegraff, K. A., & Umaña-Taylor, A. J. (2015). What can we learn from the study of Mexican-origin families in the United States? *Family Process, 54*, 205–216. https://doi.org/10.1111/famp.12135

Van Der Heijden, I., Abrahams, N., & Harries, J. (2019). Additional layers of violence: The intersections of gender and disability in the violence experiences of women with physical disabilities in South Africa. *Journal of Interpersonal Violence, 34*(4), 826–847. https://doi.org/10.1177/0886260516645818

Vasilopoulou, E., & Nisbet, J. (2016). The quality of life of parents of children with autism spectrum disorder: A systematic review. *Research in Autism Spectrum Disorders, 23*, 36–49. https://doi.org/10.1016/j.rasd.2015.11.008

Warrier, V., Greenberg, D. M., Weir, E., Buckingham, C., Smith, P., Lai, M. C., Allison, C., & Baron-Cohen, S. (2020). Elevated rates of autism, other neurodevelopmental and psychiatric diagnoses, and autistic traits in transgender and gender-diverse individuals. *Nature Communications, 11*(1), 3959. https://doi.org/10.1038/s41467-020-17794-1

Williams, E., Smith, M., Sherwood, K., Lovelace, T., & Bishop, L. (2022). Brief report: Initial evidence of depressive symptom disparities among Black and White transition age autistic youth. *Journal of Autism and Developmental Disorders, 52*, 3740–3745. https://doi.org/10.1007/s10803-021-05242-y

Wilson, N. J., Macdonald, J., Hayman, B., Bright, A. M., Frawley, P., & Gallego, G. (2018). A narrative review of the literature about people with intellectual disability who identify as lesbian, gay, bisexual, transgender, intersex or questioning. *Journal of Intellectual Disabilities: JOID, 22*(2), 171–196. https://doi-org.ezproxy.lib.utexas.edu/10.1177/1744629516682681

Wolbring, G. (2008). The politics of ableism. *Development, 51*(2), 252–258.

Xu, Y., Zeng, W., Wang, Y., & Magaña, S. (2022). Barriers to service access for immigrant families of children with developmental disabilities: A scoping review. *Intellectual and Developmental Disabilities, 60*(5), 382–404. https://doi.org/10.1352/1934-9556-60.5.382

Xu, Y., Morales, M. A., Magaña, S., & Hsieh, K. (2023). Racial and ethnic disparities in chronic disease outcomes among adults with intellectual and developmental disabilities. *Journal of Developmental and Physical Disabilities*. https://doi.org/10.1007/s10882-023-09923-7

Zablotsky, B., Ng, A. E., Black, L. I., & Blumberg, S. J. (2023). *Diagnosed developmental disabilities in children aged 3–17 years: United States, 2019–2021. NCHS Data Brief, no 473*. National Center for Health Statistics. https://doi.org/10.15620/cdc:129520

Genetic Contributions to the Expression of Traits

Christopher L. Kliethermes

Abstract This chapter provides an overview of genetic and molecular processes involved in the inheritance and expression of cognitive and behavioral traits, including intellectual and developmental disabilities. Chromosomes, the structures that house genes, are described, and an overview of some important terminology and mechanisms involved in gene expression is provided. Mendel's work with single gene traits is overlaid onto our current understanding of the segregation and assortment of chromosomes into gametes. Conditions resulting from large chromosomal abnormalities and smaller scale mutations are discussed. Mechanisms underlying the inheritance and expression of traits and conditions attributable to single genes are contrasted with those of quantitative traits, which result from complex interactions between multiple genes and the environment. The chapter concludes with an overview of approaches used to identify genetic contributions to behavior.

Keywords Inheritance · Genes · Phenotypes · Chromosomes

Genetic Contributions to Phenotypes

Offspring tend to resemble their parents in appearance, including in things like their hair color, facial features, and height. We tend to think of physical traits like this as "genetic" and (for the most part) are completely relegated to the notion that we will, on average, look something like our parents. But what about our brains? We might be dismissive of the idea that our thoughts and behavior might also resemble those of our parents—we can even voluntarily choose to *not* act like our parents—yet there is substantial evidence to suggest that, on average, we tend to think and act like our parents. More broadly, we tend to think and act like people with whom we share

C. L. Kliethermes (✉)
Department of Psychology and Neuroscience, Drake University, Des Moines, Iowa, USA
e-mail: christopher.kliethermes@drake.edu

M. G. Valdovinos (ed.), *Intellectual and Developmental Disabilities*,
Contemporary Clinical Neuroscience,
https://doi.org/10.1007/978-3-031-66932-3_4

particular gene variants, known as alleles (see Box 1), regardless of our familial relationships.

Box 1: Genes and Alleles

The word *gene* is often used when we really mean *allele.* For example, all people have the HTT *gene* which encodes a protein called *huntingtin.* However, only people with a particular mutation in the HTT *allele* develop Huntington's disease. In a population, there can be several to many such *alternative alleles* for any given gene. These alternative alleles are part of an *allelic series*. For example, human ABO blood type is determined by the inheritance of two alleles from an allelic series consisting of three alleles (A, B, and O) of the *ABO* gene. These alleles follow a Mendelian pattern of expression: A and B are dominant over O but codominant with respect to each other.

Like all physical traits, genetic contributions to the structure and function of the brain originate in the *genes.* The human genome contains about 20,000 genes which code for proteins, and each of these genes exists in one or more versions (alleles) within a population of individuals. There are two alleles for each of these 20,000 genes, one from each parent. If an individual received two of the same alleles for a gene, they are *homozygous* for this allele, and if the alleles are different, the person is *heterozygous* for those alleles. Except in the case of identical twins, the specific combination of alleles an individual inherited for each of these 20,000 genes is unique to them. These alleles guide development from the single-cell (fertilized egg) stage through the end of a person's life.

Genes, however, do not function in isolation. Both the function and expression of any given gene are regulated by other genes contained within the genome. These interactive effects of genetic material are referred to as *epistasis*. In addition, the *environment*, which in traditional genetic parlance is any factor that has a causal role in the expression of a behavior that is not genetic in origin, also regulates the expression and function of genes. Environmental regulation of gene function is exemplified by processes like long-term memory formation and long-term adaptations to stress, both of which require changes in the expression of specific genes to occur. But to a geneticist, the environment also includes everything from diet and drugs to birth order and socioeconomic status, any of which can have mild to profound effects on the expression and function of any given gene. These long-term changes to gene expression are referred to as *epigenetic effects* of the experiences. More broadly, *epigenetics* is the study of the biological mechanisms that underlie these experience-dependent changes in gene expression, as well as those that underlie cell- and tissue-specific changes in gene expression during development.

The overall goal of this chapter is to describe the relationship between *genotypes*, which are the specific alleles an individual has, and *phenotypes*, which are physical or behavioral traits. This relationship is described for two different categories of phenotypes. In *Mendelian traits*, the relationship between a genotype and a phenotype is very strong such that a single allele almost entirely explains (correlates with) the expression of within and across individuals, whereas *quantitative* or *complex traits* are the result of interactions among multiple alleles and environmental factors. The inheritance of both types of traits is described in relation to chromosomal segregation and assortment during sex cell formation, and an overview of mechanisms of gene expression is provided.

Genetics Meets Molecular Biology

Before introducing some of the terminologies developed by cellular and molecular geneticists to explain cellular processes, it is worthwhile to remember the purpose these terminologies serve for geneticists. Genetics is the study of the inheritance of traits. Genetics as a field existed quite happily and independently of cellular and molecular biology from before the rediscovery of Mendel's work in the early 1900s through the advent of modern molecular biology in the 1950s and 1960s. Geneticists in this pre-molecular time developed terms to describe the expression of inherited traits we still use today, such as dominant, recessive, and sex-linked. They invented a hypothetical construct, a *gene* (a word first used by Wilhelm Johannsen in 1909), to serve as the hypothetical carrier of hereditary information across generations, and of course, they also invented the term *genetics* (William Bateson in 1905) to describe the field that studies the inheritance of traits.

Importantly, the use of these terms by geneticists was completely agnostic with respect to the nature of the information carried across generations. A *gene*, for a geneticist during this time, was an *intervening variable* that could explain the resemblance of offspring to their parent. It was a hypothetical *thing* that existed, but what it was made of was not relevant. Just like electricians do not need to know anything about the quantal properties of an electron to install electrical circuits in a house, geneticists did not need to know anything about the physical nature of genes to explain phenotypic resemblance across generations. For historical context, chromosomes had been proposed to contain hereditary information since the early 1900s, but it was not until the experiments of Alfred Hershey and Martha Chase in the early 1950s that the DNA contained in chromosomes was established as the actual source of this information.

I make these points not to diminish the importance of cellular and molecular biology to our understanding of genetics. As described in this chapter, geneticists' contemporary understandings of everything from gene expression to mutations are based entirely on mechanisms discovered by molecular biology. Rather, my point is to introduce the idea that understanding *whether* specific genes affect the expression of a given trait is separate from the question of *how* specific genes (as defined by molecular biology) affect the expression of the trait.

Experimental approaches to answer the "*how*" question of the inheritance and expression of cognitive and behavioral traits fall into an area called *developmental neurogenetics*. Scientists working in this field study the roles of specific genes and environmental conditions in the development of the nervous system, as well as in its primary output: behavior. For example, the identification of variations in the *Mecp2* gene as the cause of Rett syndrome (Amir et al., 1999) led to a better understanding of the molecular and biochemical pathways *Mecp2* normally functions in during nervous system development, and to a number of ongoing studies that examine how variations of this gene impact the normal sequence of development.

Studies designed to determine *whether* genes affect the expression of cognitive and behavioral traits take the form of family studies, twin and adoption studies, linkage analysis, and large-scale Genome Wide Association Studies (GWAS). Unlike studies aimed at addressing the *how* question, studies asking *whether* genes affect the expression of a trait typically do not need to have a clear hypothesis about which genes might be involved in a trait of interest. Rather, they simply ask to what extent genetic similarity across individuals can explain phenotypic similarities.

Chromosomes, DNA, and Genes

Although the words chromosomes, DNA, and genes are used more or less interchangeably in common speech to refer to anything and everything "genetic," these terms refer to distinct biological entities, and the relationships between them are important toward our understanding of the inheritance and expression of traits. This section will provide an overview of chromosomal structure and function, especially in relation to the process at the heart of genetics: gametogenesis, the production of sex cells. Some of the important terms used in this section are summarized in Box 2.

Box 2: Terms and Definitions

Allele: a version of a gene

Aneuploidy: the presence of an abnormal number of chromosomes

Chromosome: structures in the nucleus of cells that contain hereditary information in the form of DNA. Chromosomes 1–22 are called the autosomes, and chromosome 23 is called the sex chromosome.

Copy number variation (CNV): a variation in the number of copies of a section of DNA. CNVs include insertions, which are repeated or translocated sections of DNA, and deletions, which are missing sections of DNA. These are commonly referred to as indel mutations.

DNA: a large molecule made of repeating sections of four nucleotide bases, adenine, guanine, cytosine, and thymine, arranged in particular sequences which carry hereditary information.

DNA marker: a section of DNA with a known location on a chromosome.

Gene: a unit of heredity; an allelic sequence that codes for a specific protein.

Haplotype: a sequence of alleles or DNA markers in a chromosome.

(continued)

Box 2 (continued)

Linkage disequilibrium (LD): the nonrandom association of two or more DNA markers in a chromosome within a population. LD forms the basis of linkage analysis.

Linkage analysis: a technique that associates DNA segments in linkage disequilibrium with the presence of a particular trait or condition. Also known as gene mapping.

Mutation: a change in a DNA sequence.

Nondisjunction: a chromosomal abnormality resulting from the improper segregation of the maternal and paternal chromosomes into gametes.

Quantitative (complex) trait: a trait which varies an expression in a population; quantitative traits are the results of interactions among many genes and the environment.

Chromosomes Except in cases of nondisjunction, people have 23 pairs of chromosomes. One chromosome from each chromosomal pair is *maternal* in origin, and the other is *paternal*, meaning that one-half of the total chromosomal makeup of a person comes from either parent. Chromosome pairs 1–22 are called *autosomes*, and their numbers are based on size: 1 is the biggest and 22 is the smallest. The 23rd pair of chromosomes is the sex chromosomes, and these chromosomes include X and Y. It is worth noting that the names "X" and "Y" *do not* refer to the shape of these chromosomes. In humans, all chromosomes look vaguely like an X when duplicated (see Fig. 1, described below). The X chromosome was initially studied and described in insects, and it was originally called the "X element" because it differed in its appearance and behavior from other chromosomes during sex cell formation (see Paliulis et al., 2023, for discussion). "Y" was chosen because it is the next letter of the alphabet. No one ever said cytologists were poetic people.

In females, each of the 23 pairs of chromosomes—that is, the autosomes and the sex chromosomes—are *homologous*: the maternally derived chromosomes are similar in appearance to the paternally derived chromosomes, including the two X chromosomes. In males, only the 22 autosomes are homologous. The pair of sex chromosomes, X and Y, are *heterologous*: they differ in appearance. This difference in the shape and structure of the X and Y chromosomes contributes to the inheritance and expression of *sex-linked traits*. Because males inherit only a single X chromosome (from the mother), they can express only the alleles for each of the genes contained in this chromosome.

Chromosomal Structure Chromosomes are made of chromatin, a chemical made of DNA and tightly associated proteins called histones. For most of a cell's life, chromatin exists in a relaxed, partially uncoiled state. Prior to cell division, the chromosomes replicate (each chromosome makes an identical copy of itself) and condense in length. Chromosomes in this replicated, condensed state display some visible features, which are useful as landmarks. Notably, the newly replicated strand

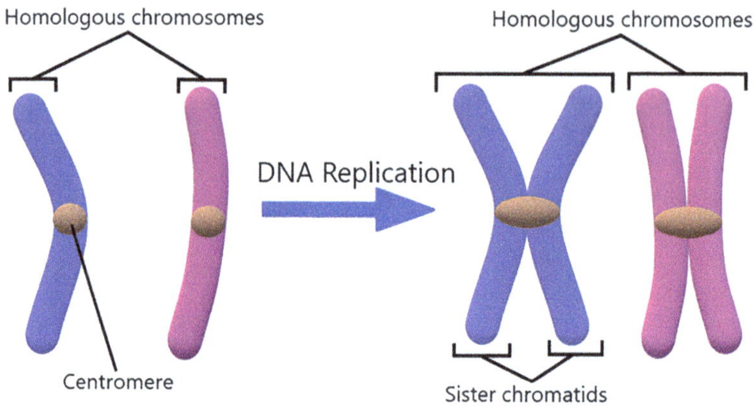

Fig. 1 A pair of unreplicated (left) and replicated (right) chromosomes. The blue chromosome is paternal in origin, and the pink chromosome is maternal in origin. The two sister chromatids that make up one replicated chromosome contain identical DNA sequences; homologous chromosomes (the pink *and* blue chromosomes) contain similar, but not identical DNA sequences (alleles). This image is distributed under the terms of the Creative Commons Attribution Non Commercial 4.0 International license (CC-BY-NC), a copy of which is available at http://creativecommons.org/licenses/by-nc/4.0/

of chromatin remains attached to the original strand of chromatin at a structure called the centromere (see Fig. 1). The centromere divides the chromosome up into two "arms"; the *p* arm (from the French *petit*, for small) and the *q* arm, which is so designated because it is the letter after "p."

The other prominent feature of chromosomes in this condensed state is the *bands*, which give chromosomes their ladder-like appearance (see Fig. 2). These bands result from applying chemical stains, such as Giemsa stain (named for Gustav Giemsa) to microscope slides containing chromosomes isolated from cells. The locations of the bands produced by Giemsa or other stains are stereotyped within a species, a feature that makes them incredibly useful as chromosomal referents across individuals. For example, "22q11.2" refers to the region around the 11th band (counting outwards from the centromere) in the "q" arm of human chromosome 22.

Chromosomal Abnormalities As indicated above, chromosomes can be visualized when they are in a condensed state. To do this, cells are first isolated from a tissue and then chemically treated to isolate the chromosomes from other cellular components. Fluid containing the isolated chromosomes is placed onto a microscope slide, and various stains are added. Different stains bind to different chemical constituents of the chromosomes, which allows for direct visualization of all the chromosomes. This visualization of all stained chromosomes is called a *karyotype* (Fig. 2).

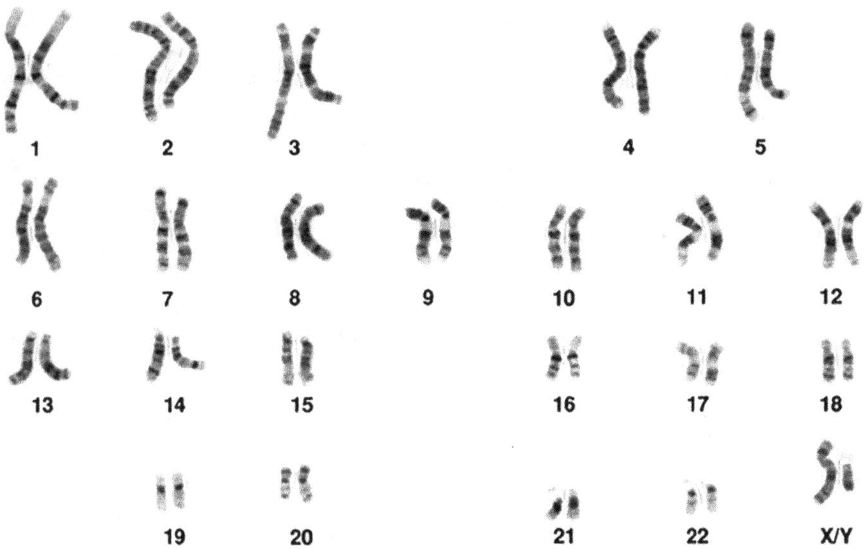

Fig. 2 Karyotype showing the 22 pairs of autosomes and the pair of sex chromosomes from a single (male) individual. These chromosomes are shown in an *unreplicated* state; the two chromosomes which comprise each pair are homologous. The bands visible in each chromosome are the result of staining. Note that the location and (approximate) size of the bands are the same in each homologous pair of chromosomes. This feature makes the bands useful as landmarks when referring to the location of genes within chromosomes. This image is distributed under the terms of the Creative Commons Attribution Non Commercial 4.0 International license (CC-BY-NC), a copy of which is available at http://creativecommons.org/licenses/by-nc/4.0/

In the clinic, karyotypes are commonly used to identify chromosomal abnormalities, including differences in chromosomal count, known as *aneuploidies*, and changes in chromosomal size resulting from *translocations*, *deletions*, or *duplications* of portions of chromosomes (Box 3). Specific changes in chromosomal size that are too small to visualize directly in karyotypes can be detected using a technique called fluorescence in situ hybridization (FISH), which uses labeled probes engineered to bind to specific DNA sequences at known locations on chromosomes (see Ozkan & Lacerda, 2024, for an overview of these techniques).

Box 3: Chromosomal Abnormalities

Aneuploidy (plural, aneuploidies): a change in chromosomal count.
Translocation: the relocation of a portion of one chromosome to another.
Deletion: the removal of a portion of one chromosome.
Duplication: the formation of an extra copy of a chromosomal region.

Most chromosomal abnormalities occur during the production of gametes, sperm, or egg cells, as described above. Aneuploidies are the result of *nondisjunction*, which occurs when homologous chromosomes fail to separate properly during meiosis (described below) and result in gametes with unequal numbers of chromosomes. For each nondisjunction event, one gamete will receive an extra copy of a chromosome while another will receive none.

When fertilization occurs, the developmental trajectory for the embryo with a chromosomal abnormality can differ mildly to significantly from the normal sequence of development. Most major abnormalities, including almost all aneuploidies, result in spontaneous miscarriage. However, some aneuploidies are survivable. The best known of these is Down syndrome, which is caused by the presence of an extra copy of chromosome 21 (trisomy 21), but others include trisomy 18 (Patau syndrome), trisomy 15 (Edwards' syndrome), and a variety of conditions resulting from missing or extra copies of the sex chromosomes. These include monosomy X (Turner syndrome), and trisomy or tetrasomy of the X and/or Y chromosome (XXY, which causes Klinefelter syndrome, as well as XXX, XYY, XXXX, XXYY). A variety of syndromes resulting from the deletion of portions of chromosomes (microdeletions) have also been identified (see Watson et al., 2014, for review), such as DiGeorge syndrome, which results from the deletion of a portion of the *q* arm of chromosome 21 (22q11.2), and Langer–Giedion syndrome, which results from the deletion of 8q24.1. These syndromes occur regardless of whether the deletion occurred in the maternally or paternally derived chromosome. In contrast, both Prader–Willi and Angelman syndromes result from the microdeletion of a portion of chromosome 15 (15q11), with the specific syndrome that develops dependent on which chromosome the deletion occurred in deletion of 15q11 in the maternally derived chromosome results in Angelman syndrome, while deletion of this segment in the paternally derived chromosome results in Prader–Willi syndrome (see Ehrhart et al., 2019; Ma et al., 2023, for review). As with chromosome-wide aneuploidies mentioned above, these microdeletion events usually occur during gametogenesis (described below).

Chromosomes and DNA The DNA in chromatin exists as a sequence (a biopolymer) of four *nucleotides*, adenine (A), guanine (G), cytosine (C) and thymine (T), strung together repetitively with sugar molecules (deoxy ribose) and phosphate groups into a long, continuous chain. The famous double helix structure of DNA is formed by the bonding of nucleotides in one strand of DNA, called the *sense* or "+" strand, with their *complements* in a second strand of DNA, called the *antisense* or "–" strand: A bonds with T and C bonds with G. Complementary nucleotides in the sense and antisense strands are called *base pairs* of DNA, and the number of base pairs ultimately determine the size of a chromosome. The largest, chromosome 1, contains approximately 249 million base pairs, or 249 megabases (Mb) of nucleotide sequences in each of the two DNA strands, while the smallest autosome, chromosome 22, is about 5 times smaller, containing about 50 Mb of nucleotides. All chromosomes together contain about 3000 Mb of DNA, or 3 billion base pairs.

DNA can be divided into two broad functional classes, coding and noncoding DNA. As these names imply, coding DNA is a sequence of nucleotides which "codes" for specific genes, while noncoding DNA does not. The human genome contains around 20,000 sections of coding DNA, and larger chromosomes generally have more of these sections: chromosome 1 contains about 3000 gene-coding regions; chromosome 10 has about 1400; and chromosome 20 has about 900. The two sex chromosomes, X and Y, contain about 1400 and 200 gene-coding regions, respectively.

By far, most of our DNA is noncoding. It accounts for 2.97 billion of our 3 billion base pairs (99%). Importantly however, noncoding DNA is not nonfunctional. It is intimately involved in the regulation of gene expression and maintenance of chromosomal structure.

Gene Expression

The basic explanation for how the information contained in a sequence of DNA nucleotides becomes expressed as a functional protein was codified (pun intended) in the early 1960s as the *Central Dogma of Molecular Biology*. This model, although incomplete and heavily appended since it was first proposed, still forms the basis for our understanding of gene expression and function. In this model, a sequence of DNA nucleotides is *transcribed* (written out, copied) as a messenger RNA (mRNA) molecule, whose nucleotide sequence is then *translated* (converted) to a protein; proteins are the molecules that underlie most cellular functions. Importantly, there is a direct correspondence between the sequence of nucleotides in a coding section of DNA, the sequence of nucleotides in the mRNA transcribed from the DNA, and both the type and sequence of amino acids in the protein coded for by the mRNA. This relationship is called the *genetic code* (see Box 4).

Box 4: The genetic code
The genetic code is the relationship between a sequence of DNA nucleotides and the ordering and type of amino acids in a protein. The genetic code is based in units called codons, with each codon comprising a sequence of 3 nucleotides in an mRNA transcript. Because there are 4 nucleotide possibilities at a given position in an mRNA transcript (A, U, C, or G) and there are 3 nucleotides in a codon, there are 64 (4×4×4) possible codons. Most of these codons 'code for' the inclusion of a specific amino acid in a protein, while others, called 'start' and 'stop' codons, signal the beginning and end, respectively, of translation. For example, the codon AUG signals the start site for translation, UCU codes for the amino acid serine, AAA codes for lysine, GAG codes for glutamate, and UAA is a stop codon. Any nucleotides in a transcript before the AUG start codon or after the UAA stop codon are not translated.

Because the function of a protein is ultimately determined by its structure, the structure of a protein is ultimately determined by the types and sequence of amino acids it contains, and the types and sequence of amino acids a protein contains are ultimately determined by a sequence of nucleotides in DNA, it follows that an understanding of DNA sequence will lead to an understanding of cellular function. But, and this is a big but, the expression of a given gene is both dynamic—the rate of transcription for a given gene varies as a function of the signals it receives—and tissue specific: not all cells express all genes.

Like all DNA, a coding strand of DNA consists of a repeating sequence of four nucleotides: A, C, T, and G. A sequence of DNA nucleotides that codes for a gene can be identified by the presence of a *transcription initiation site*, which is a specific sequence of nucleotides (TATA, to be precise) located *upstream* (before or in front of) a coding section of DNA. This upstream region, called the promoter, also includes a number of *transcription factor binding sites*. As their name implies, these sites are nucleotide sequences recognized and bound by *transcription factors*, which are proteins that either promote or repress the transcription of a coding section of DNA into mRNA. Transcription factors are the reason that pancreatic beta cells, for example, express the gene that codes for insulin, while hippocampal pyramidal neurons do not.

More importantly, transcription factors are the reason we *have* a pancreas and a hippocampus—or any tissue or organ—in the first place. Think about all the different types of cells you have in your body: neurons, skin cells, muscle cells, immune cells, bone cells, and many more. Each of these cells contains the same 23 pairs of chromosomes, yet each expresses only some of the 20,000 gene sequences contained in these chromosomes. This regulation of gene expression begins shortly after fertilization, when the developing embryo consists of only *stem cells*, which are cells capable of giving rise to all the cell types in the human body and contributes to the differentiation of these cells into lineages *fated* to become parts of various organs and tissues. Cells fated to become neurons, for example, begin to express neural-specific transcription factors which cause them to express genes necessary for the formation of cellular structures found in neurons, including genes that code for proteins that underlie the structure of axons and dendrites. As neurons form connections (synapses) with other neurons, they begin to express additional proteins necessary to maintain these synapses, and as synaptic activity changes throughout development, learning, sensory experience, and the like, yet more changes in gene expression occur.

Cell-specific regulation of gene expression also occurs at the mRNA transcript level. A newly transcribed mRNA contains sequences of nucleotides that are complementary to a coding sequence of DNA. However, not all parts of an mRNA transcript become translated into a functional protein. Many mRNA transcripts include sections called *introns* (derived from "intragenic regions") which are excised from (cut out of) the transcript before translation into protein. The remaining portions of the mRNA transcript, the *exons* (derived from "expressed regions"), are then *spliced* (joined) together and translated into a protein (Fig. 3).

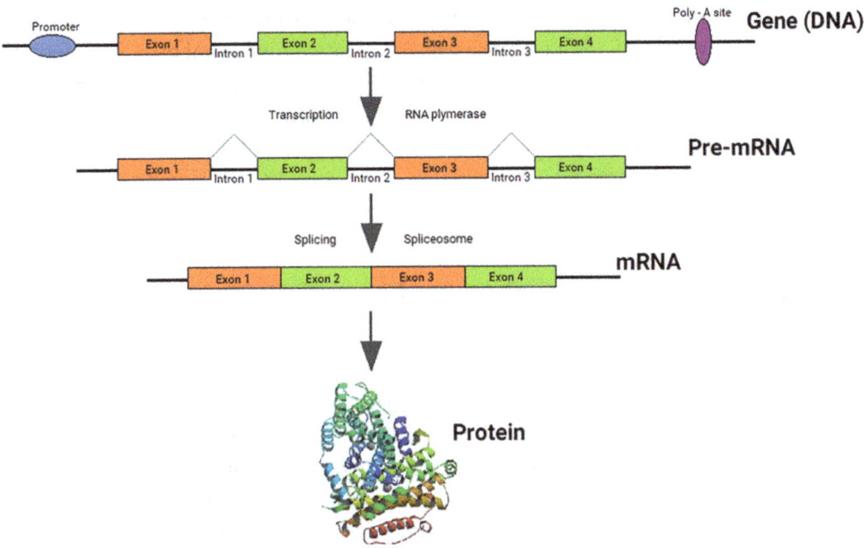

Fig. 3 An overview of the processes of transcription and translation. The promoter sequence in the recruits transcription factors and the proteins that make up the transcription complex, including RNA polymerase, to this section of DNA. This gene (allele) is initially transcribed as a full-length transcript (pre-mRNA) that includes all exons and introns. The introns are then cut out, the exons spliced together, and the final mRNA translated into a protein. Alterative splicing of the pre-mRNA will result in mRNAs with different complements of exons, and consequently, the production of a different protein. This image is distributed under the terms of the Creative Commons Attribution Non Commercial 4.0 International license (CC-BY-NC), a copy of which is available at http://creativecommons.org/licenses/by-nc/4.0/

If it seems odd to you that a cell would cut out parts from an mRNA it just synthesized, it is. Bacterial genes do not include exons and introns, and they seem to be getting along fine. However, the genes of every living thing on this planet that is not a bacteria or archaebacteria include exons and introns, leading to a very active debate about when and how they first arose in evolution (see Rogozin et al., 2012, for review).

So, while we do not know how exons and introns came to be, we do know one important thing that they do: they provide a mechanism for cells to express different mRNA transcripts from the same DNA sequence. *Alternative splicing* occurs when exons are spliced together in different combinations, resulting in the production of proteins with different functions and properties. The specific exons included are regulated by the cell type, developmental time point, and multiple epigenetic factors. For example, a type of receptor for dopamine, called DRD2, is coded for by one of the 1300 or so genes located in chromosome 11, and the mRNA transcribed from this gene contains 9 exons. In some cell types, such as medium spiny neurons in the striatum, this transcript containing all 9 exons is translated into a functional protein called the “long” form of the receptor (DRD2L). In other cells, including neurons in the substantia nigra, this same gene on chromosome 11 is transcribed as an mRNA containing only 8 exons—the transcript does not include exon number 6; it is alternatively spliced—but likewise translated into a functional protein called the "short" form of the receptor (DRD2S). These two types of receptors are products of the same gene, but the functional proteins have different properties and are found in distinct locations within different cells (see De Mei et al., 2009, for mechanisms and discussion).

Gametogenesis: Crossing Over and Mutations

Gametogenesis, the production of sex cells, is at the heart of genetics, as it is these cells that carry genetic information across generations. The gametes you produce contain shuffled versions of the chromosomes you inherited from your parents, who passed on shuffled versions of the ones inherited from their parents (i.e., your grandparents) to you. Going back (much) further in time, the gametes you produce are part of an unbroken line of genetic information going back to the origin of life as we know it on earth.

The process of sex cell formation is similar to, but distinct from, the formation of somatic (body, not sex) cells. Sex cells are produced through a process called *meiosis* (see Fig. 4). Unlike somatic cell division (mitosis), which results in the formation of two cells that are genetic clones, meiosis occurs in two stages and results in 4 genetically unique cells, each having one-half of an individual's total genetic complement—i.e., one chromosome from each of the 23 chromosomal pairs, which is to say, 23 chromosomes. Upon fertilization, the 23 chromosomes from an egg combine with the 23 chromosomes from a sperm, resulting in a fertilized egg containing the full genetic complement of 46 chromosomes, or 23 pairs.

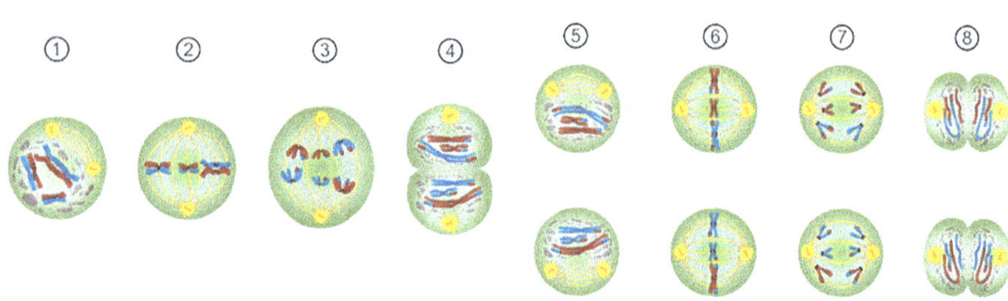

Fig. 4 Meiosis results in the reduction and reassortment of genetic material in the gametes. A single progenitor cell (panel 1) containing replicated copies of all chromosomes (only 3 pairs of homologous chromosomes are shown) undergoes two rounds of divisions (panels 2 and 7), resulting in the formation of 4 gametes (panel 8). Each gamete contains one unreplicated copy of all chromosomes, each of which consists of recombined portions of the maternal and paternal chromosomes.

Two processes very important for genetics occur during gametogenesis. The first, *homologous recombination*, known more poetically as "*crossing over*,", occurs when the maternal and paternal chromosomal pairs exchange (swap) homologous segments. This swapping of segments results in a new chromosome which contains alleles from both the maternal and paternal chromosomes. In females, homologous recombination occurs in all 23 homologous chromosomal pairs, including the two sex chromosomes (XX), while in males, recombination occurs in the 22 autosomal pairs but not the heterologous sex chromosomes (XY). In the absence of homologous recombination of chromosomes, a sperm or egg cell would contain a copy of each chromosome that is identical (barring any new mutations; see below) to that of one of the parents. Homologous recombination increases the diversity of the human genome and, as described below, makes it possible to map alleles that contribute to trait variation.

The second important process for genetics that occurs during gametogenesis is mutation, which is a change in a DNA sequence. Mutations can occur from errors introduced during DNA replication, during the continuously ongoing processes of DNA maintenance and repair, or through exposure to mutagens.

Mutations can occur in any cell, but the ones of interest to geneticists are those that occur in the gametes, as these mutations can be passed on as gene variants (alleles or polymorphisms) to the next generation (see Karki et al., 2015, for discussion and suggestions on the use of these terms).

The functional effects of a mutation depend upon its size, the gene(s) involved, and the nature of the mutation. *Copy number variants (CNVs)* result from the deletion or duplication of a chromosomal section. These mutations include changes large enough to be detected in karyotypes, such as the translocation of a chromosomal arm, and more subtle chromosomal events, including microdeletions and duplications of portions of a chromosome.

Frameshift mutations are changes to the *genetic code* (i.e., relationship between DNA and protein structure) caused by the deletion or insertion of nucleotides into a DNA sequence. Although the DNA change can be as minor as the deletion or insertion of a single nucleotide, such mutations can result in the expression of a protein with a radically different structure because of the change in the translation reading frame.

Trinucleotide expansion repeats are repetitive sequences of three DNA nucleotides that can become duplicated (expanded) during DNA replication. The nucleotide sequence upstream of the Fmr1 gene, for example, contains 5–55 repeats of the trinucleotide sequence "CGG" in most people. In people with Fragile X syndrome, this section contains more than 200 repeats of this sequence, resulting in greatly altered expression of the *Fmr1* gene. People with 55–200 repeats of the CGG sequence are said to carry a *premutation* of this gene; they are at greater risk of producing offspring with this condition.

Another important class results is the substitution of a single nucleotide in a DNA sequence. Single nucleotide polymorphisms (SNPs) can be found throughout the genome, including in coding and noncoding DNA sections. A *nonsynonymous SNP* is a nucleotide substitution in a coding section of DNA that results in a different amino acid being incorporated into the protein coded for by that gene. Depending on the amino acid substitution, the function of the protein can be affected in mild to major ways. A nonsynonymous SNP can also result in the expression of a truncated (shortened) version of the gene by introducing a *premature stop codon* into the genetic code (see Box 3), terminating the translation process for a given gene.

A *synonymous SNP*, also known as a *silent mutation*, also results from the change of a single nucleotide in a coding section of DNA. But unlike a nonsynonymous SNP, which ultimately results in an altered protein structure, a section of DNA with synonymous SNP codes for a protein with the same structure as one without the SNP. An *intergenic SNP*, as its name suggests, is a change in a single nucleotide of DNA found between coding sections of DNA. Intergenic SNPs can be found in areas immediately adjacent to genes, such as the upstream promoter sequence, where they can result in altered expression of a gene. Finally, an *intronic SNP* is found in *introns*, sections of DNA that are transcribed into mRNA but are then "cut out" of the sequence prior to translation into a protein. Because the specific sites where an mRNA is cut are determined by the sequence of nucleotides, an intronic SNP can change the site where the cut occurs, resulting ultimately in different amino acids becoming incorporated into the protein coded for by the gene.

A given SNP can be used to define an allele, and a sequence of SNPs defines a *haplotype*, which is a collection of alleles that tends to be inherited together. Family members tend to share haplotypes, as do ethnic groups. Because they are relatively easy to identify and located throughout the genome, SNPs are used extensively to identify chromosomal segments and gene variants associated with conditions (described below), and by for-profit companies such as 23andMe.

Qualitative and Quantitative Traits

The previous section provided a broad overview of some concepts and mechanisms related to the nature, inheritance, and expression of genes. This section will describe two different (but related) ways gene variations (alleles) can contribute to the expression of traits. *Mendelian traits* exist as discrete types in a population, and their expression in a population is attributable mostly or completely to a single allele. In contrast, *quantitative* or *complex traits* show continuous variation in a population, and their expression in a population is the result of multiple genes and environmental factors, as well as complex interactions between the two.

Mendelian (Qualitative) Traits

A great deal of instructional time in introductory biology courses tends to be spent describing Mendel's experiments: learning which pea trait is dominant over the other, which letter to capitalize in your Punnett squares, and more generally, using

Punnett squares to show which combination of upper- and lower-case letters goes with which particular phenotype. Once students can determine the proportion of offspring who should be capitalized or lower-cased, they go on to estimate proportions of offspring with two sets of capital letters. *Voila*, genetics.

There are a few things wrong with this.

1. *Mendel was not a genetic engineer.* Mendel—like all geneticists before and after him—conducted experiments that asked whether the appearance (phenotype) of offspring can be predicted from the phenotype of the parents. He did not arrange crosses among pea based on their genotypes. He *inferred* the genotypes of the parents based on the phenotypes of their offspring. If you have ever used a Punnett square to predict the proportions of offspring with a given phenotype given the *genotypes* of the parents, you are not doing what Mendel did.
2. *One of Mendel's "laws"—the principle of independent assortment—is wrong.* At least, in many cases. Mendel proposed that traits are inherited independently. Linkage disequilibrium occurs when gene sequences are passed on as a set (haplotype), rather than independently. This occurs when the sequences are located physically close together on a chromosome. Linkage disequilibrium forms the basis of *linkage analysis*, which is used to associate chromosomal regions with a trait of interest.
3. Mendelian traits are exceptional. Most phenotypes we can measure—and certainly those most of interest to cognitive and behavioral scientists—are not Mendelian. Things like intelligence scores, personality characteristics, and learning ability are complex (quantitative) traits, meaning they vary continuously from individual to individual. These traits are the result of multiple alleles, multiple environmental effects, and complex interactions between the two. Notably, however, a number of intellectual disabilities, including phenylketonuria (PKU), are attributable to the effects of a single allele. These conditions are inherited and expressed in a Mendelian manner. Other conditions, including Prader–Willi and Rett syndromes, are attributable to single alleles, but the conditions are not inherited in a Mendelian manner.

Mendel's Experiments Mendelian traits are those where there is a clear, consistent, and predictable relationship between the phenotypes of parents and their offspring, and this relationship is attributable to the effects of a single gene. In Mendel's original experiments with pea plants, these phenotypes included the shape and color of peas, among others. Mendel worked with plants from lines of peas that "bred true" for particular traits, such as pea plants that had either yellow or green seeds. If you are familiar with descriptors like "heirloom tomatoes" or "purebred dogs," it means the same thing. These are all euphemisms for *inbred*, meaning there is little to no genetic variation (allelic diversity) underlying the traits of interest within a population.

Mendel found that if he crossed plants from a line that "bred true" for yellow seed color with plants that similarly "bred true" for green seed color, then saved and planted the resultant seeds, all of the offspring from these plants (the first filial or F1 generation) were uniformly yellow. That is, pea plants derived from a cross of yellow- and green-seeded parents produced only yellow seeds. He saved and planted these F1 (yellow) seeds and found that plants derived from these seeds (the F2 generation) produced both yellow and green seeds, but with 3 times as many yellow-seeded plants as green in this generation.

If you have heard about these experiments before, everything in the previous paragraph should make sense. You can probably recognize the concepts of dominance and recessiveness in the F1 generation—all the peas in this generation were yellow, yet all these peas were derived from one parent from a line that produced yellow seeds and another which produced green seeds—so we infer that yellow seed color is dominant over green. Furthermore, because both yellow and green peas were found in the F2 generation, we know that the "green" allele these plants received from one of their parents was not lost but rather masked by the "yellow" allele these plants inherited from the other, yellow, parent. You could call the F1 plants "carriers" of the recessive "green" allele. Mendel's *principle of segregation* derived from these experimental results. This principle states that each plant possesses two factors that control the expression of a trait, but only one of these is passed on to the next generation.

Mendel's most famous experiment led to his formulation of the *principle of independent assortment*. This experiment was conducted identically to the one with green and yellow peas described above but included pea plants from lines that differed in their expressions of two traits: pea color and pea shape. A given pea plants can produce yellow or green seeds, as described above, and the seed colors of the F1 and F2 generations can be explained through Mendel's principle of segregation. An identical situation exists for pea shape, which can be round or wrinkled, where round is dominant over wrinkled. If you cross parents that breed true for both of these traits—let us say a pea plant that produces yellow, round peas with one that produces green, wrinkled peas—you will see in the F1 generation the expression of the dominant traits only: all peas will be yellow and round. If you cross an F1 plant with itself (or any other F1 plant), the offspring from these peas will include all possible phenotypic combinations: yellow and round; yellow and wrinkled; green and round; green and wrinkled. These offspring will be found in a 9:3:3:1 ratio. Mendel's finding of all possible combinations of the two phenotypes in the F2 generation—and in the amounts predicted by the experiments that considered only one trait at a time—suggested that inheriting one trait, pea color, did not influence the inheritance of a second trait, pea shape. Or stated differently, pea color and pea shape are inherited independently of each other.

Some Single Gene Traits in People Most traits that we can measure in people, animals, and plants (including pea plants) are not attributable to a single allele. However, there are a number of traits and conditions which are.

ABO human blood type, for example, is encoded by a gene called *alpha 1-3-N-acetylgalactosaminyltransferase* (also known as *ABO*), which has three alleles, called A, B, and O. A and B are expressed in a dominant fashion over O, but A and B are codominant with respect to each other. The alleles that code for this trait *segregate* during gametogenesis (described below) and *assort independently* from alleles that code for another important blood marker, the Rh factor.

Huntington's disease is a fatal, progressive neurological disorder that causes severe motoric, psychiatric, and cognitive problems. This disorder is expressed in a Mendelian dominant fashion through the inheritance of a mutation in the *IT-15* gene. Notably, this gene was the first to be identified as the cause of a non-sex-linked trait using the technique of *linkage analysis*, which is described below. For context, most of the genes implicated in the expression of inherited conditions (including the *ABO* gene) were identified with this technique.

Phenylketonuria (PKU) is a neurodevelopmental disorder resulting from toxic accumulation of the amino acid phenylalanine, resulting in a variety of cognitive and behavioral symptoms, including intellectual disability. This disorder is caused by one or more inherited mutations in the *PAH* gene, which normally functions to convert phenylalanine to tyrosine, another amino acid. PKU is inherited in a recessive manner, meaning that an individual must inherit two copies (one from each parent) of the mutated PAH allele. Notably, PKU provides one of the classic examples of *gene by environment interaction*, in which the effect of an allele on the expression of a phenotype depends on specific environmental conditions. If an infant with a mutation in *PAH* is placed on a phenylalanine-free diet, symptom severity is greatly reduced.

As indicated above, a number of syndromes associated with intellectual disability are also attributable to mutations in single alleles, including Rett syndrome, Fragile X syndrome, and Prader–Willi syndrome (see Ilyas et al., 2020, for review). Some of these conditions are inherited and expressed in a Mendelian manner, while others are attributable to new (*de novo*) mutations in genes during gametogenesis, as described above. That is, for some conditions, neither of the parents necessarily have the condition. This role of these genes in the expression of these conditions is very analogous to the relationship between trisomy 21 (i.e., the condition of having 3 copies of chromosome 21) and Down syndrome, for example.

Mendelian Genetics and Chromosomes Mendelian traits are rare. This is true in people, animals, and even pea plants. Had Mendel picked for his studies almost another trait, such as pea pod length, for example, he would have found that pea pods vary continuously in length and that these lengths form a normal distribution. However, what is rare about Mendelian traits is the extreme control a single allele has over the expression of a particular phenotype; the behavior of chromosomes during meiosis is decidedly Mendelian in nature.

If you were to count the chromosomes in a gamete (a sperm or egg), you would find that the gamete contains a single, unduplicated copy of each autosome (chromosomes 1–22) and a single sex chromosome (X or Y). Because there is a single copy of each chromosome in the gamete, there can be only one allele for each gene in the gamete, and this allele will be either *maternal* or *paternal* in origin. (Another way to think about this is that a given allelic sequence contained in a sperm or egg cell is derived either from the mother or father of the person who produced the egg or sperm.) Half of the gametes produced by a person will contain the maternal allele for any given gene, and the other half will contain the paternal allele for that same gene. This *segregation* of the maternal and paternal alleles among gametes means that any offspring derived from a gamete will receive only one of the two possible alleles for a gene from each parent—just like Mendel's peas.

Mendel's principle of independent assortment also applies to the behavior of chromosomes during meiosis but with a couple of very important caveats. Perhaps most importantly, independent assortment of alleles occurs when the alleles are contained in *different* chromosomes. Alleles that are parts of the *same* chromosome, i.e., alleles that are physically connected because they are a part of the same continuous strand of DNA, tend to be inherited together, with the likelihood of this co-inheritance determined by the *recombination frequency* in the area between them. The recombination frequency is the likelihood of a crossing over event (homologous recombination of the maternal and paternal chromosomes) between any two alleles contained on a chromosome. It is primarily related to the distance (in Mb of DNA) between the alleles in the chromosome. In general, the further apart two alleles are in a chromosome, the more likely a recombination event will occur between them, with the result that they wind up in different gametes. Alleles located at opposite ends of chromosome (in the *p* and *q* arms of chromosome 1, for example) are likely to be recombined in any given gamete, while alleles located within a single band of chromosome 1, such as 1q21.2, are more likely to be found together in the same gamete. Such "linked" alleles are in *linkage disequilibrium*, which forms the basis of our ability to conduct *linkage analysis*, as described below.

Another violation of the principle of independent assortment occurs in the X and Y chromosomes. Because the X and Y chromosomes in males are mostly dissimilar (heterologous), meiotic recombination between these two chromosomes can occur in only a very single small region, called the pseudoautosomal region, where there is homology between the X and Y chromosomes. This general lack of recombination in males means that alleles contained in the Y chromosome will be identical in sperm cells and somatic (non-sex) cells, but more importantly for genetic purposes, that the Y chromosome alleles contained in sperm are entirely paternal in origin. That is, the alleles contained in the Y chromosome will be identical (barring new mutations) to the father of the person whose sperm cells we are considering here.

Similarly, for half of the sperm cells that contain an X chromosome, the alleles in this chromosome will also be identical to those in the somatic cells, and these alleles are entirely *maternal* in origin. Daughters will therefore inherit an X chromosome identical to that of the father, and this will be the version found in all somatic cells. However, each of their gametes (eggs) will contain shuffled (recombined) versions of the X chromosome inherited from the father and the mother.

In contrast, the homologous X chromosomes in females act much like the other chromosomes (the autosomes). Alleles on these homologous chromosomes segregate during meiosis, and recombination between the maternal and paternal X chromosomes occurs as in the autosomes. These differences between the behavior of the XX chromosomes in females and the XY chromosomes in males contribute to the expression of *sex-linked traits*. In most cases, sex-linked traits are more commonly expressed in males than in females, and males also tend to express the trait more strongly.

Genetic Influence on Trait Variation (Quantitative)

Figure 5 shows a normal distribution. The x-axis depicts possible scores on a measure, ranging from low to high, and the *y*-axis shows the proportion of individuals with a given score. Most things we can measure in people follow distributions that are normal or nearly normal: body weight, outgoingness, height, blood pressure, blood sugar levels, dopamine turnover rate, athletic ability, visual acuity, learning rates, intelligence scores, and many others. If you measure any of these traits in a large population of people, you will find that half of the population will score higher than the mean, and the other half will score lower. In each of these halves, most people will score very close to the mean, and very few will score extremely far above or below the population mean.

Consider a trait like intelligence in 8-year-olds, which can be defined (rightly or wrongly) by a child's composite score on a standardized intelligence test, such as the WISC. Most children—68% of them; a little over two-thirds—will score between 85 and 115 on this test. This range includes the normalized average score on the test (100) plus and minus one standard deviation, which is a "standard" (pun intended…) way to depict variation in scores among individuals. On the WISC, the standard deviation is 15. Nearly all children (95% of them) will score between 70 and 130, which is the mean plus or minus two standard deviations, and 99.7% of all

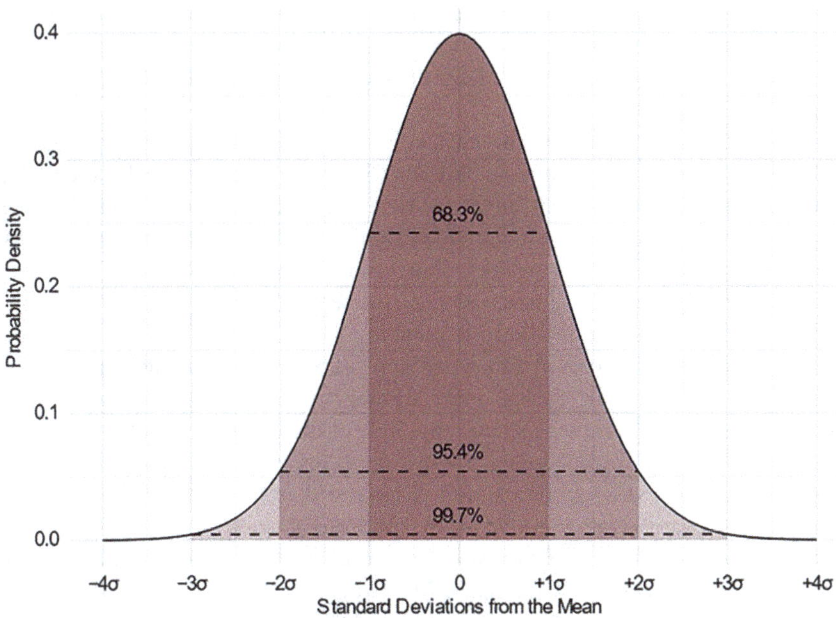

Fig. 5 A normal distribution. The X axis depicts individual scores on a measure, and the Y axis shows the proportion of the population with a given score. The score of 0 indicates the population mean, and the negative and positive sigma values represent deviations from that mean

children will score between 55 and 145. The remaining 0.3% of children will have scores greater than 145 or less than 55.

This pattern of variation in the population is clearly not the result of a gene acting in a Mendelian fashion. If such a gene existed, we would expect children to fall into discrete categories of intelligence scores, with little to no overlap between the categories. To be clear, we can apply labels of "above average" or "below average" to WISC scores in the right or left halves of the distribution (respectively), or even "mild," "moderate," or "profound" intellectual disability to first, second, and third standard deviations below the mean (again, respectively), but these labels are simply conveniences that can be used to describe similarities and differences among individuals with similar scores (again, rightly or wrongly). On average, children with WISC scores between 55 and 69 have a mild intellectual disability, yet on average, there is likely little functional difference between a child with a score of 69 and one with a score of 70, even though the latter would fall within the neurotypical range for WISC scores.

If you were going to identify factors that underlie variation in intelligence scores, how might you go about it? You could devise many studies to ask whether social and environmental factors as varied as the duration of breast-feeding during infancy, access to pre-K education, the amount of lead in the tap water, parental educational attainment, the number and ages of other children in the household, presence of poverty in the neighborhood, or the child's diet and physical activity levels might have roles. Because most of these factors are not possible, practical, or ethical to manipulate, your study will likely be correlational in design, and it will show you whether variation exists in the population in a given factor, such as children whose parents either do or do not have a college degree, can explain some of the variations in WISC scores among them.

You could also conduct correlational studies that ask whether genes contribute to the variation in WISC scores among children. The classic ways to do this are with a *twin study* or an *adoption study*, both of which estimate the degree to which genes and the environment (defined here as any factor that is not genetic in origin) contribute to the variation seen in the scores. Stated more formally, these types of studies are used to partition the total amount of variation seen in a trait into components that can be explained by genes and the environment.

Twin studies are based on the idea that any set of identical (monozygotic; MZ) twins share 100% of their alleles—they are genetic clones of each other—while fraternal (dizygotic; DZ) twins share only 50% of their alleles, which is the same degree of relatedness shared among any two same-sex siblings. If MZ twins resemble each other more strongly on the trait of interest than DZ twins, this is evidence that genetic similarity is responsible for the trait resemblance. A twin study of intelligence, as measured by the WISC, would test whether there is a stronger agreement (concordance) between WISC scores of MZ twins and DZ twins. If so, this would indicate that the greater genetic similarity in MZ twins versus DZ twins might explain the similarity in scores. However, twin studies assume that genetic similarity is the *only* real difference between MZ and DZ twins, and that the environments experienced by MZ and DZ twins are equal. If you have ever seen identical twins dressed in identical outfits, you can understand why the *equal environments assumption* can be difficult to justify (Fosse et al., 2015; Richardson & Norgate, 2005; see Felson, 2014, for discussion).

Adoption studies are another classical way to determine the degree to which genes contribute to variation in a trait. In an adoption study, scores on the measure of interest are obtained for three groups of individuals: adoptees and their biological and adoptive parents. The degree to which you can predict adoptee scores based on the scores of the (A) biological and (B) adoptive parents provides an estimate of the (A) genetic and (B) environmental influence on the scores.

Note that neither twin nor adoption studies will tell you *which* gene variants might explain the phenotypic similarity between individuals. These studies simply estimate the extent to which the phenotypic resemblance among related individuals might be the result of genetic similarities. For example, twin studies estimate that identical twins will share a diagnosis of autism spectrum disorder (ASD) up to 90% of the time, whereas this agreement is 53% in fraternal twins (Tick et al., 2016). These numbers tell us how much genes might matter in the expression of a particular trait or condition, but not the specific gene variants that might lead to the development of ASD.

To identify specific gene variants (alleles) associated with a particular trait, geneticists primarily use two approaches, *linkage analysis* and *association analysis*. These approaches are described in the context of family studies, as well as in the general population.

Linkage Analysis and Association Analysis

Family studies have a long history in genetic studies. In their simplest form, these studies ask whether there is evidence for transmission of a phenotype across generations within families. This evidence is then used to infer a mode of transmission of the trait, including whether one or more genes might be involved, and the degree to which these genes can explain the incidence of the trait. More broadly, these studies are used to determine whether family members who express a condition might have either chromosomal regions or gene variants (alleles) in common.

Linkage analysis looks for specific regions of chromosomes that are *identical by descent* in family members who express the condition (*affected* family members). These chromosomal regions are usually defined by SNPs, which can be used to differentiate between maternal and paternal sections of chromosomes in an individual. Chromosomal regions found at a higher-than-expected frequency in affected family members are said to be in *linkage disequilibrium* with the trait. Regions in linkage disequilibrium shared among affected family members (*linkage intervals*) might contain one or more gene variations (alleles) that cause the condition. It is worth noting that the *linkage* in linkage analysis refers to linked alleles in a chromosomal region, *not* to the association between that region and the condition of interest.

Association analysis looks directly for gene variants associated with the condition of interest. Association analyses are frequently used as follow-up studies to linkage analysis. For example, imagine there is linkage interval associated with an intellectual disability in a particular family that contains 8 genes. An investigator might attempt to determine whether family members with the disability differ from unaffected family members in the alleles for one or more of these 8 genes. However, because the region containing the 8 genes is in linkage disequilibrium, the problem with this approach is that all affected family members are likely to have the same allele for each of these 8 genes, making it impossible to identify a single gene variant as the culprit.

If the size of the family under study is increased to include additional generations and extended family members who also have the condition, there is a greater likelihood of finding an affected family member with a recombination in the interval. The portion of the linkage interval this affected individual shares with unaffected family members can then be excluded from the linkage interval, leaving only the portion shared among affected members as the implicated region.

A similar effect occurs with the inclusion of additional families in the analysis. This approach, however, obviously requires that the condition is not unique to the family under consideration. But more subtly, it also requires that the mechanism through which the alleles produce the intellectual disability is also not unique to the family. An allele might have a very large effect in one family—it could be both highly *penetrant* (almost all family members with the allele have the condition) and have high *expressivity* (all family members with the allele express the condition strongly)—yet have a smaller effect in other families. Alleles like this are described as being *variably penetrant* and *variably expressive* with respect to the condition of interest. The effect of the alleles depends on the presence of other, often unknown, factors.

It is important to note that the concepts of penetrance and expressivity do not apply to other possible causes of a condition, but rather refer only to the effect of a single allele (or multiple alleles) on the condition being considered. That is, an allele might have a highly penetrant effect on a condition—almost everyone with the particular allele has the condition—yet most people with the condition do not have this allele. Mutations in the *Fmr1* gene, for instance, can cause intellectual disability, which is one of the constellation of symptoms seen in Fragile X syndrome, but most people with intellectual disability do not have Fragile X syndrome.

Because of the development of high-throughput and low-cost genotyping chips, it is now feasible to obtain allelic information—in the form of SNPs—for the entire human genome in very large samples of people. Genome Wide Association (GWA) Studies ask whether there is an association within these large groups of people between specific alleles (defined by SNPs) and the condition of interest. For example, Grove et al. (2019) compared the genomes of 18,381 people diagnosed with autism spectrum disorder to 27,969 controls identified alleles localized to 5 different regions of the genome, including regions of chromosomes 8 and 20, which differed between these two groups.

Some Parting Words of Caution: Language Matters

We often speak of a disorder or condition as "genetic" in a way that makes us sound like we understand more about the condition or disorder than we actually do. And to be clear, for some of these disorders, such as Fragile X syndrome, PKU, or Rett syndrome, we do know more. Specifically, we know what causes them—a gene, or more precisely, a specific gene variation (an allele, as described above). But describing a condition as "genetic" comes with some important disclaimers.

Perhaps most importantly, the word "genetic" as it is used in this chapter does not equal "inherited." There are many disorders that are *genetic*—they are caused by genes—but not *inherited* in the sense we normally use the word. Syndromes resulting from mutations or nondisjunction events that occur during gametogenesis are clearly the result of genes, but they are not *inherited* across generations. Conversely, there are conditions that might appear to be inherited but are not, nor are these conditions necessarily *genetic* in origin. For example, a condition that appears to "run in a family," such as a mild intellectual disability, could result from affected family members' responses to an environmental condition, which could be anything from lead in the drinking water to stress resulting from domestic abuse. A linkage analysis conducted within this family might identify alleles that render individuals more or less susceptible to the effects of this environmental condition, but these alleles would likely not directly affect intellect.

Similarly, fetal alcohol spectrum disorder (FASD) results from fetal alcohol exposure during pregnancy. FASD is clearly the result of an environmental effect (alcohol exposure), yet not all fetuses exposed to alcohol develop FASD, and the severity of symptoms seen in children with the condition varies (Popova et al., 2023). Alcohol exposure causes FASD—alcohol is both necessary and sufficient for FASD to develop—so we would not likely think about this condition as "genetic" or "inherited." However, there are both protective and risk alleles which increase or decrease (respectively) the chances of FASD developing, given that fetal alcohol exposure occurred, as well as alleles that moderate both the presence and severity of specific aspects of FASD when it develops (see Kaminen-Ahola, 2020, for review), so there is clearly a genetic basis for FASD. However, much like the example in the preceding paragraph, these alleles do not themselves actually produce FASD (alcohol does that), but an individual with FASD might not otherwise have developed this condition, were it not for their genes.

My point with the above examples is not to be needlessly pedantic about correct and proper uses of *genetic* and *inherited*. But the choice of words used to describe a condition does have an effect on our perceptions of a condition, especially in regard to the blame and guilt that can occur among family members with a genetic or inherited condition (James et al., 2006). Furthermore, it is worthwhile to consider what the use of these descriptors actually does for us. Knowing that a child's intellectual disability is attributable to Fragile X syndrome, for example, does not mean that this child's prospects are better or worse than those of a child with an intellectual disability resulting from unknown (sporadic) causes, nor will there necessarily be any difference in the behavioral and cognitive interventions used with these two children. Of course, knowing that Fragile X syndrome is caused by an expanded trinucleotide repeat in the *Fmr1* gene provides a starting point for therapeutic development, and it allows for scientific literature to develop around the condition in the first place.

General Summary

Genetic influences on behavior manifest through the effects of particular alleles on the development of the nervous system. These alleles are contained within the DNA sequences in the chromosomes and are expressed as functional proteins in a manner regulated by cell type, developmental timepoint, and epigenetic factors. For named genetic conditions that cause intellectual disability, the effects of a given allele are large, but (most) cases are the result of complex interactions between multiple alleles and the environment. Some genetic conditions are the result of very large genetic changes that are visible in the chromosomes themselves, while others are the result of mutations affecting an area as small as a single nucleotide of DNA. The specific alleles associated with a trait or disorder can be identified through linkage analysis conducted within affected families, and through GWAS in large populations. Studies aimed at identifying the genetic bases of neurodevelopmental disorders might lead to better treatments for the conditions and, at a minimum, will help us understand how behavioral and cognitive traits are inherited and expressed.

References

Amir, R. E., Van den Veyver, I. B., Wan, M., Tran, C. Q., Francke, U., & Zoghbi, H. Y. (1999). Rett syndrome is caused by mutations in X-linked MECP2, encoding methyl-CpG-binding protein 2. *Nature Genetics, 23*(2), 185–188. https://doi.org/10.1038/13810

De Mei, C., Ramos, M., Iitaka, C., & Borrelli, E. (2009). Getting specialized: Presynaptic and postsynaptic dopamine D2 receptors. *Current Opinion in Pharmacology, 9*(1), 53–58. https://doi.org/10.1016/j.coph.2008.12.002

Ehrhart, F., Janssen, K. J. M., Coort, S. L., Evelo, C. T., & Curfs, L. M. G. (2019). Prader-Willi syndrome and Angelman syndrome: Visualisation of the molecular pathways for two chromosomal disorders. *The World Journal of Biological Psychiatry, 20*(9), 670–682. https://doi.org/10.1080/15622975.2018.1439594

Felson, J. (2014). What can we learn from twin studies? A comprehensive evaluation of the equal environments assumption. *Social Science Research, 43*, 184–199. https://doi.org/10.1016/j.ssresearch.2013.10.004

Fosse, R., Joseph, J., & Richardson, K. (2015). A critical assessment of the equal-environment assumption of the twin method for schizophrenia. *Frontiers in Psychiatry, 6*, 62. https://doi.org/10.3389/fpsyt.2015.00062

Grove, J., Ripke, S., Als, T. D., Mattheisen, M., Walters, R. K., Won, H., Pallesen, J., Agerbo, E., Andreassen, O. A., Anney, R., Awashti, S., Belliveau, R., Bettella, F., Buxbaum, J. D., Bybjerg-Grauholm, J., Bækvad-Hansen, M., Cerrato, F., Chambert, K., Christensen, J. H., et al. (2019). Identification of common genetic risk variants for autism spectrum disorder. *Nature Genetics, 51*(3), 431–444. https://doi.org/10.1038/s41588-019-0344-8

Ilyas, M., Mir, A., Efthymiou, S., & Houlden, H. (2020). The genetics of intellectual disability: Advancing technology and gene editing. *F1000Research, 9*, F1000 Faculty Rev-22. https://doi.org/10.12688/f1000research.16315.1

James, C. A., Hadley, D. W., Holtzman, N. A., & Winkelstein, J. A. (2006). How does the mode of inheritance of a genetic condition influence families? A study of guilt, blame, stigma, and understanding of inheritance and reproductive risks in families with X-linked and autosomal recessive diseases. *Genetics in Medicine: Official Journal of the American College of Medical Genetics, 8*(4), 234–242. https://doi.org/10.1097/01.gim.0000215177.28010.6e

Kaminen-Ahola, N. (2020). Fetal alcohol spectrum disorders: Genetic and epigenetic mechanisms. *Prenatal Diagnosis, 40*(9), 1185–1192. https://doi.org/10.1002/pd.5731

Karki, R., Pandya, D., Elston, R. C., & Ferlini, C. (2015). Defining "mutation" and "polymorphism" in the era of personal genomics. *BMC Medical Genomics, 8*, 37. https://doi.org/10.1186/s12920-015-0115-z

Ma, V. K., Mao, R., Toth, J. N., Fulmer, M. L., Egense, A. S., & Shankar, S. P. (2023). Prader-Willi and Angelman syndromes: Mechanisms and management. *The Application of Clinical Genetics, 16*, 41–52. https://doi.org/10.2147/TACG.S372708

Ozkan, E., & Lacerda, M. P. (2024). Genetics, cytogenetic testing and conventional karyotype. In *StatPearls*. StatPearls Publishing. https://www.ncbi.nlm.nih.gov/books/NBK563293/

Paliulis, L., Fabig, G., & Müller-Reichert, T. (2023). The X chromosome still has a lot to reveal—revisiting Hermann Henking's work on firebugs. *Journal of Cell Science, 136*(4), jcs260998. https://doi.org/10.1242/jcs.260998

Popova, S., Charness, M. E., Burd, L., Crawford, A., Hoyme, H. E., Mukherjee, R. A. S., Riley, E. P., & Elliott, E. J. (2023). Fetal alcohol spectrum disorders. *Nature Reviews Disease Primers, 9*(1). Article 1. https://doi.org/10.1038/s41572-023-00420-x

Richardson, K., & Norgate, S. (2005). The equal environments assumption of classical twin studies may not hold. *The British Journal of Educational Psychology, 75*(Pt 3), 339–350. https://doi.org/10.1348/000709904X24690

Rogozin, I. B., Carmel, L., Csuros, M., & Koonin, E. V. (2012). Origin and evolution of spliceosomal introns. *Biology Direct, 7*, 11. https://doi.org/10.1186/1745-6150-7-11

Tick, B., Bolton, P., Happé, F., Rutter, M., & Rijsdijk, F. (2016). Heritability of autism spectrum disorders: A meta-analysis of twin studies. *Journal of Child Psychology and Psychiatry, and Allied Disciplines, 57*(5), 585–595. https://doi.org/10.1111/jcpp.12499

Watson, C. T., Tomas, M.-B., Sharp, A. J., & Mefford, H. C. (2014). The Genetics of Microdeletion and microduplication syndromes: An update. *Annual Review of Genomics and Human Genetics, 15*(1), 215–244. https://doi.org/10.1146/annurev-genom-091212-153408

Epigenetics of Neurodevelopmental Disorders

Daniel B. Campbell

Abstract Epigenetics is the way that the environment affects the way genes work. DNA is the basis of inheritance, but DNA does not have a function of its own. Epigenetic mechanisms determine how DNA is packaged in the nucleus, how available DNA is to be transcribed into RNA and proteins, and how the food that you eat and the air that you breathe alter your genetic destiny. Epigenetics are widely believed to contribute to neurodevelopmental disorders. Strong genetic evidence indicates that genes involved in making DNA available, called chromatin regulators, are often mutated in individuals with neurodevelopmental disorders. However, there are challenges to directly studying epigenetic mechanisms. Foremost, epigenetic mechanisms are specific to individual cells at precise time points in development. Because the human brain during neurodevelopment is inaccessible, epigenetic mechanisms must be studied in a variety of proxies. Convergence of evidence in animal models, postmortem human brains, and non-brain human tissues can indicate the mechanisms underlying neurodevelopmental disorders. The study of epigenetics in neurodevelopmental disorders is in its infancy, but important clues are beginning to emerge.

Keywords DNA · Chromatin · Histone · Methylation · Noncoding RNA

What Is Epigenetics?

Epigenetics is the study of how cells control gene activity without changing the DNA sequence. "Epi-" means above, so epigenetics describes factors beyond DNA that control how genes work. Many nongenetic factors, such as your behaviors and

D. B. Campbell (✉)
Department of Pediatrics and Human Development, Michigan State University, Grand Rapids, MI, USA
e-mail: campb971@msu.edu

M. G. Valdovinos (ed.), *Intellectual and Developmental Disabilities*, Contemporary Clinical Neuroscience,
https://doi.org/10.1007/978-3-031-66932-3_5

environment, change the way that genes function. DNA contains the template for the generation of RNA and proteins. However, DNA does not itself have a function. Outside forces must act on the DNA to generate the RNA. And those outside, non-genetic factors are influenced by the environment. Epigenetics is the study of how behaviors and the environment change how the messages within DNA are read.

Epigenetic mechanisms contribute to most common neurodevelopmental disorders, such as epilepsy, autism spectrum disorder (ASD), attention deficit hyperactivity disorder (ADHD), and schizophrenia (SCZ). Each of these common neurodevelopmental disorders affects more than 1% of the population, and collectively they impact 5% of the population. Despite clear heritability and extensive genetic profiling, DNA mutations account for only ~10% of epilepsy, ASD, ADHD, or SCZ cases. Neurodevelopmental disorders are caused by a complex interplay between genetic susceptibility and environmental exposure. Epigenetics is the interface between genes and the environment.

The Piano Analogy

One way to think about epigenetics is to use the analogy of a piano and a piano player. DNA is the piano keys. When a key on the piano keyboard is pressed, a predictable note is produced. Mutations of the DNA are individual piano keys that are missing or damaged; it is often possible to still play the piano without a missing key, but some songs will not sound the same without the missing note.

Epigenetics plays several roles. Epigenetics is the pianist. That pianist can be a maestro who trained for years and has perfected the art of playing the piano. Or that pianist may be a beginner. Or the pianist may be a cat who ran across the piano keys. All of these outside forces upon the piano keys will create notes on the piano, but they will likely produce different combinations of sound that will sound more or less like an organized piece of music. Epigenetics is also the tuning of the piano and the environment in which the piano is played. The same piano will have different acoustics when it is played in a concert hall or in a basement. Epigenetics will determine how the piano sounds when the DNA keys are pressed.

Cell Type Specificity of Epigenetics

Each human body is composed of ~50 trillion cells, and every cell within an individual has the same DNA sequence. However, every one of those 50 trillion cells has a different epigenetic state. There are millions of different types of cells in the human body, each of which looks and functions very differently. Epigenetics determines how the DNA is expressed to make a neuron in the brain, an osteoclast in bones, or a T cell in the immune system. Neurons, immune cells, liver cells, and skin cells all look very different and have very different purposes. How can one DNA blueprint create so many different types of cells? The answer is epigenetics. Epigenetics expresses different genes from the DNA in different cell types to create a variety of cells.

Challenges of Studying the Epigenetics of Neurodevelopmental Disorders

The cell type specificity of epigenetics is a challenge for the direct study of the epigenetics of human neurodevelopmental disorders. The perfect tissue to study the epigenetics of neurodevelopmental disorders would be a piece of brain tissue at the onset of the neurodevelopmental disorder symptoms. For many ethical reasons, that tissue will never be available. So, instead, we must use proxies: animal models with specific genetic alterations, postmortem brain tissues, human cerebral organoids, blood samples, and cord samples. However, none of these exactly represent the epigenetics of the relevant brain regions at the onset of neurodevelopmental disorder symptoms. Instead, we can only hope to model the epigenetics of the cell type-specific and developmental time point of the onset of symptoms. This is why we use a variety of animal models and tissue samples.

Animal Models

Mice and rats are often used for epigenetics studies. The advantages of using animal models to study neurodevelopmental disorders are many. Foremost, the animal model brain develops in the context of in vivo physiology—with a growing blood supply, three-dimensional cues, and exposure to placental hormones. A mouse model can be created with a mutation in a specific gene and impacts of that genetic mutation can be determined on a host of phenotypes from behavioral to molecular, including epigenetic effects. However, there are dramatic differences in mouse and human brain development that limit the relevance of animal models in the study of epigenetics in human neurodevelopmental disorders. First, although there is a high conservation of DNA sequences within protein-coding genes, there are enormous differences between humans and rodents in the genetic sequence of noncoding regions of the genome. For example, most of the DNA sequences in the promoter regions of human protein-coding genes do not have analogous sequences in the promoters of mouse protein-coding genes. Furthermore, most of the long noncoding RNAs expressed in the human brain simply do not exist in the mouse genome. Therefore, the epigenetic control of gene expression differs intensely between humans and mice. Developing mouse brain lacks some of the transient cell types present during human brain development, precluding the study of these types of cells on human neurodevelopmental disorders. Perhaps most concerning, the changes in gene expression in response to external challenges differ dramatically—and often in diametrically opposed directions—between mouse cerebral organoids and human cerebral organoids. Thus, the advantage of the in vivo physiology of animal models is tempered by organismal differences at the level of epigenetics.

Human Cerebral Organoids

Human cerebral organoids are "mini-brains" derived from human-induced pluripotent stem cells (iPSCs) (Bilinovich et al., 2020a, b). One advantage of human cerebral organoids is that they are human cells of the tissue of interest, the brain. Experiments have shown high fidelity of human cerebral organoids with typical in vivo human brain development, particularly in the fidelity of human cerebral organoids to match the epigenetic changes known to happen in the developing human brain. Another advantage of human cerebral organoids is that they can be derived from easily accessible skin tissue of any individual, and can thus be chosen for specific neurological disorders or known genetic mutations. However, there are disadvantages to human cerebral organoids. The in vitro culture system may not reproduce important physiological aspects of in vivo brain development, such as a growing blood supply and exposure to placental hormones. Another disadvantage is that, due to a lack of blood supply to support tissues growing at the core of the human cerebral organoids, organoids typically die at about the neurodevelopmental timepoint of human birth. So, while human cerebral organoids reasonably represent human neurodevelopment up to birth, they cannot be used to study epigenetic changes during the first years outside the womb.

Postmortem Brain

Postmortem human brain provides a sample of the tissue of interest, the brain. The advantages include the diagnostic phenotypes of the individual from whom the brain was donated and in vivo development of the donated brain tissue. However, there are several disadvantages to measuring the epigenetic phenomenon in postmortem brain tissue. First, the brain tissue is harvested at the time of death, which can be years or even decades after the onset of neurodevelopmental disorder. Therefore, even if there are differences between individuals with neurodevelopmental disorder and neurotypical controls, it is impossible to discern whether those epigenetic changes are an original cause of the neurodevelopmental disorder or the result of living with the neurodevelopmental disorder for many years. Second, the circumstances of death can alter the epigenome. It is simply not known how epigenetic measures in the brain are affected by traumatic causes of death such as seizures, drowning, heart attacks, or car crashes. Epigenetic responses to the cause of death may overshadow epigenetic signals of the neurodevelopmental disorder with which the patient lived. Third, there are inherent differences in the harvest of postmortem brain samples that can affect epigenetic measures. The time from death to harvest of the brain ranges from minutes to several days, and is typically several hours. Epigenetic changes may occur over time after the death of the individual. Thus, while postmortem human brain samples provide the most direct access to the tissue of interest for neurodevelopmental disorders, the interpretations of epigenetic changes in postmortem brain samples should be cautious.

Blood Samples

Blood samples have several advantages for measuring epigenetic changes. The collection of blood samples is relatively noninvasive, in vivo, and can be performed longitudinally to determine changes during postnatal development. Conditions surrounding the blood draw can be reasonably controlled to ensure that differences between neurodevelopmental disorder cases and neurotypical controls represent epigenetic differences specific to neurodevelopmental disorder. However, the disadvantage of measuring epigenetic differences in blood is that they may not be informative of the epigenetic differences in the brain. The observed epigenetic differences may be diagnostic biomarkers of the neurodevelopmental disorders, but not be indicative of mechanisms that cause the onset and maintenance of neurodevelopmental disorder.

Cord Blood Samples

Cord blood provides an in vivo human sample at the time of birth. Because cord blood is often saved, the samples can be used to determine epigenetics of a cohort of individuals after a diagnosis of a neurodevelopmental disorder, thus providing a retrospective look at the epigenetic condition. However, there are disadvantages to epigenetic measures in cord blood. Much like blood samples, the relevance to brain epigenetics can be questioned. Moreover, cord blood is collected at a very traumatic time, immediately after birth, as the birthing process involves trauma for both the mother and the baby. The epigenetic signature of the cord blood likely represents that trauma and interpretations must include an interaction of the neurodevelopmental disorder with the trauma of birth (Table 1).

Table 1 Advantages and disadvantages of epigenetics models

Model	Advantage	Disadvantage
Animal	In vivo physiology	Dramatic differences in epigenetic mechanisms from human
Human cerebral organoids	Human epigenetics and neurodevelopment	Lack of blood supply and in vivo physiological signals
Postmortem brain	Human samples from tissue of interest	Years/decades beyond the onset of neurodevelopmental symptoms; impossible to distinguish among the (a) cause of the disorder, (b) consequence of the disorder, (c) effect of the disorder treatment, or (d) circumstances of death
Blood samples	Human patient with specific diagnosis; longitudinal collection possible	Due to cell type specificity, epigenetics measured in blood cells may not be relevant to epigenetic changes in the brain
Cord samples	Human sample at the time of birth	Due to cell type specificity, epigenetics measured in cord cells may not be relevant to epigenetic changes in the brain

DNA Structure and Chromatin Packaging

Because epigenetics is the study of how nongenetic factors influence how genes work, we need to do a brief review of DNA structure. DNA is composed of four types of nucleotides: adenine (A), cytosine (C), guanine (G), and thymidine (T). A DNA molecule is made of two strands of nucleotides wound around each other, with each strand held together by bonds between the bases. Adenine pairs with thymine and cytosine pairs with guanine. The iconic double helix of a DNA molecule consists of two strings of nucleotides that bind together and form a helix.

Each of the nucleotides has a chemical formula that can be altered. One of the major modifications to DNA is called "methylation," where a methyl group replaces a hydrogen on guanine. The methylation alters the chemical structure in a way that changes the way DNA interacts with its environment. Specifically, the methyl group changes the way that transcription factors bind to DNA. Transcription factors are proteins that recognize specific DNA sequences, and either enhance or decrease the transcription rate of a gene. A methyl group changes the transcription factors that recognize the sequence.

The string of three billion nucleotides in the human genome would occupy 2 meters of linear space. Therefore, there are more than 2 meters of DNA in every single cell in the human body. The human body contains about 50 trillion cells, which means that there are more than 100 trillion meters of DNA in each human. For context, the sun is 150 billion meters from the Earth, so the DNA in each human could reach the sun and back 300 times. How does all of this DNA fit inside the human body? DNA is compacted by proteins called histones to fit inside the nucleus.

Histones are positively charged proteins that strongly bind to negatively charged DNA to form complexes called nucleosomes. Each nucleosome consists of 166 nucleotides of double-stranded DNA and 8 histone proteins compact that segment of DNA into a small ball. Between the compacted balls of DNA are strings of DNA which are not compacted and available to transcription factor proteins. The resulting ball-and-string pattern that is either open or compacted DNA–protein complex is called the chromatin. Importantly, because the interaction between DNA and histones is based on charge and not DNA sequence, chromatin—the pattern of nucleosomes—is dynamic. Chromatin patterns can change based on the environment of the cell. If the DNA template for a gene is compacted in a nucleosome, then that gene will not be expressed in a cell. Therefore, much of epigenetic regulation is simply the folding of DNA into nucleosomes, and the cell controlling which parts of the DNA molecule are available for transcription factors to act.

Types of Epigenetic Changes

Histone Modifications and Chromatin Remodeling

Because DNA is packaged into chromatin by histone proteins to play a major role in gene regulation, there are several cellular mechanisms devoted to regulating nucleosomes (Fig. 1). First, there are groups of proteins known as chromatin remodeling complexes that displace histones from DNA and thus make the underlying DNA sequences available to transcription factors. Rare genetic mutations of these chromatin remodeling proteins play a major role in neurodevelopmental disorders, although the specific functions of the chromatin remodeling proteins are largely

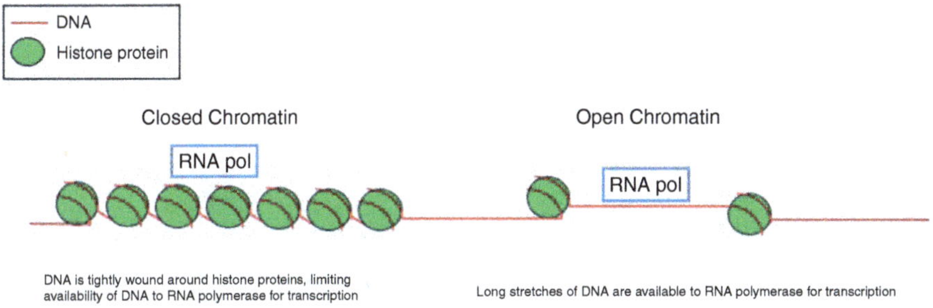

Fig. 1 Histone proteins and chromatin regulation. Histones are proteins that condense DNA so that it can be packaged into the nucleus. DNA that is wrapped around a histone protein is not available to RNA polymerase (RNA pol) for transcription. This is known as "closed chromatin," that is DNA tightly wrapped in histones and thus not available for transcription. Conversely, "open chromatin" is when histones are absent from the long stretches of DNA, and the DNA is available to RNA polymerase for transcription. The degree of histone binding to DNA is regulated by two processes. First, chromatin regulation proteins determine the regions of the genome that are wrapped in histones. Second, the histones can be chemically modified (e.g., by the addition of methylation or acetylation of specific amino acids) to alter the degree of DNA binding

unknown. Although the mutations are associated with neurodevelopmental disorders, the cell types and genes that the chromatin remodeling proteins regulate are an active area of ongoing research. Second, histones can be chemically modified by the addition of acetyl, methyl, or phosphate groups that make the histone–DNA complexes less compact. The histone modifications provide access to some genes that are within the nucleosome for transcription factors.

DNA Methylation

DNA methylation changes the chemical structure of DNA to alter the transcription factors that recognize that particular DNA sequence (Fig. 2). DNA methylation is catalyzed by a group of enzymes known as DNA methyltransferases (DNMTs) that add a methyl group to cytosine nucleotides. DNA methylation primarily occurs in regions of the DNA with enrichment of CpG dinucleotides, known as CpG islands. The CpG islands are often found in the promoter regions of genes and are critical for the regulation of gene expression. The degree of methylation of CpG islands within a gene promoter can change the accessibility of transcription factors, and thus change gene expression.

DNA methylation is dynamic. This is best described in cancer, where the DNA methylation state is a hallmark of cancer diagnosis. CpG islands are hypermethylated in the promoter regions of genes that suppress cancer and, in contrast, CpG islands are hypomethylated in the promoter regions of oncogenes. The DNMT enzymes are responsive to environmental factors such as diet, stress, and exposure to toxins. Therefore, environmental contributions to neurodevelopmental disorders may act through methylation changes.

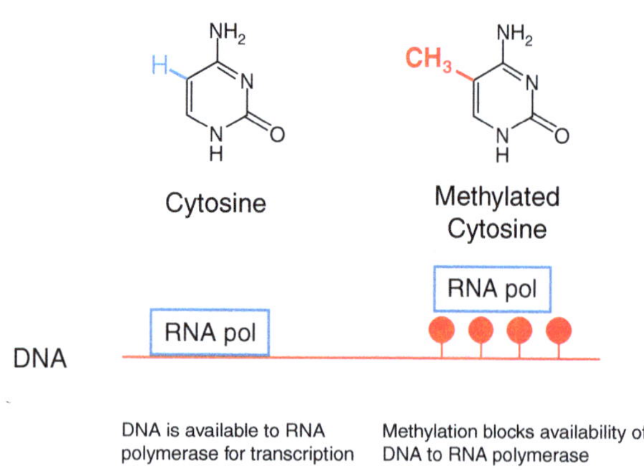

Fig. 2 DNA methylation. A specific hydrogen (blue) in the structure of the DNA nucleotide cytosine can be chemically modified to a methyl group (-CH_3; in red). The addition of a methyl group to cytosine is called DNA methylation. The methyl groups on cytosine can block the availability of DNA to RNA polymerase, and thus regulate the expression of genes

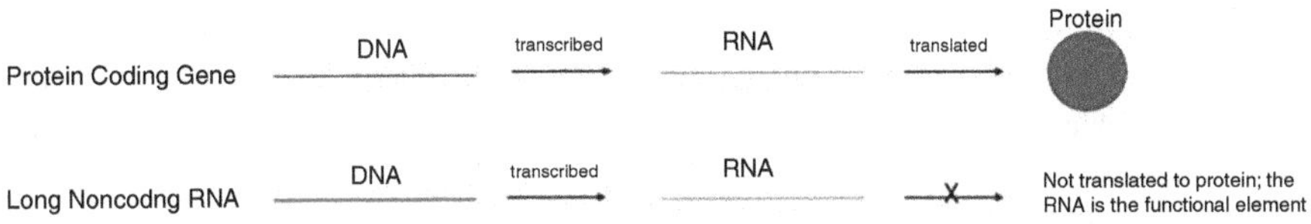

Fig. 3 Protein coding and long noncoding genes. Protein coding genes are produced in two steps: DNA is transcribed into RNA and RNA is translated into protein. There are approximately 20,000 protein-coding genes in humans. However, only 1.5% of the human DNA encodes a protein-coding gene. Much of the other 98.5% of the human genome generates long noncoding RNA, which is RNA transcribed from DNA but not translated into protein. The human brain contains more long noncoding RNAs than protein-coding RNAs, but the functions of most long noncoding RNAs in human brain development are unexplored

Noncoding RNA

Only 1.5% of the DNA in the human genome encodes a protein (Fig. 3). Until recently, it was thought that the other 98.5% of the 3 billion nucleotides in the human genome were "junk DNA," merely fillers for the protein-coding regions of the genome. The advent of next-generation sequencing revealed that 80% of the human genome is actually transcribed into RNA. If RNA is not translated into protein, then what does that RNA do? It turns out that noncoding RNAs have a number of functions in every cell in the body. They regulate the expression of genes at every level: noncoding RNAs can bind to DNA to inhibit transcription; noncoding RNAs can bind to RNA to regulate the stability of protein-coding mRNAs, and noncoding RNAs bind to proteins to alter the function of RNA–protein complexes.

Importantly for our context of neurodevelopmental disorders, the number and abundance of noncoding RNAs increases with the increasing complexity of organisms and organs. The human brain contains the most noncoding RNAs. In fact, noncoding RNAs are more abundant in the human brain than protein-coding RNAs. It is likely that the more we understand about noncoding RNAs, the more we will understand about the altered brain development that leads to neurodevelopmental disorders.

Specific Contributions of Epigenetics to Neurodevelopmental Disorders

Histone Modifications and Chromatin Remodeling Contributions to Neurodevelopmental Disorders

The most convincing evidence that chromatin remodeling contributes to neurodevelopmental disorders comes from DNA sequencing studies that indicate that chromatin remodeling genes are most often mutated in individuals with neurodevelopmental disorders (An et al., 2018; Iossifov et al., 2012; O'Roak et al., 2012b; Sanders et al., 2012; Satterstrom et al., 2020). There is a great deal of circumstantial evidence that histone modifications and chromatin remodeling contribute to neurodevelopmental disorders, but the mechanisms are poorly understood and an ongoing topic of research (Mossink et al., 2021).

Histone modifications refer to an array of post-translational modifications to histone proteins. Processes such as methylation, phosphorylation, and acetylation of specific amino acids within histone proteins are known to have biological consequences (Smith & Carregari, 2022). For example, histone acetylation has long been associated with memory formation, particularly long-term memory. However, the mechanisms by which the general biological process of adding an acetyl group to amino acids within histone proteins alters memory consolidation is poorly understood. Similarly, histone deacetylases (HDACs), enzymes that remove acetyl groups from proteins, have been implicated in neurodevelopmental disorders such as SCZ and ASD (Farrelly et al., 2022). However, the specific proteins deacetylated and the mechanisms by which they contribute to altered brain development are not known.

Methylation Contributions to Neurodevelopmental Disorders

Fragile X Syndrome Is Caused by Hypermethylation of the FMR1 Gene

Fragile X Syndrome is the most common single-gene cause of inherited intellectual disability, affecting 1 in 4000 males and 1 in 7000 females (Nobile et al., 2021). The causative gene for Fragile X Syndrome was cloned in 1991 and named *FMR1*. The *FMR1* gene encodes the FMRP (Fragile X Messenger Ribonucleoprotein), which is an mRNA-binding protein that is primarily expressed in the developing brain. However, it is not mutations in the coding region of the *FMR1* gene that cause Fragile X Syndrome. Instead, there is a DNA region just upstream of the *FMR1* gene that has a repetitive sequence of CGG. The typical number of these CGG repeats is 20–40. However, the number of CGG repeats can be expanded in some individuals to over 200 CGG repeats. The expanded repeats over 200 CGG repeats invariably lead to Fragile X Syndrome. The mechanism by which the expanded CGG repeats leads to the Fragile X Syndrome phenotype involves DNA methylation. When there are more than 200 CGG repeats, the DNA just upstream of the *FMR1* gene becomes full methylated. This hypermethylation of the DNA just upstream of the *FMR1* gene inhibits binding of transcription factors that would normally transcribe the *FMR1* mRNA. The lack of *FMR1* mRNA creates a deficit in FRMP, and causes the symptoms of Fragile X syndrome. However, the mechanisms by which the expanded CGG repeat becomes hypermethylated and how the hypermethylation silences the *FMR1* gene remain a topic of research.

Umbilical Cord Blood Methylation in ASD

Methylation status for any particular CpG island can be quantified. Because methylation is cell-type specific, one of the best tissues for neurodevelopmental disorders is the umbilical cord. Umbilical cord blood samples can be stored indefinitely and then tested for DNA methylation profiles. A study of 152 umbilical cord blood samples from individuals with ASD showed that there are DNA regions that are differentially methylated. There were 537 differentially methylated genes in males and 1762 differentially methylated genes in females. The genes were enriched for embryonic expression and X chromosome location (Mordaunt et al., 2020). While the mechanisms by which these genes are differentially methylated are not yet known, this is an ongoing line of research.

Long Noncoding RNAs (lncRNAs) in Neurodevelopmental Disorders

The abundance of lncRNAs is directly correlated to the complexity of an organism, with the complex human brain expressing the most lncRNAs (Hecht et al., 2015; Liu et al., 2013). LncRNAs are more abundant and more cell type specific in human brain than protein-coding RNAs (Hecht et al., 2015; Mattick et al., 2023). Therefore, to understand neurodevelopmental disorders, it is critical to understand the functions of lncRNAs in neurodevelopment. Genetic studies have successfully identified mutations in protein-coding genes that contribute to ASD and other neurodevelopmental disorders (An et al., 2018, 2020, p. 8; Ayhan & Konopka, 2019; Bernier et al., 2014; Iossifov et al., 2012; Neale et al., 2012; O'Roak et al., 2012a, b; Sanders et al., 2012; Satterstrom et al., 2020). However, there is substantial evidence that noncoding regions of the genome contribute to potentially more cases of ASD (An et al., 2018; Grove et al., 2019). A series of publications on gene expression in postmortem brains of individuals with ASD highlight the contributions of primate-specific lncRNAs (Gandal et al., 2018; Parikshak et al., 2016; Velmeshev et al., 2019). The Functional ANnoTation Of the Mammalian genome (FANTOM) Consortium has been a leader in providing resources to understand the transcriptome and transcriptional regulation for two decades (Abugessaisa et al., 2021). The latest edition, FANTOM6, focuses specifically on the function of lncRNAs (Abugessaisa et al., 2021; Ramilowski et al., 2020). In an effort to systematically elucidate the functions of lncRNAs, FANTOM6 focused on knockdown of 285 lncRNAs in human dermal fibroblasts (Abugessaisa et al., 2021; Ramilowski et al., 2020). FANTOM6 measured both cellular and molecular outcomes, and found that over 25% of lncRNA knockdowns result in changes in cell growth and morphology in dermal fibroblasts. The FANTOM6 project has highlighted the importance of lncRNAs in chromatin regulation, translational regulation, and cellular differentiation. The FANTOM6 project has also discovered that lncRNAs are less conserved and more cell-type specific than protein-coding genes (Abugessaisa et al., 2021; Antonov & Medvedeva, 2020; Ramilowski et al., 2020). While this is an enormous step forward in understanding the functions of lncRNAs, the roles of lncRNAs in other cell types and in neurodevelopmental disorders remain unclear. It is also clear that lncRNAs are more often over-expressed than under-expressed in neurodevelopmental disorders (Bilinovich et al., 2019; Kerin et al., 2012).

Like the evidence for chromatin regulators and DNA methylation, the evidence that lncRNAs contribute to neurodevelopmental disorders also comes from large genetic studies. A genome wide association studies (GWAS) efficiently identify genotypes at millions of nucleotides that are often different among individuals. These common genetic variants are typically found in the nonprotein-coding region of the genome. GWAS identifies genetic variants that are even more common in individuals with a specific neurodevelopmental disorder than in the general population. Because the genetic variants are common, these findings are interpreted as identifying regions of a chromosome that are implicated in neurodevelopmental disorder. For ASD, two large GWAS have been performed. The first GWAS identified had six different genetic markers that were all close to each other on chromosome 5 and significantly associated with ASD risk (Wang et al., 2009). The second GWAS identified a region of chromosome 20 that is strongly associated with ASD (Anney et al., 2010). Both of these results have been replicated (Autism Spectrum Disorders Working Group of The Psychiatric Genomics Consortium, 2017; Grove et al., 2019; Prandini et al., 2012; Wang et al., 2019). Both results were largely interpreted in the context of neighboring protein-coding genes. However, a series of publications indicated that both of these GWAS positive results showed that the genetic signal implicated lncRNAs (Bilinovich et al., 2019; DeWitt et al., 2016a, b; Kerin et al., 2012). Over-expression of both of these lncRNAs in human neural progenitor cells caused similar changes in gene expression and altered neuronal morphology, indicating impacts on neuronal differentiation (Bilinovich et al., 2019; DeWitt et al., 2016a). The functions of these lncRNAs are an active area of research.

Several publications identified additional lncRNAs that contribute to neurodevelopmental disorders (Alinejad-Rokny et al., 2020; Gudenas et al., 2017; Liu et al., 2020; Tong et al., 2020; Wang & Wang, 2020). All of these bioinformatics approaches provide strong statistical association but do not attempt to determine functions of the lncRNA in human neuronal cells. A brief outline of each of the publications are given:

1. 266 lncRNAs were identified in regions of the genome with DNA deletions or duplications (Alinejad-Rokny et al., 2020). One lncRNA (named *ENSG00000237940*) was highlighted as the only gene—coding or noncoding—that is present within one of the DNA deletion or duplication regions. The function of *ENSG00000237940* is unknown.
2. 57 lncRNAs were identified using machine learning models (Wang & Wang, 2020). Three distinct machine learning approaches were used to train models to predict protein-coding ASD risk genes based on developmental brain gene expression and transcript sequence, then applied the models to predict lncRNAs

associated with ASD. There were 57 lncRNAs predicted to be associated with ASD by all three machine learning models. None of the 57 lncRNAs have a known function in brain development.

3. 532 lncRNAs were identified by gene co-expression analyses of postmortem brains of twins (Tong et al., 2020). Weighted gene co-expression networks analysis (WGCNA) identified groups of genes with similar patterns of expression. The groups with known ASD genes also included 532 lncRNAs. There have been no further functional analyses of the 532 lncRNAs.
4. 243 lncRNAs were identified as enriched in the GWAS analyses of seven psychiatric traits (ASD, depression, bipolar disorder, SCZ, ADHD, anxiety, and neuroticism) (Liu et al., 2020). The lncRNA *LINC00461* was identified as the gene with the highest effects across the psychiatric traits. A lncRNA homologous to *LINC00461* was identified in mice, manipulated, and shown to cause changes in mouse behavior and delayed neuronal migration in the developing mouse cerebral cortex. These data indicate that a lncRNA homologous to *LINC00461* can impact brain development in mice. However, the effect of *LINC00461* on human brain development has not been tested.
5. 126 lncRNAs were identified by integrating postmortem cerebral cortex RNA sequencing data with a developmental co-expression network. These data provide a rationally data-driven set of lncRNAs that may be involved in ASD, but none of the lncRNAs have a known function.

These well-designed hypothesis-generating studies identified more than 1200 lncRNAs associated with ASD and other neurodevelopmental disorders. These more than 1200 ASD-associated lncRNAs are part of a much larger set of more than 30,000 lncRNAs expressed in the developing human brain (Hon et al., 2017). Each study provides a quantitative prioritization of the lncRNAs, but only one lncRNA is tested functionally; even then, the functionally tested lncRNA is a predicted mouse homolog of the lncRNA implicated in human neurodevelopmental disorder, not the human lncRNA itself. Thus, none of these more than 1200 neurodevelopmental disorder-relevant lncRNAs has been functionally tested in human neuronal cells.

Conclusions

Strong genetic evidence indicates that epigenetic mechanisms contribute to neurodevelopmental disorders. The genetic evidence that chromatin-regulating genes are more often mutated in individuals with neurodevelopmental disorders indicates that altered packaging of DNA contributes to neurodevelopmental disorders. The genetic evidence that longer stretches of CGG repeats cause Fragile X syndrome and subsequent evidence that longer stretches of CGG repeats are methylated at higher rates indicates that DNA methylation contributes to neurodevelopmental disorders. The genetic evidence that common genetic variants with strong associations to neurodevelopmental disorders regulate the expression of lncRNAs indicates that lncRNAs contribute to neurodevelopmental disorders. The mechanisms by which these epigenetic changes alter brain development is an active area of research. The convergence of evidence from multiple tissue sources will reveal the epigenetic mechanisms of neurodevelopment.

References

Abugessaisa, I., Ramilowski, J. A., Lizio, M., Severin, J., Hasegawa, A., Harshbarger, J., Kondo, A., Noguchi, S., Yip, C. W., Ooi, J. L. C., Tagami, M., Hori, F., Agrawal, S., Hon, C. C., Cardon, M., Ikeda, S., Ono, H., Bono, H., Kato, M., et al. (2021). FANTOM enters 20th year: Expansion of transcriptomic atlases and functional annotation of non-coding RNAs. *Nucleic Acids Research, 49*(D1), D892–D898. https://doi.org/10.1093/nar/gkaa1054

Alinejad-Rokny, H., Heng, J. I. T., & Forrest, A. R. R. (2020). Brain-enriched coding and long non-coding RNA genes are overrepresented in recurrent neurodevelopmental disorder CNVs. *Cell Reports, 33*(4), 108307. https://doi.org/10.1016/j.celrep.2020.108307

An, J.-Y., Lin, K., Zhu, L., Werling, D. M., Dong, S., Brand, H., Wang, H. Z., Zhao, X., Schwartz, G. B., Collins, R. L., Currall, B. B., Dastmalchi, C., Dea, J., Duhn, C., Gilson, M. C., Klei, L., Liang, L., Markenscoff-Papadimitriou, E., Pochareddy, S., et al. (2018). Genome-wide de novo risk score implicates promoter variation in autism spectrum disorder. *Science (New York, N.Y.), 362*(6420). https://doi.org/10.1126/science.aat6576

An, Y., Zhang, L., Liu, W., Jiang, Y., Chen, X., Lan, X., Li, G., Hang, Q., Wang, J., Gusella, J. F., Du, Y., & Shen, Y. (2020). De novo variants in the Helicase-C domain of CHD8 are associated with severe phenotypes including autism, language disability and overgrowth. *Human Genetics, 139*(4), 499–512. https://doi.org/10.1007/s00439-020-02115-9

Anney, R., Klei, L., Pinto, D., Regan, R., Conroy, J., Magalhaes, T. R., Correia, C., Abrahams, B. S., Sykes, N., Pagnamenta, A. T., Almeida, J., Bacchelli, E., Bailey, A. J., Baird, G., Battaglia, A., Berney, T., Bolshakova, N., Bölte, S., Bolton, P. F., et al. (2010). A genome-wide scan for common alleles affecting risk for autism. *Human Molecular Genetics, 19*(20), 4072–4082. https://doi.org/10.1093/hmg/ddq307

Antonov, I., & Medvedeva, Y. (2020). Direct interactions with nascent transcripts is potentially a common targeting mechanism of long non-coding RNAs. *Genes, 11*(12). https://doi.org/10.3390/genes11121483

Autism Spectrum Disorders Working Group of The Psychiatric Genomics Consortium. (2017). Meta-analysis of GWAS of over 16,000 individuals with autism spectrum disorder highlights a novel locus at 10q24.32 and a significant overlap with schizophrenia. *Molecular Autism, 8*, 21. https://doi.org/10.1186/s13229-017-0137-9

Ayhan, F., & Konopka, G. (2019). Regulatory genes and pathways disrupted in autism spectrum disorders. *Progress in Neuro-Psychopharmacology & Biological Psychiatry, 89*, 57–64. https://doi.org/10.1016/j.pnpbp.2018.08.017

Bernier, R., Golzio, C., Xiong, B., Stessman, H. A., Coe, B. P., Penn, O., Witherspoon, K., Gerdts, J., Baker, C., Vulto-van Silfhout, A. T., Schuurs-Hoeijmakers, J. H., Fichera, M., Bosco, P., Buono, S., Alberti, A., Failla, P., Peeters, H., Steyaert, J., Vissers, L. E. L. M., et al. (2014). Disruptive CHD8 mutations define a subtype of autism early in development. *Cell, 158*(2), 263–276. https://doi.org/10.1016/j.cell.2014.06.017

Bilinovich, S. M., Lewis, K., Grepo, N., & Campbell, D. B. (2019). The long noncoding RNA RPS10P2-AS1 is implicated in autism spectrum disorder risk and modulates gene expression in human neuronal progenitor cells. *Frontiers in Genetics, 10*, 970. https://doi.org/10.3389/fgene.2019.00970

Bilinovich, S. M., Lewis, K., Thompson, B. L., Prokop, J. W., & Campbell, D. B. (2020a). Environmental epigenetics of diesel particulate matter toxicogenomics. *International Journal of Environmental Research and Public Health, 17*(20). https://doi.org/10.3390/ijerph17207386

Bilinovich, S. M., Uhl, K. L., Lewis, K., Soehnlen, X., Williams, M., Vogt, D., Prokop, J. W., & Campbell, D. B. (2020b). Integrated RNA sequencing reveals epigenetic impacts of diesel particulate matter exposure in human cerebral organoids. *Developmental Neuroscience, 42*(5–6), 195–207. https://doi.org/10.1159/000513536

DeWitt, J. J., Grepo, N., Wilkinson, B., Evgrafov, O. V., Knowles, J. A., & Campbell, D. B. (2016a). Impact of the autism-associated long noncoding RNA MSNP1AS on neuronal architecture and gene expression in human neural progenitor cells. *Genes, 7*(10). https://doi.org/10.3390/genes7100076

DeWitt, J. J., Hecht, P. M., Grepo, N., Wilkinson, B., Evgrafov, O. V., Morris, K. V., Knowles, J. A., & Campbell, D. B. (2016b). Transcriptional gene silencing of the autism-associated long noncoding RNA MSNP1AS in human neural progenitor cells. *Developmental Neuroscience, 38*(5), 375–383. https://doi.org/10.1159/000453258

Farrelly, L. A., Zheng, S., Schrode, N., Topol, A., Bhanu, N. V., Bastle, R. M., Ramakrishnan, A., Chan, J. C., Cetin, B., Flaherty, E., Shen, L., Gleason, K., Tamminga, C. A., Garcia, B. A., Li, H., Brennand, K. J., & Maze, I. (2022). Chromatin profiling in human neurons reveals aberrant roles for histone acetylation and BET family proteins in schizophrenia. *Nature Communications, 13*(1), 2195. https://doi.org/10.1038/s41467-022-29922-0

Gandal, M. J., Haney, J. R., Parikshak, N. N., Leppa, V., Ramaswami, G., Hartl, C., Schork, A. J., Appadurai, V., Buil, A., Werge, T. M., Liu, C., White, K. P., CommonMind Consortium, PsychENCODE Consortium, iPSYCH-BROAD Working Group, Horvath, S., & Geschwind, D. H. (2018). Shared molecular neuropathology across major psychiatric disorders parallels polygenic overlap. *Science (New York, N.Y.), 359*(6376), 693–697. https://doi.org/10.1126/science.aad6469

Grove, J., Ripke, S., Als, T. D., Mattheisen, M., Walters, R. K., Won, H., Pallesen, J., Agerbo, E., Andreassen, O. A., Anney, R., Awashti, S., Belliveau, R., Bettella, F., Buxbaum, J. D., Bybjerg-Grauholm, J., Bækvad-Hansen, M., Cerrato, F., Chambert, K., Christensen, J. H., et al. (2019). Identification of common genetic risk variants for autism spectrum disorder. *Nature Genetics, 51*(3), 431–444. https://doi.org/10.1038/s41588-019-0344-8

Gudenas, B. L., Srivastava, A. K., & Wang, L. (2017). Integrative genomic analyses for identification and prioritization of long non-coding RNAs associated with autism. *PLoS One, 12*(5), e0178532. https://doi.org/10.1371/journal.pone.0178532

Hecht, P. M., Ballesteros-Yanez, I., Grepo, N., Knowles, J. A., & Campbell, D. B. (2015). Noncoding RNA in the transcriptional landscape of human neural progenitor cell differentiation. *Frontiers in Neuroscience, 9*, 392. https://doi.org/10.3389/fnins.2015.00392

Hon, C.-C., Ramilowski, J. A., Harshbarger, J., Bertin, N., Rackham, O. J. L., Gough, J., Denisenko, E., Schmeier, S., Poulsen, T. M., Severin, J., Lizio, M., Kawaji, H., Kasukawa, T., Itoh, M., Burroughs, A. M., Noma, S., Djebali, S., Alam, T., Medvedeva, Y. A., et al. (2017). An atlas of human long non-coding RNAs with accurate 5′ ends. *Nature, 543*(7644), 199–204. https://doi.org/10.1038/nature21374

Iossifov, I., Ronemus, M., Levy, D., Wang, Z., Hakker, I., Rosenbaum, J., Yamrom, B., Lee, Y.-H., Narzisi, G., Leotta, A., Kendall, J., Grabowska, E., Ma, B., Marks, S., Rodgers, L., Stepansky, A., Troge, J., Andrews, P., Bekritsky, M., et al. (2012). De novo gene disruptions in children on the autistic spectrum. *Neuron, 74*(2), 285–299. https://doi.org/10.1016/j.neuron.2012.04.009

Kerin, T., Ramanathan, A., Rivas, K., Grepo, N., Coetzee, G. A., & Campbell, D. B. (2012). A noncoding RNA antisense to moesin at 5p14.1 in autism. *Science Translational Medicine, 4*(128), 128ra40. https://doi.org/10.1126/scitranslmed.3003479

Liu, G., Mattick, J. S., & Taft, R. J. (2013). A meta-analysis of the genomic and transcriptomic composition of complex life. *Cell Cycle (Georgetown, Tex.), 12*(13), 2061–2072. https://doi.org/10.4161/cc.25134

Liu, S., Rao, S., Xu, Y., Li, J., Huang, H., Zhang, X., Fu, H., Wang, Q., Cao, H., Baranova, A., Jin, C., & Zhang, F. (2020). Identifying common genome-wide risk genes for major psychiatric traits. *Human Genetics, 139*(2), 185–198. https://doi.org/10.1007/s00439-019-02096-4

Mattick, J. S., Amaral, P. P., Carninci, P., Carpenter, S., Chang, H. Y., Chen, L.-L., Chen, R., Dean, C., Dinger, M. E., Fitzgerald, K. A., Gingeras, T. R., Guttman, M., Hirose, T., Huarte, M., Johnson, R., Kanduri, C., Kapranov, P., Lawrence, J. B., Lee, J. T., et al. (2023). Long non-coding RNAs: Definitions, functions, challenges and recommendations. *Nature Reviews. Molecular Cell Biology*. https://doi.org/10.1038/s41580-022-00566-8

Mordaunt, C. E., Jianu, J. M., Laufer, B. I., Zhu, Y., Hwang, H., Dunaway, K. W., Bakulski, K. M., Feinberg, J. I., Volk, H. E., Lyall, K., Croen, L. A., Newschaffer, C. J., Ozonoff, S., Hertz-Picciotto, I., Fallin, M. D., Schmidt, R. J., & LaSalle, J. M. (2020). Cord blood DNA methylome in newborns later diagnosed with autism spectrum disorder reflects early dysregulation of neurodevelopmental and X-linked genes. *Genome Medicine, 12*(1), 88. https://doi.org/10.1186/s13073-020-00785-8

Mossink, B., Negwer, M., Schubert, D., & Nadif Kasri, N. (2021). The emerging role of chromatin remodelers in neurodevelopmental disorders: A developmental perspective. *Cellular and Molecular Life Sciences: CMLS, 78*(6), 2517–2563. https://doi.org/10.1007/s00018-020-03714-5

Neale, B. M., Kou, Y., Liu, L., Ma'ayan, A., Samocha, K. E., Sabo, A., Lin, C.-F., Stevens, C., Wang, L.-S., Makarov, V., Polak, P., Yoon, S., Maguire, J., Crawford, E. L., Campbell, N. G., Geller, E. T., Valladares, O., Schafer, C., Liu, H., et al. (2012). Patterns and rates of exonic de novo mutations in autism spectrum disorders. *Nature, 485*(7397), 242–245. https://doi.org/10.1038/nature11011

Nobile, V., Pucci, C., Chiurazzi, P., Neri, G., & Tabolacci, E. (2021). DNA Methylation, Mechanisms of FMR1 inactivation and therapeutic perspectives for fragile x syndrome. *Biomolecules, 11*(2), 296. https://doi.org/10.3390/biom11020296

O'Roak, B. J., Vives, L., Fu, W., Egertson, J. D., Stanaway, I. B., Phelps, I. G., Carvill, G., Kumar, A., Lee, C., Ankenman, K., Munson, J., Hiatt, J. B., Turner, E. H., Levy, R., O'Day, D. R., Krumm, N., Coe, B. P., Martin, B. K., Borenstein, E., et al. (2012a). Multiplex targeted sequencing identifies recurrently mutated genes in autism spectrum disorders. *Science (New York, N.Y.), 338*(6114), 1619–1622. https://doi.org/10.1126/science.1227764

O'Roak, B. J., Vives, L., Girirajan, S., Karakoc, E., Krumm, N., Coe, B. P., Levy, R., Ko, A., Lee, C., Smith, J. D., Turner, E. H., Stanaway, I. B., Vernot, B., Malig, M., Baker, C., Reilly, B., Akey, J. M., Borenstein, E., Rieder, M. J., et al. (2012b). Sporadic autism exomes reveal a highly interconnected protein network of de novo mutations. *Nature, 485*(7397), 246–250. https://doi.org/10.1038/nature10989

Parikshak, N. N., Swarup, V., Belgard, T. G., Irimia, M., Ramaswami, G., Gandal, M. J., Hartl, C., Leppa, V., de la Ubieta, L. T., Huang, J., Lowe, J. K., Blencowe, B. J., Horvath, S., & Geschwind, D. H. (2016). Genome-wide changes in lncRNA, splicing, and regional gene expression patterns in autism. *Nature, 540*(7633), 423–427. https://doi.org/10.1038/nature20612

Prandini, P., Pasquali, A., Malerba, G., Marostica, A., Zusi, C., Xumerle, L., Muglia, P., Da Ros, L., Ratti, E., Trabetti, E., Pignatti, P. F., & Italian Autism Network (ITAN). (2012). The association of rs4307059 and rs35678 markers with autism spectrum disorders is replicated in Italian families. *Psychiatric Genetics, 22*(4), 177–181. https://doi.org/10.1097/YPG.0b013e32835185c9

Ramilowski, J. A., Yip, C. W., Agrawal, S., Chang, J.-C., Ciani, Y., Kulakovskiy, I. V., Mendez, M., Ooi, J. L. C., Ouyang, J. F., Parkinson, N., Petri, A., Roos, L., Severin, J., Yasuzawa, K., Abugessaisa, I., Akalin, A., Antonov, I. V., Arner, E., Bonetti, A., et al. (2020). Functional annotation of human long noncoding RNAs via molecular phenotyping. *Genome Research, 30*(7), 1060–1072. https://doi.org/10.1101/gr.254219.119

Sanders, S. J., Murtha, M. T., Gupta, A. R., Murdoch, J. D., Raubeson, M. J., Willsey, A. J., Ercan-Sencicek, A. G., DiLullo, N. M., Parikshak, N. N., Stein, J. L., Walker, M. F., Ober, G. T., Teran, N. A., Song, Y., El-Fishawy, P., Murtha, R. C., Choi, M., Overton, J. D., Bjornson, R. D., et al. (2012). De novo mutations revealed by whole-exome sequencing are strongly associated with autism. *Nature, 485*(7397), 237–241. https://doi.org/10.1038/nature10945

Satterstrom, F. K., Kosmicki, J. A., Wang, J., Breen, M. S., De Rubeis, S., An, J.-Y., Peng, M., Collins, R., Grove, J., Klei, L., Stevens, C., Reichert, J., Mulhern, M. S., Artomov, M., Gerges, S., Sheppard, B., Xu, X., Bhaduri, A., Norman, U., et al. (2020). Large-scale exome sequencing study implicates both developmental and functional changes in the neurobiology of autism. *Cell, 180*(3), 568–584.e23. https://doi.org/10.1016/j.cell.2019.12.036

Smith, B. J., & Carregari, V. C. (2022). Histone modifications in neurological disorders. *Advances in Experimental Medicine and Biology, 1382*, 95–107. https://doi.org/10.1007/978-3-031-05460-0_7

Tong, Z., Zhou, Y., & Wang, J. (2020). Identification and functional analysis of long non-coding RNAs in autism spectrum disorders. *Frontiers in Genetics, 11*, 849. https://doi.org/10.3389/fgene.2020.00849

Velmeshev, D., Schirmer, L., Jung, D., Haeussler, M., Perez, Y., Mayer, S., Bhaduri, A., Goyal, N., Rowitch, D. H., & Kriegstein, A. R. (2019). Single-cell genomics identifies cell type-specific molecular changes in autism. *Science (New York, N.Y.), 364*(6441), 685–689. https://doi.org/10.1126/science.aav8130

Wang, J., & Wang, L. (2020). Prediction and prioritization of autism-associated long non-coding RNAs using gene expression and sequence features. *BMC Bioinformatics, 21*(1), 505. https://doi.org/10.1186/s12859-020-03843-5

Wang, K., Zhang, H., Ma, D., Bucan, M., Glessner, J. T., Abrahams, B. S., Salyakina, D., Imielinski, M., Bradfield, J. P., Sleiman, P. M. A., Kim, C. E., Hou, C., Frackelton, E., Chiavacci, R., Takahashi, N., Sakurai, T., Rappaport, E., Lajonchere, C. M., Munson, J., et al. (2009). Common genetic variants on 5p14.1 associate with autism spectrum disorders. *Nature, 459*(7246), 528–533. https://doi.org/10.1038/nature07999

Wang, Z., Zhang, J., Lu, T., Zhang, T., Jia, M., Ruan, Y., Zhang, D., Li, J., & Wang, L. (2019). Replication of previous GWAS hits suggests the association between rs4307059 near MSNP1AS and autism in a Chinese Han population. *Progress in Neuro-Psychopharmacology & Biological Psychiatry, 92*, 194–198. https://doi.org/10.1016/j.pnpbp.2018.12.016

Understanding the Neurobiological Components of Neurodevelopmental Disorders

Barbara Thompson

Abstract In this chapter, we focus on the neurobiology of intellectual and developmental disabilities. We first approach the topic by providing the reader with a broad understanding of neurobiology and neurodevelopment. While it is well known that the cellular and molecular components underlying development are complex and tedious, so too are the complex circuit formations that develop and continue to refine through a product of intrinsic signaling as well as by way of additional external exposures and experiences across development. To better understand how unique exposures may differentially impact developmental processes and skew system development, we first tackle how neurobiological processes typically develop across time. In this chapter we give a broad overview of typical neurodevelopment, highlight key neurobiological components that are involved in neurodevelopment, and then turn to how atypical trajectories may arise. Additionally, an overview is provided of the tools at our current disposal that allow us to measure neurobiological processes, giving insight at the cellular, circuit, and ultimately, behavioral level. Wrapping up the chapter, we explore in depth our current understanding of two specific neurodevelopmental disorders associated with intellectual disability.

Keywords Neurobiology · Development · Circuitry · Trajectories · Heterogeneity

Typical Trajectory of Neurodevelopment

A quick review of what makes up the central nervous system (CNS) and how it develops will help set the stage for the basic components of neurodevelopment, which is critical to understanding how things may go awry. The CNS includes the

B. Thompson (✉)
Department of Pediatrics and Human Development, College of Human Medicine, Michigan State University, Grand Rapids, MI, USA
e-mail: thom1756@msu.edu

M. G. Valdovinos (ed.), *Intellectual and Developmental Disabilities*, Contemporary Clinical Neuroscience,
https://doi.org/10.1007/978-3-031-66932-3_6

brain and the spinal cord, and the primary cell relevant to this chapter is the neuron. Neurons are the cells that receive sensory information, and signal to the motor system the appropriate response given the incoming information. These signals can be either electrical or chemical in nature. In the most simplified description, neurons have three main parts, a part that receives information from other neurons, a part to process that information, and last, a part to act upon that information. In more detail, neuronal dendrites receive information from other neurons. These dendrites transmit information to the cell body which contains the nucleus of the cell. Extending from the cell body of the neuron is the axon, which is responsible for sending information to the axon terminals. Glial cells coat the axon to both insulate and protect the axon, allowing for a faster relay of information down the axon. Once information reaches the axon terminals, it is then transmitted via the synaptic cleft (gap between neurons) to the next neuron.

The typical steps of neuronal development include proliferation, migration, differentiation, and last, cell death. Proliferation of progenitor cells promotes the exponential increase of neural cells including neurons and glial cells, known as neurogenesis. As these neural cells move from their origination site, they may travel (migrate) along the scaffolding structure formed by glial cells or aided by other axons of neurons already established to reach their destination. Differentiation into a specific type of neuron begins earlier than migration and is influenced by local transcription factors but can continue to be influenced by environmental signals received during migration. Thus, what a cell encounters during migration can impact and impart a lasting signature on that cell that helps to determine what type of cell is formed. Importantly, the nervous system generates an overabundance of neurons that go through a programmed cell death to ensure a balanced ratio between innervation neurons and targets of these neurons. Once a neuron has reached its destination, it begins a process of growing axons and dendrites. Growth cones at the tip of an axon allow for the axon to grow in length. Molecular guidance cues either in the extracellular matrix or on cell surfaces help to either promote or repel the axon by way of receptors located within the growth cone. Once an axon has reached its target, then the process of forming synapses (synaptogenesis) between the two neurons occurs. These synapses serve as the location for information to pass between neurons, and while they initially occur at very specific locations along neurons as a product of molecular signaling, later in development these synapses may undergo a process of elimination, allowing for the remaining synapses to strengthen and removal of those underutilized. This strengthening of neuronal activity between neurons paves the way for circuit formation, linking of groups of neurons that are involved in a similar function, or linked by a pattern of firing. One of the last steps of neuronal development is the insulation of the established axonal tracts, known as myelination. This insulation is provided by glial cells that wrap around the axon, and allow the message, the action potential, to be transmitted quickly by way of gaps in the myelin, known as nodes of Ranvier.

Additionally, development is shaped by experiences, and these changes can be grouped into two different categories. Those that are experience expectant and those that are experience dependent. Experience expectant changes are those developmental events that are common to all members of the species, and generally occur during very specific and circumscribed periods of development. In contrast, experience-dependent changes are brought about due to specific experiences, can be different for each individual, and can occur at any point in life (Greenough et al., 1987). This understanding sets the stage for how different experiences throughout life can lead to significant heterogeneity in development (Fox et al., 2010; Rakic, 1995; Thompson & Levitt, 2010; Thompson et al., 2009).

Neurotransmitters

Neurons communicate by way of firing action potentials (electrical signals) down the length of the neuron. Once the signal reaches the axon terminal, synaptic vesicles, which are holding neurotransmitters, then bind to the cell membrane and release neurotransmitters into the synapse, the gap between neurons. These neurotransmitters serve as the chemical communicators between neurons. Neurotransmitters can be classified or grouped into different families corresponding to their chemical nature or according to their primary action. The first are the amino acids, containing GABA and glutamate, aspartate, and glycine. These amino acid neurotransmitters are either excitatory (glutamate and aspartate) or inhibitory (GABA and glycine). GABA is the predominant inhibitory neurotransmitter, used by inhibitory interneurons, and once released binds to GABA-A or GABA-B receptors found distributed throughout the brain. Given how early it is expressed in development, it plays role in the early steps of neuronal development including migration and proliferation of other neurons. Similar to GABA, glutamatergic neurons are expressed throughout the brain. There are more than 20 different types of glutamatergic receptors, and those are distributed in different areas throughout the brain. Glutamate serves as the main excitatory neurotransmitter for the central nervous system (CNS), and when normally expressed, is balanced by GABAergic activity allowing for a balanced ratio of excitation and inhibition. If there is too much glutamatergic activity, or not enough GABAergic activity, then this imbalance in excitation and inhibition leads to hyperactivity, and can promote seizure activities, which are prominent in intellectual and developmental disabilities (IDD) and neurodevelopmental disorders (NDDs).

The second family is the Monoamines, synthesized from amino acids. This family includes the catecholamines of dopamine (DA), norepinephrine (NE), epinephrine (also thought of as a hormone), the indolamine serotonin (5HT), and lastly, acetylcholine. All the catecholamines are synthesized from the amino acid tyrosine. Dopamine is found in the midbrain including the substantia nigra and the ventral tegmentum. It is known to be involved in emotion regulation, reward, movement, and drug use disorders. Dopamine has five known receptors, broken into two main families. Norepinephrine is found in the locus coeruleus in the CNS, and is involved in emotional arousal, reward, drug use disorders, and in regulating sleep and mood. There are three norepinephrine receptors. Serotonin is found in the raphe nuclei in the pons and is involved in temperature regulation, sensory perception, sleep onset, and mood. There are 15 known serotonin receptors, categorized into three main subfamilies. In the CNS, there are two main regions of expression of acetylcholine. These include the upper midbrain that targets the thalamus, brainstem, and hypothalamus. The second region includes the anterior hypothalamus and ventral septum which innervates the cortex, hippocampus, and olfactory bulbs. There are two families of cholinergic receptors, known as muscarinic and nicotinic families. Given the wide range of activities and behaviors that are mediated by all of the monoamines, and the number of monoaminergic receptors, it is not surprising that numerous pharmacotherapies have been developed targeting the monoaminergic receptors and reuptake transporters. Therapeutics that act to increase levels of monoamines in the synapse by blocking the reuptake of serotonin, norepinephrine, or both are routinely prescribed for depression and anxiety, antipsychotics act by blocking dopamine activity, and stimulants act by increasing norepinephrine and dopamine levels. Additionally, these psychotropics are routinely prescribed to individuals with IDD, with comorbid anxiety, depression, attention deficit hyperactivity disorder, or the presence of challenging behaviors that may threaten the safety of the individual or those commonly surrounding in the immediate presence of the individual.

Last, the final family of neurotransmitters we present belong to the peptide group, those that are a product of at least two amino acids linking together. Some well-known peptides include arginine vasopressin (AVP), glucagon-related peptide, opioids, cholecystokinin, substance P, and oxytocin. Peptides are also not exclusively found in the CNS, they are found in the pituitary gland, additional endocrine glands, and throughout the immune system. Peptide concentrations in the CNS are far lower than amino acid or monoamine transmitters. While both vasopressin and oxytocin are released directly into the blood stream, and have roles peripherally, both are thought to contribute to maternal–infant attachment and other prosocial behaviors. AVP binds to three receptor subtypes, located throughout the cortex, hippocampus, and hypothalamus. Similarly, the endogenous opioids enkephalin and dynorphin are distributed throughout the brain, as are the five different types of opioid receptors. It is important to also note that all these neurotransmitters and some of their accompanying receptors have been found external to the CNS.

Circuits

Once the different anatomical substrates have formed, connections between the substrates begin to refine, and circuits begin to build. These circuits are built by strengthening connections by adding additional neuronal connections, as well as modifications to synaptic strength in a hierarchical fashion. Additionally, refinement includes weakening or removing altogether those less-needed neuronal connections. Simple systems develop first, and then more integrated and complex circuits develop. Building on the successful development of sensory systems like vision and audition, located in the visual or auditory cortex, then more complex systems like language and speech can be built. This leads then to higher cognitive functions developing. This can also be used to describe how skill sets may build, including those related to social and emotional regulation, in which more basic skills build upon each other to allow for complex and integrated behaviors to emerge, develop, and become refined (Thompson & Levitt, 2010). Importantly, white matter tracts (bundles of myelinated axons) in the prefrontal cortex continue to build throughout early adulthood. Thus, while neural development begins early in utero, it continues to develop and be refined throughout early adulthood, allowing then for an extended period of time in which neurodevelopment may be influenced by an expected trajectory, but also providing an expansive window in which neurodevelopment may be able to be righted back onto the expected trajectory.

Visualizing Neurobiology

At the heart of dissecting apart neurobiological structures and functions that may underlie intellectual and developmental disabilities, one must be able to probe or visualize that which is contained within the central nervous system. Limited access points, ethical considerations, and developmental expression necessitate creative probes to visualize that which cannot be visualized by the unaided eye. Creative tools like animal models, neural imaging, and behavioral assays have resulted in allowing scientists a view not otherwise visible.

Animal Models

Animal models serve a uniquely powerful role in understanding typical and atypical neurodevelopment. Importantly, the selection of species used must be guided by consideration of the specific brain regions, circuits, or behaviors one wants to probe (Thompson & Levitt, 2010). Once the appropriate species is selected, the ability to closely model specific behavioral disruptions that closely mimic the human disorder allows for a powerful understanding of the underlying neurobiology that contributes to IDD and NDD risk. With advances in microscopic and histology techniques as well as tissue culture, detailed cellular and molecular studies can be performed in vitro, in vivo, or ex vivo. Carefully controlled and manipulated genetic and environmental challenges allow for a precise understanding of both typical development and functional consequences of disruptions at the level of cells, circuits, or behavioral output.

Neural Imaging

Given the limitations on in vivo studies that can be performed in the human, the most useful protocols allow for the intact identification of substrates, and their connections, both structural and functional to additional substrates located further away. Luckily, modern medical science has continued to discover and build an assortment of neural imaging tools to be used to probe both structure and function of human neural substrates and circuits. These tools allow a view of the inner workings of the brain from the outside looking in, without requiring extraction of the brain, and in most instances without needing sedation. As protocols for these tools have been refined to ensure quality data collection, so too have the data processing steps continued to evolve to allow for improved signal to noise data collection. Pipelines for data analyses created and universally shared allow for broader consistency and the merging of databases allowing for larger sample sizes to be analyzed, permitting both more refined conclusions and more statistical power to grouping of subcategories. These processing steps have taken advantage of advances in high computing technology and have begun to lasso the strength of machine learning tools. Of course, we would be remiss if we did not remind the reader that discovery moves quickly, and current high computing machine learning technology may be rendered obsolete with newer discoveries.

Magnetic Resonance Imaging Magnetic resonance imaging (MRI) serves as the most utilized tool for capturing exquisitely detailed spatial resolution scans of the brain. MRI uses high-powered magnets to drive protons off their natural alignment with earth's magnetic field, and the signal that is recorded is the amount of time it takes the proton to realign. MRI has a superior spatial resolution (1 mm^3) and can noninvasively probe deep into subcortical regions. Structural images gathered from MRI studies can differentiate gray and white matter, as well as ventricular space and information can include size, shape, volume, position, and even integrity of structures located across the entire brain. When combining MRI scans while performing a task, functional MRI (fMRI) then measures changes as a response to the task demands and performance. Specifically, fMRI measures blood oxygenation level dependent (BOLD) signals, as a product of the magnetic properties of deoxyhemoglobin. Thus, changes in hemodynamic response provide insight into the substrates that are active during a specific task. The temporal resolution is on the order of seconds, so it is not considered a sensitive measurement for the temporal resolution of brain activation. Interestingly, changes in BOLD signal are not large, giving rise to the awareness of the level of neural activity even at rest. Additionally, fMRI studies have shown that there is incredible demand on the brain as measured by fluctuating BOLD activity even at rest (resting state activity). This has given rise to the discovery of resting state networks including salience networks, default mode networks, language networks, visual networks, and sensorimotor networks. Additionally, fMRI can probe the functionality of specific circuits by coupling changes in different areas of interest to each other as a function of evoked network activity. This functional connectivity allows for a better understanding of the breadth of cortical and subcortical activation during tasks.

A few features of MRI protocols make it a complicated, but not impossible, tool to use in children, as well as those with neurodevelopmental or intellectual disabilities. First is the need for the participant to remain still during an imaging scan. Movement artifacts can interfere with the spatial resolution of a structural MRI, or even by creating noise in the BOLD signal that could be misattributed to the task demands. All types of movement can interfere, whether they are gross sporadic motor behaviors, motor tics, or even repetitive rhythmic movements such as breathing. Post collection processing protocols that both identify artifacts, as well as clean artifacts from the raw imaging data, are now routine steps in data analyses. A second complication of MRI is the audible noise created during data collection. An MRI machine can generate 110 decibels of noise, that are heard as loud bangs emanating from the machine as the gradient coils receive electrical stimulation and vibrate to create the magnetic field. For this reason, ear protection is provided for individuals undergoing MRI. This, in combination with the confined space of the imaging scanner, makes it even more difficult for individuals with sensory sensitivities to participate. Improved protocols for scanning young children, or individuals with sensory issues incorporate steps to reduce parent and child anxiety for scans. These protocols may include pre-visit visual stories that allow participants to view actors portraying each step of the protocol, mock scans that allow the individual to lie in a similar scanner with a similar noise sequence, for a similar length of time, audio feedback to the participant during the scan, foam cushions and pillows to minimize jostling, weighted blankets, and child-friendly décor and visual stimuli.

Diffusion Imaging While magnetic resonance imaging provides high-resolution images of cerebral gray matter, white matter details, like directionality of fiber tracts, remain elusive. Instead, if one is interested in specific white matter details, diffusion tensor imaging (DTI) is a relatively newer technique that relies on magnetic pulses to drive water diffusion. Water diffusion in the presence of a magnetic field gradient will de-phase the magnetic resonance signal, and the signal decrease is related to the diffusivity of water along the direction of the applied gradient. Since water prefers to diffuse along, and not against, white matter fibers, the fastest direction of diffusion indicates the overall orientation of the fiber. These scans are particularly good for scanning the integrity of white matter tracts, and for measuring changes in white matter tracts across time. Of course, not all white matter tracts are lined up neat and orderly, and in fact, white matter fibers cross and touch each other, which is difficult for DTI to quantify. An even more recent technique, diffusion spectrum imaging (DSI) allows for the imaging of several different orientations within a single voxel. This then allows for a more comprehensive understanding of fiber tract directionality and paths. The trade-off for using DSI is the increased time needed for scanning.

Electroencephalogram Electroencephalogram (EEG) is a tool used to capture changes in the conductance of electrical activity by capturing the activity of groups of neurons that are firing in synchrony. This is done by placing electrodes onto an individual's head, making contact with the surface of the scalp, to collect electrical activity changes at the surface of the scalp. Electroencephalogram excels in temporal resolution of neural activity, capable of measuring sub-millisecond changes in electrical activity. Thus, for those instances in which details related to the speed of neuronal activity are needed, EEG is the more exact methodology. Not surprisingly, the spatial resolution of EEG is more limited compared to MRI, as it is dependent on groups of neurons firing together, thereby reducing specificity in the location of activity. Spatial resolution can be improved by increasing the density of electrodes on the scalp surface, but even then, resolution is still limited. The output of EEG is a continuous measurement of electrical activity, that is further grouped together as oscillations in activity, ranging in frequency from 1 to 100+ Hz. These frequency oscillations are then sub-grouped into delta, theta, alpha, beta, and gamma bands, with each band having been associated with specific neural tasks or activities. For example, delta bands, as the lowest frequency bands, are typically found during sleep, whereas gamma bands, representing the highest frequencies, predominate during higher order cognitive tasks like attention and object recognition (Pizzagalli, 2007)

Event-Related Potentials Event-related potentials (ERP) are a shorter window of recording of EEG, utilized to measure brain responses to a specific stimulus. Importantly, specific evoked potentials have been recorded in response to auditory, visual, and somatosensory stimuli, as well as novelty, thought to tap into attention. ERPs can occur within 100 milliseconds of stimulus presentation, and hundreds of milliseconds following stimulus presentation. While the manual setup of the electrodes to allow for the recording of electrical activity is the same, the recording of the electrical activity is importantly time locked to the presentation of a specific stimulus. Due to the short nature of the response to the onset of the stimulus, and the background noise of the rest of the ongoing spontaneous EEG, high numbers of presentation trials are necessary to accurately capture the specific response to the noted stimulus.

Magnetoencephalography (MEG) A more recent technique, MEG shares temporal resolution properties with EEG, on the order of sub-millisecond changes, and similarly excels in temporal over spatial resolution. Different from EEG, MEG measures changes in magnetic fields generated by neural events. The individual is placed in a chair and pre-arranged sensors in the shape of a helmet are positioned directly above the head, while the individual sits or lays down. As magnetic fields are not prone to disruptions by the scalp or skull, MEG can provide better spatial resolution compared to EEG. MEG has been used to provide unique insight into sensory responsiveness, and functional network activity in a variety of populations, including those with IDDs.

Behavior

Behavior is often overlooked when describing tools for measuring neurobiological processes. Understandably, complex technological tools that allow the visualization of cellular and molecular activity, and other microscopic insights are sought out to better understand neurobiology at the level of the cells or even cellular contents. However, we could argue that behavior itself should be classified as the most insightful tool to measure neurobiological function. As behavior is a direct output of neurobiological functions, disruptions to behavior give direct insight into the cellular, molecular, and circuitwide activities that are occurring, as well as into how different neurodevelopmental disorders may differ in their underlying neurobiology from each other. As some NDDs with unknown etiology like autism spectrum disorders (ASD) rely on a formal behavioral diagnosis for the classification of the disorder, the importance of reliable and valid measurements of behavioral constructs cannot be understated. Here, we take a look at some of the more commonly disrupted behaviors in NDD and IDD, and the current knowledge related to the underlying neurobiology that plays a role in the behaviors. Of course, understanding the underlying neurobiology has come about with the help of animal models, unique imaging tools that have evolved, and post-mortem studies.

Higher-order cognitive functions can be robustly probed in both animal models as well as in the human condition. Innovative protocols, whether they are observational, task based, performed with pen and paper, or computerized, can reliably measure cognition, attention, working memory, executive function, and even creativity. Task demands and expectations differ based on the age of the subject, and the testing or observational environment. Additionally, adaptations to the innovative protocols allow for simultaneous collection of task performance, as well as monitoring of brain activity, be that physiology or imaging based. These collaborative efforts have propelled the field forward into a better understanding of the development of higher-order cognition and the underlying neurobiological constructs. Extensive imaging, physiology, animal model experiments, and postmortem human studies have revealed that it is a complex circuitry that is involved. This circuitry includes the frontal lobe, and four main regions of the prefrontal cortex (dorsolateral, ventrolateral, anterior, and medial). Just as important as the prefrontal cortex (PFC) are the areas that connect with the PFC, either by projecting directly to the PFC, or by receiving projects from the PFC. Confirming the importance of the PFC is the knowledge that it connects to motor, limbic, sensory, and arousal systems, essentially demonstrating its involvement in all activities. Importantly too, we know that the development of the PFC occurs across a prolonged amount of time, with refinements in connectivity occurring well into early adulthood.

Fear and anxiety are two behavioral constructs that have been extensively studied in animal models and in humans. Studies incorporating behavioral observations, lesion, and direct stimulation (animals), imaging, and postmortem dissections have revealed that emotion processing is centered in the limbic circuit. The amygdala is a substrate with several different nuclei that in totem has been established as the main headquarters for fear behaviors. The amygdala receives information from sensory systems and then coordinates an outgoing response proportional to the incoming stimulus. Alterations in connections between the different nuclei that connect with the amygdala, the amygdaloid nuclei connections, and the substrates to which the amygdaloid nuclei project have all been implicated in disruptions in typical fear processing. Connections with the thalamus, prefrontal cortex, hippocampus, bed nucleus of the stria terminalis, hypothalamus, and brainstem position the amygdala as a key substrate whose typical function can easily be disrupted leading to significant alterations in fear behaviors. While the function of fear is to allow for survival of the species, and to ensure an appropriate response during danger and direct threats to survival, anxiety can help guide an individual toward avoiding an actual or perceived dangerous situation. Measuring fear and anxiety in humans can range from surveys and questionnaires to detailed research protocols developed in animal models, then translated for human use like fear-potentiated startle. Autonomic measures like galvanic skin response can document levels and changes in arousal.

Disruptions in language and language processing are a prominent piece of NDD and IDDs and are often some of the earliest signs of delays in neurodevelopment. However, there is great variability in language acquisition, that can be even further confounded by experiences afforded by the environment including multiplicity of languages in the immediate environment and quantity of exposure to quality language early in development (Golinkoff et al., 2019). The developmental timeline on which language develops is broad and extended and there is considerable heterogeneity for typical neurodevelopmental trajectories. Even so, milestones and recommendations are provided by the Centers for Disease Control and Prevention and the American Academy of Pediatrics for monitoring the foundational steps that allow full language capacity to develop. These include orienting and responding to sounds, babbling, and joint attention to name a few. Language is a more easily demonstrated, observed, and quantified capacity than other cognitive skills, and can be easily captured at home, in the naturalistic setting with basic video and audio tapes if not elicited in a direct testing situation. Additionally, parent, caregiver and teacher reports can also provide insight into capacity. The original understanding of the neural substrates involved in language separated it into the production area and the perception or understanding area. Broca's area was originally thought to serve as the site for the production of speech, and Wernickes area served as the site for speech comprehension. What is well understood now however is that language is scattered across a much broader area, involves both motor and sensory pieces of the cortex, and is intimately tied to the prefrontal cortex and temporal lobes.

Social behaviors are complex, species-centered, and increase in complexity as circuits and systems develop across time. As social species, humans engage in a variety of daily social interactions, beginning in early infancy. Additionally, those social interactions change as the infant develops, and different biological needs emerge. Social behaviors then are refined across development as the child engages in different social contexts and becomes more proficient in navigating their social worlds. Familiarity with social partners and environment can dramatically impact observed social behaviors, as can external environmental stimuli (loudness or brightness of room) or individual states (hungry, tired, scared). Thus, great consideration must be given to the design of social tasks and to the interpretation of elicited social behaviors. Measuring social behaviors can be done through direct observation (in the field, or an assessment), and they can also be reported by caregivers and teachers. Human imaging studies have revealed several areas of activation involved in social behaviors, including the fusiform gyrus, amygdala, superior temporal sulcus, insular cortex, and prefrontal cortex.

Sensory and motor behaviors are commonly assayed either by direct observation or combined using physiology and neural imaging to reveal the precise location and timing of neural activity in response to the stimulus or initiation of the motor behavior. Sensory functions can be measured in a variety of ways, including parent or teacher reports, or direct observation and probing of the individual. As they are the first systems to develop, sensory and motor functions can be tested from a very early age, and of course, can be probed easily in animal models as well, though species differences in importance of each sensory system, and species-specific motor responses must be taken into consideration. Basic psychophysical assays can probe for detection and discrimination thresholds across any of the eight sensory domains. These eight sensory domains include auditory, visual, gustatory, tactile, olfaction, vestibular, proprioception, and interoception. Building on that, as individuals age, sensory systems continue to refine and they also enable multisensory learning, and assays can then probe multisensory integration and responsiveness. Animal models and neural imaging reveal specific cortical locations of the brain for each sense, the thalamus serving as the relay center for all sensory information except for olfaction which has a direct connection from the olfactory bulbs to the olfactory cortex, and the cranial nerves as the critical communicators for both sensory perception and motor activity. Motor behaviors can also be assayed early, including reflexes and some gross motor activity, that are further shaped and refined through experiences and development. As the individual ages, more fine motor skills, and highly complex gross motor skills begin to emerge and can be assayed. Animal models, neural imaging, and lesion studies have allowed for identifying the location of the motor cortex, as well as the involvement of the basal ganglia and cerebellum in motor activities.

Limitations to all these behavioral assays include the amount of time it takes to collect the data, as well as to analyze the collected data. Additionally, there are limitations in the reliability and validation of questionnaires and surveys. The high demands on data collection and analysis are due to a number of factors, including the need for specialized training for individuals collecting the data to ensure reliable data collection. Additionally, there is substantial training required for the analysis of data, especially the data that serves as a quantification of qualitative states. Taken together, human behavior studies are not considered a high-throughput assay but are nonetheless one of the most critical tools toward our understanding of functional disruptions in neurobiology.

Disrupting Neurodevelopment

Before we dive into the numerous ways that neurodevelopment can be disrupted, we will briefly discuss genetics. Knowing now the many different components involved in typical neurodevelopment, it is not difficult to imagine that any small or large perturbation in genetic code could significantly alter neurodevelopment and shift it toward an atypical trajectory. Our prior discussion of the many cell signaling pathways, neurotransmitters, windows of specific development events, timelines on which the different circuits develop, and then subsequent communication between circuits demonstrate the multitude of different events or ways in which neurodevelopment could be genetically perturbed (Fox et al., 2010; Rakic, 1995; Thompson et al., 2009). Unexpectedly, however, we know that there is significant redundancy in genetic code, and all perturbations do not necessarily result in altered neurodevelopmental outcomes. Rather, some genetic perturbations may result in endophenotypes, or a trait common to the disorder, but not the disorder. Detailed characterization of behaviors within each individual across a population allows then for a better understanding of the consequences of each genetic disruption beyond a diagnosis of NDD or IDD.

Visualizing neurodevelopment as a trajectory allows one to understand how different experienced stimuli may alter the path away from the typical trajectory, resulting in an atypical trajectory. Those different stimuli, or stressors, when encountered, provide "stress" onto the typical path, and may result in an altered development. However, just as there are challenges, there are also built in guardrails that help buffer these disruptions. As such, depending on the timing of the exposures, the strength of the stressors, and the capacity of the buffers, not all stressors will disrupt neurodevelopment (Fig. 1). We will now turn to discussing how some of the more commonly experienced stressors may impact neurodevelopment.

Age of Exposures

Given our understanding of the timeline of development provided above, we know that the age of exposure partially determines the intensity and permanency of consequences, as well as downstream effects, following exposure to stimuli. A once popular theme of critical periods has been expanded to include sensitive windows with the understanding that if an exposure occurs outside a sensitive window time frame, then the stimulus must be increased in either duration or intensity to have a similar effect as when presented during the sensitive window.

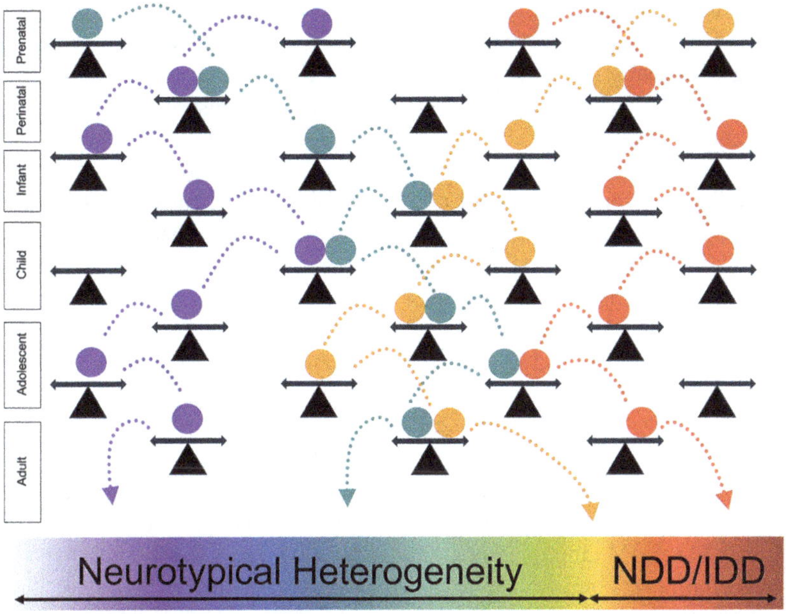

Fig. 1 Impact of experienced stressors on neurodevelopmental trajectories. Stressors can occur at all developmental time points and can alter neurodevelopment depending on the intensity or timing of the stressor as well as whether there are buffers to help minimize the impact of the stressors. Therefore, while each starting location for neurodevelopment may lie on a continuum of risk for NDD/IDD, additional lived experiences may shift an individual trajectory away or towards the range of neurotypical heterogeneity.

Prenatal and Perinatal Exposures

A variety of prenatal exposures have been linked to altering typical neurodevelopment. They run the gamut from psychosocial challenges such as poverty, racism, and food insecurity, to physiological stressors such as direct or indirect cause of injury to the embryo. Some of these physiological stressors may be a result of direct challenge to the embryo, or, instead, early challenges to the pregnant host or to the placenta that can then significantly impact the development of the embryo. Physiological stressors can include infection and immune activation, extreme temperatures, extreme changes in blood pressure, exposures to exogenous substances, or stimuli that may generate a pain response. Challenges to the pregnant host can produce a response that may not be limited to just the pregnant host, but rather, set up a cascade of responses in the embryo. Depending on the age of the embryo when the challenge occurs, the impact on the embryo can range from no impact to altered neurodevelopmental trajectory, dependent on which signaling systems and pathways were already built, or were in the process of being built when the challenge occurs (Thompson & Stanwood, 2009). For example, mid gestational in utero cocaine exposure results in significant disruptions to the developing brain, whereas early gestational exposure does not have the same significant impact on the developing brain. Detailed studies revealed these differences are due to the timing of developmental expression of dopamine receptors, whereby early gestational exposures cannot impact dopamine receptors as they are not yet expressed.

Investigations probing the relationship between the physiological stressor of maternal immune activation and the development of NDD and IDD in the offspring have also begun to finely tease apart the specific mechanistic causes, and subsequent neurobiological disruptions. Immune activation by way of either infection or environmental triggers during pregnancy has been shown to increase the risk of neurodevelopmental disorders in the offspring (Brown, 2012; Estes & McAllister, 2016). Exquisitely designed animal models of maternal immune activation significantly impact the proliferation and migration of neurons as well as significantly alter the genes linked to immune responsivity of the offspring (McEwan et al., 2023; Canales et al., 2021).

Similar to prenatal challenges, perinatal and postnatal challenges and exposures are also diverse and can also lead to long-lasting impacts on neurodevelopmental trajectories. Significant complications during birth delivery, early postnatal infections, and additional environmental challenges have been associated with altered neurodevelopmental trajectories in the offspring (Atladottir et al., 2010; Dalman et al., 2008; Kariuki et al., 2016).

Susceptibility and Resilience

It is important to note that as was stated for the genetic associations with neurodevelopmental disorders, environmental disruptions also increase the risk, but do not necessarily confer or guarantee a diagnosis of a neurodevelopmental disorder. Rather, we think of each of these challenges as increasing the susceptibility of an individual. That increased susceptibility ultimately means that there is a stronger likelihood when challenged a second, or third, or fourth, time, either by the same or a different exposure, to then confer a diagnosis. Additionally, this increased risk conferred by environmental challenges can interact with an already established susceptibility due to a genetic disruption. Importantly, part of our shared scientific goals within the field of neurodevelopmental disorders is understanding this susceptibility, and the timing of this plasticity to better design intervention approaches that may reduce this susceptibility or remove any additional challenges that may increase the probability of a neurodevelopmental disorder. By detailing the exact timing and dosing of each challenge, combined with the understanding of the timing of neurodevelopmental circuits and key developmental changes that are occurring, we can then provide detailed timelines that explain exactly when interventions or therapies may work best, and detailed maps that illustrate exactly which circuit and which developmental milestone should be targeted by an intervention for maximum impact on neurodevelopmental trajectory.

Specific Intellectual and Developmental Disabilities

While findings specific to each disorder will be presented below in each relevant subsection, we mention here some of the known disruptions shared amongst the disorders. It has been repeatedly shown that disruptions to the synapse are common between neurodevelopmental disorders, especially the known single gene mutations that result in Fragile X, Rett syndrome, neurofibromatosis, and tuberous sclerosis. Each of these single gene mutations is also comorbid with autism. The underlying genes known to be responsible for these disorders have been found to alter the functional capacity of the synapse, leading NDD and IDDs to be characterized as disorders of the synapse. These functional disruptions may occur at several levels of the synapse, either as a result of disrupting the development of the synapse, activity of the synapse, or even transcriptional regulation of the synapse (Wang et al., 2018; West & Greenberg, 2011; Zoghbi & Bear, 2012). As intellectual and developmental disabilities are heterogeneous in their descriptive characteristics, so too are the individual phenotypes. This likely is a product of the numerous paths that can be taken to disrupt a neurotransmitter, anatomical substrate, and circuit. However, variability within the group generally does not exceed that for distinguishing the different groups from each other.

Fragile X Syndrome

Fragile X syndrome is caused by changes in the gene Fragile X Messenger Ribonucleoprotein 1 (*FMR1*), found on the X chromosome, which is responsible for coding Fragile X Messenger Ribonucleoprotein 1 (FMRP). FMR1 has a CGG trinucleotide repeat in the 5′ untranslated region that is unstable, meaning that in the general population, the number of repeats of CGG can range up to 55 repeats. However, in individuals with a diagnosis of FXS, there are at least 200 repeats of this trinucleotide. With these 200 repeats, the gene is no longer able to make the protein, leading to an absence of FMRP. FMRP is known to generally negatively regulate protein synthesis, so, without FMRP, there is an upregulation of protein synthesis. The absence of FMRP and subsequent increased protein synthesis then disrupts a multitude of cellular functions and activities. These occur very early, as FMRP is necessary for neurogenesis and neuronal differentiation. Additionally, FMRP has been shown to inhibit the activity of metabotropic glutamate receptors 1 and 5, which led to clinical trials to test the ability of mGluR inhibitors and GABA agonists to improve symptoms in Fragile X.

Specific neurobiological disruptions found in FX have included brain atrophy, white matter differences, dendritic spine abnormalities, and hyperexcitability. Neural imaging studies have revealed an enlarged caudate nucleus, thought likely to be involved in the repetitive movements exhibited by individuals with Fragile X syndrome, in addition to increased volume of globus pallidus and putamen as early as 6 months of age (Shen et al., 2022). Functional imaging has demonstrated differences in neural activity in the insula, cingulate cortex, and superior temporal sulcus in response to facial stimuli. Tasks probing executive functioning have revealed differences in performance, correlated with a reduction in brain areas recruited during more complex tasks (Kwon et al., 2001; Menon et al., 2000; Rivera et al., 2002). Moreover, decreased functional connectivity has been found across a variety of resting state networks including visuospatial and affective networks (Hall et al., 2013).

Furthermore, coexisting symptoms and disorders that are found with FX include anxiety, hyperactivity and hyperarousal, depression, OCD, and social deficits (Bear et al., 2004; Bhakar et al., 2012; Hall et al., 2013; Leigh & Hagerman, 2020; Lightbody & Reiss, 2009; Willemsen et al., 2011), though few studies have tried to tease apart the neurobiological components of each of these comorbid disorders from FX.

Autism Spectrum Disorders

As a behaviorally defined disorder, the behavioral repertoire for each individual with a diagnosis of ASD is unique and varied, though generally fitting into a group with a combination of altered social communication and repetitive behaviors. As varied and unique as the behaviors, so too are the underlying neurobiological disruptions that have been reported. Autism spectrum disorder is a highly heritable disorder, numerous reports have implicated genes with functional contributions ranging from neurotransmitters, scaffolding proteins, cell signaling molecules, and even noncoding RNAs. Importantly as well, we have moved beyond the initial discussions of whether genes or environmental exposures paved the way for ASD. Thankfully, we now have a more comprehensive understanding of how genetic predispositions and environmental exposures may set a baseline from which additional exposures or challenges may alter the developmental trajectory even further. That is, it is no longer an either-or question to be posed, but rather, either-and that also takes into account the nuances of dosing and timing of these challenges (Geschwind, 2009; Happe et al., 2006). Some of the most convincing environmental exposures that have been studied include maternal immune activation and prenatal exposure to diesel particulate (Volk et al., 2014, 2013). While neither guarantees a diagnosis of autism, they both increase the risk of a diagnosis.

A core feature of ASD, social communication differences are posited to arise from complex combinations of theory of mind, social motivation, and communication disruptions. It has been proposed that social disruptions occur as a disruption in the theory of mind (Baron-Cohen et al., 1985). Disruptions in the theory of mind have been demonstrated in individuals with a diagnosis of ASD. Imaging studies probing neural activity during theory of mind tasks reveal a network of activity including the medial prefrontal cortex, anterior cingulate cortex, medial posterior parietal cortex, and temporal parietal junction, and disruptions in these circuits in individuals with a diagnosis of ASD (Castelli et al., 2002; Mason et al., 2008).

Disruptions in social motivation, specifically a reduction in reward from social interactions, have also been proposed as a driver of disruptions in social behavior in individuals with ASD (Chevallier et al., 2012). While similar neural substrates are involved in social motivation as those for theory of mind and social cognition in general, social motivation circuitry also incorporates the amygdala for attention and the ventral striatum for reward (Clements et al., 2018). Still a popular theory, recent findings illustrate the confounding effects of anxiety and sex on social motivation, as well as findings that self-reported interest in social interactions is high in some individuals with a diagnosis of ASD, but communication disruptions may interfere with successful interactions (Bagg et al., 2023; Kapp et al., 2019).

Prior to the revision in diagnostic criteria published in the DSM5, social and communication skills were categorically separate. As such, there was a large body of literature generated to better understand the specific communicative skills and underlying neurobiology supporting those skills. While specific language skills and qualities were found to be disrupted in children with autism, including echolalia and perseverative speech patterns, the most general description for language disruptions reveals that individuals with a diagnosis of ASD do not use appropriate communication in social contexts, that is, pragmatic use of language is disrupted (Mody & Belliveau, 2013; Volden et al., 2009). As with all symptoms, the heterogeneity of the disruptions in language is vast. Due to the lack of responding to verbal communication in toddlers, and reduced vocalizations very early in development, significant effort was made to identify earlier precursors to language to help identify children at greater risk for a later diagnosis, ultimately promoting earlier interventions. Talented researchers identified joint attention as a reliable precursor to language capacity at later ages, and that those with greater difficulties with joint attention were at greater risk for communication deficits as well as elevated autism symptoms at later ages (Mundy et al., 2007; Thorup et al., 2018). Neuroimaging studies have shown differences in brain activation in response to voice stimuli, which is hypothesized to contribute to social differences in children with ASD (Gervais et al., 2004), as well as differences in auditory cortical processing of speech sounds, resulting in altered language development. Additionally, electrophysiology studies have also found differences in sound and speech processing in individuals with ASD (Kuhl et al., 2005).

The second diagnostic feature for ASD includes restrictive and repetitive behaviors. These behaviors occur frequently and may increase in frequency in new or novel environments, as well as during periods of greater stress. Measurement of repetitive behaviors can be done through direct observation (in the field, or in a direct assessment), and they can also be reported by caregivers and teachers if not directly observed. Behaviors can be elicited at the level of fine motor disruptions, like repetitive wheel spinning, or more global motor behaviors including rocking, clapping, and pacing. Disruptions in the corticostriatal circuitry have been shown to underly repetitive behaviors. As the name implies, this circuitry allows for communication between the striatum and premotor, motor, and prefrontal cortical areas in a feedback loop. Disruption in either the striatum or cortical areas results in disrupted signaling, leading to repetitive behaviors, as well as disrupted feedback, preventing the inhibition of the repetitive behaviors. Structural imaging studies have focused on the basal ganglia, and relationships between repetitive behaviors and striatal volumes have been found. Functional imaging studies have found roles for the caudate nucleus, as well as both the anterior cingulate and posterior parietal cortex (Kohls et al., 2014; Langen et al., 2014; Shafritz et al., 2008; Turner et al., 2006). Due to the elicitation of these behaviors during stressful or novel situations, hypotheses related to the calming and self-soothing effects of these repetitive behaviors implicate the role of the reward pathways in supporting these behaviors. Here, animal models have provided insight, suggesting that motor areas in the brainstem that have shown importance to repetitive behaviors in animals also project to multiple reward areas, and are sensitive to selective serotonin reuptake inhibitors. This further supports the hypothesis that these repetitive behaviors may provide emotional relief (Sun et al., 2022; Zhang et al., 2022).

A large number of individuals with a diagnosis of ASD also have high levels of anxiety, and anxiety disorders are comorbid with ASD. Imaging studies have revealed enlarged amygdala in children with ASD (Schumann et al., 2004), structural scans reveal overgrowth of the amygdala early, even before ASD is diagnosed (Shen et al., 2022), and it has been reported that children with both a diagnosis of ASD and anxiety have an enlarged right amygdala volume (Andrews et al., 2022). In comparison, functional MRI scans have revealed both positive and negative associations between amygdala activation in individuals with ASD. When individuals with a diagnosis of ASD performed a facial discrimination task, enhanced amygdala activation during gaze fixation was reported (Dalton et al., 2005; Tottenham et al., 2014). In comparison, blunted amygdala activation during social viewing has also been reported (Grelotti et al., 2005; Pierce et al., 2001). Though none of these specific studies assayed anxiety in these participants, it is possible that differences in amygdala activation in these earlier studies may have been driven by differences in anxiety in the autism groups. Beyond the limbic structures, several investigations have also probed hypothalamic–pituitary–adrenal activity to better understand stress in individuals with ASD. However, no consistent differences in daily patterns of fluctuations or total cortisol levels have been found in individuals with ASD. Additionally, in studies probing both anxiety and salivary cortisol levels, no association between emotional stress and the biological output of stress has been found (Simon & Corbett, 2013; van der Linden et al., 2022).

Disrupted sensory responding and sensory processing have been reported as highly prevalent in individuals with ASD, and occupational therapy remains a highly requested therapy by families to help with sensory processing. The early differences in sensory processing have given rise to the "sensory first" description of how ASD may develop (Falck-Ytter & Bussu, 2023). Numerous studies employing both psychophysical and survey techniques have revealed that individuals with ASD show hypersensitivity, hyposensitivity, and sometimes even a combination of both to varied sensory stimuli (Baranek et al., 2006; Foss-Feig et al., 2012). All sensory modalities have been reported to be altered (Blanche et al., 2012; Oster & Zhou, 2022; Robertson & Baron-Cohen, 2017), and the integration and processing of multi-sensory information have also been shown to be disrupted (Foss-Feig et al.,

2010; Stevenson et al., 2014). Electroencephalogram and magnetoencephalogram recordings have shown group differences across a wide range of sensory stimuli (Cascio et al., 2015; Roberts et al., 2021). Diffusion-weighted imaging has shown relationships between brainstem microstructure and sensory features in individuals with ASD diagnosis compared to typically developing individuals (Surgent et al., 2022). Functional MRI scans have revealed differences in thalamolimbic connectivity that begins as hyperconnectivity at 1.5 months old, which then transitions to thalamocortical hypoconnectivity by 9 months of age in a population of infants with high familial risk for ASD (Wagner et al., 2023). During a visual processing task, the TD group activated more prefrontal cortex during perception tasks, whereas the ASD group engaged more the extrastriate V2 region (Jassim et al., 2021). Importantly, relationships between sensory over responsivity and anxiety in individuals with ASD have been demonstrated, again highlighting the overlapping mechanisms by which behavioral phenotypes may emerge (Green & Ben-Sasson, 2010).

Summary

In summary, this overview of the underlying neurobiology of intellectual and developmental disabilities gives a brief insight into the numerous avenues by which altered neurobiological development and subsequent behaviors may arise. While the plasticity of the brain and circuits allows for new skills and behaviors to be learned, it also allows for systems and circuits to be misaligned by challenges, sending a system toward altered neurodevelopment that may result in an intellectual or neurodevelopmental disorders. Neurodevelopmental and intellectual disorders may arise from any number of challenges, either at the genetic or environmental level. The variability of environmental challenges can differentially impact specific circuits and result in different levels of severity of responses to those challenges, dependent upon the timing of exposures, the age of exposures, the intensity and duration of the exposures, and the systems already in place to buffer the challenges. Importantly, the plethora of common genetic mutations has revealed the redundancy in the central nervous system and has revealed the heterogeneity in typical neurodevelopment. Without a doubt, neurodevelopment is a sensitive period in which a complexity of intertwined circuits, cellular, and molecular signaling systems can each be impacted by different challenges. Importantly, detailed studies revealing when in development each stage occurs, provides insight into when and specifically what, could possibly be targeted as an intervention target to right the path back toward a typical trajectory. The promise and hope for early intervention is directly informed by the translational neuroscience studies and emphasizes the need for collaborative science and discourse amongst professionals serving those individuals and their families with IDD/NDD, and those scientists teasing apart the details related to neurobiological development.

References

Andrews, D. S., Aksman, L., Kerns, C. M., Lee, J. K., Winder-Patel, B. M., Harvey, D. J., Waizbard-Bartov, E., Heath, B., Solomon, M., Rogers, S. J., Altmann, A., Nordahl, C. W., & Amaral, D. G. (2022). Association of amygdala development with different forms of anxiety in autism spectrum disorder. *Biological Psychiatry, 91*(11), 977–987. https://doi.org/10.1016/j.biopsych.2022.01.016

Atladottir, H. O., Thorsen, P., Schendel, D. E., Ostergaard, L., Lemcke, S., & Parner, E. T. (2010). Association of hospitalization for infection in childhood with diagnosis of autism spectrum disorders: A Danish cohort study. *Archives of Pediatrics and Adolescent Medicine, 164*(5), 470–477. https://doi.org/10.1001/archpediatrics.2010.9

Bagg, E., Pickard, H., Tan, M., Smith, T. J., Simonoff, E., Pickles, A., Carter Leno, V., & Bedford, R. (2023). Testing the social motivation theory of autism: The role of co-occurring anxiety. *Journal of Child Psychology and Psychiatry*. https://doi.org/10.1111/jcpp.13925

Baranek, G. T., David, F. J., Poe, M. D., Stone, W. L., & Watson, L. R. (2006). Sensory experiences questionnaire: Discriminating sensory features in young children with autism, developmental delays, and typical development. *Journal of Child Psychology and Psychiatry, 47*(6), 591–601. https://doi.org/10.1111/j.1469-7610.2005.01546.x

Baron-Cohen, S., Leslie, A. M., & Frith, U. (1985). Does the autistic child have a "theory of mind"? *Cognition, 21*(1), 37–46. https://doi.org/10.1016/0010-0277(85)90022-8

Bear, M. F., Huber, K. M., & Warren, S. T. (2004). The mGluR theory of fragile X mental retardation. *Trends in Neurosciences, 27*(7), 370–377. https://doi.org/10.1016/j.tins.2004.04.009

Bhakar, A. L., Dolen, G., & Bear, M. F. (2012). The pathophysiology of fragile X (and what it teaches us about synapses). *Annual Review of Neuroscience, 35*, 417–443. https://doi.org/10.1146/annurev-neuro-060909-153138

Blanche, E. I., Reinoso, G., Chang, M. C., & Bodison, S. (2012). Proprioceptive processing difficulties among children with autism spectrum disorders and developmental disabilities. *American Journal of Occupational Therapy, 66*(5), 621–624. https://doi.org/10.5014/ajot.2012.004234

Brown, A. S. (2012). Epidemiologic studies of exposure to prenatal infection and risk of schizophrenia and autism. *Developmental Neurobiology, 72*(10), 1272–1276. https://doi.org/10.1002/dneu.22024

Canales, C. P., Estes, M. L., Cichewicz, K., Angara, K., Aboubechara, J. P., Cameron, S., Prendergast, K., Su-Feher, L., Zdilar, I., Kreun, E. J., Connolly, E. C., Seo, J. M., Goon, J. B., Farrelly, K., Stradleigh, T. W., van der List, D., Haapanen, L., Van de Water, J., Vogt, D., et al. (2021). Sequential perturbations to mouse corticogenesis following in utero maternal immune activation. *eLife, 10*. https://doi.org/10.7554/eLife.60100

Cascio, C. J., Gu, C., Schauder, K. B., Key, A. P., & Yoder, P. (2015). Somatosensory event-related potentials and association with tactile behavioral responsiveness patterns in children with ASD. *Brain Topography, 28*(6), 895–903. https://doi.org/10.1007/s10548-015-0439-1

Castelli, F., Frith, C., Happe, F., & Frith, U. (2002). Autism, Asperger syndrome and brain mechanisms for the attribution of mental states to animated shapes. *Brain, 125*(Pt 8), 1839–1849. https://doi.org/10.1093/brain/awf189

Chevallier, C., Kohls, G., Troiani, V., Brodkin, E. S., & Schultz, R. T. (2012). The social motivation theory of autism. *Trends in Cognitive Sciences, 16*(4), 231–239. https://doi.org/10.1016/j.tics.2012.02.007

Clements, C. C., Zoltowski, A. R., Yankowitz, L. D., Yerys, B. E., Schultz, R. T., & Herrington, J. D. (2018). Evaluation of the social motivation hypothesis of autism: A systematic review and meta-analysis. *JAMA Psychiatry, 75*(8), 797–808. https://doi.org/10.1001/jamapsychiatry.2018.1100

Dalman, C., Allebeck, P., Gunnell, D., Harrison, G., Kristensson, K., Lewis, G., Lofving, S., Rasmussen, F., Wicks, S., & Karlsson, H. (2008). Infections in the CNS during childhood and the risk of subsequent psychotic illness: A cohort study of more than one million Swedish subjects. *American Journal of Psychiatry, 165*(1), 59–65. https://doi.org/10.1176/appi.ajp.2007.07050740

Dalton, K. M., Nacewicz, B. M., Johnstone, T., Schaefer, H. S., Gernsbacher, M. A., Goldsmith, H. H., Alexander, A. L., & Davidson, R. J. (2005). Gaze fixation and the neural circuitry of face processing in autism. *Nature Neuroscience, 8*(4), 519–526. https://doi.org/10.1038/nn1421

Estes, M. L., & McAllister, A. K. (2016). Maternal immune activation: Implications for neuropsychiatric disorders. *Science, 353*(6301), 772–777. https://doi.org/10.1126/science.aag3194

Falck-Ytter, T., & Bussu, G. (2023). The sensory-first account of autism. *Neuroscience and Biobehavioral Reviews, 153*, 105405. https://doi.org/10.1016/j.neubiorev.2023.105405

Foss-Feig, J. H., Kwakye, L. D., Cascio, C. J., Burnette, C. P., Kadivar, H., Stone, W. L., & Wallace, M. T. (2010). An extended multisensory temporal binding window in autism spectrum disorders. *Experimental Brain Research, 203*(2), 381–389. https://doi.org/10.1007/s00221-010-2240-4

Foss-Feig, J. H., Heacock, J. L., & Cascio, C. J. (2012). Tactile responsiveness patterns and their association with core features in autism spectrum disorders. *Research in Autism Spectrum Disorders, 6*(1), 337–344. https://doi.org/10.1016/j.rasd.2011.06.007

Fox, S. E., Levitt, P., & Nelson, C. A., 3rd. (2010). How the timing and quality of early experiences influence the development of brain architecture. *Child Development, 81*(1), 28–40. https://doi.org/10.1111/j.1467-8624.2009.01380.x

Gervais, H., Belin, P., Boddaert, N., Leboyer, M., Coez, A., Sfaello, I., Barthelemy, C., Brunelle, F., Samson, Y., & Zilbovicius, M. (2004). Abnormal cortical voice processing in autism. *Nature Neuroscience, 7*(8), 801–802. https://doi.org/10.1038/nn1291

Geschwind, D. H. (2009). Advances in autism. *Annual Review of Medicine, 60*, 367–380. https://doi.org/10.1146/annurev.med.60.053107.121225

Golinkoff, R. M., Hoff, E., Rowe, M. L., Tamis-LeMonda, C. S., & Hirsh-Pasek, K. (2019). Language matters: Denying the existence of the 30-million-word gap has serious consequences. *Child Development, 90*(3), 985–992. https://doi.org/10.1111/cdev.13128

Green, S. A., & Ben-Sasson, A. (2010). Anxiety disorders and sensory over-responsivity in children with autism spectrum disorders: Is there a causal relationship? *Journal of Autism and Developmental Disorders, 40*(12), 1495–1504. https://doi.org/10.1007/s10803-010-1007-x

Greenough, W. T., Black, J. E., & Wallace, C. S. (1987). Experience and brain development. *Child Development, 58*(3), 539–559. https://www.ncbi.nlm.nih.gov/pubmed/3038480

Grelotti, D. J., Klin, A. J., Gauthier, I., Skudlarski, P., Cohen, D. J., Gore, J. C., Volkmar, F. R., & Schultz, R. T. (2005). fMRI activation of the fusiform gyrus and amygdala to cartoon characters but not to faces in a boy with autism. *Neuropsychologia, 43*(3), 373–385. https://doi.org/10.1016/j.neuropsychologia.2004.06.015

Hall, S. S., Jiang, H., Reiss, A. L., & Greicius, M. D. (2013). Identifying large-scale brain networks in fragile X syndrome. *JAMA Psychiatry, 70*(11), 1215–1223. https://doi.org/10.1001/jamapsychiatry.2013.247

Happe, F., Ronald, A., & Plomin, R. (2006). Time to give up on a single explanation for autism. *Nature Neuroscience, 9*(10), 1218–1220. https://doi.org/10.1038/nn1770

Jassim, N., Baron-Cohen, S., & Suckling, J. (2021). Meta-analytic evidence of differential prefrontal and early sensory cortex activity during non-social sensory perception in autism. *Neuroscience and Biobehavioral Reviews, 127*, 146–157. https://doi.org/10.1016/j.neubiorev.2021.04.014

Kapp, S. K., Goldknopf, E., Brooks, P. J., Kofner, B., & Hossain, M. (2019). Expanding the critique of the social motivation theory of autism with participatory and developmental research. *Behavioral and Brain Sciences, 42*, e94. Article e94. https://doi.org/10.1017/S0140525X18002479

Kariuki, M., Raudino, A., Green, M. J., Laurens, K. R., Dean, K., Brinkman, S. A., Lenroot, R. K., Liu, E., Harris, F., Luo, L., & Carr, V. J. (2016). Hospital admission for infection during early childhood influences developmental vulnerabilities at age 5 years. *Journal of Paediatrics and Child Health, 52*(9), 882–888. https://doi.org/10.1111/jpc.13239

Kohls, G., Yerys, B. E., & Schultz, R. T. (2014). Striatal development in autism: Repetitive behaviors and the reward circuitry. *Biological Psychiatry, 76*(5), 358–359. https://doi.org/10.1016/j.biopsych.2014.07.010

Kuhl, P. K., Coffey-Corina, S., Padden, D., & Dawson, G. (2005). Links between social and linguistic processing of speech in preschool children with autism: Behavioral and electrophysiological measures. *Developmental Science, 8*(1), F1–F12. https://doi.org/10.1111/j.1467-7687.2004.00384.x

Kwon, H., Menon, V., Eliez, S., Warsofsky, I. S., White, C. D., Dyer-Friedman, J., Taylor, A. K., Glover, G. H., & Reiss, A. L. (2001). Functional neuroanatomy of visuospatial working memory in fragile X syndrome: Relation to behavioral and molecular measures. *American Journal of Psychiatry, 158*(7), 1040–1051. https://doi.org/10.1176/appi.ajp.158.7.1040

Langen, M., Bos, D., Noordermeer, S. D., Nederveen, H., van Engeland, H., & Durston, S. (2014). Changes in the development of striatum are involved in repetitive behavior in autism. *Biological Psychiatry, 76*(5), 405–411. https://doi.org/10.1016/j.biopsych.2013.08.013

Leigh, M. J., & Hagerman, R. J. (2020). Chapter 15 - Fragile X clinical features and neurobiology. In J. Rubenstein, P. Rakic, B. Chen, & K. Y. Kwan (Eds.), *Neurodevelopmental disorders* (pp. 351–375). Academic Press. https://doi.org/10.1016/B978-0-12-814409-1.00015-X

Lightbody, A. A., & Reiss, A. L. (2009). Gene, brain, and behavior relationships in fragile X syndrome: Evidence from neuroimaging studies. *Developmental Disabilities Research Reviews, 15*(4), 343–352. https://doi.org/10.1002/ddrr.77

Mason, R. A., Williams, D. L., Kana, R. K., Minshew, N., & Just, M. A. (2008). Theory of mind disruption and recruitment of the right hemisphere during narrative comprehension in autism. *Neuropsychologia, 46*(1), 269–280. https://doi.org/10.1016/j.neuropsychologia.2007.07.018

McEwan, F., Glazier, J. D., & Hager, R. (2023). The impact of maternal immune activation on embryonic brain development. *Frontiers in Neuroscience, 17*, 1146710. https://doi.org/10.3389/fnins.2023.1146710

Menon, V., Kwon, H., Eliez, S., Taylor, A. K., & Reiss, A. L. (2000). Functional brain activation during cognition is related to FMR1 gene expression. *Brain Research, 877*(2), 367–370. https://doi.org/10.1016/s0006-8993(00)02617-2

Mody, M., & Belliveau, J. W. (2013). Speech and language impairments in autism: Insights from behavior and neuroimaging. *North American Journal of Medicine and Science (Boston), 5*(3), 157–161. https://doi.org/10.7156/v5i3p157

Mundy, P., Block, J., Delgado, C., Pomares, Y., Van Hecke, A. V., & Parlade, M. V. (2007). Individual differences and the development of joint attention in infancy. *Child Development, 78*(3), 938–954. https://doi.org/10.1111/j.1467-8624.2007.01042.x

Oster, L. M., & Zhou, G. (2022). Balance and vestibular deficits in pediatric patients with autism spectrum disorder: An underappreciated clinical aspect. *Autism Research and Treatment, 2022*, 7568572. https://doi.org/10.1155/2022/7568572

Pierce, K., Muller, R. A., Ambrose, J., Allen, G., & Courchesne, E. (2001). Face processing occurs outside the fusiform 'face area' in autism: Evidence from functional MRI. *Brain, 124*(Pt 10), 2059–2073. https://doi.org/10.1093/brain/124.10.2059

Pizzagalli, D. A. (2007). Electroencephalography and high-density electrophysiological source localization. In G. Berntson, J. T. Cacioppo, & L. G. Tassinary (Eds.), *Handbook of Psychophysiology* (3rd ed., pp. 56–84). Cambridge University Press. https://www.cambridge.org/core/product/F39CD1186B84D725412AFC986B6DA32A

Rakic, P. (1995). A small step for the cell, a giant leap for mankind: A hypothesis of neocortical expansion during evolution. *Trends in Neurosciences, 18*(9), 383–388. https://doi.org/10.1016/0166-2236(95)93934-p

Rivera, S. M., Menon, V., White, C. D., Glaser, B., & Reiss, A. L. (2002). Functional brain activation during arithmetic processing in females with fragile X Syndrome is related to FMR1 protein expression. *Human Brain Mapping, 16*(4), 206–218. https://doi.org/10.1002/hbm.10048

Roberts, T. P. L., Kuschner, E. S., & Edgar, J. C. (2021). Biomarkers for autism spectrum disorder: Opportunities for magnetoencephalography (MEG). *Journal of Neurodevelopmental Disorders, 13*(1), 34. https://doi.org/10.1186/s11689-021-09385-y

Robertson, C. E., & Baron-Cohen, S. (2017). Sensory perception in autism. *Nature Reviews. Neuroscience, 18*(11), 671–684. https://doi.org/10.1038/nrn.2017.112

Schumann, C. M., Hamstra, J., Goodlin-Jones, B. L., Lotspeich, L. J., Kwon, H., Buonocore, M. H., Lammers, C. R., Reiss, A. L., & Amaral, D. G. (2004). The amygdala is enlarged in children but not adolescents with autism; The hippocampus is enlarged at all ages. *Journal of Neuroscience, 24*(28), 6392–6401. https://doi.org/10.1523/JNEUROSCI.1297-04.2004

Shafritz, K. M., Dichter, G. S., Baranek, G. T., & Belger, A. (2008). The neural circuitry mediating shifts in behavioral response and cognitive set in autism. *Biological Psychiatry, 63*(10), 974–980. https://doi.org/10.1016/j.biopsych.2007.06.028

Shen, M. D., Swanson, M. R., Wolff, J. J., Elison, J. T., Girault, J. B., Kim, S. H., Smith, R. G., Graves, M. M., Weisenfeld, L. A. H., Flake, L., MacIntyre, L., Gross, J. L., Burrows, C. A., Fonov, V. S., Collins, D. L., Evans, A. C., Gerig, G., McKinstry, R. C., Pandey, J., et al. (2022). Subcortical brain development in autism and fragile x syndrome: Evidence for dynamic, age- and disorder-specific trajectories in infancy. *American Journal of Psychiatry, 179*(8), 562–572. https://doi.org/10.1176/appi.ajp.21090896

Simon, D. M., & Corbett, B. A. (2013). Examining associations between anxiety and cortisol in high functioning male children with autism. *Journal of Neurodevelopmental Disorders, 5*(1), 32. https://doi.org/10.1186/1866-1955-5-32

Stevenson, R. A., Siemann, J. K., Schneider, B. C., Eberly, H. E., Woynaroski, T. G., Camarata, S. M., & Wallace, M. T. (2014). Multisensory temporal integration in autism spectrum disorders. *Journal of Neuroscience, 34*(3), 691–697. https://doi.org/10.1523/JNEUROSCI.3615-13.2014

Sun, J., Yuan, Y., Wu, X., Liu, A., Wang, J., Yang, S., Liu, B., Kong, Y., Wang, L., Zhang, K., Li, Q., Zhang, S., Yuan, T., Xu, T. L., & Huang, J. (2022). Excitatory SST neurons in the medial paralemniscal nucleus control repetitive self-grooming and encode reward. *Neuron, 110*(20), 3356–3373. https://doi.org/10.1016/j.neuron.2022.08.010

Surgent, O., Riaz, A., Ausderau, K. K., Adluru, N., Kirk, G. R., Guerrero-Gonzalez, J., Skaletski, E. C., Kecskemeti, S. R., Dean Iii, D. C., Weismer, S. E., Alexander, A. L., & Travers, B. G. (2022). Brainstem white matter microstructure is associated with hyporesponsiveness and overall sensory features in autistic children. *Molecular Autism, 13*(1), 48. https://doi.org/10.1186/s13229-022-00524-3

Thompson, B. L., & Levitt, P. (2010). The clinical-basic interface in defining pathogenesis in disorders of neurodevelopmental origin. *Neuron, 67*(5), 702–712. https://doi.org/10.1016/j.neuron.2010.08.037

Thompson, B. L., & Stanwood, G. D. (2009). Pleiotropic effects of neurotransmission during development: Modulators of modularity. *Journal of Autism and Developmental Disorders, 39*(2), 260–268. https://doi.org/10.1007/s10803-008-0624-0

Thompson, B. L., Levitt, P., & Stanwood, G. D. (2009). Prenatal exposure to drugs: Effects on brain development and implications for policy and education. *Nature Reviews Neuroscience, 10*(4), 303–312. https://doi.org/10.1038/nrn2598

Thorup, E., Nystrom, P., Gredeback, G., Bolte, S., Falck-Ytter, T., & Team, E. (2018). Reduced alternating gaze during social interaction in infancy is associated with elevated symptoms of autism in toddlerhood. *Journal of Abnormal Child Psychology, 46*(7), 1547–1561. https://doi.org/10.1007/s10802-017-0388-0

Tottenham, N., Hertzig, M. E., Gillespie-Lynch, K., Gilhooly, T., Millner, A. J., & Casey, B. J. (2014). Elevated amygdala response to faces and gaze aversion in autism spectrum disorder. *Social Cognitive and Affective Neuroscience, 9*(1), 106–117. https://doi.org/10.1093/scan/nst050

Turner, K. C., Frost, L., Linsenbardt, D., McIlroy, J. R., & Muller, R. A. (2006). Atypically diffuse functional connectivity between caudate nuclei and cerebral cortex in autism. *Behavioral and Brain Functions, 2*, 34. https://doi.org/10.1186/1744-9081-2-34

van der Linden, K., Simons, C., van Amelsvoort, T., & Marcelis, M. (2022). Emotional stress, cortisol response, and cortisol rhythm in autism spectrum disorders: A systematic review. *Research in Autism Spectrum Disorders, 98*, 102039. https://doi.org/10.1016/j.rasd.2022.102039

Volden, J., Coolican, J., Garon, N., White, J., & Bryson, S. (2009). Brief report: Pragmatic language in autism spectrum disorder: Relationships to measures of ability and disability. *Journal of Autism and Developmental Disorders, 39*(2), 388–393. https://doi.org/10.1007/s10803-008-0618-y

Volk, H. E., Lurmann, F., Penfold, B., Hertz-Picciotto, I., & McConnell, R. (2013). Traffic-related air pollution, particulate matter, and autism. *JAMA Psychiatry, 70*(1), 71–77. https://doi.org/10.1001/jamapsychiatry.2013.266

Volk, H. E., Kerin, T., Lurmann, F., Hertz-Picciotto, I., McConnell, R., & Campbell, D. B. (2014). Autism spectrum disorder: Interaction of air pollution with the MET receptor tyrosine kinase gene. *Epidemiology, 25*(1), 44–47. https://doi.org/10.1097/EDE.0000000000000030

Wagner, L., Banchik, M., Okada, N. J., McDonald, N., Jeste, S. S., Bookheimer, S. Y., Green, S. A., & Dapretto, M. (2023). Associations between thalamocortical functional connectivity and sensory over-responsivity in infants at high likelihood for ASD. *Cerebral Cortex, 33*(12), 8075–8086. https://doi.org/10.1093/cercor/bhad100

Wang, X., Christian, K. M., Song, H., & Ming, G. L. (2018). Synaptic dysfunction in complex psychiatric disorders: From genetics to mechanisms. *Genome Medicine, 10*(1), 9. https://doi.org/10.1186/s13073-018-0518-5

West, A. E., & Greenberg, M. E. (2011). Neuronal activity-regulated gene transcription in synapse development and cognitive function. *Cold Spring Harbor Perspectives in Biology, 3*(6). https://doi.org/10.1101/cshperspect.a005744

Willemsen, R., Levenga, J., & Oostra, B. A. (2011). CG G repeat in the FMR1 gene: Size matters. *Clinical Genetics, 80*(3), 214–225. https://doi.org/10.1111/j.1399-0004.2011.01723.x

Zhang, Y., Roy, D. S., & Feng, G. (2022). Finding pleasure in repetitive behaviors. *Neuron, 110*(20), 3225–3227. https://doi.org/10.1016/j.neuron.2022.09.022

Zoghbi, H. Y., & Bear, M. F. (2012). Synaptic dysfunction in neurodevelopmental disorders associated with autism and intellectual disabilities. *Cold Spring Harbor Perspectives in Biology, 4*(3). https://doi.org/10.1101/cshperspect.a009886

The Role of Inborn Errors of Metabolism in Intellectual Disability

Samantha A. Schrier Vergano

Abstract As genetic diagnostic technology advances, clinicians are able to identify more individuals with intellectual disability who have underlying genetic, or in some cases, hereditary forms of disease. Inborn errors of metabolism (metabolic disorders) represent a class of genetic conditions that involve congenital abnormalities in the biochemical processes of the body. The advent of newborn screening in the United States has enabled the fast and accurate detection of many inborn errors that are potentially life-threatening but may still result in significant intellectual disability. Many of these disorders are managed with diet, medications, and in some cases, solid-organ or stem cell transplantation. There are many more metabolic conditions for which available treatment is not readily available or does not yet exist. This chapter will cover many of the most recognizable inborn errors and their role in learning and developmental delay.

Keywords Amino acid disorders · Urea cycle disorders · Storage disorders · Purine and pyrimidine disorders · Mitochondrial disease

Inborn errors of metabolism are a group of genetic conditions that refer to some derangement in the normal biochemical functioning of the body at the cellular level. There are hundreds of inborn errors of metabolism that exist, but only a certain percentage are likely easily identifiable and treatable. Many of those that are treatable are found on newborn screening programs, to pick up asymptomatic infants and provide the best chance at a typical life. This chapter discusses some of the more common inborn errors of metabolism that have an impact on learning and

---

S. A. Schrier Vergano (✉)
Division of Genetic Medicine, Seattle Children's Hospital, Seattle, WA, USA
e-mail: samantha.vergano@seattlechildrens.org

M. G. Valdovinos (ed.), *Intellectual and Developmental Disabilities*, Contemporary Clinical Neuroscience,
https://doi.org/10.1007/978-3-031-66932-3_7

development. Some disorders have milder presentations that can be relatively easily managed with dietary restrictions or medication. Other disorders are felt to significantly affect an individual's cognitive growth despite tight adherence to recommended treatments and therapies. Conversely, metabolic disorders that may not have learning and developmental differences as a primary feature still may affect cognition if individuals present with encephalopathy (derangement in brain function), seizures, hyperammonemia (elevated ammonia levels), or metabolic decompensation (low blood sugar, acidosis, respiratory failure) in the setting of intercurrent illness.

Categories of inborn errors have been organized by the macromolecules primarily affected: namely, proteins (amino acid disorders and urea cycle disorders) and carbohydrates (galactosemia), as well as storage disorders including mucopolysaccharidoses, sphingolipidoses, GM gangliosidoses, peroxisomal disorders, and mitochondrial diseases.

Amino Acid Disorders

Amino acids, the "building blocks" of protein, are crucial for all organ systems and play a vital role in the biochemical processes of the cell. While the human body can produce nonessential amino acids, essential amino acids are only available from the food we consume. All amino acids are important for the proper functioning of the Krebs cycle and urea cycle and for the production of other nitrogen-containing compounds including but not limited to neurotransmitters and other peptides. Disorders involving the production or breakdown of amino acid–related compounds will be discussed.

Phenylalanine Hydroxylase Deficiency (PAH)

Amino acid disorders comprise a number of conditions in which the body is not able to utilize protein, either taken in as food or broken down by the body in times of catabolism. Phenylalanine hydroxylase deficiency (PAH, often known as phenylketonuria) is arguably the most well-recognized inborn error of metabolism, as it was the first disorder on the United States' newborn screen. Prior to the institution of newborn screening in 1960, many of the inborn errors discussed in this chapter went undetected and untreated, resulting not only in mortality but in irreversible intellectual impairment.

Phenylalanine hydroxylase deficiency is caused by deficiency of the enzyme in its name (abbreviated PAH). The enzyme is responsible for the conversion of phenylalanine (PHE) to tyrosine (TYR) and is assisted by the cofactor tetrahydrobiopterin (BH4). In PKU, the accumulation of phenylalanine is neurotoxic, and early and prolonged buildup of PHE leads to profound intellectual disability, autism, and seizures. Prior to newborn screening, individuals with untreated PKU were often relegated to institutions, unable to function as independent members of society. Now, screening of these individuals in all 50 states in the United States identifies these newborns early in order to initiate treatment and enables these individuals to live healthy, independent lives. The estimated prevalence of PKU worldwide is just under 1 in 24,000 births (Hillert et al., 2020) with approximately just under half a million individuals affected throughout the globe (Regier & Greene, 1993).

The neurologic impact of elevated phenylalanine levels is felt to be due to injury to the white matter from abnormal myelination. There is direct evidence of the risk of neurologic impairment as correlated with phenylalanine levels. It is recommended that phenylalanine levels are maintained between 120 and 360 mg/dL in order to minimize the neurologic impact of phenylalanine on the brain. Despite good control, individuals may still have some learning and developmental differences, including ADHD, and may be at increased risk for anxiety and depression. Infants picked up on newborn screening can be expected to have good development and learning with levels between 120 and 360 μmol/L. As they grow and begin to eat typical food (not PKU formula), levels may fluctuate especially in times of growth, puberty, or increased activity. All individuals with PKU should at some point in childhood (ideally early school age) have a comprehensive neuropsychological assessment. Working with a specialized metabolic dietician throughout life is vital for individuals with PKU in order to determine protein requirements as their metabolism changes over time.

Treatment for PKU, as with most amino acid disorders, is the removal of the offensive amino acid from the diet in order to reduce levels of phenylalanine and its metabolites in the brain. While there is no cure, there are several different types of medication that have been used, in addition to a low-phenylalanine diet. Treated levels of phenylalanine ideally should range between 120 and 360 μmol/L (Adams et al., 2023; Rink et al., 2020). Most individuals adhere to typically a vegan-style diet, with a limited amount of natural protein (complete protein found in food). Protein void of phenylalanine is supplied through medical formula (medical protein). Adhering to a diet, in general, is challenging, and individuals who may have difficulty maintaining the recommended diet run into difficulty as their levels rise, which can result in forgetfulness or disorganization; this compounds the adherence challenges. Although phenylalanine is found in a number of nonprotein sources (aspartame, as in sodas), it is likely not a huge contributor to elevated levels if the individual otherwise avoids larger sources of nonmedical protein.

One of the earliest medications, sapropterin dihydrochloride, acts functionally as a synthetic form of BH4 (tetrahydrobiopterin) to aid in the conversion of phenylalanine to tyrosine and thereby reduces the substrate. Other medications include large neutral amino acids that compete for uptake with phenylalanine at the level of the blood–brain barrier. This enables individuals who may have high blood levels of phenylalanine to minimize the effect of Phe on the brain. Pegvialase, developed in 2018, is a pegylated form of phenlyalanine ammonia lyase (PAL) and is an injectable form of medication that brings down phenylalanine levels and enables individuals to liberalize their diet (Lichter-Konecki & Vockley, 2019; Vockley et al., 2014).

Phenylalanine also acts as a teratogen to the developing fetus. Women who themselves have PAH are at risk of having fetuses exposed to elevated levels of phenylalanine, resulting in intrauterine growth retardation, cardiac anomalies, microcephaly, and significant learning and developmental differences (referred to as “maternal PKU”). Women of child-bearing age are recommended to maintain even lower levels of phenylalanine in order to reduce the risk of the fetus early in pregnancy. It is recommended that any woman with PKU of child-bearing age maintain compliance with diet in order to keep levels ideally at 360 µmol/L or lower to reduce the chance of having a child with maternal PKU (Adams et al., 2023; Rink et al., 2020). This can be difficult given the challenges in pregnancy with food avoidance and cravings. Establishing a low pre-conception baseline and identifying low-Phe options in times of nausea or cravings can help with these challenges.

The vast majority of individuals in the United States diagnosed with PAH are picked up on newborn screening; however, the US population is not stagnant. It is important to consider PAH on the differential of learning and developmental differences, especially in individuals who have migrated to the United States from countries where there may be no or limited newborn screening technologies. Although biochemical testing including plasma amino acids may be diagnostic, definitive molecular sequencing of the *PAH* gene is typically recommended for improved understanding of genotype–phenotype correlations and to inform regarding family planning.

Maple Syrup Urine Disease (MSUD)

Although the buildup of phenylalanine and its derivatives is neurotoxic, newborns do not present with a catastrophic metabolic decompensation, which can be seen in other amino acid disorders. Maple syrup urine disease (MSUD) is caused by deficiency in the branched-chain alpha-keto acid dehydrogenase enzyme (BCKDH) within the leucine degradation pathway. MSUD is one of these; classic and intermittent forms exist. Like phenylalanine, the amino acid leucine is neurotoxic. Newborns with the classic form may have a maple syrup odor first in their cerumen (ear wax) and later in their urine. They may present with an overwhelming encephalopathy within the first week of life, lethargy, anorexia, ketosis, irritability, “bicycling,” or other seizure-like movements, and, if left untreated, eventually cerebral edema and death (Amaral & Wajner, 2022; Strauss et al., 1993). Biochemical testing reveals elevations in the branched-chain amino acids (BCAA, leucine, isoleucine, and valine), as well as alloisoleucine and increased beta-hydroxybutyrate levels. Treatment involves restriction of leucine in the diet by limiting complete natural protein as well as supplementation with a BCAA-free formula. Sequencing of the genes known to be associated with MSUD is the gold standard for obtaining a definitive diagnosis, although there are likely cases of MSUD that do not have a known molecular etiology (personal communication).

Intermittent forms of MSUD may become evident in older children who may only have hyperleucinosis in times of intercurrent illnesses; biochemical testing in times of wellness may show normal levels of leucine and absent alloisoleucine. There appears to be a higher prevalence of MSUD in certain populations with founder mutations (e.g., Mennonites; Blackburn et al., 2017). Overall prevalence is estimated at 1 in every 185,000 (Nellis et al., 2003).

Treatment is focused on providing sufficient protein sources with limited amounts of branched-chain amino acids and providing supportive care in times of intercurrent illness. When catabolic, individuals are most at risk for elevations of leucine, which is neurotoxic. This can lead to altered mental status, seizures, and most notably, cerebral edema with abnormalities in sodium levels. Although the exact mechanism is not entirely known, cerebral herniation may occur, leading to death (Amaral & Wajner, 2022). Acute treatment for hyperleucinosis includes reducing catabolism with leucine-free protein such as either metabolic formula or MSUD-total parenteral nutrition (TPN), high-dextrose-containing IV fluids, and close monitoring of electrolytes. As restriction of leucine may also lead to low levels of the other branched-chain amino acids (valine and isoleucine), supplementation of these other BCAA may be needed (Frazier et al., 2014; Strauss et al., 2020).

As intermittent MSUD may not be picked up on newborn screening, individuals may only come to medical attention during times of metabolic stress (e.g., illness, pregnancy, surgery, anesthesia). A normal newborn screen, therefore, should not be used to "rule out" MSUD if other concerning signs are present. There are additional "forms" of MSUD that may have different periods of presentation or genes involved (Amaral & Wajner, 2022). Some forms of MSUD are treated with thiamine and are considered "thiamine-responsive." Any individual suspected of having MSUD should undergo a trial with thiamine. Milder forms with thiamine responsiveness may not present with a neonatal illness but may be managed with thiamine and BCAA restriction.

Liver transplant for MSUD, along with several other IEMs, may be "curative" in that it essentially eliminates the risk of hyperleucinosis in times of catabolism and a restricted diet is typically no longer necessary. However, it does not undo any neurologic damage or developmental delay that may have already occurred (Deon et al., 2023; Froese & Gravel, 2010).

Organic Acidurias

Organic acidurias, including propionic aciduria (PA) and methylmalonic aciduria (MMA), are two disorders of amino acid metabolism that arguably can have the most impact on an individual's cognitive potential. Most individuals with PA and MMA, even if asymptomatic when picked up on newborn screening, are still at risk for learning and developmental differences. Both propionyl-CoA carboxylase and methylmalonyl mutase are instrumental in moving analytes such as succinyl-CoA toward the Krebs cycle for the production of ATP to be used for energy.

Propionic aciduria, caused by deficiency of propionyl-coA carboxylase, often presents with overwhelming neonatal illness, including lethargy, encephalopathy, coma, and subsequently, death. Lab abnormalities include metabolic acidosis, hypoglycemia, neutropenia, and hyperammonemia. Some asymptomatic infants may first be picked up on newborn screening, but many infants may present prior to when screening results return. Biochemical testing shows elevated C3 on the acylcarnitine profile and elevated propionyl species on urine organic acids (Shchelochkov et al., 1993). MMA, similarly, may also present with catastrophic neonatal decompensation, including bone marrow suppression, encephalopathy, hyperammonemia, and metabolic acidosis. Biochemical testing is likely to reveal elevated C3 levels on acylcarnitine profile and elevated methylmalonate on urine organic acids. Individuals with MMA, in addition to the risk for decompensation in the setting of intercurrent illness (as with PA and many other AA disorders), are at increased risk for recurrent pancreatitis, renal failure, and bone marrow suppression (Manoli et al., 1993).

Treatment of both PA and MMA involves dietary restriction of protein, including supplementation with medical protein free of the following, which can create toxic accumulation in these individuals (for both MMA and PA, these components spell V.O.M.I.T: Valine, Odd-chain fatty acids, Methionine, Isoleucine, and Threonine). Despite good control, individuals with PA may still have extra-hepatic manifestations including cardiomyopathy, neutropenia (low neutrophils, a type of white blood cell), retinal abnormalities, and learning and developmental differences.

Some genotypes causing MMA may be responsive to treatment with hydroxycobalamin (HCbl, B12) which may help reduce MMA levels in blood. Propionate is produced in the gut as a result of intestinal bacteria; administration of metronidazole may be able to reduce these levels in individuals with these disorders.

There are many other amino acid disorders derived from enzyme deficiencies within the breakdown of the branched-chain amino acids, including HMG-CoA lyase, HMG-CoA synthase, 3-methylcrotonylcarboxylase, and others, which in the setting of severe metabolic decompensation puts individuals at risk for learning and developmental impairments, but it does not appear that the majority of individuals have significant cognitive deficits at baseline. Treatment for these disorders may include protein restriction with or without supplementation with medical formula. All individuals require vigilance around intercurrent illness and other metabolic stressors in order to reduce episodes of catabolism and minimize the breakdown of the body's complete sources of protein. Most of these disorders are generally picked up by the majority of states' newborn screening programs.

Elevations in methylmalonic acid may also occur as a result of cobalaminopathies (disorders of intracellular cobalamin/B12) as adenosylcobalamin (AdoBcl) is the B12-derived coenzyme for methylmalonyl mutase. Cobalamin plays a number of roles in the proper biochemical functioning of the cell, including acting as a coenzyme for several different enzymes. There are eight different disorders of cobalamin recognized (cblA, cblB, cblC, cblD-combined, cblD-homocystinuria, cblE, cblF, cblG, cblJ, and cblX.). Deficiencies in cblC (the most common cobalaminopathy), cblA, and cblB, all show elevations of MMA (in addition to homocystinuria, see below). Cobalaminopathies may have varying presentations, including neonatal encephalopathy, retinal degeneration, seizures, cardiomyopathy, microcephaly, global developmental delay, and other neurologic findings. Individuals may display megaloblastic anemia (as one might see in B12 deficiency) and have other laboratory abnormalities, including hyperammonemia, metabolic acidosis, and evidence of bone marrow suppression. Adults may display psychiatric symptoms as well (Froese & Gravel, 2010).

The types of cobalaminopathies that result in isolated elevations of MMA (cblA and cblB) may present with the same signs and symptoms that one would see in MMA caused by deficiency of the enzyme alone, including overwhelming neonatal encephalopathy, neutropenia, seizures, and growth and developmental delay. However, elevations in MMA caused by cobalamin defects may be responsive to hydroxycobalamin (B12) administered intramuscularly (Sloan et al., 1993).

The cobalaminopathies that result in the elevation of homocysteine (cblD-homocystine, cblE, and cblG) are caused by deficiencies in methylcobalamin, the coenzyme for the enzyme methionine synthase. Elevations in homocysteine put an individual at risk for thromboembolic events, such as strokes. They may also demonstrate similar symptoms to the other types of cobalaminopathies and may include (in infancy and childhood) lethargy, encephalopathy, seizures, hypotonia, developmental delay, and intellectual disability. Intramuscular B12 administration is typically the mainstay of treatment for most of the cobalaminopathies and may include additional medications depending on the specific cobalaminopathy (Huemer et al., 2017).

Levels of homocysteine may also be elevated in cystathione-beta-synthase deficiency (CBS, classical homocystinuria). Cystathione-beta-synthase is an enzyme responsible for the conversion of homocysteine to cystathionine. Methionine is a precursor to homocysteine and therefore CBS may also result in elevations in methionine. Individuals who have homocystinuria from CBS deficiency typically are tall and may have skeletal abnormalities, including dolichostenomelia and arachnodactyly; the phenotype may resemble that of individuals with Marfan syndrome. A specific feature also may include downward dislocation of the lens of the eye (ectopia lentis). Elevations in homocysteine also result in vascular/thromboembolic events including strokes, and individuals often have intellectual disability. The diagnosis is made through biochemical testing and typically shows elevated levels of homocysteine and methionine on plasma amino acids and is included in the newborn screening for all 50 states. Clinicians may also confirm the diagnosis through sequencing of the *CBS* gene (Sacharow et al., 1993). Treatment involves administration of B6 (pyridoxine), betaine, folate, and B12 (hydroxycobalamin) administration. Additionally, individuals should be put on a low-methioninc diet with the use of medical formulas and low-protein foods.

In general, the life expectancy of individuals with any of the organic acidurias, cobalaminopathies, or other amino acid disorders generally depends on the severity of presentation during the neonatal period and the degree of any residual enzyme or genotype. Long-term survival has been improved as supportive care has improved and organ transplantation for these disorders is becoming more widely performed.

Urea Cycle Disorders

Urea cycle disorders (UCD) are caused by enzyme deficiencies within the urea cycle, the mechanism by which the body is able to process protein and safely excrete nitrogen and ammonia in the form of urea. As a general rule, urea cycle disorders result in the elevation of ammonia levels in the blood; ammonia is neurotoxic, and individuals are at risk for severe encephalopathy, seizures, coma, and death. Depending on the degree of enzyme deficiency as determined by mutations in the particular gene affected, individuals may present with catastrophic neonatal illness or may present as children or even adults after a metabolic stressor, including surgery or pregnancy.

The majority of enzymes within the urea cycle have correlating disorders due to deficiencies. All urea cycle disorders are autosomal recessive, except for ornithine transcarbamylase deficiency (OTC), which is X-linked. OTC is also the most common UCD, occurring in approximately 1 in every 57,000 births (Ah Mew et al., 1993).

As many of the symptoms of ammonia intoxication may be mistaken for more common conditions in neonates, predominately sepsis, and given the rarity of urea cycle disorders in general, they are often misdiagnosed or diagnosed too late to provide useful intervention. Studies have shown that there is a significant loss of IQ points per 24 h of hyperammonemic encephalopathy (Msall et al., 1984). Even with early recognition, neonates who present with elevated ammonia levels are at high risk for significant learning and developmental delay, cerebral palsy, and other neurologic morbidities. First-line clinicians who may see these infants, including neonatologists and ER physicians, should strive to continue to put UCDs on their differential for the infant with altered mentation or poor feeding as obtaining an ammonia level early gives these infants the best chance at treatment.

Depending on the enzyme deficiency within the cycle, various amino acids may be elevated or profoundly low; proximal urea cycle defects may be distinguished from more distal defects by the presence or absence of elevated orotic acid. Urine organic acids typically show elevated levels of glutamine, alanine, and arginine.

Ammonia, when elevated, freely crosses the blood–brain barrier and is converted to glutamine. Elevations in glutamine may cause cellular swelling and brain edema. Clinical manifestations, particularly in neonates with absent enzyme activity, may include respiratory alkalosis (blood pH greater than 7.45), poor feeding, poikilothermia (inability to maintain a normal body temperature), seizures, and cerebral edema.

Acute treatment of urea cycle defects requires the removal of protein sources and supplementation with high-dose dextrose and lipids in order to reduce catabolism in the form of nonprotein calories. Nitrogen scavenging medication (sodium phenylacetate and sodium benzoate) helps bind glutamine and glycine and excretes the excess ammonia in the urine. Often, ammonia levels may be so high that hemodialysis is needed to aid in their reduction. Outside of the initial presentation, individuals may continue to have periods of hyperammonemia during various times of metabolic stressors. They are at risk during intercurrent illnesses, surgeries, pregnancies, steroid use, or increased protein intake that overwhelms the capacity of the enzyme.

Ongoing treatment of individuals with UCDs involves, akin to the amino acid disorders, protein restriction and supplementation with specialized medical formula, as well as administration of nitrogen scavengers either orally or via gastrostomy-tube. Particular care is taken during times of illness or poor oral intake in order to prevent a catabolic state. Arginase deficiency, while a urea cycle disorder, may present as progressive cerebral palsy with developmental delay, as opposed to an acute neurologic presentation.

Hepatic transplant for UCDs also is "curative" in the sense that the majority of the urea cycle takes place in the liver and transplanting an organ from an unaffected individual essentially eliminates the need for protein restriction as well as the risk for recurrent hyperammonemic episodes (Foschi et al., 2015). However, individuals with urea cycle defects may still have intellectual disability, ADHD, or autism despite a successful liver transplant. Neurocognitive outcomes appear to approximate individuals who are medically managed (Posset et al., 2023). Argininosuccinate acid lyase deficiency caused by pathogenic variants in *ASL* is a slightly more distal defect that can cause significant developmental challenges even if an individual has been transplanted or has not had a prior episode of hyperammonemic encephalopathy (Ah Mew et al., 1993).

Galactosemia

Classic galactosemia, a disorder of carbohydrate metabolism, results from the deficiency of galactose-1-phosphate-uridyltransferase (GALT), which converts galactose-1-phosphate (G1P) to glucose-1-phosphate. Symptoms of infants presenting with galactosemia are the result of buildup of the substrate galactose-1-phosphate. Babies develop rapid and fulminant liver failure, vomiting, lethargy, and *E. coli* sepsis. Urine-reducing substances are positive for galactose, GALT levels are profoundly reduced, and G1P levels are elevated. Cataracts develop over time due to the buildup of G1P and may still develop even if treated with dietary interventions. Prevalence in the United States is approximately 1 in 48,000 births and is higher in Ireland (1 in 16,476 births; Coss et al., 2013).

Treatment for classic galactosemia is the removal of lactose-containing foods, which is often easier in the first 6 months of life with the ability to switch to soy-based formulas. There are a number of foods that have been shown to contain galactose that are not dairy based, including tomato sauce and garbanzo beans. There are several types of hard cheeses that have less lactose and therefore are "safe" for individuals with galactosemia. Current treatment recommendations do not include restricting some of these nondairy foods for the most part since their amounts of galactose are not particularly significant and help individuals have a more varied diet.

However, the body also creates endogenous galactose which is not able to be measured or eliminated. Individuals with galactosemia therefore, despite adequate restriction of lactose, are still at risk for various comorbidities. In women, this includes premature ovarian failure. In both men and women, studies have shown that many individuals have some degree of cognitive impairment. Several studies have shown that more than half of individuals with classic galactosemia have speech problems as well as motor abnormalities, including tremors. IQs are often below average, but as with many disorders affecting cognition, there is a wide range, including individuals who have above-average IQs (Doyle et al., 2010). Cohort studies in Ireland have shown that 30% of individuals over 6 years of age have IQs less than 70% (Coss et al., 2013).

There are other enzyme deficiencies within the galactosemia pathway, including galactokinase deficiency and epimerase deficiency. Epimerase deficiency may cause some developmental delay and therefore treatment is emphasized.

Storage Disorders

The term "lysosomal storage disorders" refers broadly to a group of inborn errors that are caused by enzyme deficiencies usually resulting in the buildup of substance disorders that may be subcategorized depending on the product buildup or the organelle where the defect occurs. Examples include mucopolysaccharidoses, sphingolipidoses, oligosaccharidosis, and glycoprotein disorders. This section will discuss the more common lysosomal disorders and their role in intellectual disability. While treatment for several disorders exists, morbidities still occur, and mortality is generally within the first decade of life if untreated.

Mucopolysaccharidoses are a group of LSDs that result in the buildup of glycosaminoglycans (GAGs), a sugar moiety, in multiple tissues of the body. There are seven known MPS disorders, the majority of which cause some degree of intellectual disability in affected individuals. All MPS disorders are autosomal recessive, with the exception of Hunter syndrome (iduronate-2-sulfatase deficiency), which is X-linked. The disorders are named after the enzyme affected, but most also have an eponym.

Hurler syndrome (alpha-L-iduronidase deficiency, MPS1) is now on many states' newborn screens. Left untreated, build up of heparan and dermatan sulfate can result in multi-organ dysfunction, including hepatosplenomegaly, coarse facial features, gingival hypertrophy, cardiac valvular disease, pulmonary fibrosis, hydrocephalus, and corneal clouding (Clarke, 1993). GAGs may cross the blood–brain barrier (BBB) and result in significant intellectual disability, worsening over time. Enzyme replacement therapy (ERT) with laronidase may be given intravenously and can help slow down the accumulation of GAGs in various organs. However, it does not cross the BBB and therefore cannot affect the cognitive impact of the disease. Hematopoietic stem cell transplant (HSCT) is considered the gold standard and, although not without risks itself, is the only current treatment that is able to eliminate the risk of further progression of the disease.

Hurler-Scheie (MPS 1a and Scheie syndrome) are milder forms of Hurler syndrome, and individuals may not have as significant medical complications and may also not have any or as severe intellectual impairments.

Hunter syndrome (iduronate-2 sulfatase deficiency, MPS II) has also been proposed for inclusion in newborn screening. Boys affected with this disorder can have similar symptoms to those of individuals with Hurler syndrome but may also exhibit more behavioral disturbances than those seen in Hurler syndrome. Additionally, corneal clouding is not a feature and can distinguish MPS II from MPS 1. As with MPS 1, intravenous enzyme replacement therapy is available but cannot reverse or slow the cognitive symptoms; individuals may be candidates for HSCT if symptoms have not significantly affected the heart, lungs, or central nervous system (Scarpa, 1993).

Sanfilippo disease (MPS III) is caused by a deficiency in one of four different enzymes (types A–D) and may be more difficult to pick up than Hunter and Hurler as symptoms may be more subtle. Individuals with Sanfilippo syndrome may have significant sleep disturbances, behavioral concerns, developmental regression, and seizures. There is currently no treatment for MPS III (Wagner & Northrup, 1993).

MPS IV and MPS VI (Morquio syndrome and Maroteaux-Lamy syndrome) are unique among the MPS disorders in that cognitive impairment is not the rule. Multi-organ involvement can still occur without treatment but may be milder or slowed in individuals who receive intravenous ERT. Maroteaux-Lamy may have mild, moderate, or severe progression in terms of organ-system involvement but typically, learning and developmental differences are not as significant as in MPS I and II (Khan et al., 2017).

A number of lysosomal storage disorders may present during the first year of life or shortly after and are so severe that individuals die in early childhood. For many of these disorders, there is no treatment or treatment is not yet standard of treatment. These include Tay-Sachs (a GM Gangliosidosis), Niemann-Pick A/B (NP A/B), Krabbe disease, and Metachromatic Leukodystrophy (MLD; sphingolipidoses). However, there are some later forms of these disorders that may present in late childhood, adolescence, or in adulthood, depending on the genotype and the degree of residual enzyme. Some of these later-onset conditions may present with neurologic abnormalities or intellectual disability.

There are a few reports of adult-onset Krabbe disease which may manifest as progressive neurologic symptoms, including muscle weakness and mental and physical degeneration. Niemann-Pick type C (NPC) which is a different condition than NP A and B, may manifest as dysarthria, ataxia, and supranuclear gaze palsy in older children or adolescents. As the disease progresses, bulbar signs such as dysphagia worsen, and learning and developmental differences may become more evident (Walterfang et al., 2012).

NPC has a wide range of ages of presentation, from early infantile (individuals under 2 years of age) with more hepatic involvement to adolescent/adult onset (>15 years) with more neurologic features, and everything in between. Depending on the age, individuals may present with early hypotonia or developmental delay, whereas older children may first be noticed to be becoming clumsier, with ataxia and some academic difficulties. Cognitive decline and psychiatric symptoms may become evident, including misdiagnosed ADHD, psychosis, schizophrenia, or bipolar disorder (Josephs et al., 2003). One classic but atypical finding is supranuclear gaze palsy in which individuals are unable to "look up" (the examiner sees their eyes staring straight ahead when they are asked to look upwards).

The definitive diagnosis is made by sequencing the *NPC1* and *NPC2* genes and detecting biallelic pathogenic variants. Treatment currently remains primarily supportive, although trials with Cyclodextrin, a type of oligosaccharide, continue (Kirkegaard et al., 2016). Miglustat, an iminosugar, is primarily a drug approved to treat Gaucher disease Type 1 but has been used off-label for individuals with NPC and may slow the progression of the disease (Geberhiwot et al., 2018). Additionally, there are trials utilizing the drug Arimoclomol, which improves lysosomal function, in order to slow the progression of NPC (Mengel et al., 2021).

Purine and Pyrimidine Disorders

Arguably the most widely recognized purine disorder is hypoxanthine–guanine–phosphoribosyltransferase deficiency (HGPRT), an X-linked disorder. Deficiencies in this enzyme cause elevations of uric acid in the blood; symptoms may range from gout, with uric acid buildup in the joints, to nephrolithiasis with uric acid stones in the kidney, and to neurologic involvement with profound intellectual disability. Mutations in the *HPRT-1* gene are felt to be causative. Depending on the percent of the residual enzyme, individuals may have hyperuricemia alone (*HPRT1*-related hyperuricemia, HRH), *HPRT1*-related neurologic dysfunction, with moderate intellectual impairment, or Lesch-Nyhan disease (LND; Nyhan, 1973) with profound neurobehavioral abnormalities, cerebral palsy and motor disorders, seizures, and severe self-injurious behavior (SIB; Sampat et al., 2011; Torres et al., 2016).

Individuals with LND may present initially with hypotonia and developmental delays, and then typically develop severe dystonia. About one-third of individuals develop seizures. Cognitive impairment is significant but intellectual disability is felt to be in the mild–moderate range. SIB is typically universal. Individuals who have attenuated versions of the disease may exhibit clumsiness or more significant motor abnormalities. Cognitive dysfunction appears to be mild–moderate (Jinnah et al., 2010). Treatment for LND is primarily supportive. Restraints or bite-guards may need to be utilized for individuals who demonstrate SIB. Individuals may live for several decades, and mortality appears primarily as a result of aspiration pneumonia or sepsis (Neychev & Jinnah, 2006).

Disorders of Creatine Metabolism

There are several treatable inborn errors of metabolism responsible for the metabolism and transport of creatine in the body. Creatine is primarily stored and utilized within the muscles for energy, however deficiencies in creatine can cause significant neurologic symptoms including severe intellectual impairment and developmental delays as cerebral levels of creatine are low. Creatine has been shown to function possibly as a neuromodulator for proper brain function (Stöckler et al., 1996). Inborn errors can occur involving the production of creatine, as seen in Arginine:glycine amidinotransferase (AGAT) deficiency, Guanidinoacetate methyltransferase (GAMT) deficiency (both autosomal recessive), and deficiencies of a creatine transporter caused by pathogenic variants in the *SLC6A8* gene (X-linked).

Diagnosis may be achieved by initial biochemical testing including urine and plasma levels of guanidinoacetate or plasma levels of creatine. Brain imaging (magnetic resonance spectroscopy) may show a low peak of creatine. Confirmation of the diagnosis is made by sequencing *GAMT* or *SLC6A8*.

Individuals with creatine disorders may present with learning and developmental delays and largely go undiagnosed given the rarity of the disorders and nonspecific features on history and physical. There also may be behavioral challenges, seizures, and occasionally movement disorders such as dystonia. Intellectual disability may range from mild to severe. Recently, GAMT was added to the Recommended Uniform Screening Panel (RUSP) in addition to states' newborn screens given the available treatments (Mercimek-Mahmutoglu et al., 2012; Pasquali et al., 2014).

Recommended treatment for these individuals includes supplementation with creatine monohydrate to replete low cerebral levels of creatine; for individuals with GAMT, dietary restriction of arginine/low protein diet as well as the supplementation of ornithine is recommended (Khaikin et al., 2018). A study looking at the various treatment regimens found that treatment showed improvement in all individuals, but only those who were younger than 9 months had normal developmental outcomes (Stockler-Ipsiroglu et al., 2014).

Mitochondrial Disease

The mitochondria are organelles in cells of the body that are responsible for generating energy in the form of adenosine-tri-phosphate (ATP) through the respiratory chain. Individuals often remember learning the phrase in school that the mitochondria are the "powerhouse of the cell." Mitochondria are unique in that they have their own genomic DNA which is passed down from one's mother. Diseases caused by pathogenic variants in mitochondrial DNA are passed down from mother to child, but not from father. However, there are genes within the nucleus of each cell that control the mitochondria, and diseases caused by these genes may be inherited in any Mendelian pattern (autosomal dominant, autosomal recessive, X-linked).

For the purposes of this chapter, "mitochondrial disease" will refer to disorders caused by pathogenic variants in both nuclear and mitochondrial DNA. Mitochondrial diseases are difficult to both diagnose and treat as they may manifest in many different tissues of the body, or in a single organ, they may present at any age, with different patterns of progression, and with different degrees of severity. This chapter will include some of the more well-recognized primary mitochondrial diseases, including MELAS (mitochondrial encephalopathy, lactic acidosis, and stroke-like episodes), pyruvate dehydrogenase deficiency (PDH), and Leigh syndrome. It is important to keep in mind that there are a multitude of other mitochondrial diseases that may have components of intellectual disability as a feature.

Diagnosis of mitochondrial disease, as mentioned previously, can be difficult given the various presentations and the ages and tissues affected. Prior to the development of advanced next-generation sequencing, the gold standard for diagnosis of mitochondrial disease tended to be the biopsy of various tissues. This is still used today in situations where various mitochondrial enzymes may need to be measured or electron microscopy is helpful in aiding with diagnosis. Blood, buccal, and urine samples may all be submitted for mitochondrial DNA analysis when molecular testing may be unclear. Typically, whole exome or whole genome sequencing with mitochondrial DNA sequencing is utilized currently as the most rapid testing available for the diagnosis of mitochondrial disease.

MELAS is a complex and life-threatening disorder that may present at any age and may have a waxing and waning course. Individuals may have persistently elevated lactate levels in the blood and cerebrospinal fluid and are at risk of "stroke-like" episodes/metabolic strokes. These do not have the same pathogenesis as an ischemic or hemorrhagic stroke, as they do not present along the same distribution as the cerebral vasculature, but present with similar neurologic deficits and may be life-threatening. Although symptoms of acute stroke may improve over time, individuals may have persistent neurologic impairment. They are also at risk for seizures and dementia, as well as intellectual impairment. The latter certainly can occur after stroke-like episodes but may exist at baseline as well depending on the specific genotype or the underlying pathophysiology of the condition (El-Hattab et al., 1993). Other comorbidities include the risk of arrhythmias, diabetes, and hearing loss. Although some literature recommends the use of a "mitochondrial cocktail" which includes various vitamins (coenzyme Q10, riboflavin, biotin, alpha-lipoic acid, carnitine, and others), treatment with these supplements is in no way curative (Parikh et al., 2015). There is good evidence, however, to suggest that prophylactic treatment with Arginine and Taurine can reduce the risk of stroke in these individuals (El-Hattab et al., 2017; Ohsawa et al., 2019). Acute management of strokes in individuals with MELAS includes intravenous arginine which increases the production of nitrous oxide (NO) and improves perfusion to the brain (El-Hattab et al., 2017).

Pyruvate dehydrogenase complex deficiency (PDHCD) is a mitochondrial disease that can present in both males and females and may present with varying severity at varying ages, somewhat dependent on the gene and also the genotype. There are several different genes within the pyruvate dehydrogenase complex, all of which can contribute to disease with variable features. Individuals may be detected prenatally with microcephaly, ventriculomegaly, and corpus callosal dysgenesis (Ganetzky et al., 1993). At birth, individuals may have profound lactic acidosis and elevated pyruvate levels in the blood or cerebrospinal fluid. Individuals may have features of encephalopathy or seizures; ketogenic diets have been used with some success in individuals. Intellectual impairment is common. Variable neurologic phenotypes may be seen in older individuals or those diagnosed in later childhood. Individuals with mutations in a gene in the pyruvate dehydrogenase complex may also develop Leigh's syndrome/Leigh encephalopathy. Treatment is primarily supportive, although there is some literature to suggest that the use of dichloroacetic acid (DCA) may be used for PDH and some additional mitochondrial diseases by lowering lactic acid levels (Schoenmann et al., 2023; Karissa et al., 2022).

Leigh syndrome (subacute necrotizing encephalomyopathy) is a term typically used to describe changes to the basal ganglia and/or brainstem on brain MRI with accompanying neurologic regression that occurs after an intercurrent illness. The onset is typically in infancy and is associated with a shortened lifespan. Biochemical findings include elevated lactate levels in the blood and cerebrospinal fluid (Thorburn et al., 1993). Other neurologic abnormalities may include seizures, spasticity, movement disorders, or peripheral neuropathy. While there are certain nuclear genes that can cause Leigh syndrome, there may be others that cause a Leigh-like presentation that is either in the mitochondrial DNA or for which a molecular etiology cannot be established. As with most mitochondrial disorders, various supplements may be utilized but do not stop the progression of disease or lengthen lifespan.

Peroxisomal Disorders

The peroxisome is an organelle in cells that is responsible for the production of various enzymes which aid in, among other things, oxidative reactions within the cell. There are a number of different peroxisomal disorders, including the Zellweger syndrome spectrum, chondrodyplasia punctata, and adult Refsum disease, but this chapter will focus on X-linked adrenoleukodystrophy (X-ALD). One of the hallmark biochemical findings in X-ALD is elevated levels of very long chain fatty acids (VLCFA). These are found throughout the body but accumulate at higher levels in the brain, adrenal cortex, and testes. Elevations of VLCFAs (C26:0) are used to detect cases on newborn screening for the 41 states that include this condition.

X-ALD is becoming more widely diagnosed as more states are including it on their newborn screens. Prior to this, however, individuals were often not diagnosed until they became symptomatic. Individuals with X-ALD typically are at risk for both endocrinologic and neurologic manifestations. Adrenocortical insufficiency puts individuals at risk for cortisol deficiency and often requires regular cortisol supplementation or stress-dosing during times of illness or metabolic stress (Raymond et al., 1993). The neurologic manifestations of X-ALD (cCALD, childhood cerebral adrenoleukodystrophy) may vary but typically present between the ages of 3 and 11. Males may present with worsening school performance or features of ADHD that do not appear to improve with medication administration. Neurologic symptoms progress and ultimately lead to profound disability and death due to cerebral demyelination.

Diagnosis of XALD is ultimately made by sequencing of the *ABCD1* gene; however, individuals may have biochemical evidence of peroxisomal dysfunction that may flag on newborn screening, but molecular testing may be inconclusive. Additionally, other types of peroxisomal disorders may be picked up as a result of this screening; false negative rates may be as high as 20%. It is not uncommon for the diagnosis of cCALD in an infant to prompt the diagnosis of additional family members with different presentations of XALD such as adrenomyeloneuropathy (AMN) in young adult men or carrier females, or primary adrenocortical insufficiency (Turk et al., 2020).

Treatment for boys with the cALD form of XALD is hematopoietic stem cell transplantation (HSCT). As with all HSCT, there is a risk of morbidity and mortality with this method. There has previously been use of dietary restriction of VLCFAs and supplementation with what has been termed "Lorenzo's Oil," a mixture of oleic and erucic acid triglycerides. This treatment however has not reliably been shown to halt neurologic progression (Moser et al., 2007).

Conclusion

This chapter has focused on a number of inborn errors of metabolism and their role in contributing to the differential diagnoses of individuals with intellectual delay or cognitive impairment. This group of disorders should always be considered in individuals with intellectual and developmental disabilities even if they had a normal newborn screen at birth; unfortunately, there will likely always be conditions that are not good candidates for public health initiatives such as newborn screening, and diagnosis can be delayed. Early recognition of these disorders enables patients to have better access to available treatments and therapies. With the advent of gene therapy, inborn errors continue to be the target of various treatment modalities utilizing gene editing or gene additions. Additionally, research continues to aim for improvements in the administration of certain treatments to improve compliance and outcomes.

References

Adams, A. D., Fiesco-Roa, M., Wong, L., Jenkins, G. P., Malinowski, J., Demarest, O. M., et al. (2023). Phenylalanine hydroxylase deficiency treatment and management: A systematic evidence review of the American College of Medical Genetics and Genomics (ACMG). *Genetics in Medicine, 25*(9). https://doi.org/10.1016/j.gim.2022.12.005

Ah Mew, N., Simpson, K. L., Gropman, A. L., Lanpher, B. C., Chapman, K. A., & Summar, M. L. (1993). Urea cycle disorders overview. In *GeneReviews® [Internet]*. University of Washington, Seattle.

Amaral, A. U., & Wajner, M. (2022, July 1). Pathophysiology of maple syrup urine disease: Focus on the neurotoxic role of the accumulated branched-chain amino acids and branched-chain α-keto acids. *Neurochemistry International*. Elsevier Ltd. https://doi.org/10.1016/j.neuint.2022.105360

Blackburn, P. R., Gass, J. M., Pinto e Vairo, F., Farnham, K. M., Atwal, H. K., Macklin, S., et al. (2017, September 6). Maple syrup urine disease: Mechanisms and management. *Application of Clinical Genetics*. Dove Medical Press Ltd. https://doi.org/10.2147/TACG.S125962

Clarke, L. A. (1993). Mucopolysaccharidosis type I. In *GeneReviews® [Internet]*. University of Washington, Seattle.

Coss, K. P., Doran, P. P., Owoeye, C., Codd, M. B., Hamid, N., Mayne, P. D., et al. (2013). Classical galactosaemia in Ireland: Incidence, complications and outcomes of treatment. *Journal of Inherited Metabolic Disease, 36*(1), 21–27. https://doi.org/10.1007/s10545-012-9507-9

Deon, M., Guerreiro, G., Girardi, J., Ribas, G., & Vargas, C. R. (2023, October 1). Treatment of maple syrup urine disease: Benefits, risks, and challenges of liver transplantation. *International Journal of Developmental Neuroscience*. John Wiley and Sons Inc. https://doi.org/10.1002/jdn.10283

Doyle, C. M., Channon, S., Orlowska, D., & Lee, P. J. (2010). The neuropsychological profile of galactosaemia. *Journal of Inherited Metabolic Disease, 33*(5), 603–609. https://doi.org/10.1007/s10545-010-9154-y

El-Hattab, A. W., Almannai, M., & Scaglia, F. (1993). MELAS. In *GeneReviews® [Internet]*. University of Washington, Seattle.

El-Hattab, A. W., Zarante, A. M., Almannai, M., & Scaglia, F. (2017). Therapies for mitochondrial diseases and current clinical trials. *Molecular Genetics and Metabolism, 122*(3), 1–9. https://doi.org/10.1016/j.ymgme.2017.09.009

Foschi, F. G., Morelli, M. C., Savini, S., Dall'Aglio, A. C., Lanzi, A., Cescon, M., et al. (2015). Urea cycle disorders: A case report of a successful treatment with liver transplant and a literature review. *World Journal of Gastroenterology, 21*(13), 4063–4068. https://doi.org/10.3748/wjg.v21.i13.4063

Frazier, D. M., Allgeier, C., Homer, C., Marriage, B. J., Ogata, B., Rohr, F., et al. (2014). Nutrition management guideline for maple syrup urine disease: An evidence- and consensus-based approach. *Molecular Genetics and Metabolism, 112*(3), 210–217. https://doi.org/10.1016/j.ymgme.2014.05.006

Froese, D. S., & Gravel, R. A. (2010, February). Genetic disorders of vitamin B 12 metabolism: Eight complementation groups - Eight genes. *Expert Reviews in Molecular Medicine*. https://doi.org/10.1017/S1462399410001651

Ganetzky, R., McCormick, E. M., & Falk, M. J. (1993). Primary pyruvate dehydrogenase complex deficiency overview. In *GeneReviews® [Internet]*. University of Washington, Seattle.

Geberhiwot, T., Moro, A., Dardis, A., Ramaswami, U., Sirrs, S., Marfa, M. P., et al. (2018, April 6). Consensus clinical management guidelines for Niemann-Pick disease type C. *Orphanet Journal of Rare Diseases*. BioMed Central Ltd. https://doi.org/10.1186/s13023-018-0785-7

Hillert, A., Anikster, Y., Belanger-Quintana, A., Burlina, A., Burton, B. K., Carducci, C., et al. (2020). The genetic landscape and epidemiology of phenylketonuria. *American Journal of Human Genetics, 107*(2), 234–250. https://doi.org/10.1016/j.ajhg.2020.06.006

Huemer, M., Diodato, D., Schwahn, B., Schiff, M., Bandeira, A., Benoist, J. F., et al. (2017). Guidelines for diagnosis and management of the cobalamin-related remethylation disorders cblC, cblD, cblE, cblF, cblG, cblJ and MTHFR deficiency. *Journal of Inherited Metabolic Disease, 40*(1), 21–48. https://doi.org/10.1007/s10545-016-9991-4

Jinnah, H. A., Ceballos-Picot, I., Torres, R. J., Visser, J. E., Schretlen, D. J., Verdu, A., et al. (2010). Attenuated variants of Lesch-Nyhan disease. *Brain, 133*(3), 671–689. https://doi.org/10.1093/brain/awq013

Josephs, K. A., Van Gerpen, M. W., & Van Gerpen, J. A. (2003). Adult onset Niemann-Pick disease type C presenting with psychosis. *Journal of Neurology, Neurosurgery, and Psychiatry, 74*. https://www.jnnp.com

Karissa, P., Simpson, T., Dawson, S. P., Low, T. Y., Tay, S. H., Nordin, F. D. A., et al. (2022, May 19). Comparison between dichloroacetate and phenylbutyrate treatment for pyruvate dehydrogenase deficiency. *British Journal of Biomedical Science*. Institute of Biomedical Science (IBMS). https://doi.org/10.3389/bjbs.2022.10382

Khaikin, Y., Sidky, S., Abdenur, J., Anastasi, A., Ballhausen, D., Buoni, S., et al. (2018). Treatment outcome of twenty-two patients with guanidinoacetate methyltransferase deficiency: An international retrospective cohort study. *European Journal of Paediatric Neurology, 22*(3), 369–379. https://doi.org/10.1016/j.ejpn.2018.02.007

Khan, S. A., Peracha, H., Ballhausen, D., Wiesbauer, A., Rohrbach, M., Gautschi, M., et al. (2017). Epidemiology of mucopolysaccharidoses. *Molecular Genetics and Metabolism, 121*(3), 227–240. https://doi.org/10.1016/j.ymgme.2017.05.016

Kirkegaard, T., Gray, J., Priestman, D. A., Wallom, K. L., Atkins, J., Olsen, O. D., et al. (2016). Heat shock protein-based therapy as a potential candidate for treating the sphingolipidoses. *Science Translational Medicine, 8*(355). https://doi.org/10.1126/scitranslmed.aad9823

Lichter-Konecki, U., & Vockley, J. (2019). Phenylketonuria: Current treatments and future developments. *Drugs, 79*(5), 495–500. https://doi.org/10.1007/s40265-019-01079-z

Manoli, I., Sloan, J. L., & Venditti, C. P. (1993). Isolated methylmalonic acidemia. In *GeneReviews® [Internet]*. University of Washington, Seattle.

Mengel, E., Patterson, M. C., Da Riol, R. M., Del Toro, M., Deodato, F., Gautschi, M., et al. (2021). Efficacy and safety of arimoclomol in Niemann-Pick disease type C: Results from a double-blind, randomised, placebo-controlled, multinational phase 2/3 trial of a novel treatment. *Journal of Inherited Metabolic Disease, 44*(6), 1463–1480. https://doi.org/10.1002/jimd.12428

Mercimek-Mahmutoglu, S., Sinclair, G., van Dooren, S. J. M., Kanhai, W., Ashcraft, P., Michel, O. J., et al. (2012). Guanidinoacetate methyltransferase deficiency: First steps to newborn screening for a treatable neurometabolic disease. *Molecular Genetics and Metabolism, 107*(3), 433–437. https://doi.org/10.1016/j.ymgme.2012.07.022

Moser, H. W., Moser, A. B., Hollandsworth, K., Brereton, N. H., & Raymond, G. V. (2007, September). "Lorenzo's oil" therapy for X-linked adrenoleukodystrophy: Rationale and current assessment of efficacy. *Journal of Molecular Neuroscience*. https://doi.org/10.1007/s12031-007-0041-4

Msall, M., Batshaw, M. L., Suss, R., Brusilow, S. W., & Mellits, E. D. (1984). Neurologic outcome in children with inborn errors of urea synthesis. Outcome of urea-cycle enzymopathies. *The New England Journal of Medicine, 310*(23), 1500–1505. https://doi.org/10.1056/NEJM198406073102304

Nellis, M. M., Kasinski, A., Carlson, M., Allen, R., Schaefer, A. M., Schwartz, E. M., & Danner, D. J. (2003). Relationship of causative genetic mutations in maple syrup urine disease with their clinical expression. *Molecular Genetics and Metabolism, 80*(1–2), 189–195. https://doi.org/10.1016/s1096-7192(03)00144-6

Neychev, V. K., & Jinnah, H. (2006). Sudden death in Lesch-Nyhan disease. *Developmental Medicine & Child Neurology, 48*(11), 923. https://doi.org/10.1017/S0012162206002015

Nyhan, W. L. (1973). *The Lesch-Nyhan Syndrome*. https://www.annualreviews.org

Ohsawa, Y., Hagiwara, H., Nishimatsu, S.-I., Hirakawa, A., Kamimura, N., Ohtsubo, H., et al. (2019). Taurine supplementation for prevention of stroke-like episodes in MELAS: A multicentre, open-label, 52-week phase III trial. *Journal of Neurology, Neurosurgery, and Psychiatry, 90*(5), 529–536. https://doi.org/10.1136/jnnp-2018-317964

Parikh, S., Goldstein, A., Koenig, M. K., Scaglia, F., Enns, G. M., Saneto, R., et al. (2015, September 2). Diagnosis and management of mitochondrial disease: A consensus statement from the Mitochondrial Medicine Society. *Genetics in Medicine*. Nature Publishing Group. https://doi.org/10.1038/gim.2014.177

Pasquali, M., Schwarz, E., Jensen, M., Yuzyuk, T., Debiase, I., Randall, H., & Longo, N. (2014). Feasibility of newborn screening for guanidinoacetate methyltransferase (GAMT) deficiency. *Journal of Inherited Metabolic Disease, 37*(2), 231–236. https://doi.org/10.1007/s10545-013-9662-7

Posset, R., Garbade, S. F., Gleich, F., Scharre, S., Okun, J. G., Gropman, A. L., et al. (2023). Severity-adjusted evaluation of liver transplantation on health outcomes in urea cycle disorders. *Genetics in Medicine, 26*(4), 101039. https://doi.org/10.1016/j.gim.2023.101039

Raymond, G. V., Moser, A. B., & Fatemi, A. (1993). X-linked adrenoleukodystrophy. In *GeneReviews® [Internet]*. University of Washington, Seattle.

Regier, D. S., & Greene, C. L. (1993). Phenylalanine Hydroxylase Deficiency. In *GeneReviews® [Internet]*. University of Washington, Seattle.

Rink, B., Dukhovny, S., & Timofeev, J. (2020). ACOG COMMITTEE OPINION Number 802 Committee on Genetics This Committee Opinion was developed by the Committee on Genetics in collaboration with committee members Management of Women With Phenylalanine Hydroxylase Deficiency (Phenylketonuria) (Vol. 135).

Sacharow, S. J., Picker, J. D., & Levy, H. L. (1993). Homocystinuria caused by cystathionine beta-synthase deficiency. In *GeneReviews® [Internet]*. University of Washington, Seattle.

Sampat, R., Fu, R., Larovere, L. E., Torres, R. J., Ceballos-Picot, I., Fischbach, M., et al. (2011). Mechanisms for phenotypic variation in Lesch-Nyhan disease and its variants. *Human Genetics, 129*(1), 71–78. https://doi.org/10.1007/s00439-010-0901-9

Scarpa, M. (1993). Mucopolysaccharidosis type II. In *GeneReviews® [Internet]*. University of Washington, Seattle.

Schoenmann, N., Tannenbaum, N., Hodgeman, R. M., & Raju, R. P. (2023, October 1). Regulating mitochondrial metabolism by targeting pyruvate dehydrogenase with dichloroacetate, a metabolic messenger. *Biochim Biophys Acta - Mol Basis Dis*. Elsevier B.V. https://doi.org/10.1016/j.bbadis.2023.166769

Shchelochkov, O. A., Carrillo, N., & Venditti, C. (1993). Propionic acidemia. In *GeneReviews® [Internet]*. University of Washington, Seattle.

Sloan, J. L., Carrillo, N., Adams, D., & Venditti, C. P. (1993). Disorders of intracellular cobalamin metabolism. In *GeneReviews® [Internet]*. University of Washington, Seattle.

Stöckler, S., Isbrandt, D., Hanefeld, F., Schmidt, B., & von Figura, K. (1996). Guanidinoacetate methyltransferase deficiency: The first inborn error of creatine metabolism in man. *American Journal of Human Genetics, 58*(5), 914–922.

Stockler-Ipsiroglu, S., van Karnebeek, C., Longo, N., Korenke, G. C., Mercimek-Mahmutoglu, S., Marquart, I., et al. (2014). Guanidinoacetate methyltransferase (GAMT) deficiency: Outcomes in 48 individuals and recommendations for diagnosis, treatment and monitoring. *Molecular Genetics and Metabolism, 111*(1), 16–25. https://doi.org/10.1016/j.ymgme.2013.10.018

Strauss, K. A., Puffenberger, E. G., & Carson, V. J. (1993). Maple syrup urine disease. In *GeneReviews® [Internet]*. University of Washington, Seattle.

Strauss, K. A., Carson, V. J., Soltys, K., Young, M. E., Bowser, L. E., Puffenberger, E. G., et al. (2020). Branched-chain α-ketoacid dehydrogenase deficiency (maple syrup urine disease): Treatment, biomarkers, and outcomes. *Molecular Genetics and Metabolism, 129*(3), 193–206. https://doi.org/10.1016/j.ymgme.2020.01.006

Thorburn, D. R., Rahman, J., & Rahman, S. (1993). Mitochondrial DNA-associated leigh syndrome and NARP. In *GeneReviews® [Internet]*. University of Washington, Seattle.

Torres, R. J., Prior, C., Garcia, M. G., & Puig, J. G. (2016, December 1). A review of the implication of hypoxanthine excess in the physiopathology of Lesch–Nyhan disease. *Nucleosides, Nucleotides and Nucleic Acids*. Taylor and Francis. https://doi.org/10.1080/15257770.2016.1147579

Turk, B. R., Theda, C., Fatemi, A., & Moser, A. B. (2020). X-linked adrenoleukodystrophy: Pathology, pathophysiology, diagnostic testing, newborn screening and therapies. *International Journal of Developmental Neuroscience, 80*(1), 52–72. https://doi.org/10.1002/jdn.10003

Vockley, J., Andersson, H. C., Antshel, K. M., Braverman, N. E., Burton, B. K., Frazier, D. M., et al. (2014). Phenylalanine hydroxylase deficiency: Diagnosis and management guideline. *Genetics in Medicine, 16*(2), 188–200. https://doi.org/10.1038/gim.2013.157

Wagner, V. F., & Northrup, H. (1993). Mucopolysaccharidosis type III. In *GeneReviews®*. University of Washington, Seattle.

Walterfang, M., Chien, Y. H., Imrie, J., Rushton, D., Schubiger, D., & Patterson, M. C. (2012). Dysphagia as a risk factor for mortality in Niemann-Pick disease type C: Systematic literature review and evidence from studies with miglustat. *Orphanet Journal of Rare Diseases, 7*(1). https://doi.org/10.1186/1750-1172-7-76

Prenatal Risk and Later Intellectual and Developmental Disabilities

Johanna Bick, Tara M. Rutter, Cristin M. Holland, Isabelle Mueller, Venus Mahmoodi, Angela Gigliotti-Manessis, Livia Merrill, M. Catalina Camacho, Sanjana Inala, Andrea Ortiz-Jimenez, Dana DeMaster, and Marisa N. Spann

Abstract The fetal period is marked by heightened vulnerability to environmental insults. The complex and patterned organization of the fetal brain lays the foundation for later maturation and behavior. Prenatal influences play a critical role in shaping the developmental trajectory of intellectual and developmental disabilities (IDD). This chapter explores prenatal risk factors, including maternal physical and mental health and exposure to adversity, teratogens, environmental hazards, and systemic factors. We consider how exposure to these factors shapes long-term neurodevelopmental processes that lead to functional abnormalities, subsequently contributing to IDD. We discuss these factors to underscore the importance of public health interventions and targeted strategies to foster healthy pregnancies, support optimal neurodevelopment, and mitigate the long-term consequences of early life adversities.

Keywords Prenatal exposures · Maternal healthy · Infant development · IDD · Sensitive periods

---

J. Bick (✉) · L. Merrill · A. Ortiz-Jimenez
Department of Psychology, University of Houston, Houston, TX, USA
e-mail: jrbick@uh.edu

T. M. Rutter
Department of Psychiatry, Oregon Health & Science University, Portland, OR, USA

C. M. Holland · I. Mueller · V. Mahmoodi · S. Inala · M. N. Spann (✉)
Department of Psychiatry, Vagelos College of Physicians and Surgeons, Columbia University, New York, NY, USA
e-mail: mns2125@cumc.columbia.edu

A. Gigliotti-Manessis
Teachers College, Columbia University, New York, NY, USA

M. C. Camacho
Washington University in St. Louis, St. Louis, MO, USA

D. DeMaster
University of Texas Health Houston, Houston, TX, USA

M. G. Valdovinos (ed.), *Intellectual and Developmental Disabilities*,
Contemporary Clinical Neuroscience,
https://doi.org/10.1007/978-3-031-66932-3_8

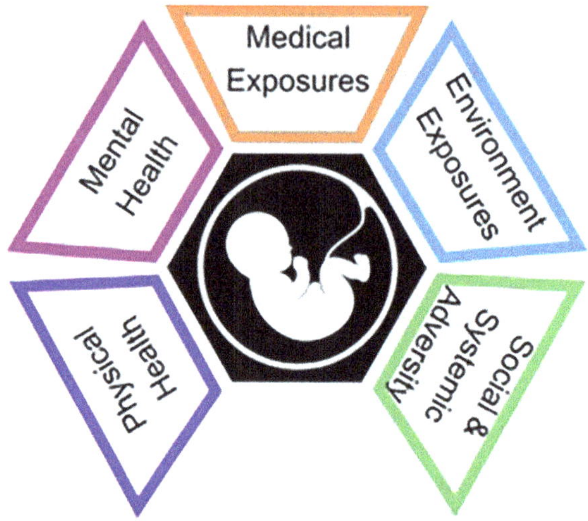

Fig. 1 Schematic of prenatal factors that influence IDD risk and development

Intrauterine exposures have long been associated with risk for later intellectual and developmental disabilities (IDD). Increasing evidence points to formative alterations in the developing fetal brain as pathways to postnatal developmental risk. Critically, influences of risk are multifactorial and cut across individual, intrauterine, and environmental influences. Not surprisingly, similar exposures can have diverse consequences for later cognitive and/or social development. Beyond genetics, environmental factors include maternal health characteristics, psychosocial factors, and environmental exposures, which may be exacerbated by institutional and socioeconomic structures that increase propensity for risks at all levels (See Fig. 1). The goal of this chapter is to review evidence on prenatal influences on IDD risk in childhood. First, we consider the fetal period as a period of heightened vulnerability due to complex and patterned organization of the fetal brain that lays the foundation for later maturation. We review various prenatal exposures as risk factors for IDD and discuss potential disturbances in early fetal brain development that may mediate later risk. Our goal is to synthesize understanding on the prenatal pathways to IDD risk and highlight future directions that have critical implications for public policy and prevention.

Fetal Brain Development: Sensitive Periods and Vulnerabilities

Sensitive periods are phases in which the developing brain is particularly impacted by environmental insults (Knudsen, 2004). It is thought that sensitive periods encompass both prenatal and postnatal phases, serving a fundamental role in shaping later infant developmental outcomes. Advances in fetal and infant imaging have made it possible to examine how perinatal exposures might influence trajectories of fetal and infant brain development in ways that elucidate connections between perinatal environments and the later emergence of behavioral and cognitive delays that comprise various syndromes and IDD.

A substantial portion of brain development occurs during the prenatal period, including gastrulation, neural induction and neurulation, neurogenesis, neuronal migration, development of axons and dendrites, synaptogenesis and pruning, myelination, and formation of neural circuitries (Linderkamp et al., 2009). Primary neurulation, the process in which the neural tube is formed, begins during the first trimester. Neural tube defects, generally assessed through prenatal screening using maternal blood, detect abnormalities of neural folding, incomplete or premature dysfunction, defective gastrulation, disordered secondary neurulation, and sacral agenesis (Singh & Munakomi, 2019). These defects are known to be a cause of stillbirths and palsies in offspring (Chandran et al., 2021). Neurogenesis—the formation of neurons derived from neuroepithelial cells of the neural tube—begins during the late embryonic period and is largely completed during the second trimester (Leibovitz et al., 2022).

Atypical neurogenesis is thought to cause brain growth abnormalities, in which rates of symmetric and asymmetric cell divisions or changes to differentiations of neuronal cells may lead to cell cycle dysregulation. Between 12- and 20-week gestation, the developing brain undergoes neuronal migration, with post-migration neuronal development and cortical organization beginning at 22-week gestation and continuing through the postnatal period (Leibovitz et al., 2022). Atypical neuronal migration, which may be a particularly important mechanistic pathway of IDD (Pan et al., 2019), leads to the malformation and atypical functioning of brain circuits. However, the brain also undergoes rapid development during the early postnatal period (Gilmore et al., 2018). Thus, postnatal neural plasticity and environmental supports have the potential to buffer relations between environmental insults and offspring brain development, potentially reducing later risk for IDD.

Maternal Physical Health

In utero exposures to maternal infection, inflammation, and chronic illness are increasingly shown to increase risk for a broad range of IDD in offspring. While pathways are currently unclear, emerging work suggests that there are likely direct effects of maternal immune responses that compromise the healthy or typical development of the fetus and, specifically, alter foundational neurodevelopmental pathways that increase risk for IDD and neuropsychiatric disorders in later life (Amgalan, et al., 2021; Estes et al., 2016). We review evidence on various common maternal health conditions related to hypertension, diabetes, obesity, and individual risk factors that can occur during pregnancy that may disrupt fetal brain organization and enhance risk for IDD.

Hypertension

Hypertension is a major risk factor for cardiovascular disease, which is the leading cause of death in the United States. The prevalence of hypertension in individuals of reproductive age is almost 8% and rising, particularly in people of color (Ford, 2022). While hypertensive disorders of pregnancy occur outside of the context of prepregnancy hypertension, pregnant individuals with chronic hypertension are at high risk for developing hypertensive disorders of pregnancy (HDPs). HDPs affect approximately 10% of pregnancies and are the most common medical comorbidity seen by obstetricians (Gupta et al., 2023).

Chronic hypertension, now included in the definition of HDPs, has been associated with neurodevelopmental outcomes in offspring. For example, in a study using a Swedish medical registry, HDPs were associated with a hazard ratio (HR) of 1.22 for autism spectrum disorder (ASD), 1.10 for attention-deficit/ hyperactivity disorder (ADHD), and 1.39 for intellectual disability (ID) in addition to slightly lower cognitive scores (Brand et al., 2021). Similarly, a large Taiwanese study showed offspring exposed to HDPs, either pregnancy-induced hypertension or preeclampsia, had an increased HR of 1.27 for ASD, 1.23 for ADHD, 1.29 for developmental delay, and 1.53 for ID compared to offspring not exposed to HDPs (Chen et al., 2021). Offspring exposed to chronic hypertension during pregnancy, compared to those of unexposed mothers, had similar a greater risk for ID (HR = 1.67) compared to other neurodevelopmental disorders including developmental delay (HR = 1.29) and ADHD (HR = 1.22). Neurodevelopmental outcomes at 12 months of age for preterm infants exposed to HDPs did not differ from preterm infants who were not exposed to HDPs in utero (Avorgbedor et al., 2019).

A meta-analysis reported a pooled, adjusted odds ratio of 1.35 for ASD from 11 studies and a pooled, adjusted odds ratio of 1.29 for ADHD from six studies, indicating an elevated likelihood for these neurodevelopmental outcomes from exposure to HDPs (Maher et al., 2018). In another meta-analysis of 21 studies, offspring who were born prematurely due to HDP had higher risk for ASD than matched controls (Xu et al., 2017). Additionally, offspring exposed to preeclampsia and chronic hypertension complicating pregnancy had higher elevated likelihood for ASD than those exposed to gestational hypertension. In summary, numerous studies suggest a connection between HDP and preeclampsia has a higher association with IDD compared to other neurodevelopmental disorders.

Diabetes

The prevalence of type 1 and type 2 diabetes mellitus (T1DM, T2DM, respectively) has increased in individuals of reproductive age and affects about one percent of all pregnancies (Ornoy et al., 2021). Meanwhile, gestational diabetes mellitus (GDM) also has steadily increased across the last 20 years and affects anywhere from 1.7 to 15.7 percent of pregnancies. Incidence of gestational diabetes can be dependent on ethnicity, age, and diagnostic criteria (Kong et al., 2018; Schneider et al., 2012).

A large retrospective cohort study assessed the elevated likelihood of neurodevelopmental disorders for offspring exposed to T1DM, T2DM, and GDM. Adjusted for demographic variables, maternal hypertensive disorders, and preterm delivery, offspring exposed to T1DM had the greatest elevated likelihood for developmental delay (HR = 1.80) and IDD (HR = 2.69), while those exposed to T2DM had increased likelihood of ASD (HR = 1.28), ADHD (HR = 1.29), developmental delay (HR = 1.34), and ID (HR = 1.58). Offspring exposed to GDM had an elevated likelihood of ASD (HR = 1.14), ADHD (HR = 1.07), and developmental delay (HR = 1.03). Another population-based cohort study in Sweden similarly investigated the effects of maternal T1DM, pre-gestational T2DM, and GDM on risk of neurodevelopmental disorders in offspring, also finding an association with all types of maternal diabetes and an increased odds of neurodevelopmental disorders in offspring (Chen et al., 2021). T2DM had the greatest elevated likelihood for neurodevelopmental disorders risk including IDD (OR = 2.09), ASD (OR = 1.27), and ADHD (OR = 1.43). A diagnosis of GDM between 27- and 30-week gestation was associated with the greatest elevated likelihood of neurodevelopmental disorders compared with a diagnosis of GDM at other stages of gestation. In a study following up with children exposed to GDM at a median age of 18, GDM was associated with offspring ADHD (HR = 1.64), but not ASD (HR = 1.46; Perea et al., 2022). Additionally, there were higher rates of ADHD for children whose mothers went on to develop T2DM. A meta-analysis demonstrated an elevated likelihood of ASD and ADHD for those exposed to GDM, with pooled OR of 1.42 and 1.01, respectively (Rowland & Wilson, 2021).

Exposure to diabetes and the developing fetus can be complicated by other factors, such as air pollution, socioeconomic factors, and obesity. For example, there is an increased likelihood of ASD when there is exposure to ozone in the first trimester and a diagnosis of maternal GDM before 24 weeks of gestation (Jo et al., 2019). Elevated likelihood of ADHD increased 14-fold when offspring of mothers with GDM were also exposed to low socioeconomic status (SES) (Nomura et al., 2012). The likelihood of ADHD and ASD, as well as mood disorders, was increased for offspring exposed to non-insulin-treated T2DM in severely obese mothers compared to offspring of normal weight mothers without diabetes, though offspring exposed to GDM in severely obese mothers had lower overall effect sizes (Kong et al., 2018). These studies indicate that additional factors should be considered when exploring the impact of maternal diabetes on offspring IDD development. In summary, numerous studies suggest a higher association of GDM with IDD compared to other neurodevelopmental disorders, with ratios around a two-fold increased risk.

Obesity and Body Mass Index

Over 40% of reproductive-age women from 2014 in the United States have a body mass index (BMI) that is considered obese (Flegal et al., 2016). Accumulating evidence from epidemiological studies suggests that infants' exposure to maternal prenatal obesity has vast impacts on later neurodevelopmental outcomes, including IDD (Edlow, 2017). A majority of cohort and case-control studies have examined prepregnancy BMI as a potential risk factor, linking maternal obesity to offspring cognitive outcomes including lower IQ at age 5 (Bliddal et al., 2014; Eriksen et al., 2013) and at age 7 (Huang et al., 2014), as well as lower reading scores (Tanda et al., 2013). Further, one U.S. cohort study of 2734 children found that prepregnancy maternal BMI was significantly associated with a diagnosis of IDD in offspring (Li et al., 2016).

Maternal obesity has also been linked to other neurodevelopmental outcomes in offspring, including ASD, ADHD, and schizophrenia. One U.S. study of 62 mother-offspring pairs found that maternal obesity, measured via BMI during the first trimester, was significantly associated with increased odds of a positive result or ASD via the Modified Checklist for Autism in Toddlers (M-CHAT), a parent report used to screen for ASD in toddlers aged 16 to 30 months (Robins et al., 2001), at age 2 among preterm infants (Reynolds et al., 2014). Lending further credence to this association, a case-control study of 1004 children in the United States found that maternal obesity (measured by prepregnancy BMI) was associated with increased odds of a medical diagnosis of ASD among young children (Krakowiak et al., 2012). However, results among cohorts have been mixed. Several studies suggest that maternal obesity alone is not a risk factor; rather, maternal obesity *with* gestational diabetes or maternal obesity *with* excess gestational weight gain demonstrates associations with increased risk for ASD diagnosis in offspring (Gardner et al., 2015.; Li et al., 2016). Numerous cohort studies in Sweden, Denmark, and Finland have also demonstrated links between maternal obesity and increased risk of ADHD symptoms (Buss et al., 2012; Rodriguez et al., 2008; Rodriguez, 2010), although a majority have found these relations to be unique to teacher-reported ADHD symptoms rather than primary caregiver report. Offspring schizophrenia has also been linked to maternal prepregnancy BMI in both a US cohort (Schaefer et al., 2000) and a Finnish cohort (Jones et al., 1998), with the former study achieving statistical significance while the latter did not.

Relations between offspring neurodevelopmental outcomes and maternal obesity during the prenatal period are thought to be confounded by both biological and nonbiological factors including genetics and environment (Brion et al., 2011; Heikura et al., 2008), underscoring the importance of employing animal models in elucidating causal mechanisms. Critically, the extant evidence from nonhuman primate models and rodent models suggests that dietary interventions that induce maternal obesity, such as high-fat diets, result in offspring behavior changes (Patel et al., 2015) and cognition (memory and learning; White et al., 2009), consistent with neurodevelopmental differences such as ID. However, several preclinical studies using affective measures in rodents show mixed results, where maternal obesity may lead to increased or decreased anxiety-like behaviors and sociability (Hasebe et al., 2021), correlates of behavioral abnormalities noted in offspring of obese women.

Parental Age

Several other key factors during pregnancy are linked to later offspring IDD. For example, advanced maternal age has been associated with higher incidence of ASD (Durkin et al., 2008), Down syndrome (Yoon et al., 1996), and schizophrenia (Fountoulakis et al., 2018). Advanced maternal age has been well studied as a potential risk factor for IDD. A meta-analysis by Sandin et al. (2012) found an increased risk for ASD in offspring of mothers who were aged 35 years or older compared to offspring of younger mothers (between the ages of 25 and 29 years) during their pregnancy. Other medical risk factors, such as preeclampsia (Burstyn et al., 2010; Mann et al., 2010) and maternal diabetes (Gardener et al., 2009) may potentially mediate or moderate the association between advanced maternal age during pregnancy and IDD. Additionally, factors such as socioeconomic disadvantage and maternal multiparity have been implicated as a risk factor for IDD (Heikura et al., 2008). Of note, advanced parental age has also been linked to increased risk for neurodevelopmental disorders, including ASD and schizophrenia. Importantly, recent genomics research indicates that advanced paternal age increases the likelihood of sperm vulnerability to de novo mutations, thereby influencing vulnerability to neurodevelopmental disorders (Crow, 2000).

Both maternal and paternal age are risk factors associated with offspring IDD (Croen, 2007; Durkin & Rubenstein, 2021; Huang et al., 2016; Khandwala et al., 2018; Sandin et al., 2016; Wu et al., 2017). Research focusing on maternal age has been extensive, but studies focusing on paternal age have been scarce, despite both significantly influencing risk of offspring IDD (Croen, 2007; Khandwala et al., 2018). Lower maternal age has been associated with a 10% reduction in risk of ASD (Wu et al., 2017). However, both advanced and very young maternal age were independently associated with increased risk of ASD (Croen, 2007). For very young mothers, income and adversity play a significant role in lower education and diminished access to care and education, which delay diagnosis of IDD (Delobel-Ayoub et al., 2015; Sandin et al., 2016). Prior to the twenty-first century, advanced maternal age was considered for women over 30 (Huang et al., 2016); however, more recently women have delayed building families, so the definition of advanced maternal age has been shifted to women over 35 (Durkin & Rubenstein, 2021; Huang et al., 2016). For women over 35, the risk of ASD diagnosis is increased by 41% (Wu et al., 2017).

In recent years, paternal age has also been associated with IDD, namely ASD (Khandwala et al., 2018; Wu et al., 2017). Research has shown infants born to fathers aged 45 and older experienced more adverse perinatal outcomes such as lower gestational age, lower birthweight, increased risk of morbidity, and more likely to have a stay in the neonatal intensive care unit (Khandwala et al., 2018). Among fathers younger than 35, risk of ASD in offspring reduces by 20%; however, risk increases by 55% among offspring with fathers over 45 (Khandwala et al., 2018; Wu et al., 2017). Sandin et al. (2016) found the highest risk of an IDD diagnosis for offspring was when both parents were of advanced age. Furthermore, when there is a moderate to large difference between maternal and paternal age, risk of ASD in offspring is also higher than younger parental age by 18 to 21% (Sandin et al., 2016; Wu et al., 2017).

Understanding the mechanism of how advanced parental age is an increased risk of ASD is associated with several factors. Maternal age factors are associated with increased rates of genomic alterations, higher rates of obstetric and perinatal complications, and more time for exposure to environmental stressors (Durkin & Rubenstein, 2021; Wu et al., 2017). Male gametes are vulnerable to de novo mutations and epigenetic alterations (Croen, 2007; Sandin et al., 2016; Wu et al., 2017). Advanced maternal and paternal age has been associated with mutations on the X chromosome, increasing the likelihood to affect girls than boys, since both mothers and fathers contribute an X chromosome (Croen, 2007).

Maternal Mental Health

Prenatal maternal mental health can have a bidirectional relationship with IDD. The presence of mental health challenges may lead to delays in seeking diagnosis and intervention for developmental concerns in their child (Petterson & Albers, 2001). Conversely, the combination of the parent's mental health conditions and the inherent challenges of raising a child with IDD creates a compounding effect on both the well-being of the parent and the developmental trajectory of the child. Neuroimaging techniques have provided insights into the association between maternal depressive symptoms and the structural alterations in the developing brains of children (Lebel et al., 2016), thereby influencing the neurodevelopment of those with IDD. The following sections will explore specific effects of maternal mental health (and its treatments) as possible influencing factors in the development of IDD.

Prenatal Stress

Research has shown prenatal stress before and during pregnancy to be an independent factor in the elevated likelihood of ASD and ADHD. Several studies have demonstrated that maternal stress stimulates release of glucocorticoids that can pass through the placenta and impact the development of the fetal hypothalamic-pituitary-adrenal (HPA) axis (Matthews, 2000). This is important in the context of ASD and ADHD development because altered HPA activity, specifically hyperactivity, has been associated with increased ASD diagnoses, symptomology, and abnormal patterns of behavioral development (Corbett et al., 2006, Zelazo 2001). Additionally, factors related to mental health diagnosis, prenatal stress, lower social support, and trouble paying for basic needs elevated the likelihood of ASD in offspring at 3 years old. Maternal stressful events during pregnancy are associated with increased autistic traits and ADHD behaviors in offspring at 2 years of age (Ronald et al., 2011), in addition to increased rates of child ADHD (Robinson et al., 2023). In a predominantly urban, low-income, and minority prospective sample, children were 1.45 times more likely to receive an ADHD diagnosis if their mother experienced a major stressful event and 3.03 times more likely if mothers had reported a high level of perceived stress during pregnancy (Okano et al., 2019). The risk of ADHD also increased with the number of family adversities. Among children with ASD, exposure to prenatal stressful life events significantly predicted ASD-related symptom severity and communication abilities (Varcin et al., 2017). However, higher ASD symptom severity was only demonstrated in the context of exposure to two or more prenatal stressful life events.

Depression and Anxiety

Estimates of prenatal depression vary depending on the criteria used, with reports reaching approximately 50%, particularly in low-income samples (Chung et al., 2004). Children born to mothers with unipolar major depression are significantly more prone to IDD. This relationship seems to be even stronger with those with disabilities with a genetic basis (i.e., chromosomal, autosomal, and X-linked; Di Prinzio et al., 2018). Specifically, maternal depression during or prior to pregnancy is associated with an elevated likelihood of ASD (Kodesh et al., 2022). In a large cohort study, both maternal and paternal depression during prepregnancy and prenatal periods were associated with ADHD and ASD in offspring (HR = 1.42–2.25; Chen et al., 2020). Though chronicity of depression and additive effects of maternal and paternal depression increased elevated likelihood of ADHD and ASD for offspring, the effects of maternal depression were stronger than paternal depression effects. For ADHD, children were three times more likely to be diagnosed if birthed by mothers who had depression within one year prior to giving birth (Nidey et al., 2021).

Maternal prenatal depression can also indirectly elevate the risk of IDD by contributing to issues such as inadequate nutrition, unemployment, and insufficient prenatal care (Burd et al., 1999; Price & Caughey, 2022; Sidebottom et al., 2017). While there is a link between parental psychiatric history and the diagnosis of IDD, it is plausible that children born to parents with a diagnosed psychiatric disorder are more likely to be identified and diagnosed compared to children whose parents do not have a psychiatric diagnosis (Larsson et al., 2005).

Like depression rates, the prevalence of anxiety disorders in pregnancy varies according to studies, though it has been reported that over half of women reported anxiety symptoms during at least one trimester (Lee et al., 2007). While rates seem to be inconsistent, a significant portion of women experience prenatal anxiety both in general and about their pregnancy (Ross et al., 2006). Prenatal anxiety is an independent contributor to child ADHD symptoms, particularly in males, and social communication challenges, but child neurodevelopment also may be impacted by maternal prenatal anxiety and its coupling with other exposures and gene-environment interactions (O'Connor et al., 2022; O'Donnell et al., 2017; Shao et al., 2020). Prenatal anxiety additively contributes to ADHD symptomology with maternal immune activation (O'Connor et al., 2022). For example, maternal anxiety-induced elevation of C-reactive protein (CRP), a marker of systemic inflammation and part of the complement immune system, may have neurodevelopment consequences for the fetus, particularly for boys. Elevated CRP levels have been linked to alterations in brain development and function, which may increase the risk of ADHD symptoms later in childhood (Shao et al., 2020). Emerging evidence also suggests that the influence of prenatal maternal anxiety on early childhood risk for ADHD may also be mediated by other key biological pathways and/or genetic variants, such as catechol-o-methyltransferase (COMT), an enzyme involved in differences in neurotransmitter metabolism, which, in turn, can influence various cognitive and behavioral processes (O'Donnell et al., 2017).

Importantly, depression and anxiety are highly comorbid; women who report high prenatal anxiety also had high scores of depression (Field et al., 2003). Maternal mood disorders such as these have been shown to activate the maternal HPA axis and program the HPA axis and physiology of the developing fetus (Charil et al., 2010; Dickens & Pawluski, 2018; Glover et al., 2010). Alterations in this development in turn can raise the risk of IDD. Furthermore, prenatal anxiety and depression have been linked to preterm birth, gestational age, and birth complications (Accortt et al., 2015; Ding et al., 2014), factors associated with an increased risk of child IDD—as discussed in previous sections. These indirect associations make it challenging to disentangle causality and are an important consideration for future work (Segenreich et al., 2015).

Medications for Depression and Anxiety

Given that poor maternal mental health is shown to enhance risk for offspring disability, there is increasing research on the risk and benefits of using medications to treat neuropsychiatric and mental health disorders in pregnant women. It is now appreciated that prior standards of care, involving mood stabilizers and antiepileptic medications, may have adverse effects on offspring development and increase risk for cognitive delays and lower IQ (Bromley et al., 2014).

Antidepressant medications, especially serotonin selective reuptake inhibitors (SSRIs), are now the most common form of pharmacological treatment for depression and anxiety in pregnancy, and past decades have seen the rise in use of these medications in pregnant women and the general population (Olfson & Marcus, 2009; Andrade et al., 2008; Bakker et al., 2008). Because antidepressants cross the placenta, research explores if and to what extent these forms of medications alter fetal and infant outcomes in the short and long term. Animal work shows that exposure to antidepressants and high levels of serotonin can influence fetal brain growth (Gaspar et al., 2003). Observational work in humans has drawn links between prenatal antidepressant exposures and pregnancy and birth risks (Lund et al., 2009; Oberlander et al., 2006; Victorin et al., 2016; Wen et al., 2006), reduced birth weight, and neurodevelopmental disorders, such as ASD following birth (Andalib et al., 2017; Seebeck et al., 2023). The use of SSRIs in the second or third trimester of pregnancy increases likelihood of ASD, even after accounting for depression history (Boukhris et al., 2016). However, the associations, especially with later risk for IDD (and ASD), in offspring are inconsistent. Some of the more rigorous large-scale studies show that associations may be better explained by confounding factors that underlie the need for treatment rather than due to effects of antidepressant medications (Victorin et al., 2017). These include non-mutually exclusive influences, including causal effects of exposure, genetics, environmental causes, and indications for antidepressant use (Sujan et al., 2019). While antidepressant exposure may increase the likelihood of neurodevelopmental disorders, the risk associated may be less than expected and due to interactive influences that need further study (Ackerman et al., 2017).

Traumatic Stress Exposure

Posttraumatic stress disorder (PTSD) occurs after experiencing or witnessing a traumatic event or set of circumstances, marked by disturbing thoughts and feelings that endure long after the event has passed. One review reported that the prevalence of PTSD during pregnancy is documented at 3.3% in community samples and rises to 18% in high-risk samples. Four percent of individuals report symptoms of PTSD during the postpartum period, primarily associated with traumatic events experienced during childbirth (Yildiz et al., 2017). The majority of pregnant women who met diagnostic criteria for PTSD from a traumatic event were reported to be due to intimate partner violence and domestic violence (see section below; Harris-Britt et al., 2004). PTSD during pregnancy is linked to neuroendocrine alterations (Yehuda, 2005), potentially predisposing individuals to birth complications and infants to a biological vulnerability (Friedman & McEwen, 2004; Marshall & Garakani, 2002). Additionally, maternal exposure to environmental traumatic triggers could induce in utero epigenetic alterations, a theorized mechanism of the development of offspring psychiatric disorders (Varela et al., 2022). PTSD symptoms have been associated with 2–3 times greater risk of child's diagnosis of ASD (Roberts et al., 2014). It has been theorized that maternal PTSD and child ASD may have a shared genetic risk, although few shared genetic variants have been identified (Miller et al., 2009). Importantly, PTSD is also associated with high rates of psychiatric comorbidities, such as anxiety and depression (see previous sections for relationship of anxiety and depression and child IDD; Keane et al., 1997). While general maternal PTSD diagnosis has implications for child outcomes, the following sections will delve into specific instances of trauma, such as childhood maltreatment and domestic violence.

Childhood Maltreatment Childhood maltreatment (CM) is defined by any type of abuse (i.e., physical, emotional, and sexual) and neglect experienced during childhood and has deleterious consequences on mental and physical health persisting throughout the lifespan (Afifi et al., 2009; Bryce & Collier, 2022; Hansen & Olff, 2017). Research indicates that exposure to CM not only affects the individuals directly subjected to it but also has implications on the subsequent generations (Arditti & Strat, 2020; Rijlaarsdam et al., 2014). As it pertains to IDD, maternal exposure to childhood abuse has been associated with increased risk for ASD in their children. This relationship has been shown to increase monotonically, such that higher exposure of combined types of childhood abuse exposure raises the likelihood of having a child with ASD (Roberts et al., 2013). Notably, the impact of maternal CM on child outcomes appears to be more pronounced if the maternal exposure occurred during her early childhood rather than at a later stage in life (Thompson, 2007). Maternal CM may also contribute to the development of depression (Heim, 2018) and PTSD, as assessed through behavioral and neuroimaging measures (Killion & Weyandt, 2020; Wieck et al., 2014), in addition to increased genetic loading for ADHD and ASD (Ratanatharathorn et al., 2021). As highlighted in other sections of this chapter, these maternal conditions may increase the risk of IDD for the child. This body of research underscores the complexity of the intergenerational and lifelong repercussions of CM on maternal and child outcomes, emphasizing the need for comprehensive understanding and intervention strategies.

Domestic Violence With a lifetime prevalence of approximately 25% for women in general and almost 10% for pregnant people specifically, intimate partner violence (IPV) and domestic violence (DV) are significant public health problems (Hahn et al., 2018; Truman & Morgan, 2014). Cases of violence against women are rising globally, and the World Health Organization has declared it an urgent concern (Garcia-Moreno & Watts, 2011). Despite this urgency, there is a scarcity of research on DV and IPV, especially with respect to the long-term consequences of perinatal exposure to violence (Mueller & Tronick, 2019). Women affected by violence often fear the consequences of speaking out or participating in research, such as retaliation from their abuser and the societal stigma that comes with IPV. During pregnancy or if children are involved, the potential involvement of official agencies, such as Child Protective Services, is of concern as well.

Despite these challenges, a growing body of literature shows that exposure to DV or IPV during pregnancy is associated with poor outcomes for the fetus, outcomes that overlap with known major risk factors for IDD (Boy & Salihu, 2004; Coker et al., 2004; Cokkinides et al., 1999; Rosen et al., 2007). Across studies, violence during pregnancy has been found to substantially increase the risk for low birth weight or very-low birth weight in the offspring (Lipsky et al., 2003; Rosen et al., 2007; Silverman et al., 2006), which is also significantly connected to the risk to develop ASD, developmental delays, and cognitive deficits, as shown in large population-based studies worldwide, such as Sweden (Hultman et al., 2002), Denmark (Larsson et al., 2005), Finland (Lampi et al., 2012), and the United States (Schendel & Bhasin, 2008). Similarly, a population-based study in the State of Washington reported a three-fold increased risk for preterm or very-preterm birth after exposure to IPV during pregnancy (Lipsky et al., 2003), which has been shown to be associated with an increased likelihood of developing ASD (Chang et al., 2023), and conditions associated with neurodevelopment, such as sensory impairment and IDD (Schieve et al., 2016; Tammela, 2022). The only study to date that directly investigated the impact of IPV during pregnancy on offspring neurodevelopmental disorders reported a moderate association between long-term exposure to violence before birth and an elevated risk for an ASD diagnosis in a population-based cohort study that assessed exposure to IPV via a retrospective questionnaire (Roberts et al., 2016).

While obstetric outcomes can be directly influenced by IPV and successively increase the risk for later IDD of the child, exposure to violence may also lead to a chronic increase of psychosocial stress levels, which have been shown to have a potentially negative affect on fetal development through neuroendocrine or immuno-metabolic pathways (Sarkar, 2008). Next to current experience of violence, a history of childhood DV has also been reported to be associated with increased stress levels during pregnancy and a rise in subsequent risk for developmental disorders, such as ASD or ADHD (Roberts et al., 2013, 2018). Even short-term stressors, such as exposure to a hurricane, have been reported to elevate the ASD diagnosis in the affected population; hence, a chronic stressor such as DV or IPV is assumed to have the potential for a similar effect, despite a scarcity of research studies (Kinney et al., 2008). The psychobiological pathways underlying adverse pregnancy outcomes and developmental delays are not well understood (Mueller & Tronick, 2020; Yim & Kofmann, 2019). Chronic psychosocial stress, such as exposure to violence, has been reported to promote immune activation through stress-induced cytokine production and glucocorticoid resistance (Maes et al., 1999; Schmidt et al., 2010). This hypothesis aligns a body of literature indicating that experiences of childhood DV and adult IPV are connected to persistent activation of inflammatory markers (Danese et al., 2007; Fernandez-Botran et al., 2011; Newton et al., 2011; Robertson Blackmore et al., 2016). Thus, immune activation and inflammation during pregnancy may be a key mechanism through which exposure to violence is able to affect fetal development and the offspring's susceptibility to developmental disorders (Braunschweig et al., 2013).

Prenatal Drug Exposure

Prenatal drug exposure is long known to increase risk for neurodevelopmental delay and IDD. In the United States, approximately 5% of fetuses are exposed to legal or illicit substances in utero, with tobacco and alcohol shown to be more prevalent than other substances (Behnke & Smith, 2013). Animal and human work suggest that exposure to various substances can have both direct and indirect effects on fetal brain maturation. For example, direct effects of psychoactive drugs appear to include alterations in monoaminergic neurotransmitter signaling and circuit formation, disruption in neural proliferation and migration, axonal growth, and dendritic branching. Direct exposure may have neurotoxic effects on the developing brain by altering the supply of nutrients and oxygen (Ross et al., 2015). Recent longitudinal work has demonstrated alterations in functional connectivity patterns in neonates who were prenatally exposed to cocaine and marijuana, with some evidence for specificity in effects. For example, cocaine exposure has been uniquely associated with alteration in frontal limbic circuitry and thalamo-cortical connections (Salzwedel et al., 2015), areas that seem to be relevant to the expression of psychopathology (Kalin, 2019). Exposure to marijuana has been associated with disruptions in regions that have high expression of certain cannabinoid receptors, specifically in striatal connectivity, which modulates fetal neurodevelopment (Grewen et al., 2015). Fetal exposure to multiple substances is common and has been associated with early hyper- and hypoconnectivity patterns throughout the brain in neonates that often subserve behaviors critical for later success (Grewen et al., 2015; Salzwedel et al., 2015, 2016).

Tobacco Exposure

Tobacco exposure is a modifiable risk factor for poor health outcomes. A systemic review and meta-analysis of smoking prevalence across all countries by Lange et al. (2018) found that global prevalence of smoking during pregnancy was approximately 1.7% (95% CI 0.0–4.5), with the European region having the highest prevalence rates (8.1%). Importantly, among pregnant people globally, approximately half of women who smoked daily prior to pregnancy continued to smoke during pregnancy (Lange et al., 2018), constituting a potential avenue of targeted cessation intervention. Prenatal exposure to tobacco—through either smoking during pregnancy or secondhand smoke—demonstrates robust effects on offspring brain alterations and neurobehavior as early as infancy (Froggat et al., 2020). The human fetal brain is particularly sensitive to nicotine, affecting synaptic activity (Dempsey & Benowitz, 2001), thereby potentially affecting offspring behavioral and neurocognitive outcomes. Tobacco exposure during pregnancy has been linked to behavior problems, poorer academic achievement (Clifford et al., 2012), and ADHD (Huang et al., 2018) in offspring. Prenatal tobacco exposure has also been linked to variation in offspring IQ, although literature examining other aspects of neurocognition, such as working memory, demonstrates mixed findings (Madley-Dowd et al., 2022). More recent evidence also suggests that tobacco exposure is not associated with offspring risk for ASD (Jung et al., 2017; Rosen et al., 2015), thus marking a specificity of potential risk for neurocognitive differences. While findings seem to be mixed regarding the association between prenatal maternal tobacco exposure and ID, a 2016 meta-analysis points to an increase in ID risk (Huang et al., 2016). One U.S. cohort study examining these associations found biological sex differences in

IDD risk while controlling for maternal education, such that male children exposed to 20 or more cigarettes per day during gestation were at greater risk of ID than unexposed males (Braun et al., 2009). Interestingly, females did not demonstrate an elevated risk (Braun et al., 2009), constituting a potential female protective effect. However, this study has been criticized for not controlling for other known confounds, and recent findings from a Danish cohort study suggest that associations between maternal smoking and increased odds for ID are attenuated to null effects following adjustment for confounds (Madley-Dowd et al., 2022). Given these confounds in human studies, experimental investigations with animal models have been employed to ascertain causal mechanisms. Rodents prenatally exposed to nicotine have demonstrated atypical neurobehavior including deficits in attention and increased motor activity (Ernst et al., 2001), consistent with ADHD symptomology in humans. Experimental animal models examining other IDD diagnoses, studies examining subgroups more affected by tobacco exposure, and the potential critical windows that may influence dose-response remain to be fully explored.

Alcohol Exposure

Globally, approximately 9.8% of birthing parents consume alcohol during pregnancy (Popova et al., 2017). Exposure to alcohol during pregnancy is well-documented to have poor offspring health outcomes. Fetal alcohol spectrum disorder (FASD) is an umbrella term for a group of neurodevelopmental conditions including fetal alcohol syndrome, alcohol-related birth defects, neurobehavioral disorder associated with prenatal alcohol exposure, and alcohol-related neurodevelopmental disorder. These conditions describe offspring neurodevelopmental and health outcomes associated with prenatal exposure to alcohol, with varying phenotypes. FASD is diagnosed by the following: (1) a history of prenatal alcohol exposure—although confirmation is not required to make a diagnosis, (2) height/weight growth deficiency during the prenatal or postnatal period, (3) dysmorphic cardinal facial features, and (4) central nervous system impairment including neurocognitive, behavioral, or adaptive deficits (Hoyme et al., 2016). The mechanistic pathway of FASD is due entirely to prenatal exposure to alcohol (Larkby, 1997) and thus constitutes the most preventable cause of IDD in offspring. The amount of alcohol exposure and timing of exposure (i.e., during specific trimesters or throughout the pregnancy) demonstrates regional variation and global subpopulations demonstrating a significantly higher prevalence of FASD rates compared to the general population (Popova et al., 2019). The risk for FASD in offspring is most likely among pregnant people who binge drink, typically defined as consuming four or more drinks per occasion or having a blood alcohol content (BAC) level of 0.08 grams percent or above (NIAAA, 2004). In some studies of FASD, binge drinking has been defined as three or more drinks per occasion in much of the FASD literature due to high correlations with child outcomes (May et al., 2007, 2008). Currently, there is no known safe amount of alcohol an individual can drink while pregnant (Williams & Smith, 2015).

Opioid Use

Opioid use during pregnancy can lead to neonatal opioid withdrawal syndrome (NOWS) in newborn infants, which manifests as hyperactivity, feeding difficulties, vomiting, and diarrhea. Treatment for NOWS involves a combination of pharmacological interventions (morphine and buprenorphine; De Young et al., 2020) and behavioral interventions focusing on skin-to-skin contact and breastfeeding. Prenatal opioid exposure enhances risk for premature birth and lower birthweight, (Graeve et al., 2022), which may contribute to higher likelihood of cognitive and motor delays, though long-term consequences are not yet fully established, with heterogeneity in findings (Yeoh et al., 2019; Balalian et al., 2023). Although there is heterogeneity among studies, research indicates that higher prenatal opioid exposure and exposure to opioids for longer durations or higher doses overtime (Wen et al., 2021) are generally associated with neurodevelopmental problems (Balalian et al., 2023). Specifically, prenatal opioid prescriptions were associated with increased odds of ASD and developmental delays with ASD features, though this relationship was not observed when the opioid exposure occurred in the first trimester (Rubenstein et al., 2019). These findings suggest not just the importance of exploring opioid exposure (both prescribed and recreational), but also the timing and chronicity of exposure.

Environmental Factors

Pregnant people and their unborn children have an increased vulnerability to environmental exposures compared to the nonpregnant population, and environmental pollutants have been shown to increase the risk for adverse birth outcomes in general and the likelihood of developmental disorders specifically. The following sections will explore the relationship between child outcomes and air pollution and chemical exposures. However, these hazards represent only a small part of potential environmental toxins that may hinder development and impact rates of IDD.

Air Pollution

A growing body of evidence associates prenatal exposure to air pollution with significantly decreased birthweight, preterm and very-preterm birth, and intrauterine growth restriction (Sram et al., 2005; Ritz et al., 2000), all of which are critical risk factors for subsequent neurodevelopmental disorders, such as ASD (Lampi et al., 2012; Sacchi et al., 2021). Several studies have aimed to investigate in detail how in utero exposure to fine and ultrafine particles can have a possible neurodegenerative effect. One study in Poland found that prenatal exposure to polycyclic aromatic hydrocarbon (PAH), an air pollutant released by diesel- and gasoline-powered vehicles, heating, cooking, and secondhand smoke, is associated with a substantial reduction in intelligence and cognitive skills at school entry (Edwards et al., 2010). PAH exposure during pregnancy in New York City was similarly related to considerably greater odds of cognitive delay in children exposed to higher levels of PAH (Perera et al., 2006). Similar results were found in a Boston-based cohort after evaluating their exposure to black carbon, a traffic-related air pollutant (Suglia et al., 2008). The authors found that even after adjusting for known risk factors such as birthweight, black carbon exposure was significantly associated with a decreased vocabulary and lower scores on a composite intelligence measurement. Moreover, they found a dose-dependent effect, with higher levels of black carbon exposure predicting lower levels of cognitive functioning across development. A pioneering study from Mexico City reported a considerable connection between exposure to air pollution during pregnancy and cognitive deficits in children, with a subset of affected children having visible prefrontal white matter lesions (Calderon-Garciduenas et al., 2008). The authors replicated the lesions found in children by scanning dogs from highly polluted areas in Mexico City, showing frontal lesions to a similar percentage as the children of the study and no lesions in dogs from low pollution areas. The conclusions of this study include that possible effects can be observed across species and that the prefrontal cortex may be one of the brain regions most vulnerable to air pollutant exposure during pregnancy and the associated cognitive and intellectual deficits observed in children.

However, different air pollutants are not just affecting intellectual and cognitive development in general—an increasing body of literature connects air pollution distinctly to ASD. Becerra and colleagues (2013) found in a retrospective study of over 7000 children with ASD an up to 15% rise in odds for an ASD diagnosis per interquartile increase particulate matter exposed to during pregnancy. This result held, even when controlling for birth weight, gestational age, sex, and SES. Similar results were found in a study looking at ambient air pollution around the birth residences of children diagnosed with ASD and found a predictive relationship between metal and solvent concentrations in air and subsequent diagnosis of ASD (Windham et al., 2006). Another study did not find a significant relationship between metal pollutants, but an increased odds ratio for methylene chloride, quinoline, and styrene and ASD diagnosis by age 8 (Kalkbrenner et al., 2010), and comparable findings are mounting in recent literature (Kalkbrenner et al., 2015; Klepac et al., 2018; Raz et al., 2015; von Ehrenstein et al., 2014).

While the underlying mechanism connecting air pollution with developmental delays and ASD is not clear, it is assumed the immune system may play a central role, as PAH and fine particle exposure during pregnancy have been linked to altered immunoglobin E levels and lymphocyte immunophenotypic disruptions in cord blood (Hertz-Picciotto et al., 2022). Prenatal exposure to air pollutants has also been associated with fetal lymphocyte maturation, a critical part of immune activation and antibody production (Hertz-Picciotto et al., 2022). In addition, translational studies in rodents and humans have identified a polymorphism in the tyrosine receptor gene as a potential interaction between genes and the environment contributing to an increased susceptibility for adverse developmental outcomes after prenatal exposure to air pollution (Volk et al., 2014). Individuals with the CC genotype of the MET receptor tyrosine K gene had a significantly increased risk for an autism-related endophenotype, but only when they were exposed to high concentrations of air pollutants. These results indicate that vulnerability to cognitive delays and developmental disorders after prenatal exposure to pollutants is a complex interplay of genetic susceptibility and air pollutants in a dose-dependent manner.

Other Chemical Exposures

Environmental agents have also been known to alter neurodevelopmental and cognitive functioning. It is increasingly evident that fetal exposure to environmental toxins increases risk for IDD and other neurodevelopmental disabilities, including ASD, schizophrenia, and ADHD. In fact, over 200 industrial applied or produced chemicals have been identified as neurotoxic, especially when exposure occurs during sensitive periods of brain development (Grandjean & Landrigan, 2006). Sources of environmental toxicant exposure include air pollution, contaminated drinking water, soil, and pesticides on food. When exposed pre- and perinatally, these chemicals pass through the fetal blood-brain barrier (BBB) or postnatally through breast-milk (Grandjean & Landrigan, 2006) to influence fetal and infant brain and physiological development. Common neurotoxicants of concern include polychlorinated biphenyls (PCBs), methylmercury (MeHg), lead, arsenic, pesticides, and bisphenol A (BPA). Etiological pathways connecting neurotoxicants to risk for IDD are varied and complex. For example, PCBs are common toxicants that are produced industrially and accumulate in the food chain (Winneke, 2011). Prenatal exposure to PCBs via food contamination has been linked with intellectual impairment and related IDD (Jacobson & Jacobson, 1996; Lai et al., 2002), via disruptions in neurotransmitter systems (i.e., dopamine, glutamate) that support synaptic transmission and plasticity (Gilbert, 2003). PCB also appears to disrupt maternal thyroid hormone production, which is necessary for healthy fetal brain development (Porterfield, 2000; Wise et al., 2012). Prenatal MeHg is another common exposure that occurs primarily via maternal consumption of contaminated fish and can have neurotoxic effects on fetal brain development, as it easily crosses the placenta and fetal BBB, where it accumulates in brain tissue (Grandjean & Landrigan, 2014; Mendola et al., 2002). Mechanistically, MeHg is thought to impair neurotransmitter systems, alter excitatory/inhibitory balance and associated synaptic plasticity, and lead to thyroid disruption (Castoldi et al., 2001; Johansson et al., 2007; Newland et al., 2015). Prenatal exposure to lead is one of the longest-studied neurotoxicants with significantly historical widespread exposure due to its use in various household products (Grandjean & Landrigan, 2006). Lead exposure during pre- peri-, and postnatal periods similarly increased risk for IDD (Symeonides et al., 2014). Mechanistically, lead exposure crosses the BBB and is demonstrated to interfere with normative neuronal differentiation, myelination and synaptogenesis, and plasticity by affected glutamate release and NMDA receptors (Costa et al., 2004; Dietrich et al., 2001; Froehlich et al., 2009). Arsenic exposure can occur from industrial pollution or contaminated drinking water from naturally occurring arsenic in groundwater and wells. Although prenatal exposure to arsenic is related to risk for IDD, longer-term exposure over childhood tends to be associated with greater risk (Tolins et al., 2014; Tyler & Allan, 2014). Arsenic appears to have a neurotoxic effect by altering levels of various neurotransmitters (i.e., glutamate, GABA, dopamine, serotonin) and altering plasticity and synaptogenesis (Tyler & Allen, 2014). Prenatal exposure to pesticides of varying types has been associated with risk for IDD and related neurodevelopmental problems including ASD (Cheslack-Postava et al., 2013; Robert et al., 2007; Shelton et al., 2014). Effects appear to depend on the timing of exposure of some pesticides. For example, exposure to chlorpyrifos during the second trimester or organophosphate and pyrethroid during the third trimester of pregnancy are more strongly linked with IDD risk than other pre- or postnatal phases of life (Shelton et al., 2014). Neurotoxicants appear to disrupt numerous processes necessary for healthy fetal brain maturation, including disrupting maternal thyroid levels (necessary for fetal neurodevelopment), increase neuroinflammation and immuno-suppression, induce oxidative stress (Fonnum et al., 2006; Rossignol & Frye, 2012), alter fetal neurotransmitter systems (i.e., dopamine and glutamate), and disrupt synaptic transmission and plasticity that occurs prior to and following birth (Gafni et al., 2004).

Systemic and Socioeconomic Factors

In recent years, sociodemographic risk factors have been identified as playing a significant role in understanding the causes of IDD through identifying determinants of health and quality of life (Durkin & Yeargin-Allsopp, 2018). Very little is known about the exact causes of many IDD, including ASD; yet there is limited research focused on SES as a cause of IDD and the studies that do exist tend to have significant methodological flaws, including small sample sizes and confounders (Croen, 2007; Rai, 2012). Yet despite these challenges, there are some key themes that emerge from the literature that elucidate the sociodemographic factors, including parental age, family income, level of education, and race/ethnicity (Camp et al., 1998; Croen, 2007; Durkin & Rubenstein, 2021; Huang et al., 2016; Rai, 2012).

This interaction between sociodemographic factors and biological determinants contributes to the development of IDD. Biological risk factors such as genetics and perinatal adverse outcomes, such as preterm birth and low birthweight, are strengthened due to environmental factors, which may be associated with sociodemographic factors (Durkin & Yeargin-Allsopp, 2018). Low SES mothers have decreased access to education, less knowledge of medical treatments and screenings, and less prenatal care (Durkin & Yeargin-Allsopp, 2018). These inequities influence decisions about terminating pregnancies in the case of fetal genetic anomalies and using in vitro fertilization to select healthy embryos, as well as diminished access to nutritious foods and macronutrients, medical care and vaccines, and exposure to toxins due to poor home conditions (Durkin & Yeargin-Allsopp, 2018).

Regarding associations between SES and IDD, there are mixed results in the literature. In the United States and Australia, the majority of studies show a positive relationship between SES and ASD diagnosis, meaning that families with higher SES and education had higher risk for an ASD diagnosis (Delobel-Ayoub et al., 2015; Durkin et al., 2010; Rai, 2012). Conversely, in Europe, studies found higher ASD diagnoses among families with low SES, including studies in Sweden and France (Delobel-Ayoub et al., 2015; Rai, 2012). Delobel-Ayoub et al. (2015) identified that the highest prevalence of ASD in France was areas of the highest deprivation, highest adult unemployment, lower education, immigrants, and single-parent households. In a Swedish study focused on children with ASD, a disadvantaged background (i.e., lower household income, parental manual labor, and parental low education) was associated with higher prevalence of ASD, even after controlling for all other potential causes (Rai, 2012). Explanation for the significant differences between study results in these different regions might be due to access to patient medical records in countries with universal healthcare, allowing researchers to assess entire populations for ASD diagnoses (Rai, 2012). Furthermore, systemic struggles associated with lower income and access to qualified healthcare providers prevent timely diagnoses of IDD, which might contribute to the lower prevalence rates of IDD in lower SES families in the United States (Huang et al., 2016; Rai, 2012; Thomas et al., 2012).

Another significant demographic factor influencing the prevalence rates of IDD is race and ethnicity (Camp et al., 1998; Durkin & Rubenstein, 2021; Emerson, 2012; Thomas et al., 2012). Some genetic mutations are specific to certain ethnic or racial groups due to isolation, consanguinity, and the founder effects (Durkin & Rubenstein, 2021). Studies focused on this association are limited, but there are some data that provide an impression of how race and ethnicity affect prevalence rates of IDD. Camp et al., (1998) identified that Black mothers, with low SES, as having offspring with higher levels of developmental disorder compared to White mothers. Among middle-to-high SES Black mothers, APGAR scores of less than two were associated with increased prevalence of IDD, with an even higher prevalence among Black mothers with low SES (Camp et al., 1998). However, once prevalence rates are stratified by SES, rates of IDD slightly change. Sex of the offspring and race/ethnicity were associated with mild and severe forms of IDD; specifically, among Black children, mild IDD were associated with economic disadvantage and higher likelihood of adverse perinatal factors (Durkin & Rubenstein, 2021). Other studies found that Black and White families had similar rate of IDD diagnosis; however, the differences arose in regards to diagnosis, where Black children were diagnosed later than white children (Gourdine et al., 2011).

Interestingly, some studies found that when SES was stratified, the rates of IDD were similar across ethnic and racial groups (Durkin et al., 2010; Emerson, 2012; Gourdine et al., 2011; Thomas et al., 2012). For ASD diagnoses in the United States, SES stratification was also influenced by the spectrum of severity of symptoms (Leonard et al., 2011). Among all families with ASD diagnosis, the lowest prevalence was found among those in the lowest SES group, whereas the highest SES groups had the highest prevalence of ASD diagnosis (Durkin et al., 2010). These stratifications were possible due to the SES diversity in the United States. In countries where diversity does not exist, there are no associations between ASD and SES (Durkin et al., 2010). In contrast, stratification of SES yielded divergent results compared to the U.S. results. Emerson (2012) found that among the lowest SES stratum, prevalence of IDD was the highest, including higher levels of deprivation and disadvantage.

Sociodemographic factors play an instrumental role in the prevalence of IDD. Although they are not the direct cause of IDD, SES affects how families receive care, diagnosis, or education about these symptoms (Thomas et al., 2012). Furthermore, exposure to adversity, violence, and increased parental stress also contributed to IDD diagnoses, creating barriers to care, less recognition of IDD symptoms by parents and teachers, and interventions to support offspring with IDD (Dodds, 2020). Therefore, parental age, SES, and minoritized status are important considerations when assessing risk factors of IDD.

Summary and Conclusion Status

This chapter reviews the intricate complexity of environmental insult during the prenatal and perinatal period of fetal brain development. Exposure to maternal infection, inflammation, chronic illnesses, and emergence of new maternal mental health conditions highlight the potential for complex interactions when multiple risk factors coincide during pregnancy. During the fetal months, the consequences of environmental exposure underscore the urgent need to address environmental risks to ensure favorable developmental outcomes. Furthermore, the lasting impact of traumatic stressors, childhood maltreatment, and domestic violence underscores the cyclic nature of violence and its profound effects on mental health trajectories across generations. Prenatal exposure to substances such as alcohol, nicotine, and marijuana emphasizes the necessity for a nuanced understanding due to their differential effects on neurodevelopmental pathways and resulting variations in phenotypes. Importantly, preventing prenatal exposure to teratogens, including illicit drugs, alcohol, and certain prescribed medications, is feasible. However, this requires addressing underlying issues such as women feeling they have limited coping options for stress, trauma, or mental health challenges. Pregnant women with untreated mental health disorders may be at increased risk of substance use to alleviate symptoms or numb emotional pain. Increasing support from healthcare providers, family, and partners to address substance use during pregnancy can empower women to make healthier choices for themselves and their babies.

Disparities in developmental outcomes disproportionately affect children from low SES families, perpetuating inequalities in academic achievement, career opportunities, and overall quality of life. These delays often manifest within the first year of life, a critical period of rapid brain development where targeted interventions can have a significant impact on child outcomes. Recognizing the multifaceted influences and vulnerabilities underscores the importance of comprehensive, interdisciplinary approaches to promote optimal neurodevelopment and mitigate the long-term consequences of early life adversities. Ultimately, this research underscores the critical role of maternal care in shaping fetal development and subsequent outcomes.

References

Accortt, E. E., Cheadle, A. C. D., & Dunkel Schetter, C. (2015). Prenatal depression and adverse birth outcomes: An updated systematic review. *Maternal and Child Health Journal, 19*(6), 1306–1337. https://doi.org/10.1007/s10995-014-1637-2

Ackerman, S., Schoenbrun, S., Hudac, C., et al. (2017). Interactive effects of prenatal antidepressant exposure and likely gene disrupting mutations on the severity of autism spectrum disorder. *Journal of Autism and Developmental Disorders, 47*, 3489–3496. https://doi.org/10.1007/s10803-017-3246-6

Afifi, T. O., Boman, J., Fleisher, W., & Sareen, J. (2009). The relationship between child abuse, parental divorce, and lifetime mental disorders and suicidality in a nationally representative adult sample. *Child Abuse and Neglect, 33*(3), 139–147. https://doi.org/10.1016/j.chiabu.2008.12.009

Amgalan, A., Andescavage, N., & Limperopoulos, C. (2021). Prenatal origins of neuropsychiatric diseases. *Acta Paediatrica, 110*(6), 1741–1749. https://doi.org/10.1111/apa.15766

Andalib, S., Emamhadi, M. R., Yousefzadeh-Chabok, S., Shakouri, S. K., Høilund-Carlsen, P. F., Vafaee, M. S., & Michel, T. M. (2017). Maternal SSRI exposure increases the risk of autistic offspring: A meta-analysis and systematic review. *European Psychiatry, 45*, 161–166.

Andrade, S. E., Raebel, M. A., Brown, J., Lane, K., Livingston, J., Boudreau, D., Rolnick, S. J., Roblin, D., Smith, D. H., Willy, M. E., Staffa, J. A., & Platt, R. (2008). Use of antidepressant medications during pregnancy: A multisite study. *American Journal of Obstetrics and Gynecology, 198*(2), 194.e1–194.e5. https://doi.org/10.1016/j.ajog.2007.07.036

Arditti, R., & Strat, Y. L. E. (2020). A traumatic life experience in childhood increases the risk of a psychiatric disorder in the offspring. *Psychiatry Research, 290*, 113101. https://doi.org/10.1016/j.psychres.2020.113101

Avorgbedor, F., Silva, S., Merwin, E., Blumenthal, J. A., & Holditch-Davis, D. (2019). Health, physical growth, and neurodevelopmental outcomes in preterm infants of women with hypertensive disorders of pregnancy. *JOGNN - Journal of Obstetric, Gynecologic, and Neonatal Nursing, 48*(1), 69–77. https://doi.org/10.1016/j.jogn.2018.10.003

Bakker, M. K., Kölling, P., Van Den Berg, P. B., De Walle, H. E. K., & De Jong Van Den Berg, L. T. W. (2008). Increase in use of selective serotonin reuptake inhibitors in pregnancy during the last decade, a population-based cohort study from The Netherlands. *British Journal of Clinical Pharmacology, 65*(4), 600–606. https://doi.org/10.1111/j.1365-2125.2007.03048.x

Balalian, A. A., Graeve, R., Richter, M., Fink, A., Kielstein, H., Martins, S. S., Philbin, M. M., & Factor-Litvak, P. (2023). Prenatal exposure to opioids and neurodevelopment in infancy and childhood: A systematic review. *Frontiers in Pediatrics, 11*, 1071889. https://doi.org/10.3389/fped.2023.1071889

Behnke, M., Smith, V. C., & Committee on Substance Abuse, & Committee on Fetus and Newborn. (2013). Prenatal substance abuse: Short- and long-term effects on the exposed fetus. *Pediatrics, 131*, e1009–e1024. https://doi.org/10.1542/peds.2012-3931

Bliddal, M., Olsen, J., Støvring, H., Eriksen, H. L. F., Kesmodel, U. S., Sørensen, T. I. A., & Nøhr, E. A. (2014). Maternal pre-pregnancy BMI and intelligence quotient (IQ) in 5-year-old children: A cohort-based study. *PLoS ONE, 9*(4), e94498. https://doi.org/10.1371/journal.pone.0094498

Brand, J. S., Lawlor, D. A., Larsson, H., & Montgomery, S. (2021). Association between hypertensive disorders of pregnancy and neurodevelopmental outcomes among offspring. *JAMA Pediatrics, 175*(6), 577–585. https://doi.org/10.1001/jamapediatrics.2020.6856

Braun, J. M., Daniels, J. L., Kalkbrenner, A., Zimmerman, J., & Nicholas, J. S. (2009). The effect of maternal smoking during pregnancy on intellectual disabilities among 8-year-old children. *Paediatric and Perinatal Epidemiology, 23*(5), 482–491. https://doi.org/10.1111/j.1365-3016.2009.01056.x

Braunschweig, D., Krakowiak, P., Duncanson, P., Bpyce, R., Hansen, R. L., Ashwood, P., Hertz-Picciotto, J., Pessah, J. N., & Van de Water, J. (2013). Autism-specific maternal autoantibodies recognize critical proteins in developing brain. *Translational Psychiatry, 3*, e277. https://doi.org/10.1038/tp.2013.50

Brion, M. J., Zeegers, M., Jaddoe, V., Verhulst, F., Tiemeier, H., Lawlor, D. A., & Smith, G. D. (2011). Intrauterine effects of maternal prepregnancy overweight on child cognition and behavior in 2 cohorts. *Pediatrics, 127*(1), e202–e211. https://doi.org/10.1542/peds.2010-0651

Bromley, R., Weston, J., Adab, N., Greenhalgh, J., Sanniti, A., Mckay, A. J., Tudur Smith, C., & Marson, A. G. (2014). Treatment for epilepsy in pregnancy: Neurodevelopmental outcomes in the child. *Cochrane Database of Systematic Reviews, 10*, CD010236. https://doi.org/10.1002/14651858.CD010236.pub2

Bryce, I., & Collier, S. (2022). A systematic literature review of the contribution accumulation makes to psychological and physical trauma sustained through childhood maltreatment. *Trauma Care, 2*(2), 307–329. https://doi.org/10.3390/traumacare2020026

Boukhris, T., Sheehy, O., Mottron, L., & Bérard, A. (2016). Antidepressant use during pregnancy and the risk of autism spectrum disorder in children. *JAMA Pediatrics, 170*(2), 117–124.

Boy, A., & Salihu, H. M. (2004). Intimate partner violence and birth outcomes: A systematic review. *International Journal of Fertility and Women's Medicine, 49*(4), 159–164.

Burd, L., Severud, R., Kerbeshian, J., & Klug, M. G. (1999). Prenatal and perinatal risk factors for autism. *Journal of Perinatal Medicine, 27*(6), 441–450. https://doi.org/10.1515/JPM.1999.059

Burstyn, I., Sithole, F., & Zwaigenbaum, L. (2010). Autism spectrum disorders, maternal characteristics and obstetric complications among singletons born in Alberta, Canada. *Chronic Diseases in Canada, 30*(4), 125–134. https://doi.org/10.24095/hpcdp.30.4.04

Buss, C., Entringer, S., Davis, E. P., Hobel, C. J., Swanson, J. M., Wadhwa, P. D., & Sandman, C. A. (2012). Impaired executive function mediates the association between maternal pre-pregnancy body mass index and child ADHD symptoms. *PLoS ONE, 7*(6), e37758. https://doi.org/10.1371/journal.pone.0037758

Calderón-Garcidueñas, L., Mora-Tiscareño, A., Ontiveros, E., Gómez-Garza, G., Barragán-Mejía, G., Broadway, J., Chapman, S., Valencia-Salazar, G., Jewells, V., Maronpot, R. R., Henríquez-Roldán, C., Pérez-Guillé, B., Torres-Jardón, R., Herrit, L., Brooks, D., Osnaya-Brizuela, N., Monroy, M. E., González-Maciel, A., Reynoso-Robles, R., Villarreal-Calderon, R., et al. (2008). Air pollution, cognitive deficits and brain abnormalities: A pilot study with children and dogs. *Brain and Cognition, 68*(2), 117–127. https://doi.org/10.1016/j.bandc.2008.04.008

Camp, B. W., Broman, S. H., Nichols, P. L., & Leff, M. (1998). Maternal and neonatal risk factors for mental retardation: Defining the 'at-risk' child. *Early Human Development, 50*(2), 159–173.

Castoldi, A. F., Coccini, T., Ceccatelli, S., & Manzo, L. (2001). Neurotoxicity and molecular effects of methylmercury. *Brain Research Bulletin, 55*, 197–203. https://doi.org/10.1016/s0361-9230(01)00458-0

Chandran, S., Sreeraj, V. S., Venkatasubramanian, G., Sathyaprabha, T. N., & Murthy, P. (2021). Corpus callosum morphometry in children with prenatal alcohol exposure. *Psychiatry Research - Neuroimaging, 318*, 111405. https://doi.org/10.1016/j.pscychresns.2021.111405

Chang, Y. S., Chen, L. W., Yu, T., Lin, S. H., & Kuo, P. L. (2023). Preterm birth and weight-for-gestational age for risks of autism spectrum disorder and intellectual disability: A nationwide population-based cohort study. *Journal of the Formosan Medical Association, 122*(6), 493–504.

Charil, A., Laplante, D. P., Vaillancourt, C., & King, S. (2010). Prenatal stress and brain development. *Brain Research Reviews, 65*(1), 56–79.

Chen, L. C., Chen, M. H., Hsu, J. W., Huang, K. L., Bai, Y. M., Chen, T. J., Wang, P. W., Pan, T. L., & Su, T. P. (2020). Association of parental depression with offspring attention deficit hyperactivity disorder and autism spectrum disorder: A nationwide birth cohort study. *Journal of Affective Disorders, 277*, 109–114. https://doi.org/10.1016/j.jad.2020.07.059

Chen, K. R., Yu, T., Kang, L., Lien, Y. J., & Kuo, P. L. (2021). Childhood neurodevelopmental disorders and maternal hypertensive disorders of pregnancy. *Developmental Medicine and Child Neurology, 63*(9), 1107–1113. https://doi.org/10.1111/dmcn.14893

Cheslack-Postava, K., Rantakokko, P. V., Hinkka-Yli-Salomäki, S., Surcel, H. M., McKeague, I. W., Kiviranta, H. A., Sourander, A., & Brown, A. S. (2013). Maternal serum persistent organic pollutants in the Finnish Prenatal Study of Autism: A pilot study. *Neurotoxicology and Teratology, 38*, 1–5. https://doi.org/10.1016/j.ntt.2013.04.001

Chung, E. K., McCollum, K. F., Elo, I. T., Lee, H. J., & Culhane, J. F. (2004). Maternal depressive symptoms and infant health practices among low-income women. *Pediatrics, 113*(6), e523–e529. https://doi.org/10.1542/peds.113.6.e523

Clifford, A., Lang, L., & Chen, R. (2012). Effects of maternal cigarette smoking during pregnancy on cognitive parameters of children and young adults: A literature review. *Neurotoxicology and Teratology, 34*(6), 560–570. https://doi.org/10.1016/j.ntt.2012.09.004

Coker, A. L., Sanderson, M., & Dong, B. (2004). Partner violence during pregnancy and risk of adverse pregnancy outcomes. *Paediatric and Perinatal Epidemiology, 18*(4), 260–269.

Cokkinides, V. E., Coker, A. L., Sanderson, M., Addy, C., & Bethea, L. (1999). Physical violence during pregnancy: Maternal complications and birth outcomes. *Obstetrics & Gynecology, 93*(5), 661–666.

Corbett, B., Mendoza, S., Abdullah, M., Wegelin, J., & Levine, S. (2006). Cortisol circadian rhythms and response to stress in children with autism. *Psychoneuroendocrinology, 31*(1), 59–68. https://doi.org/10.1016/j.psyneuen.2005.05.011

Costa, L. G., Aschner, M., Vitalone, A., Syversen, T., & Soldin, O. P. (2004). Developmental neuropathology of environmental agents. *Annual Review of Pharmacology and Toxicology, 44*, 87–110. https://doi.org/10.1146/annurev.pharmtox.44.101802.121424

Croen, L. A., Najjar, D. V., Fireman, B., & Grether, J. K. (2007). Maternal and paternal age and risk of autism spectrum disorders. *Archives of Pediatrics & Adolescent Medicine, 161*(4), 334–340. https://doi.org/10.1001/archpedi.161.4.334

Crow, J. F. (2000). The origins, patterns, and implications of human spontaneous mutation. *Nature Reviews Genetics, 1*, 40–47. https://doi.org/10.1038/35049558

Danese, A., Pariante, C. M., Caspi, A., Taylor, A., & Poulton, R. (2007). Childhood maltreatment predicts adult inflammation in a life-course study. *Proceedings of the National Academy of Sciences, 104*(4), 1319–1324.

De Young, C. G. (2020). Intelligence and personality. In R. J. Sternberg (Ed.), *The Cambridge handbook of intelligence. Cambridge handbooks in psychology* (2nd ed., pp. 1011–1047). Cambridge University Press.

Delobel-Ayoub, M., Ehlinger, V., Klapouszczak, D., Maffre, T., Raynaud, J. P., Delpierre, C., & Arnaud, C. (2015). Socioeconomic disparities and prevalence of autism spectrum disorders and intellectual disability. *PLoS One, 10*(11), e0141964.

Dempsey, D. A., & Benowitz, N. L. (2001). Risks and benefits of nicotine to aid smoking cessation in pregnancy. *Drug Safety, 24*(4), 277–322. https://doi.org/10.2165/00002018-200124040-00005

Di Prinzio, P., Morgan, V. A., Björk, J., Croft, M., Lin, A., Jablensky, A., & McNeil, T. F. (2018). Intellectual disability and psychotic disorders in children: Association with maternal severe mental illness and exposure to obstetric complications in a whole-population cohort. *American Journal of Psychiatry, 175*(12), 1232–1242. https://doi.org/10.1176/appi.ajp.2018.17101153

Dickens, M. J., & Pawluski, J. L. (2018). The HPA axis during the perinatal period: Implications for perinatal depression. *Endocrinology, 159*(11), 3737–3746. https://doi.org/10.1210/en.2018-00677

Dietrich, K. N., Ris, M. D., Succop, P. A., Berger, O. G., & Bornschein, R. L. (2001). Early exposure to lead and juvenile delinquency. *Neurotoxicology and Teratology, 23*, 511–518. https://doi.org/10.1016/s0892-0362(01)00184-2

Ding, X. X., Wu, Y. L., Xu, S. J., Zhu, R. P., Jia, X. M., Zhang, S. F., et al. (2014). Maternal anxiety during pregnancy and adverse birth outcomes: A systematic review and meta-analysis of prospective cohort studies. *Journal of Affective Disorders, 159*, 103–110.

Dodds, R. L. (2020). An Exploratory Review of the Associations between Adverse Experiences and Autism. *Journal of Aggression, Maltreatment & Trauma, 30*(8), 1093–1112. https://doi.org/10.1080/10926771.2020.1783736

Durkin, M. S., Maenner, M. J., Newschaffer, C. J., Lee, L. C., Cunniff, C. M., Daniels, J. L., Kirby, R. S., Leavitt, L., Miller, L., Zahorodny, W., & Schieve, L. A. (2008). Advanced parental age and the risk of autism spectrum disorder. *American Journal of Epidemiology, 168*(11), 1268–1276. https://doi.org/10.1093/aje/kwn250

Durkin, M. S., Maenner, M. J., Meaney, F. J., Levy, S. E., DiGuiseppi, C., Nicholas, J. S., et al. (2010). Socioeconomic inequality in the prevalence of autism spectrum disorder: Evidence from a US cross-sectional study. *PLoS One, 5*(7), e11551.

Durkin, M. S., & Yeargin-Allsopp, M. (2018, October). Socioeconomic status and pediatric neurologic disorders: Current evidence. In *In seminars in pediatric neurology* (Vol. 27, pp. 16–25). WB Saunders.

Durkin, M. S., & Rubenstein, E. (2021). Epidemiology of intellectual and developmental disabilities. In L. M. Glidden, L. Abbeduto, L. L. McIntyre, & M. J. Tassé (Eds.), *APA handbook of intellectual and developmental disabilities: Foundations* (pp. 73–87). American Psychological Association. https://doi.org/10.1037/0000194-004

Edlow, A. G. (2017). Maternal obesity and neurodevelopmental and psychiatric disorders in offspring. *Prenatal Diagnosis, 37*(1), 95–110. https://doi.org/10.1002/pd.4932

Edwards, S. C., Jedrychowski, W., Butscher, M., Camann, D., Kieltyka, A., Mroz, E., et al. (2010). Prenatal exposure to airborne polycyclic aromatic hydrocarbons and children's intelligence at 5 years of age in a prospective cohort study in Poland. *Environmental Health Perspectives, 118*(9), 1326–1331.

Emerson, E. (2012). Deprivation, ethnicity and the prevalence of intellectual and developmental disabilities. *Journal of Epidemiology and Community Health, 66*(3), 218–224. https://doi.org/10.1136/jech.2010.111773

Eriksen, H. L. F., Kesmodel, U. S., Underbjerg, M., Kilburn, T. R., Bertrand, J., & Mortensen, E. L. (2013). Predictors of intelligence at the age of 5: Family, pregnancy and birth characteristics, postnatal influences, and postnatal growth. *PLoS ONE, 8*(11), 79200. https://doi.org/10.1371/journal.pone.0079200

Ernst, M., Moolchan, E. T., & Robinson, M. L. (2001). Behavioral and neural consequences of prenatal exposure to nicotine. *Journal of the American Academy of Child and Adolescent Psychiatry, 40*(6), 630–641. https://doi.org/10.1097/00004583-200106000-00007

Estes, M. L., & McAllister, A. K. (2016). Maternal immune activation: Implications for neuropsychiatric disorders. *Science, 353*(6301), 772–777. https://doi.org/10.1126/science.aag3194

Fernandez-Botran, R., Miller, J. J., Burns, V. E., & Newton, T. L. (2011). Correlations among inflammatory markers in plasma, saliva and oral mucosal transudate in post-menopausal women with past intimate partner violence. *Brain, Behavior, and Immunity, 25*(2), 314–321.

Field, T., Diego, M., Hernandez-Reif, M., Schanberg, S., Kuhn, C., Yando, R., & Bendell, D. (2003). Pregnancy anxiety and comorbid depression and anger: Effects on the fetus and neonate. *Depression and Anxiety, 17*(3), 140–151. https://doi.org/10.1002/da.10071

Flegal, K. M., Kruszon-Moran, D., Carroll, M. D., Fryar, C. D., & Ogden, C. L. (2016). Trends in obesity among adults in the United States, 2005 to 2014. *JAMA, 315*(21), 2284–2291.

Ford, N. D., Cox, S., Ko, J. Y., Ouyang, L., Romero, L., Colarusso, T., Ferre, C. D., Kroelinger, C. D., Hayes, D. K., & Barfield, W. D. (2022). Hypertensive disorders in pregnancy and mortality at delivery hospitalization — United States, 2017–2019. *MMWR. Morbidity and Mortality Weekly Report, 71*(17), 585–591. https://doi.org/10.15585/mmwr.mm7117a1

Fonnum, F., Mariussen, E., & Reistad, T. (2006). Molecular mechanisms involved in the toxic effects of polychlorinated biphenyls (PCBs) and brominated flame retardants (BFRs). *Journal of Toxicology and Environmental Health. Part A, 69*, 21–35. https://doi.org/10.1080/15287390500259020

Fountoulakis, K. N., Gonda, X., Siamouli, M., Panagiotidis, P., Moutou, K., Nimatoudis, I., & Kasper, S. (2018). Paternal and maternal age as risk factors for schizophrenia: A case–control study. *International Journal of Psychiatry in Clinical Practice, 22*(3), 170–176. https://doi.org/10.1080/13651501.2017.1391292

Friedman, M. J., & McEwen, B. S. (2004). Posttraumatic stress disorder, allostatic load, and medical illness. In P.P. Schnurr & B.L. Green (Eds.), *Trauma and health: Physical health consequences of exposure to extreme stress* (pp. 157–188). https://doi.org/10.1037/10723-007

Froehlich, T. E., Lanphear, B. P., Auinger, P., Hornung, R., Epstein, J. N., Braun, J., & Kahn, R. S. (2009). Association of tobacco and lead exposures with attention-deficit/hyperactivity disorder. *Pediatrics, 124*, e1054–e1063. https://doi.org/10.1542/peds.2009-0738

Froggatt, S., Covey, J., & Reissland, N. (2020). Infant neurobehavioural consequences of prenatal cigarette exposure: A systematic review and meta-analysis. *Acta Paediatrica, 109*(6), 1112–1124. https://doi.org/10.1111/apa.15132

Gafni, J., Wong, P. W., & Pessah, I. N. (2004). Non-coplanar 2,2′,3,5′,6-pentachlorobiphenyl (PCB 95) amplifies ionotropic glutamate receptor signaling in embryonic cerebellar granule neurons by a mechanism involving ryanodine receptors. *Toxicological Sciences, 77*, 72–82. https://doi.org/10.1093/toxsci/kfh004

Garcia-Moreno, C., & Watts, C. (2011). Violence against women: An urgent public health priority. *Bulletin of the World Health Organization, 89*, 2–2.

Gardener, H., Spiegelman, D., & Buka, S. L. (2009). Prenatal risk factors for autism: Comprehensive meta-analysis. *British Journal of Psychiatry, 195*(1), 7–14. https://doi.org/10.1192/bjp.bp.108.051672

Gardner, R. M., Lee, B. K., Magnusson, C., Rai, D., Frisell, T., Karlsson, H., Idring, S., & Dalman, C. (2015). Maternal body mass index during early pregnancy, gestational weight gain, and risk of autism spectrum disorders: Results from a Swedish total population and discordant sibling study. *International Journal of Epidemiology, 44*(3), 870–883. https://doi.org/10.1093/ije/dyv081

Gaspar, P., Cases, O., & Maroteaux, L. (2003). The developmental role of serotonin: News from mouse molecular genetics. *Nature Reviews Neuroscience, 4*(12), 1002–1012. https://doi.org/10.1038/nrn1256

Gilbert, M. E. (2003). Perinatal exposure to polychlorinated biphenyls alters excitatory synaptic transmission and short-term plasticity in the hippocampus of the adult rat. *Neurotoxicology, 24*, 851–860. https://doi.org/10.1016/S0161-813X(03)00073-1

Gilmore, J. H., Knickmeyer, R. C., & Gao, W. (2018). Imaging structural and functional brain development in early childhood. *Nature Reviews Neuroscience, 19*(3), 123–137. https://doi.org/10.1038/nrn.2018.1

Glover, V., O'Connor, T. G. O., & O'Donnell, K. (2010). Prenatal stress and the programming of the HPA axis. *Neuroscience & Biobehavioral Reviews, 35*(1), 17–22. https://doi.org/10.1016/j.neubiorev.2009.11.008

Gourdine, R. M., Baffour, T. D., & Teasley, M. (2011). Autism and the African American community. *Social Work in Public Health, 26*(4), 454–470.

Graeve, R., Balalian, A. A., Richter, M., Kielstein, H., Fink, A., Martins, S. S., Philbin, M. M., & Factor-Litvak, P. (2022). Infants' prenatal exposure to opioids and the association with birth outcomes: A systematic review and meta-analysis. *Paediatric and Perinatal Epidemiology, 36*(1), 125–143. https://doi.org/10.1111/ppe.12805

Grandjean, P., & Landrigan, P. J. (2006). Developmental neurotoxicity of industrial chemicals. *Lancet, 368*(9553), 2167–2178. https://doi.org/10.1016/S0140-6736(06)69665-7

Grandjean, P., & Landrigan, P. J. (2014). Neurobehavioural effects of developmental toxicity. *The Lancet. Neurology, 13*, 330–338. https://doi.org/10.1016/S1474-4422(13)70278-3

Grewen, K., Salzwedel, A. P., & Gao, W. (2015). Functional connectivity disruption in neonates with prenatal marijuana exposure. *Frontiers in Human Neuroscience, 9*, 601. https://doi.org/10.3389/fnhum.2015.00601

Gupta, S., Petras, L., Tufail, M. U., Rodriguez Salazar, J. D., & Jim, B. (2023). Hypertension in pregnancy: What we now know. *Current Opinion in Nephrology and Hypertension, 32*(2), 153–164. https://doi.org/10.1097/MNH.0000000000000857

Hahn, C. K., Gilmore, A. K., Aguayo, R. O., & Rheingold, A. A. (2018). Perinatal intimate partner violence. *Obstetrics and Gynecology Clinics, 45*(3), 535–547.

Hansen, M., & Olff, M. (2017). Child maltreatment across the lifespan. *European Journal of Psychotraumatology, 8*(sup7), 1483165. https://doi.org/10.1080/20008198.2018.1483165

Harris-Britt, A., Martin, S. L., Li, Y., Casanueva, C., & Kupper, L. L. (2004). Posttraumatic stress disorder and associated functional impairments during pregnancy: Some consequences of violence against women. *Journal of Clinical Psychology in Medical Settings, 11*(4), 253–264. https://doi.org/10.1023/B:JOCS.0000045345.72671.5e

Hasebe, K., Kendig, M. D., & Morris, M. J. (2021). Mechanisms underlying the cognitive and behavioural effects of maternal obesity. *Nutrients, 13*(1), 240. https://doi.org/10.3390/nu13010240

Heikura, U., Taanila, A., Hartikainen, A. L., Olsen, P., Linna, S. L., Von Wendt, L., & Järvelin, M. R. (2008). Variations in prenatal sociodemographic factors associated with intellectual disability: A study of the 20-year interval between two birth cohorts in northern Finland. *American Journal of Epidemiology, 167*(2), 169–177. https://doi.org/10.1093/aje/kwm291

Heim, C. (2018). Psychobiological consequences of child maltreatment, In J. Noll & I. Shalev (Eds.), *The biology of early life stress* (pp. 15–30). https://doi.org/10.1007/978-3-319-72589-5_2

Hertz-Picciotto, I., Korrick, S. A., Ladd-Acosta, C., Karagas, M. R., Lyall, K., Schmidt, R. J., Dunlop, A. L., Croen, L. A., Dabelea, D., Daniels, J. L., Duarte, C. S., Fallin, M. D., Karr, C. J., Lester, B., Leve, L. D., Li, Y., McGrath, M., Ning, X., Oken, E., et al. (2022). Maternal tobacco smoking and offspring autism spectrum disorder or traits in ECHO cohorts. *Autism Research, 15*(3), 551–569. https://doi.org/10.1002/aur.2665

Hoyme, H. E., Kalberg, W. O., Elliott, A. J., Blankenship, J., Buckley, D., Marais, A. S., Manning, M. A., Robinson, L. K., Adam, M. P., Abdul-Rahman, O., Jewett, T., Coles, C. D., Chambers, C., Jones, K. L., Adnams, C. M., Shah, P. E., Riley, E. P., Charness, M. E., Warren, K. R., & May, P. A. (2016). Updated clinical guidelines for diagnosing fetal alcohol spectrum disorders. *Pediatrics, 138*(2), e20154256. https://doi.org/10.1542/peds.2015-4256

Huang, L., Wang, Y., Zhang, L., Zheng, Z., Zhu, T., Qu, Y., & Mu, D. (2018). Maternal smoking and attention-deficit/hyperactivity disorder in offspring: A meta-analysis. *Pediatrics, 141*(1), e20172465. https://doi.org/10.1542/peds.2017-2465

Huang, L., Yu, X., Keim, S., Li, L., Zhang, L., & Zhang, J. (2014). Maternal pre-pregnancy obesity and child neurodevelopment in the Collaborative Perinatal Project. *International Journal of Epidemiology, 43*(3), 783–792.

Huang, J., Zhu, T., Qu, Y., & Mu, D. (2016). Prenatal, perinatal, and neonatal risk factors for intellectual disability: A systemic review and meta-analysis. *PLoS ONE, 11*(4), e0153655. https://doi.org/10.1371/journal.pone.0153655

Hultman, C. M., Sparén, P., & Cnattingius, S. (2002). Perinatal risk factors for infantile autism. *Epidemiology, 13*(4), 417–423.

Jacobson, J. L., & Jacobson, S. W. (1996). Intellectual impairment in children exposed to polychlorinated biphenyls in utero. *The New England Journal of Medicine, 335*, 783–789. https://doi.org/10.1056/NEJM199609123351104

Jo, H., Eckel, S. P., Chen, J. C., Cockburn, M., Martinez, M. P., Chow, T., Lurmann, F. W., Funk, W. E., Xiang, A. H., & McConnell, R. (2019). Gestational diabetes mellitus, prenatal air pollution exposure, and autism spectrum disorder. *Environment International, 133*, 105110. https://doi.org/10.1016/j.envint.2019.105110

Johansson, C., Castoldi, A. F., Onishchenko, N., Manzo, L., Vahter, M., & Ceccatelli, S. (2007). Neurobehavioural and molecular changes induced by methylmercury exposure during development. *Neurotoxicity Research, 11*(3–4), 241–260. https://doi.org/10.1007/BF03033570

Jones, P. B., Rantakallio, P., Hartikainen, A. L., Isohanni, M., & Sipila, P. (1998). Schizophrenia as a long-term outcome of pregnancy, delivery, and perinatal complications: A 28-year follow-up of the 1966 North Finland general population birth cohort. *American Journal of Psychiatry, 155*(3), 355–364. https://doi.org/10.1176/ajp.155.3.355

Jung, Y., Lee, A. M., McKee, S. A., & Picciotto, M. R. (2017). Maternal smoking and autism spectrum disorder: Meta-analysis with population smoking metrics as moderators. *Scientific Reports, 7*, 4315. https://doi.org/10.1038/s41598-017-04413-1

Kalin, N. H. (2019). Prefrontal cortical and limbic circuit alterations in psychopathology. *American Journal of Psychiatry, 176*(12), 971–973. https://doi.org/10.1176/appi.ajp.2019.19101036

Kalkbrenner, A. E., Daniels, J. L., Chen, J. C., Poole, C., Emch, M., & Morrissey, J. (2010). Perinatal exposure to hazardous air pollutants and autism spectrum disorders at age 8. *Epidemiology, 21*(5), 631–641. https://doi.org/10.1097/EDE.0b013e3181e65d76

Kalkbrenner, A. E., Windham, G. C., Serre, M. L., Akita, Y., Wang, X., Hoffman, K., Thayer, B. P., & Daniels, J. L. (2015). Particulate matter exposure, prenatal and postnatal windows of susceptibility, and autism spectrum disorders. *Epidemiology, 26*(1), 30–42. https://doi.org/10.1097/EDE.0000000000000173

Keane, T. M., Taylor, K. L., & Penk, W. E. (1997). Differentiating Post-Traumatic Stress Disorder (PTSD) from Major Depression (MDD) and Generalized Anxiety Sisorder (GAD). *Journal of Anxiety Disorders, 11*(3), 317–328. https://doi.org/10.1016/S0887-6185(97)00013-3

Khandwala, Y. S., Baker, V. L., Shaw, G. M., Stevenson, D. K., Lu, Y., & Eisenberg, M. L. (2018). Association of paternal age with perinatal outcomes between 2007 and 2016 in the United States: Population based cohort study. *BMJ, 363*, k4372.

Killion, B. E., & Weyandt, L. L. (2020). Brain structure in childhood maltreatment-related PTSD across the lifespan: A systematic review. *Applied Neuropsychology: Child, 9*(1), 68–82. https://doi.org/10.1080/21622965.2018.1515076

Kinney, D. K., Munir, K. M., Crowley, D. J., & Miller, A. M. (2008). Prenatal stress and risk for autism. *Neuroscience & Biobehavioral Reviews, 32*(8), 1519–1532.

Klepac, P., Locatelli, I., Korošec, S., Künzli, N., & Kukec, A. (2018). Ambient air pollution and pregnancy outcomes: A comprehensive review and identification of environmental public health challenges. *Environmental Research, 167*, 144–159.

Knudsen, E. I. (2004). Sensitive periods in the development of the brain and behavior. *Journal of Cognitive Neuroscience, 16*(8), 1412–1425. https://doi.org/10.1162/0898929042304796

Kodesh, A., Levine, S. Z., Khachadourian, V., Rahman, R., Schlessinger, A., O'Reilly, P. F., Grove, J., Schendel, D., Buxbaum, J. D., Croen, L., Reichenberg, A., Sandin, S., & Janecka, M. (2022). Maternal health around pregnancy and autism risk: A diagnosis-wide, population-based study. *Psychological Medicine, 52*(16), 4076–4084. https://doi.org/10.1017/S0033291721001021

Kong, L., Norstedt, G., Schalling, M., Gissler, M., & Lavebratt, C. (2018). The risk of offspring psychiatric disorders in the setting of maternal obesity and diabetes. *Pediatrics, 142*(3), e20180776. https://doi.org/10.1542/peds.2018-0776

Krakowiak, P., Walker, C. K., Bremer, A. A., Baker, A. S., Ozonoff, S., Hansen, R. L., & Hertz-Picciotto, I. (2012). Maternal metabolic conditions and risk for autism and other neurodevelopmental disorders. *Pediatrics, 129*(5), e1121–e1128. https://doi.org/10.1542/peds.2011-2583

Lai, T. J., Liu, X., Guo, Y. L., Guo, N. W., Yu, M. L., Hsu, C. C., & Rogan, W. J. (2002). A cohort study of behavioral problems and intelligence in children with high prenatal polychlorinated biphenyl exposure. *Archives of General Psychiatry, 59*, 1061–1066. https://doi.org/10.1001/archpsyc.59.11.1061

Lampi, K. M., Lehtonen, L., Tran, P. L., Suominen, A., Lehti, V., Banerjee, P. N., Gissler, M., Brown, A. S., & Sourander, A. (2012). Risk of autism spectrum disorders in low birth weight and small for gestational age infants. *The Journal of Pediatrics, 161*, 830–836. https://doi.org/10.1016/j.jpeds.2012.04.058

Lange, S., Probst, C., Rehm, J., & Popova, S. (2018). National, regional, and global prevalence of smoking during pregnancy in the general population: A systematic review and meta-analysis. *The Lancet Global Health, 6*(7), e769–e776. https://doi.org/10.1016/S2214-109X(18)30223-7
Larkby, C. (1997). The effects of prenatal alcohol exposure. *Alcohol Research and Health, 21*(3), 192–198.
Larsson, H. J., Eaton, W. W., Madsen, K. M., Vestergaard, M., Olesen, A. V., Agerbo, E., Schendel, D., Thorsen, P., & Mortensen, P. B. (2005). Risk factors for autism: Perinatal factors, parental psychiatric history, and socioeconomic status. *American Journal of Epidemiology, 161*(10), 916–925. https://doi.org/10.1093/aje/kwi123
Lebel, C., Walton, M., Letourneau, N., Giesbrecht, G. F., Kaplan, B. J., & Dewey, D. (2016). Prepartum and postpartum maternal depressive symptoms are related to children's brain structure in preschool. *Biological Psychiatry, 80*, 859–868. https://doi.org/10.1016/j.biopsych.2015.12.004
Lee, A. M., Lam, S. K., Sze Mun Lau, S. M., Chong, C. S. Y., Chui, H. W., & Fong, D. Y. T. (2007). Prevalence, course, and risk factors for antenatal anxiety and depression. *Obstetrics and Gynecology, 110*(5), 1102–1112. https://doi.org/10.1097/01.AOG.0000287065.59491.70
Leibovitz, Z., Lerman-Sagie, T., & Haddad, L. (2022). Fetal brain development: Regulating processes and related malformations. *Life, 12*(6), 809. https://doi.org/10.3390/life12060809
Leonard, H., Glasson, E., Nassar, N., Whitehouse, A., Bebbington, A., Bourke, J., et al. (2011). Autism and intellectual disability are differentially related to sociodemographic background at birth. *PLoS One, 6*(3), e17875.
Li, M., Fallin, M. D., Riley, A., Landa, R., Walker, S. O., Silverstein, M., Caruso, D., Pearson, C., Kiang, S., Dahm, J. L., Hong, X., Wang, G., Wang, M. C., Zuckerman, B., & Wang, X. (2016). The association of maternal obesity and diabetes with autism and other developmental disabilities. *Pediatrics, 137*(2), e20152206. https://doi.org/10.1542/peds.2015-2206
Linderkamp, O., Janus, L., Linder, R., & Skoruppa, D. B. (2009). Timetable of normal foetal brain development. *International Journal Prenatal and Perinatal Psychology and Medicine, 21*(1/2), 4–16.
Lipsky, S., Holt, V. L., Easterling, T. R., & Critchlow, C. W. (2003). Impact of police-reported intimate partner violence during pregnancy on birth outcomes. *Obstetrics & Gynecology, 102*(3), 557–564.
Lund, N., Pedersen, L. H., & Henriksen, T. B. (2009). Selective serotonin reuptake inhibitor exposure in utero and pregnancy outcomes. *Archives of Pediatrics and Adolescent Medicine, 163*(10), 949–954. https://doi.org/10.1001/archpediatrics.2009.164
Madley-Dowd, P., Kalkbrenner, A. E., Heuvelman, H., Heron, J., Zammit, S., Rai, D., & Schendel, D. (2022). Maternal smoking during pregnancy and offspring intellectual disability: Sibling analysis in an intergenerational Danish cohort. *Psychological Medicine, 52*(10), 1847–1856.
Maes, M., Van Bockstaele, D. R., Gastel, A. V., Song, C., Schotte, C., Neels, H., et al. (1999). The effects of psychological stress on leukocyte subset distribution in humans: Evidence of immune activation. *Neuropsychobiology, 39*(1), 1–9.
Maher, G. M., O'Keeffe, G. W., Kearney, P. M., Kenny, L. C., Dinan, T. G., Mattsson, M., & Khashan, A. S. (2018). Association of hypertensive disorders of pregnancy with risk of neurodevelopmental disorders in offspring: A systematic review and meta-analysis. *JAMA Psychiatry, 75*(8), 809–819. https://doi.org/10.1001/jamapsychiatry.2018.0854
Mann, J. R., McDermott, S., Bao, H., Hardin, J., & Gregg, A. (2010). Pre-eclampsia, birth weight, and autism spectrum disorders. *Journal of Autism and Developmental Disorders, 40*(5), 548–554. https://doi.org/10.1007/s10803-009-0903-4
Marshall, R. D., & Garakani, A. (2002). Psychobiology of the acute stress response and its relationship to the psychobiology of post-traumatic stress disorder. *Psychiatric Clinics of North America, 25*(2), 385–395. https://doi.org/10.1016/S0193-953X(01)00005-3
Matthews, S. G. (2000). Antenatal glucocorticoids and programming of the developing CNS. *Pediatric Research, 47*(3), 291–300. https://doi.org/10.1203/00006450-200003000-00003

May, P. A., Gossage, J. P., Marais, A. S., Adnams, C. M., Hoyme, H. E., Jones, K. L., Robinson, L. K., Khaole, N. C. O., Snell, C., Kalberg, W. O., Hendricks, L., Brooke, L., Stellavato, C., & Viljoen, D. L. (2007). The epidemiology of fetal alcohol syndrome and partial FAS in a South African community. *Drug and Alcohol Dependence, 88*(2–3), 259–271. https://doi.org/10.1016/j.drugalcdep.2006.11.007

May, P. A., Gossage, J. P., Marais, A. S., Hendricks, L. S., Snell, C. L., Tabachnick, B. G., et al. (2008). Maternal risk factors for fetal alcohol syndrome and partial fetal alcohol syndrome in South Africa: A third study. *Alcoholism: Clinical and Experimental Research, 32*(5), 738–753.

Mendola, P., Selevan, S. G., Gutter, S., & Rice, D. (2002). Environmental factors associated with a spectrum of neurodevelopmental deficits. *Mental Retardation and Developmental Disabilities Research Reviews, 8*, 188–197. https://doi.org/10.1002/mrdd.10033

Miller, D. T., Shen, Y., Weiss, L. A., Korn, J., Anselm, I., Bridgemohan, C., Cox, G. F., Dickinson, H., Gentile, J., Harris, D. J., Hegde, V., Hundley, R., Khwaja, O., Kothare, S., Luedke, C., Nasir, R., Poduri, A., Prasad, K., Raffalli, P., et al. (2009). Microdeletion/duplication at 15q13.2q13.3 among individuals with features of autism and other neuropsychiatric disorders. *Journal of Medical Genetics, 46*(4), 242–248. https://doi.org/10.1136/jmg.2008.059907

Mueller, I., & Tronick, E. (2019). Early life exposure to violence: Developmental consequences on brain and behavior. *Frontiers in Behavioral Neuroscience, 13*, 156.

Mueller, I., & Tronick, E. (2020). The long shadow of violence: The impact of exposure to intimate partner violence in infancy and early childhood. *International Journal of Applied Psychoanalytic Studies, 17*(3), 232–245.

Newton, T. L., Fernandez-Botran, R., Miller, J. J., Lorenz, D. J., Burns, V. E., & Fleming, K. N. (2011). Markers of inflammation in midlife women with intimate partner violence histories. *Journal of Women's Health, 20*(12), 1871–1880.

Newland, M. C., Reed, M. N., & Rasmussen, E. (2015). A hypothesis about how early developmental methylmercury exposure disrupts behavior in adulthood. *Behavioural Processes, 114*, 41–51. https://doi.org/10.1016/j.beproc.2015.03.007

NIAAA. (2004). Council approves definition of binge drinking. *NIAAA Newsletter, 3*, 3. https://doi.org/10.1037/e306662005-004

Nidey, N. L., Momany, A. M., Strathearn, L., Carter, K. D., Wehby, G. L., Bao, W., Xu, G., Scheiber, F. A., Tabb, K., Froehlich, T. E., & Ryckman, K. (2021). Association between perinatal depression and risk of attention deficit hyperactivity disorder among children: A retrospective cohort study. *Annals of Epidemiology, 63*, 1–6. https://doi.org/10.1016/j.annepidem.2021.06.005

Nomura, Y., Marks, D. J., Grossman, B., Yoon, M., Loudon, H., Stone, J., & Halperin, J. M. (2012). Exposure to gestational diabetes mellitus and low socioeconomic status: Effects on neurocognitive development and risk of attention-deficit/ hyperactivity disorder in offspring. *Archives of Pediatrics and Adolescent Medicine, 166*(4), 337–343. https://doi.org/10.1001/archpediatrics.2011.784

O'Connor, T. G., Ciesla, A. A., Sefair, A. V., Thornburg, L. L., Brown, A. S., Glover, V., & O'Donnell, K. J. (2022). Maternal prenatal infection and anxiety predict neurodevelopmental outcomes in middle childhood. *Journal of Psychopathology and Clinical Science, 131*(4), 422–434. https://doi.org/10.1037/abn0000746

O'Donnell, K. J., Glover, V., Lahti, J., Lahti, M., Edgar, R. D., Räikkönen, K., & O'Connor, T. G. (2017). Maternal prenatal anxiety and child COMT genotype predict working memory and symptoms of ADHD. *PLoS One, 12*(60), e0177506. https://doi.org/10.1371/journal.pone.0177506

Oberlander, T. F., Warburton, W., Misri, S., Aghajanian, J., & Hertzman, C. (2006). Neonatal outcomes after prenatal exposure to selective serotonin reuptake inhibitor antidepressants and maternal depression using population-based linked health data. *Archives of General Psychiatry, 63*(8), 898–906. https://doi.org/10.1001/archpsyc.63.8.898

Olfson, M., & Marcus, S. C. (2009). National patterns in antidepressant medication treatment. *Archives of General Psychiatry, 66*(8), 848–856. https://doi.org/10.1001/archgenpsychiatry.2009.81

Okano, L., Ji, Y., Riley, A. W., & Wang, X. (2019). Maternal psychosocial stress and children's ADHD diagnosis: A prospective birth cohort study. *Journal of Psychosomatic Obstetrics & Gynecology, 40*(3), 217–225.

Ornoy, A., Becker, M., Weinstein-Fudim, L., & Ergaz, Z. (2021). Diabetes during pregnancy: A maternal disease complicating the course of pregnancy with long-term deleterious effects on the offspring. A clinical review. *International Journal of Molecular Sciences, 22*(6), 2965. https://doi.org/10.3390/ijms22062965

Pan, Y. H., Wu, N., & Yuan, X. B. (2019). Toward a better understanding of neuronal migration deficits in autism spectrum disorders. *Frontiers in Cell and Developmental Biology, 7*, 205. https://doi.org/10.3389/fcell.2019.00205

Patel, N., Pasupathy, D., & Poston, L. (2015). Determining the consequences of maternal obesity for offspring health. *Experimental Physiology, 100*(12), 1421–1428. https://doi.org/10.1113/EP085132

Perera, F. P., Rauh, V., Whyatt, R. M., Tsai, W. Y., Tang, D., Diaz, D., et al. (2006). Effect of prenatal exposure to airborne polycyclic aromatic hydrocarbons on neurodevelopment in the first 3 years of life among inner-city children. *Environmental Health Perspectives, 114*(8), 1287–1292.

Perea, V., Simó-Servat, A., Quirós, C., Alonso-Carril, N., Valverde, M., Urquizu, X., et al. (2022). Role of excessive weight gain during gestation in the risk of ADHD in offspring of women with gestational diabetes. *The Journal of Clinical Endocrinology & Metabolism, 107*(10), e4203–e4211.

Petterson, S. M., & Albers, A. B. (2001). Effects of poverty and maternal depression on early child development. *Child Development, 72*, 1794–1813. https://doi.org/10.1111/1467-8624.00379

Popova, S., Lange, S., Probst, C., Gmel, G., & Rehm, J. (2017). Estimation of national, regional, and global prevalence of alcohol use during pregnancy and fetal alcohol syndrome: A systematic review and meta-analysis. *The Lancet Global Health, 5*(3), e290–e299. https://doi.org/10.1016/S2214-109X(17)30021-9

Popova, S., Lange, S., Shield, K., Burd, L., & Rehm, J. (2019). Prevalence of fetal alcohol spectrum disorder among special subpopulations: A systematic review and meta-analysis. *Addiction, 114*(7), 1150–1172. https://doi.org/10.1111/add.14598

Porterfield, S. P. (2000). Thyroidal dysfunction and environmental chemicals—Potential impact on brain development. *Environmental Health Perspectives, 108*(Suppl 3), 433–438. https://doi.org/10.1289/ehp.00108s3433

Price, S. M., & Caughey, A. B. (2022). The impact of prenatal care on pregnancy outcomes in women with depression. *Journal of Maternal-Fetal and Neonatal Medicine, 35*(20), 3948–3954. https://doi.org/10.1080/14767058.2020.1844655

Rai, D., Lewis, G., Lundberg, M., Araya, R., Svensson, A., Dalman, C., et al. (2012). Parental socioeconomic status and risk of offspring autism spectrum disorders in a Swedish population-based study. *Journal of the American Academy of Child & Adolescent Psychiatry, 51*(5), 467–476.

Ratanatharathorn, A., Koenen, K. C., Chibnik, L. B., Weisskopf, M. G., Rich-Edwards, J. W., & Roberts, A. L. (2021). Polygenic risk for autism, attention-deficit hyperactivity disorder, schizophrenia, major depressive disorder, and neuroticism is associated with the experience of childhood abuse. *Molecular Psychiatry, 26*(5), 1696–1705. https://doi.org/10.1038/s41380-020-00996-w

Raz, R., Weisskopf, M. G., Davidovitch, M., Pinto, O., & Levine, H. (2015). Differences in autism spectrum disorders incidence by sub-populations in Israel 1992–2009: A total population study. *Journal of Autism and Developmental Disorders, 45*, 1062–1069.

Reynolds, L. C., Inder, T. E., Neil, J. J., Pineda, R. G., & Rogers, C. E. (2014). Maternal obesity and increased risk for autism and developmental delay among very preterm infants. *Journal of Perinatology, 34*(9), 688–692. https://doi.org/10.1038/jp.2014.80

Rijlaarsdam, J., Stevens, G. W. J. M., Jansen, P. W., Ringoot, A. P., Jaddoe, V. W. V., Hofman, A., Ayer, L., Verhulst, F. C., Hudziak, J. J., & Tiemeier, H. (2014). Maternal childhood maltreatment and offspring emotional and behavioral problems: Maternal and paternal mechanisms of risk transmission. *Child Maltreatment, 19*(2), 67–78. https://doi.org/10.1177/1077559514527639

Ritz, B., Yu, F., Chapa, G., & Fruin, S. (2000). Effect of air pollution on preterm birth among children born in Southern California between 1989 and 1993. *Epidemiology, 11*(5), 502–511.

Roberts, E. M., English, P. B., Grether, J. K., Windham, G. C., Somberg, L., & Wolff, C. (2007). Maternal residence near agricultural pesticide applications and autism spectrum disorders among children in the California Central Valley. *Environmental Health Perspectives, 115*, 1482–1489. https://doi.org/10.1289/ehp.10168

Roberts, A. L., Lyall, K., Rich-Edwards, J. W., Ascherio, A., & Weisskopf, M. G. (2013). Association of maternal exposure to childhood abuse with elevated risk for autism in offspring. *JAMA Psychiatry, 70*(5), 508–515. https://doi.org/10.1001/jamapsychiatry.2013.447

Roberts, A. L., Koenen, K. C., Lyall, K., Ascherio, A., & Weisskopf, M. G. (2014). Women's posttraumatic stress symptoms and autism spectrum disorder in their children. *Research in Autism Spectrum Disorders, 8*(6), 608–616. https://doi.org/10.1016/j.rasd.2014.02.004

Roberts, A. L., Lyall, K., Rich-Edwards, J. W., Ascherio, A., & Weisskopf, M. G. (2016). Maternal exposure to intimate partner abuse before birth is associated with autism spectrum disorder in offspring. *Autism, 20*(1), 26–36.

Roberts, A. L., Liew, Z., Lyall, K., Ascherio, A., & Weisskopf, M. G. (2018). Association of maternal exposure to childhood abuse with elevated risk for attention deficit hyperactivity disorder in offspring. *American Journal of Epidemiology, 187*(9), 1896–1906.

Robertson Blackmore, E., Mittal, M., Cai, X., Moynihan, J. A., Matthieu, M. M., & O'Connor, T. G. (2016). Lifetime exposure to intimate partner violence and proinflammatory cytokine levels across the perinatal period. *Journal of Women's Health, 25*(10), 1004–1013.

Robins, D. L., Fein, D., Barton, M. L., & Green, J. A. (2001). The modified checklist for autism in toddlers: An initial study investigating the early detection of autism and pervasive developmental disorders. *Journal of Autism and Developmental Disorders, 31*, 131–144.

Robinson, R., Girchenko, P., Pulakka, A., Heinonen, K., Lähdepuro, A., Lahti-Pulkkinen, M., et al. (2023). ADHD symptoms and diagnosis in adult preterms: Systematic review, IPD meta-analysis, and register linkage study. *Pediatric Research, 93*(5), 1399–1409.

Rodriguez, A. (2010). Maternal pre-pregnancy obesity and risk for inattention and negative emotionality in children. *Journal of Child Psychology and Psychiatry and Allied Disciplines, 51*(2), 134–143. https://doi.org/10.1111/j.1469-7610.2009.02133.x

Rodriguez, A., Miettunen, J., Henriksen, T. B., Olsen, J., Obel, C., Taanila, A., Ebeling, H., Linnet, K. M., Moilanen, I., & Järvelin, M. R. (2008). Maternal adiposity prior to pregnancy is associated with ADHD symptoms in offspring: Evidence from three prospective pregnancy cohorts. *International Journal of Obesity, 32*(3), 550–557. https://doi.org/10.1038/sj.ijo.0803741

Ronald, A., Pennell, C. E., & Whitehouse, A. J. (2011). Prenatal maternal stress associated with ADHD and autistic traits in early childhood. *Frontiers in Psychology, 1*, 223.

Rosen, D., Seng, J. S., Tolman, R. M., & Mallinger, G. (2007). Intimate partner violence, depression, and posttraumatic stress disorder as additional predictors of low birth weight infants among low-income mothers. *Journal of Interpersonal Violence, 22*(10), 1305–1314.

Rosen, B. N., Lee, B. K., Lee, N. L., Yang, Y., & Burstyn, I. (2015). Maternal smoking and autism spectrum disorder: A meta-analysis. *Journal of Autism and Developmental Disorders, 45*, 1689–1698.

Ross, L. E., McLean, L. M., & Psych, C. (2006). Anxiety disorders during pregnancy and the postpartum period: A systematic review. *Depression, 6*(9), 1–14.

Ross, E. J., Graham, D. L., Money, K. M., & Stanwood, G. D. (2015). Developmental consequences of fetal exposure to drugs: What we know and what we still must learn. *Neuropsychopharmacology: Official publication of the American College of Neuropsychopharmacology, 40*, 61–87. https://doi.org/10.1038/npp.2014.147

Rossignol, D. A., & Frye, R. E. (2012). A review of research trends in physiological abnormalities in autism spectrum disorders: Immune dysregulation, inflammation, oxidative stress, mitochondrial dysfunction and environmental toxicant exposures. *Molecular Psychiatry, 17*(4), 389–401. https://doi.org/10.1038/mp.2011.165

Rowland, J., & Wilson, C. A. (2021). The association between gestational diabetes and ASD and ADHD: A systematic review and meta-analysis. *Scientific Reports, 11*(1), 5136. https://doi.org/10.1038/s41598-021-84573-3

Rubenstein, E., Young, J. C., Croen, L. A., DiGuiseppi, C., Dowling, N. F., Lee, L.-C., Schieve, L., Wiggins, L. D., & Daniels, J. (2019). Brief report: Maternal opioid prescription from preconception through pregnancy and the odds of autism spectrum disorder and autism features in children. *Journal of Autism and Developmental Disorders, 49*(1), 376–382. https://doi.org/10.1007/s10803-018-3721-8

Sacchi, C., O'Muircheartaigh, J., Batalle, D., Counsell, S. J., Simonelli, A., Cesano, M., et al. (2021). Neurodevelopmental outcomes following intrauterine growth restriction and very preterm birth. *The Journal of Pediatrics, 238*, 135–144.

Salzwedel, A. P., Grewen, K. M., Vachet, C., Gerig, G., Lin, W., & Gao, W. (2015). Prenatal drug exposure affects neonatal brain functional connectivity. *The Journal of Neuroscience: The Official Journal of the Society for Neuroscience, 35*(14), 5860–5869. https://doi.org/10.1523/JNEUROSCI.4333-14.2015

Salzwedel, A. P., Grewen, K. M., Goldman, B. D., & Gao, W. (2016). Thalamocortical functional connectivity and behavioral disruptions in neonates with prenatal cocaine exposure. *Neurotoxicology and Teratology, 56*, 16–25. https://doi.org/10.1016/j.ntt.2016.05.009

Sandin, S., Hultman, C. M., Kolevzon, A., Gross, R., MacCabe, J. H., & Reichenberg, A. (2012). Advancing maternal age is associated with increasing risk for autism: A review and meta-analysis. *Journal of the American Academy of Child and Adolescent Psychiatry, 51*(5), 477–486.e1. https://doi.org/10.1016/j.jaac.2012.02.018

Sandin, S., Schendel, D., Magnusson, P., Hultman, C., Surén, P., Susser, E., Grønborg, T., Gissler, M., Gunnes, N., Gross, R., Henning, M., Bresnahan, M., Sourander, A., Hornig, M., Carter, K., Francis, R., Parner, E., Leonard, H., Rosanoff, M., Stoltenberg, C., et al. (2016). Autism risk associated with parental age and with increasing difference in age between the parents. *Molecular Psychiatry, 21*(5), 693–700. https://doi.org/10.1038/mp.2015.70

Sarkar, P., Bergman, K., O'Connor, T. G., & Glover, V. (2008). Maternal antenatal anxiety and amniotic fluid cortisol and testosterone: Possible implications for foetal programming. *Journal of Neuroendocrinology, 20*(4), 489–496.

Schaefer, C. A., Brown, A. S., Wyatt, R. J., Kline, J., Begg, M. D., Bresnahan, M. A., & Susser, E. S. (2000). Maternal prepregnant body mass and risk of schizophrenia in adult offspring. *Schizophrenia Bulletin, 26*(2), 275–286. https://doi.org/10.1093/oxfordjournals.schbul.a033452

Schendel, D., & Bhasin, T. K. (2008). Birth weight and gestational age characteristics of children with autism, including a comparison with other developmental disabilities. *Pediatrics, 121*(6), 1155–1164.

Schieve, L. A., Tian, L. H., Rankin, K., Kogan, M. D., Yeargin-Allsopp, M., Visser, S., & Rosenberg, D. (2016). Population impact of preterm birth and low birth weight on developmental disabilities in US children. *Annals of Epidemiology, 26*(4), 267–274. https://doi.org/10.1016/j.annepidem.2016.02.012

Schmidt, D., Reber, S. O., Botteron, C., Barth, T., Peterlik, D., Uschold, N., et al. (2010). Chronic psychosocial stress promotes systemic immune activation and the development of inflammatory Th cell responses. *Brain, Behavior, and Immunity, 24*(7), 1097–1104.

Schneider, S., Freerksen, N., Röhrig, S., Hoeft, B., & Maul, H. (2012). Gestational diabetes and preeclampsia - Similar risk factor profiles? *Early Human Development, 88*(3), 179–184. https://doi.org/10.1016/j.earlhumdev.2011.08.004

Seebeck, J., Sznajder, K. K., & Kjerulff, K. H. (2023). The association between prenatal psychosocial factors and autism spectrum disorder in offspring at 3 years: A prospective cohort study. *Social Psychiatry and Psychiatric Epidemiology*, 1–11.

Segenreich, D., Paez, M. S., Regalla, M. A., Fortes, D., Faraone, S. V., Sergeant, J., & Mattos, P. (2015). Multilevel analysis of ADHD, anxiety and depression symptoms aggregation in families. *European Child and Adolescent Psychiatry, 24*(5), 525–536. https://doi.org/10.1007/s00787-014-0604-1

Shao, S., Wang, J., Huang, K., Wang, S., Liu, H., Wan, S., Yan, S., Hao, J., Zhu, P., & Tao, F. (2020). Prenatal pregnancy-related anxiety predicts boys' ADHD symptoms via placental C-reactive protein. *Psychoneuroendocrinology, 120*, 104797. https://doi.org/10.1016/j.psyneuen.2020.104797

Shelton, J. F., Geraghty, E. M., Tancredi, D. J., Delwiche, L. D., Schmidt, R. J., Ritz, B., Hansen, R. L., & Hertz-Picciotto, I. (2014). Neurodevelopmental disorders and prenatal residential proximity to agricultural pesticides: The CHARGE study. *Environmental Health Perspectives, 122*, 1103–1109. https://doi.org/10.1289/ehp.1307044

Sidebottom, A. C., Hellerstedt, W. L., Harrison, P. A., & Jones-Webb, R. J. (2017). Prenatal care: Associations with prenatal depressive symptoms and social support in low-income urban women. *Archives of Women's Mental Health, 20*(5), 633–644. https://doi.org/10.1007/s00737-017-0730-0

Silverman, J. G., Decker, M. R., Reed, E., & Raj, A. (2006). Intimate partner violence victimization prior to and during pregnancy among women residing in 26 US states: Associations with maternal and neonatal health. *American Journal of Obstetrics and Gynecology, 195*(1), 140–148.

Singh, R., & Munakomi, S. (2019). *Embryology, neural tube*. StatPearls Publishing.

Šrám, R. J., Binková, B., Dejmek, J., & Bobak, M. (2005). Ambient air pollution and pregnancy outcomes: A review of the literature. *Environmental Health Perspectives, 113*(4), 375–382.

Sujan, A. C., Öberg, A. S., Quinn, P. D., & D'Onofrio, B. M. (2019). Annual Research Review: Maternal antidepressant use during pregnancy and offspring neurodevelopmental problems–a critical review and recommendations for future research. *Journal of Child Psychology and Psychiatry, 60*(4), 356–376.

Suglia, S. F., Gryparis, A., Wright, R. O., Schwartz, J., & Wright, R. J. (2008). Association of black carbon with cognition among children in a prospective birth cohort study. *American Journal of Epidemiology, 167*(3), 280–286.

Symeonides, C., Ponsonby, A. L., Vuillermin, P., Anderson, V., & Sly, P. (2014). Environmental chemical contributions to ADHD and the externalizing disorders of childhood—A review of epidemiological evidence. *Journal of Environmental Immunology and Toxicology, 1*, 92–104. https://doi.org/10.7178/jeit.21

Tammela, O. K. (2022). Neurodevelopmental impairments in childhood after moderate and late preterm birth in Finland: A summary of national register studies. *Medical Research Archives, 10*(10) https://doi.org/10.18103/mra.v10i10.3205

Tanda, R., Salsberry, P. J., Reagan, P. B., & Fang, M. Z. (2013). The impact of prepregnancy obesity on children's cognitive test scores. *Maternal and Child Health Journal, 17*(2), 222–229. https://doi.org/10.1007/s10995-012-0964-4

Thomas, P., Zahorodny, W., Peng, B., Kim, S., Jani, N., Halperin, W., & Brimacombe, M. (2012). The association of autism diagnosis with socioeconomic status. *Autism, 16*(2), 201–213. https://doi.org/10.1177/1362361311413397

Thompson, R. (2007). Mothers' violence victimization and child behavior problems: Examining the link. *American Journal of Orthopsychiatry, 77*(2), 306–315. https://doi.org/10.1037/0002-9432.77.2.306

Tolins, M., Ruchirawat, M., & Landrigan, P. (2014). The developmental neurotoxicity of arsenic: Cognitive and behavioral consequences of early life exposure. *Annals of Global Health, 80*, 303–314. https://doi.org/10.1016/j.aogh.2014.09.005

Truman, J. L., & Morgan, R. E. (2014). *Nonfatal domestic violence*. US Department of Justice, Bureau of Justice Statistics.

Tyler, C. R., & Allan, A. M. (2014a). The effects of arsenic exposure on neurological and cognitive dysfunction in human and rodent studies: A review. *Current Environmental Health Reports, 1*, 132–147. https://doi.org/10.1007/s40572-014-0012-1

Varcin, K. J., & Jeste, S. S. (2017). The emergence of autism spectrum disorder: Insights gained from studies of brain and behaviour in high-risk infants. *Current Opinion in Psychiatry, 30*(2), 85–91.

Varela, R. B., Cararo, J. H., Tye, S. J., Carvalho, A. F., Valvassori, S. S., Fries, G. R., & Quevedo, J. (2022). Contributions of epigenetic inheritance to the predisposition of major psychiatric disorders: Theoretical framework, evidence, and implications. *Neuroscience and Biobehavioral Reviews, 135*, 104579. https://doi.org/10.1016/j.neubiorev.2022.104579

Viktorin, A., Lichtenstein, P., Lundholm, C., Almqvist, C., D'Onofrio, B. M., Larsson, H., Landén, M., & Magnusson, P. K. E. (2016). Selective serotonin re-uptake inhibitor use during pregnancy: Association with offspring birth size and gestational age. *International Journal of Epidemiology, 45*(1), 170–177. https://doi.org/10.1093/ije/dyv351

Viktorin, A., Uher, R., Kolevzon, A., Reichenberg, A., Levine, S. Z., & Sandin, S. (2017). Association of antidepressant medication use during pregnancy with intellectual disability in offspring. *JAMA Psychiatry, 74*(10), 1031–1038. https://doi.org/10.1001/jamapsychiatry.2017.1727

Volk, H. E., Kerin, T., Lurmann, F., Hertz-Picciotto, I., McConnell, R., & Campbell, D. B. (2014). Autism spectrum disorder: Interaction of air pollution with the MET receptor tyrosine kinase gene. *Epidemiology, 25*(1), 44–47.

von Ehrenstein, O. S., Aralis, H., Cockburn, M., & Ritz, B. (2014). In utero exposure to toxic air pollutants and risk of childhood autism. *Epidemiology, 25*(6), 851–858.

Wen, S. W., Yang, Q., Garner, P., Fraser, W., Olatunbosun, O., Nimrod, C., & Walker, M. (2006). Selective serotonin reuptake inhibitors and adverse pregnancy outcomes. *American Journal of Obstetrics and Gynecology, 194*(4), 961–966. https://doi.org/10.1016/j.ajog.2006.02.019

Wen, X., Lawal, O. D., Belviso, N., Matson, K. L., Wang, S., Quilliam, B. J., & Meador, K. J. (2021). Association between prenatal opioid exposure and neurodevelopmental outcomes in early childhood: A retrospective cohort study. *Drug Safety, 44*(8), 863–875. https://doi.org/10.1007/s40264-021-01080-0

White, C. L., Pistell, P. J., Purpera, M. N., Gupta, S., Fernandez-Kim, S. O., Hise, T. L., Keller, J. N., Ingram, D. K., Morrison, C. D., & Bruce-Keller, A. J. (2009). Effects of high fat diet on Morris maze performance, oxidative stress, and inflammation in rats: Contributions of maternal diet. *Neurobiology of Disease, 35*(1), 3–13. https://doi.org/10.1016/j.nbd.2009.04.002

Wieck, A., Grassi-Oliveira, R., Hartmann Do Prado, C., Teixeira, A. L., & Bauer, M. E. (2014). Neuroimmunoendocrine interactions in post-traumatic stress disorder: Focus on long-term implications of childhood maltreatment. *Neuroimmunomodulation, 21*(2–3), 145–151. https://doi.org/10.1159/000356552

Williams, J. F., & Smith, V. C. (2015). Fetal alcohol spectrum disorders. *Pediatrics, 136*(5), e20153113. https://doi.org/10.1542/peds.2015-3113

Windham, G. C., Zhang, L., Gunier, R., Croen, L. A., & Grether, J. K. (2006). Autism spectrum disorders in relation to distribution of hazardous air pollutants in the San Francisco Bay area. *Environmental Health Perspectives, 114*(9), 1438–1444.

Winneke, G. (2011). Developmental aspects of environmental neurotoxicology: Lessons from lead and polychlorinated biphenyls. *Journal of Neurology Sciences, 308*, 9–15. https://doi.org/10.1016/j.jns.2011.05.020

Wise, A., Parham, F., Axelrad, D. A., Guyton, K. Z., Portier, C., Zeise, L., Zoeller, R. T., & Woodruff, T. J. (2012). Upstream adverse effects in risk assessment: A model of polychlorinated biphenyls, thyroid hormone disruption and neurological outcomes in humans. *Environmental Research, 117*, 90–99. https://doi.org/10.1016/j.envres.2012.05.013

Wu, S., Wu, F., Ding, Y., Hou, J., Bi, J., & Zhang, Z. (2017). Advanced parental age and autism risk in children: A systematic review and meta-analysis. *Acta Psychiatrica Scandinavica, 135*(1), 29–41.

Xu, R. T., Chang, Q. X., Wang, Q. Q., Zhang, J., Xia, L. X., Zhong, N., Yu, Y. H., Zhong, M., & Huang, Q. T. (2017). Association between hypertensive disorders of pregnancy and risk of autism in offspring: A systematic review and meta-analysis of observational studies. *Oncotarget, 9*(1), 1291–1301. https://doi.org/10.18632/oncotarget.23030

Yehuda, R. (2005). Neuroendocrine aspects of PTSD. In F. Holsboer & A. Ströhle (Eds.), *Anxiety and anxiolytic drugs* (Handbook of Experimental Pharmacology, vol 169). Springer. https://doi.org/10.1007/3-540-28082-0_13

Yeoh, S. L., Eastwood, J., Wright, I. M., Morton, R., Melhuish, E., Ward, M., & Oei, J. L. (2019). Cognitive and motor outcomes of children with prenatal opioid exposure: A systematic review and meta-analysis. *JAMA Network Open, 2*(7), e197025. https://doi.org/10.1001/jamanetworkopen.2019.7025

Yildiz, P. D., Ayers, S., & Phillips, L. (2017). The prevalence of posttraumatic stress disorder in pregnancy and after birth: A systematic review and meta-analysis. *Journal of Affective Disorders, 208*, 634–645. https://doi.org/10.1016/j.jad.2016.10.009

Yim, I. S., & Kofman, Y. B. (2019). The psychobiology of stress and intimate partner violence. *Psychoneuroendocrinology, 105*, 9–24.

Yoon, P. W., Freeman, S. B., Sherman, S. L., Taft, L. F., Gu, Y., Pettay, D., Dana Flanders, W., Khoury, M. J., & Hassold, T. J. (1996). Advanced maternal age and the risk of down syndrome characterized by the meiotic stage of the chromosomal error: A population-based study. *American Journal of Human Genetics, 58*(3), 628–633.

Zelazo, P. R. (2001). A developmental perspective on early autism: Affective, behavioral, and cognitive factors. In J. A. Burack, T. Charman, N. Yirmiya, & P. R. Zelazo (Eds.), *The development of autism: Perspectives from theory and research* (pp. 39–60). Laurence Erlbaum Associates. https://doi.org/10.4324/9781410600196

Perinatal Risk and Later Intellectual and Developmental Disabilities

Angela Gigliotti-Manessis, Tara M. Rutter, Kevin Cook, Livia Merrill, Marta Korom, Thiviya Selvanathan, Maria Catalina Camacho, Fetal, Infant, and Toddler Neuroimaging Group (FIT'NG), and Johanna Bick

Abstract Intellectual and developmental disabilities (IDD) are lifelong medical conditions that affect neurodevelopmental trajectories. Numerous risk factors have been linked to IDD, including biomedical and environmental influences. However, specific underlying etiology is not necessarily easily identified. Nongenetic pathways are dependent upon sensitive periods of neurodevelopment, including the perinatal period. In this chapter, we review nongenetic perinatal risk factors that influence fetal brain development and confer risk to later IDD diagnosis in childhood. Following this, we outline factors that mitigate risk for future offspring development of IDD. Given that environmental factors are inherently modifiable, we discuss research implications for future public health policy and advocacy.

Authors Angela Gigliotti-Manessis and Tara M. Rutter are the joint first-authorship to this chapter.

A. Gigliotti-Manessis
Teachers College, Columbia University, New York, NY, USA

T. M. Rutter
Department of Psychiatry, Oregon Health & Science University, Portland, OR, USA

K. Cook
Developing Brain Institute, Children's National Hospital, Washington, DC, USA

L. Merrill
Department of Psychology, University of Houston, Houston, TX, USA

M. Korom
Department of Psychological and Brain Science, University of Delaware, Newark, DE, USA

T. Selvanathan
Department of Pediatrics, BC Children's Hospital and University of British Columbia, Vancouver, BC, Canada

Department of Pediatrics, The Hospital for Sick Children, Toronto, ON, Canada

M. G. Valdovinos (ed.), *Intellectual and Developmental Disabilities*, Contemporary Clinical Neuroscience,
https://doi.org/10.1007/978-3-031-66932-3_9

Keywords Perinatal exposures · Neurodevelopment · IDD · Sensitive periods · Infant development · Environmental exposure

Intellectual and developmental disabilities (IDD) are characterized by impairments or altered developmental trajectories of intellectual and adaptive functioning (American Psychiatric Association [APA], 2013) and represent neurodevelopmental disorders that begin prior to 18 years of age. In terms of prevalence rates, 1 out of 100 children will develop an IDD following birth (Maulik et al., 2011), which can emerge from both genetic and nongenetic sources. Genetically linked IDD represent 30% to 50% of all IDD and result from various genetic abnormalities, including chromosomal alterations (e.g., trisomy 21), inherited genetic signatures (e.g., Fragile X syndrome), and/or single gene traits (e.g., Prader-Willi syndrome) (Kaufman et al., 2010; Rauch et al., 2006). Nongenetically linked IDD arise from a

Fetal, Infant, and Toddler Neuroimaging Group (FIT'NG).
FIT'NG Group: Sahar Ahmad (The University of North Carolina at Chapel Hill, USA), Banu Ahtam (Division of Newborn Medicine, Boston Children's Hospital, Department of Pediatrics, Harvard Medical School), Laurie Bayet (Department of Neuroscience, American University), Kaitlyn Campbell (Temple University), Bosi Chen (New York University, Langone Health), Fahimeh Darki (Department of Clinical Neuroscience, Karolinska Institutet, Sweden), Angela Easterling (Florida Atlantic University), Sonja J. Fenske (Newcastle University, United Kingdom), Collin Gregg (Virginia Tech), Cassandra L. Hendrix (New York University, Langone Health), Brittany R. Howell (Fralin Biomedical Research Institute at VTC, Department of Human Development and Family Science, Virginia Tech), Janelle Liu (Cedars-Sinai Medical Center), Gina Marie Mason (Department of Psychiatry and Human Behavior, Alpert Medical School of Brown University, Sleep Research Laboratory, E.P. Bradley Hospital, Providence, RI, USA), Roberta Pineda (University of Southern California), Caroline D. Rae (School of Psychology, The University of New South Wales and Neuroscience Research Australia, Sydney), Dustin Scheinost (Department of Radiology & Biomedical Imaging, Yale School of Medicine), Sarah Shultz (Department of Pediatrics, Emory University School of Medicine), Marisa Spann (Columbia University), Chloe M. Sobolewski (Washington University in St. Louis, School of Medicine), Moriah E. Thomason (New York University, Langone Health), Ted K. Turesky (Harvard Graduate School of Education), Alena Uus (King's College London), Lauren Wagner (The University of California, Los Angeles), Pew-Thian Yap (University of North Carolina at Chapel Hill), Elizabeth Yen (Tufts Medicine Pediatrics-Boston Children's Hospital).

M. C. Camacho
Department of Psychiatry, Washington University in St. Louis, St. Louis, MO, USA

J. Bick (✉)
University of Houston, Houston, TX, USA
e-mail: jrbick@uh.edu

variety of risk factors or events that occur pre- and perinatally. Mechanistic pathways that connect nongenetic exposures with later IDD are less understood, despite the high prevalence. This chapter will review the current understanding of common exposures known to increase risk for IDD and elucidate key directions for future work.

IDD include a range of co-occurring neurodevelopmental conditions, such as autism spectrum disorder (autism), cerebral palsy, Down syndrome, fetal alcohol spectrum disorders, schizophrenia, and attention-deficit hyperactivity disorder (ADHD). IDD populations experience a myriad of health disparities, namely, decreased life expectancy (Anderson et al., 2013). In the United States, approximately 17% of children aged 3–17 years are diagnosed with an IDD (Cogswell et al., 2022). Data from the National Health Interview Survey suggests that the prevalence of *any* diagnosed developmental disability increased from 2019 to 2021 among children 3–17 years, from 7.40% to 8.56% in the United States (Zablotsky et al., 2023), denoting an increase in prevalence. IDD conditions manifest in the early childhood period, altering an individual's neurodevelopmental trajectory compared to their typically developing peers. The lifelong impacts associated with an IDD diagnosis underscore the need for research and healthcare support to improve the well-being of individuals with IDD and their families.

Perinatal exposures have long been shown to increase the risk for IDD in later childhood, although pathways that causally explain adverse outcomes are only beginning to be understood. Progress in this area depends on careful consideration of the multiple and diverse exposures and complex interactions that increase IDD risk via immediate, short-term, and longer-term effects that unfold across childhood. *Developmental equifinality* (Cicchetti & Rogosch, 1996) refers to the idea that exposure to diverse sets of perinatal exposures can have similar phenotypic expressions (in the case of IDD, related to cognitive, adaptive, or motor delays). Thus, identifying mechanistic pathways leading to IDD will require an understanding of how a wide range of perinatal exposures (preterm birth, infection, neurotoxin exposure, etc.) influence key developmental systems. *Developmental multifinality* (Cicchetti & Rogosch, 1996) refers to the idea that any one exposure can lead to a heterogeneous range of outcomes, which is also relevant for understanding pathways that connect perinatal exposures to IDD risk. For example, preterm birth is not only associated with risk for IDD but a wide range of neuropsychiatric disorders and/or growth delays, and despite increased risk, many children also show no long-term consequences (van Baar et al., 2005). Adding to the complexity, perinatal risk exposures often co-occur, making it difficult to disentangle the individual versus interactive influences of these adversities on developmental risk. Timing or severity of exposure, infant characteristics (male sex, low birth weight), and neuroprotective factors relate to the relative risk for IDD. Neonatal factors (male sex, low birth weight), postnatal environmental risk, and neuroprotective factors may influence the likelihood of later risk and resilience (Huang et al., 2016).

Multiple theories from various disciplines consider how and why perinatal factors may influence later IDD risk. The most prominent or widely cited theories include the "fetal programming hypothesis" (Seckl & Holmes, 2007), the "developmental programming hypotheses" (Barker, 2004; Langley-Evans, 2006, 2015), and the "developmental origins of health and disease" (DOHAD) (Barker, 1990, 2004). These models generally hypothesize that intrauterine environments, events, or injury/disturbances influence developmental trajectories at a sensitive point in development. Programmatically, these exposures influence neurocognitive, behavioral, and physical development across neonatal, infant, toddler, later childhood, and even adult stages of life. A discussion of the differences in theoretical perspective goes beyond the scope of the chapter. However, these theoretical viewpoints offer a framework for understanding how and why such a diverse set of exposures can have similar or vast consequences related to IDD (and comorbid developmental problems, which we discuss later in this chapter). For example, perinatal stressors, including nutrient or oxygen deprivation, teratogen exposure, and psychosocial adversity, are hypothesized to causally influence epigenetic and hypoxic changes in the fetal brain and body, which then influence hormones, metabolic activity, and brain development. In combination, these cascading influences predispose offspring to a spectrum of adverse outcomes. Pathways may involve myriad disturbances including altered inflammatory, glucocorticoid, or HPA axis programming (Packard et al., 2016; Xiong & Zhang, 2013), GABA signaling (Braat & Kooy, 2015; Cherubini & Ben-Ari, 2023; Deidda et al., 2014; Sgadó et al., 2011; Tang et al., 2021), and macro-level changes to the brain structure and functional organization (Bohlken et al., 2014). Further, the risk that any one exposure will lead to later IDD will depend on timing, severity, co-occurring risk factors, and neuroprotective factors (Fig. 1).

In what follows, we consider risk factors during the perinatal period that have implications for IDD in offspring. These are potential consequences of one's genetic profile, environmental exposures, and birth experiences. Some known perinatal exposures that impact developmental outcomes are term and preterm brain injury,

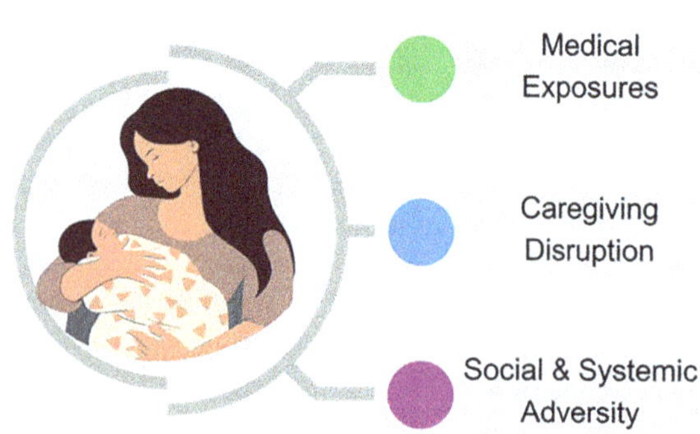

Fig. 1 Overview of perinatal factors that influence IDD risk and development

such as intraventricular hemorrhage, white matter injury, birth trauma—subpial hemorrhage, large subarachnoid hemorrhage, and perinatal stroke. Perinatal infections such as bacterial meningitis and viral meningoencephalitis present known risk factors. Other risk factors that will be addressed include infants born prematurely and infants requiring intensive neonatal interventions. First, we will define various adversities that occur during the perinatal period. Second, we will connect these exposures with various IDD that emerge in later childhood. Third, we will discuss evidence for prevention and intervention in neonatal intensive care units (NICUs) and in early development, shown to effectively mitigate risks associated with perinatal adversities and IDD in later childhood. Finally, we will discuss key gaps in understanding and addressing structural barriers that inform policy, and future directions for research.

Defining Perinatal Exposures That Pose Risk for IDD

There are many etiological pathways during brain development implicated in the perinatal environment that can consequently increase risk of IDD. We will review some common perinatal exposures that are known to undermine healthy brain development and increase risk for IDD. They are grouped into two overarching themes: medical exposures (perinatal brain injury, infection, and prematurity) and social adversities (maternal-infant separation in the NICU, and maternal mental health affecting bonding and responsiveness).

Medical Exposures

Perinatal Brain Injury The early postnatal period is a time of significant risk for brain injury and can take the form of numerous etiologies. Most commonly, 1.5 in 1000 term-born infants experience rates of hypoxic-ischemic encephalopathy (HIE) experience rates of hypoxic-ischemic encephalopathy (HIE) in 1.5 in 1000 births (Kurinczuk et al., 2010) and stroke occurs in stroke in 0.6 in out of 1000 births (Laugesaar et al., 2007). Across causes, brain injury during the perinatal period has been associated with later neurodevelopmental impairments including motor, language, and cognitive delays (Novak et al., 2018). While there is no single cause for many of these types of injury, there are multiple contributing factors that play a role in increased prevalence.

One of the most cited causes of neonatal brain injury relates to birth trauma or damage caused as a direct result of labor and/or delivery (Gupta & Cabacungan, 2021). Rather than a specific cause, birth trauma constitutes a wide range of insults across the spectrum, from minor cuts and scrapes to severe and life-threatening damage to the brain and other organs. Of factors that make birth trauma more likely is the use of vacuums and forceps during delivery (Cieplinski & Bhutani, 1996; Linder et al., 2013) and infants who are large for gestational age, which can be common in mothers who are obese or experience gestational diabetes (Boulet et al., 2003; Persson et al. (2013); Shuffrey et al., 2023). Beyond direct mechanical causes, numerous other factors can contribute to significant brain injury, with the most common being hypoxic events. These events represent instances where insufficient oxygen reaches the brain and are associated with greater neuro developmental difficulties, especially in motor domains such as those implicated with cerebral palsy (Graham et al., 2008). Importantly, while brain injury can impact infants born at any age, it is more common in infants born prematurely due to an increased vulnerability of the immature brain to perinatal insults and a greater prevalence of prematurity-related complications with lower birth gestational age (Inder et al., 2023).

Perinatal Infections One of the other major risks for later neurodevelopmental concerns is perinatal infections, including neonatal sepsis, meningitis, and various viruses that cause high fevers and respiratory infections such as pneumonia and bronchiolitis. Due to the relatively immature immune system at birth and before, neonates are particularly susceptible to infection and vulnerable to longer-term impacts on development (Ygberg & Nilsson 2012). This is especially true for infants born prematurely who have even less mature immune responses compared to term-born infants—although differences quickly minimize between term- and preterm-born infants shortly after birth (Olin et al., 2018).

Of the most common infections, neonatal sepsis, which is an extreme response to an infection, often occurs within the first days of life and impacts as many as 2.2% of infants. Although often treatable, long-lasting neurological effects can be observed throughout childhood (Glaser et al., 2021). Like sepsis, neonatal meningitis can arise from a diverse set of causes; however, it ultimately results in severe inflammation of the meninges, the membranes surrounding the brain. Importantly, more severe cases are associated with significant motor and cognitive disabilities later in childhood (Ku et al., 2015). In both sepsis and meningitis, inflammation has been reported as the primary cause of subsequent impairments likely due to its impacts on the brain (Humberg et al., 2020). Conversely, respiratory disorders in the neonatal period are common, including pneumonia and bronchiolitis, but rather than just inflammation, these disorders can also lead to severe respiratory distress which impacts oxygenation saturation (Gill et al., 2022; Nissen, 2007). Further, studies have reported that two known viruses such as Zika (Mlakar et al., 2016) and Herpes Simplex (Fa et al., 2020) when contracted during pregnancy leading to infant in utero exposure have caused structural cortical changes. Regardless of the cause, however, the presence of neonatal infection is a significant risk factor for later neurodevelopmental impairments including cerebral palsy and intellectual disability (Rand et al., 2016; Mitha et al., 2013).

Prematurity and Related Complications Premature birth is one of the most common risk factors for later IDD. While most infants are born at term around 40 gestational weeks, approximately 1 in 10 births are infants born preterm before 36 gestational weeks (Osterman et al., 2023; Walani, 2020). While those rates have remained consistent over the past two decades (Blencowe et al., 2012; Gyamfi-Bannerman & Ananth, 2014), advancements in neonatal care have dramatically increased the survival rate (Bell et al., 2022; Younge et al., 2017) and also the age at which ex utero survival for extremely premature infants is possible. However, there is consistent evidence that infants born preterm are at greater risk for a wide range of developmental delays and impairments, including pronounced sensorimotor deficits (Cabral et al., 2016; Hee Chung et al., 2020), language delays (Barre et al., 2011; Woodward et al., 2009), and broad cognitive impairments (Hee Chung et al., 2020; Kerr-Wilson et al., 2012) alongside higher rates of neurodevelopmental disorders (Bhutta et al., 2002; Joseph et al., 2017; Larroque et al, 2008; Sucksdorff et al., 2015). IDD is a potential risk factor for very prematurely born babies, as studies have found that prematurity may cause future development of neurobehavioral needs related to lower IQ, inattention, learning problems, and mental health problems (Perlman, 2001; Hack & Taylor, 2000).

The preterm period, between birth and term equivalent age, represents a particular period of neuro-vulnerability. The period typically comprising the late second through third trimesters is a period of rapid brain development. During this period, in utero fetuses experience rapid increases in brain volume (Andescavage et al., 2017), strengthening of the brain's white matter tracts (Jaimes et al., 2020), and development of the earliest functional proto-networks (Turk et al., 2019; van den Heuvel & Thomason, 2016). Perturbations to these developmental processes likely influence later neurodevelopmental sequelae. In terms of impact on the developing brain, preterm birth has been associated with structural and functional alterations that are unique and not related to focal brain injury. For example, prematurity has been connected with specific reductions in cortical and subcortical gray matter, diminished cortical gyrification, and delayed maturation in gray and white matter structures (Ball et al., 2012; Keunen et al., 2012; Rathbone et al., 2011; Shimony et al., 2016; Volpe, 2009; Thompson et al., 2007). Recent work indicates that premature birth influences connectivity patterns across the whole brain (Doria et al., 2010; Smyser et al., 2016; Scheinost et al., 2017) and has a pronounced effect on connectivity that supports thalamo-cortical circuitry, both structurally (Ball et al., 2013, 2014) and functionally (Doria et al., 2010; Smyser et al., 2010; Toulmin et al., 2015). Reductions in inter-hemispheric connectivity (Smyser et al., 2010, 2013) and altered lateralization of language regions (Kwon et al., 2015) have also been associated with premature birth. Recent work shows that prematurity may increase connectivity between superior parietal regions and motor areas, which may explain risk for developmental coordination problems that emerge later in life (Eyre et al., 2021). These disruptions in long-range and intrahemispheric connectivity seem to be candidate pathways that increase risk for long-term cognitive deficits and IDD risk that emerges in later childhood. White matter injury related to intraventricular hemorrhage is also common in high-risk preterm babies and can manifest as diminished inter and intrahemispheric connectivity on functional levels (Grotheer et al., 2023; Omidvarnia et al., 2015; Smyser et al., 2013).

IDD risk may stem from myriad maturational differences and injuries to the developing brain of prematurely born children. The most prevalent differences and injuries include perinatal brain injury, co-occurring medical diagnoses that alter circulatory or pulmonary function, and changes arising from development occurring in the extrauterine rather than intrauterine environment (Polglase et al., 2014; Sarda et al., 2021; Volpe, 2009; Yates et al., 2021). Beyond those injuries, however, many infants experience comorbid medical conditions in tandem with their premature birth. While all forms of medical comorbidities are of concern in preterm infants for later neurodevelopment in particular are those that can impair either blood flow to the brain or the oxygenation of blood being carried to the brain. These can include brief complications such as a short-term hypoxic event or longer-term medical diagnoses. Among the list of common co-occurring medical conditions in premature infants, those such as patent ductus arteriosus (PDA) or bronchopulmonary dysplasia (BPD) are associated with neurodevelopmental outcomes in childhood (Gudmundsdottir et al., 2021; Hee Chung et al., 2020; Schmidt et al., 2003). Even with infants without significant injury or comorbidities, altered developmental trajectories can be observed. One likely explanation is the stark difference between the intrauterine and extrauterine environments. The intrauterine environment radically differs from the extrauterine, especially for younger-born infants who are admitted to the NICU. During intrauterine development, fetuses are in a heavily shielded environment protected from noxious stimuli such as bright lights, loud noises, and intense tactile experiences. Daily life in a NICU is significantly more sensorially intense with loud noises—ranging from constant machine hum to alarm alerts, which can reach as loud as 85 decibels (Slevin et al., 2000; Williams et al., 2007)—and bright, intense lights (White, 2020), which have been associated with poorer outcomes (Ream & Lehwald, 2018). Efforts to reduce sensory stimulation and stress in the NICU are burgeoning, which may reduce risk for atypical development.

Social Input and Caregiving Adversities

Thus far, this chapter has focused on medical hazards that cause IDD. However, burgeoning evidence over the last decades has connected psychosocial adversity, and toxic stress, with risk for IDD and other neurodevelopmental and mental health problems. The experience of extreme stress prior to conception through the postpartum period is increasingly shown to shape developmental outcomes. Effects can occur independently or interactively with other stressors. We focus on two common stressors that infants are likely to face in the NICU or immediately following birth which can increase the risk for IDD in later childhood.

Separations and Disruptions in Caregiving Following typical birth, newborns spend much of their time in close contact with their primary caregiver, which serves to support the regulatory capabilities of the infant and promote bonding. Infants cared for in the NICU are often deprived of the same level of close physical contact with primary caregivers, which is increasingly shown to act as an additional stress exposure at this formative phase of life. Duration of separations may last up to weeks or months for the highest risk cases, adding an additional challenge to healthy development on top of myriad other stressors related to medical procedures, overstimulating conditions, pain, and instability. Not surprisingly, stress exposure during NICU, including separations from caregivers, has been linked with altered functional connectivity and decreased brain size in infancy (Smith et al., 2011) and risk for IDD or related neurodevelopmental delays when in toddlerhood (Chau et al., 2013). Painful procedures in the NICU have been linked to altered early brain development. This may be more relevant to the section above on preterm birth—explaining some of the within-group variance in outcomes among babies born preterm (Brummelte et al., 2012). Later in this chapter, we will discuss growing support for interventions that promote close physical contact and minimal separation between mothers and their babies cared for in the NICU.

Systemic Factors, Poverty, and Institutional Racism

Mental illness and poverty can exacerbate the challenges associated with IDD by the limited access to healthcare, education, and support services, in addition to adverse environmental conditions that can affect brain development. Addressing poverty and its associated disparities is critical for improving outcomes for individuals with IDD. Factors that provide support can often buffer against ill effects. By offering affordable services and collaborative treatment with medical professionals, educational systems, and family members, it is possible to provide early diagnosis and intervention, improving long-term outcomes for these individuals. Specialized resources and quality education in policy reformation not only scaffold the development and learning of those affected but also contribute to breaking the cycle of poverty that can exacerbate rates of IDD.

Racism and Discrimination Racism and discrimination escalate liability for IDD risk. Disability status including IDD is known to impact socioeconomic status, affecting SES outcomes such as educational achievement, income, and employment (Gage et al., 2021). Individuals with IDD demonstrate higher unemployment rates than their nondisabled peers and employment opportunities are, on average, lower paying (Quierós et al., 2015). Systemic, institutional, and personal experiences of discrimination, such as those related to one's sexual orientation/identity, ethnicity, and religious and cultural practices, can marginalize women and increase stress during pregnancy. For example, women predominantly from Black and Latinx communities report higher at-risk adverse mental health outcomes associated with systematic racism and preterm birth are reportedly higher for these communities (Alhusen et al., 2016; Bower et al., 2018, 2023). From the extant literature, social inequities demonstrate strong relations with an individual's IDD status.

Poverty Approximately 719 million people around the world live in impoverished conditions, with the COVID-19 pandemic recently exacerbating this issue (CDC, 2023). Poverty exerts a significant and detrimental effect, particularly on those with IDD. Epidemiological research has consistently demonstrated an association between poverty and IDD across countries, with the prevalence of IDD higher among those in lower socioeconomic positions (Emerson, 2007). Restricted access to healthcare services due to poverty can result in delayed diagnosis and treatment, hindering early intervention efforts. This limitation coupled with reduced awareness and opportunities for educational enrichment can compound the challenges faced by those with these disorders (Graham, 2005). Furthermore, inadequate nutrition and food insecurity can impair healthy brain function and development (Prado & Dewey, 2014), potentially worsening the impact of these disorders and escalating the incidence of IDD (Rose-Jacobs et al., 2008). Cross-sectional research from the US National Survey of Children's Health found that among preterm children, children living in households that could not afford nutritious meals were more likely to have a learning disability compared to households who could afford nutritious meals (Okoli et al., 2022). The chronic stressors and adverse environmental factors prevalent in impoverished households may increase the severity of IDD (Hertzman & Boyce, 2010).

Consequences of Perinatal Risk Related to IDD

Thus far, we have discussed how environmental risks during the perinatal period can consequently cause IDD, reviewing common perinatal exposures that are known to undermine healthy brain development. Below, we focus on IDD, mapping the phenotypes related to an IDD, with implications in a facet of developmental domains—cognitive function, adaptive function, motor disabilities, learning disabilities, and language delays. We also discuss the association with risk for autism, attention disorders, and other known psychiatric comorbidities.

Defining Intellectual and Developmental Disabilities

Diagnostic criteria for IDD are defined by multiple organizations and bodies, including the International Classification of Disease (ICD-11) from the World Health Organization (WHO), the American Association on Intellectual and

Developmental Disabilities (AAIDD; Schalock et al., 2010), and the APA Diagnostic Statistical Manual (5th ed., text rev.; *DSM-5-TR*; American Psychiatric Association, 2022). Definitions across these organizations define IDD based on three criteria involving (1) cognitive functioning including intellectual abilities such as reasoning, problem-solving, memory, and learning; (2) adaptive functioning, encompassing the practical skills required for everyday life, such as communication, self-care, social interactions, and the ability to live independently; and (3) a manifestation of symptoms of intellectual and adaptive disability occurring in childhood. The severity of intellectual developmental disorder (intellectual disability) is categorized into different levels through the DSM-5-TR, ranging from mild, moderate, severe, and profound. Diagnosis is usually made in childhood, based on standardized assessments and clinical evaluations. The process is often expensive and quite involved, so there are likely many individuals, especially with mild IDD, who go undiagnosed. IDD may occur in isolation or in combination with a number of diagnoses or impairments, including psychomotor disability, language delay, autism, developmental coordination disorder, ADHD, and other forms of developmental psychopathology including depression, anxiety, psychotic disorders, and somatic conditions. Throughout childhood, children with IDD often show various learning difficulties (in math, reading, etc.) and neurocognitive difficulties in memory, attention, and executive function. Below, we review the various functional domains associated with IDD understood to be influenced by perinatal adversities.

Cognitive Disability Intelligence is measured by different conventional psychometric instruments, such as the Kaufman Assessment Battery for Children (Singer et al., 2012), Wechsler intelligence scales (Wechsler, 2014), and Stanford-Binet intelligence scales (Roid & Pomplun, 2012), among others, which result in the intelligence quotient (IQ). IQ, measured typically as a standard score with an average of 100 (standard deviation of 15), allows for comparison of the performance of a child with established age-graded norms. IQ tests measure cognitive skills. Although the definition of intelligence and the use of psychometric intelligence tests are controversial (Eysenck, 1971; Ganuthula & Sinha, 2019), most psychologists agree that intelligence involves a combination of many mental processes to learn from situations, apply knowledge and skills, and think abstractly. While some research has indicated that the age of reaching developmental milestones was associated with IQ in later years (Murray et al., 2007), early language skills seem to be more predictive of later IQ than other domains (Peyre et al., 2017).

Motor Disabilities A wide range of IDD entails delays or difficulties in gross motor functions such as sitting or standing, fine movements such as grabbing an object using an index finger and thumb, and general motor coordination necessary for complex systematized movements such as walking or running (Biotteau et al., 2020; Lucas et al., 2016). Rather than a single diagnosis, motor disabilities constitute a large umbrella of disorders, which in early life include cerebral palsy, ataxia, tremors, and various tic disorders. Many motor disorders, such as cerebral palsy, are thought to arise prior to birth, although diagnosis often does not occur until at least around 6-months of age (Novak et al., 2017). Typically, diagnosis occurs once early motor milestones (such as rolling from front to back, sitting independently for brief periods, or passing toys from one hand to another) do not develop when expected. Early interventions are often thought to be key in mitigating the long-term impacts of motor disorders, with earlier intervention yielding more efficacious outcomes throughout the lifespan (Hadders-Algra, 2021). For example, premature infants are frequently referred for motor interventions prior to formal diagnosis (Spittle et al., 2015).

Language and Communication Disabilities Language and communication disabilities represent a broad umbrella of symptoms and disorders that are primarily marked by difficulty expressing oneself and can include delays in reaching normative speech milestones and difficulties engaging in age-appropriate expressive language to the absence of speech entirely (APA, 2013). By the age of 3, it is believed that language and communication disabilities affect approximately 6% of children (Boyle et al., 1996). Beyond the direct issues associated with communication difficulties, speech and language disabilities at early ages can often be an early indicator of later concerns. Toddlers with language and communication difficulties are at greater risk for later developmental concerns including cognitive delays and intellectual disabilities (Marrus & Hall 2017), broader social impairments (St Clair et al., 2011), and elevated risk for psychiatric disorders (Snowling et al., 2006) including autism spectrum disorder (Luyster et al., 2007). Using advanced neuroimaging techniques, such as electroencephalogram (EEG), magnetic resonance imaging (MRI), and functional near-infrared spectroscopy (fNIRS), neural signatures of complex interactions of brain systems have been implicated in language development (Kuhl, 2010). These neuroimaging techniques are also used to investigate the neural correlates of nonverbal communication skills in infants and children, including those with difficulties in nonverbal communication (e.g., interpreting social cues, facial expressions; Bayet & Nelson, 2019).

Learning Disabilities Learning disabilities (LDs) with rates that suggest a global impact on about 5% of school-aged children, encompass several disorders that affect the acquisition, retention, comprehension, or application of verbal and nonverbal information, and interfere with specific aspects of school achievement (Lagae, 2008). The DSM-5 categorizes LDs into three major academic domains: reading, writing, and mathematics, with difficulties lasting for at least 6 months (APA, 2013). Specific LDs include dyscalculia, dysgraphia, and dyslexia, the latter accounting for approximately 80% of LDs (Shaywitz et al., 1998). While IQ may correlate with LD, it is important to note that they are separate constructs. LDs can mimic or be comorbid with other neurodevelopmental disorders, making it imperative to distinguish symptoms for proper diagnosis. While early detection can be challenging, typical detection occurs around third grade, after the child is in the academic setting for several years. Family history seems to be the best parameter to select those at risk. Once diagnosed, it is essential for the treatment to encompass a collaborative effort, involving the family, school, medical professionals, and psychologists (Cortiella & Horowitz, 2014).

Related Neurodevelopmental Conditions Autism, attention disorders, and schizophrenia can co-occur with IDD, implicating various neurological domains such as cognitive flexibility, social impairment, attention, memory, and emotion regulation. Symptoms emerge over time, due to similar pathways (changes in gray and white matter, ventricular enlargement) or distinct pathways and processes that are altered during brain development. Associations with perinatal risk are reviewed in the section that follows.

Autism IDD frequently co-occur with autism spectrum disorder, which is a pervasive developmental condition characterized by impaired communication and social interaction as well as restricted, repetitive patterns of behavior, interests, or activities. The term "autism" is used to refer to the autism spectrum, which affects approximately 1 in 36 children (Maenner et al., 2023). Population-based case-control studies have provided evidence that autism is related to adverse pregnancy and delivery processes (Larsson et al., 2005). While there is an overlap between risk factors for ID and autism, not all autistic individuals are also diagnosed with ID. Perinatal risk factors implicated in the subsequent development of autism and overlap with IDD include the following: *Maternal autoimmune and inflammatory disorders* are also associated with a child receiving an autism diagnosis (Meltzer & van de Water, 2017; Chen et al., 2016). *Premature birth* (Allen et al., 2020; Crump et al., 2021) and/or *low birth weight* (Lampi et al., 2012; Talmi et al., 2020) are also at an increased likelihood for autism; similarly, *low APGAR scores at birth* have been linked to subsequent autism diagnosis (Modabbernia et al., 2019). Large cohort studies have found that *advanced maternal* and *paternal age* are independently associated with a higher likelihood of offspring receiving an autism diagnosis (Sandin et al., 2016). Additionally, recent studies have found a link between *a family history of psychiatric problems* and a subsequent diagnosis of autism in offspring (Xie et al., 2019). Taking all these findings together, it is important to explore these factors as well as those discussed in this chapter as potential risks for a co-diagnosis of autism and IDD.

Attention-Deficit/Hyperactivity Disorder (ADHD) Attention-deficit/hyperactivity disorder (ADHD) is characterized as a persistent pattern of inattention and/or hyperactivity-impulsivity that interferes with daily functioning or development (APA, 2013). Youth diagnosed with ADHD often experience symptoms related to deficits in executive function and challenges with behavioral self-regulation (APA, 2013). While not all children with ADHD have a co-diagnosis of IDD, the same perinatal experiences conferring risk for IDD are linked to attention disorders, including prenatal stress, exposure to toxicants and teratogens, and maternal mental health depression disturbance. These developmentally manifest as symptoms of inattention and/or hyperactivity-impulsivity, often in combination with other disorders (Wiggs et al., 2016).

Schizophrenia Childhood-onset schizophrenia is a psychiatric disorder that affects emotions, temperament, and behavior. A child with this disorder has aberrant behavior and emotional manifestations, such as psychotic symptoms (i.e., having strange ideas, thoughts, or feelings that are not based on reality; Keepers et al., 2020). Schizophrenia is not typically diagnosed in pediatric populations younger than age 12. Often, the psychotic symptoms start in adolescence and are more common in males earlier in development (Albert & McCaig, 2015). Schizophrenia is often linked to IDD, and they are not mutually inclusive; nonetheless, a schizophrenia diagnosis does not necessarily imply an IDD. Schizophrenia is also a consequence of genetic and environmental exposure, discussed in this chapter. The causes of schizophrenia are well known and include brain pathology, genetics, environmental factors, and gene-environmental interactions. The DOHAD model provides robust evidence toward a neurodevelopmental model, wherein developmental complications as early as the late first or early second trimester lead to the activation of pathologic neural circuits during adolescence or young adulthood (Fatemi & Folsom, 2009).

Prevention and Intervention

So far, our review has focused on the risk for IDD stemming from multiple adverse exposures and experiences. It is equally important to emphasize how positive environments or interventions can mitigate risk for IDD and support healthy development in the short and long term. Considering the significant neuroplasticity during the neonatal period, early interventions prove to be the most effective in mitigating potential risk factors for IDD (Cioni et al., 2016). Therapeutic input can range from basic application of neuroprotective practices that support brain health during fetal and perinatal phases of development, to higher-level nutritional interventions, to supporting bonding and reducing stressful separations between caregivers and infants around the time of birth. A nonexhaustive list of interventions that will be reviewed below includes kangaroo care, nutritional interventions, family-integrated care, pain reduction techniques, and medical prevention.

Medical, Pain Management, and Nutritional Intervention

Invasive Medical Intervention While any infant may receive early medical interventions, infants born preterm and those who are term born with genetic or congenital disorders are at greater likelihood of receiving both noninvasive and invasive medical procedures. For all infants, major interventions such as surgeries during the perinatal period can confer significant risks for later neurodevelopment, while even noninvasive procedures can confer risk for some infants, such as those born preterm.

For all infants, major invasive procedures are typically only undertaken when it is deemed medically essential given the particular vulnerability of long-term impacts on later development. For example, infants born with congenital heart disease are often at risk for serious health outcomes without surgical interventions. However, it's becoming increasingly apparent that receiving those life-saving interventions often is associated with an elevated risk for later neurodevelopmental impairments (Howell et al., 2019). Importantly, this effect is not limited to cardiac disorders, as surgical interventions for noncardiac disorders see similar increases in neurodevelopmental knock-on effects in childhood (Moran et al., 2019). There are many perioperative factors across cardiac and noncardiac disorders that may account for these observed differences. First, vulnerability to brain injury remains a persistent concern throughout the perinatal period with surgery often associated with subsequent injury related to hypoxia and inflammation (Abella et al., 2015; Pironkova et al., 2017). Moreover, administration of anesthetics during surgery is associated with greater apoptosis of nervous system cells and greater risk for neurodevelopmental delays (Sanders et al., 2013).

NICU-Based Pain Reduction Techniques Infants born at younger ages often experience extended NICU stays and, even among the healthier infants, often experience repeated exposures to painful events such as heel sticks to collect routine blood draws. These can add up quickly; however, very preterm infants can often experience over 100 skin break procedures over the standard course of care before they are discharged, with some reaching as high as 600 (Carbajal et al., 2008; Cong et al., 2017). While these heel sticks, venipuncture, and similar skin break procedures are routine in adults, they may pose a significantly larger risk in younger-born preterm infants. Specifically, pain sensation pathways are considered mature by 20 weeks gestation (Anand & Hickey, 1987; Fitzgerald, 1993), while the pain modulatory response is not fully developed until after birth and nociceptive blocking signaling may not be available until nearing term age (Fitzgerald, 1991). As a result, earlier-born infants may be at unique risk of insufficiently moderate pain response. Importantly, the relationship between pain exposure and neurodevelopment has been consistently observed in preterm infants, with greater exposures associated with broad neurodevelopmental impairments during infancy and early childhood (Valeri et al., 2015; Vinall et al., 2014).

Pain reduction techniques in the NICU include strategies to minimize the discomfort experienced by premature infants during medical procedures, such as intubation or needle pricks. While these techniques do not directly target IDD, reducing pain and stress during the NICU stay may have an indirect effect on long-term neurodevelopment. Studies have shown that interventions such as sucrose administration or nonpharmacological measures, such as facilitated tucking and gentle human touch, can reduce pain and stress in preterm infants (Stevens et al., 2016). Extending this finding, one RCT has also shown that skin-to-skin care can reduce biobehavioral pain responses in preterm infants (Cong et al., 2011), further supporting the idea that skin-to-skin contact may be one of the most effective, nonpharmacological pain reduction techniques. In sum, the evidence for a direct link between pain reduction techniques in the NICU and later IDD outcomes is not robust yet, but given the promising initial findings, future research is highly encouraged.

NICU-Based Nutritional Interventions NICUs play a pivotal role in providing the necessary care for at-risk infants during the perinatal period and are ideal locations for implementing interventions that can be continued at home (Aucott et al., 2002; Soleimani et al., 2020). Nutritional interventions in the NICU and following discharge from the hospital aim to optimize the overall health of preterm infants by ensuring adequate nutrition through parenteral and enteral feeding. Although breastfeeding may lead to slower weight gain in preterm infants than when receiving formula, breastfeeding has been associated with reduced risk for poor neurodevelopmental outcomes and better odds for a larger head circumference at 6 months and 2 and 5 years of age (Rozé et al., 2012). Correlational research also suggests that breastfeeding may reduce the risk of certain developmental disorders, such as autism spectrum disorders (Xiang et al., 2023) and attention-deficit/hyperactivity disorder (Oddy et al., 2010). While these interventions are essential for neonatal survival and growth, more evidence is needed to measure the impact on reducing developmental and intellectual disability risk.

Medical Treatments as Prevention Medical prevention and interventions are commonly provided in NICUs for respiratory support and infection management to promote infant survival and long-term health (Ho et al., 2023). Effective respiratory support can reduce the risk of brain injury and improve neurodevelopmental outcomes in at-risk infants (Polin & Sahni, 2017). For instance, surfactant therapy is commonly used to treat respiratory distress syndrome in preterm infants, as it improves lung function and oxygenation while simultaneously reducing the risk of hypoxia-related brain injury and improving neurodevelopmental outcomes (Polin & Sahni, 2017). Other medical interventions focus on neuroprotection, by minimizing brain injury and improving long-term neurodevelopmental outcomes with medications, including caffeine citrate for apnea of prematurity (Schmidt & Davis, 2016). Such interventions have the potential to indirectly reduce risk for IDD, but their causal effects are yet to be fully understood.

Maternal-Fetal Dyadic Interventions

Family-based prevention and intervention programs prior to birth and in the NICU not only contribute to the overall well-being of the infant but also empower parents to become active participants in their children's care and development. These interventions can help reduce the risk of developmental disorders by creating a supportive, nurturing, and cognitively stimulating environment for infants. We discuss two examples below.

Kangaroo Care Kangaroo care (KC), also known as skin-to-skin care, involves placing the infant on their parents' chest in a diaper and a hat, maximizing skin-to-skin contact. Randomized controlled trials (RCT) suggest that KC has causal benefits for infant neurocognitive health. For example, El-Farrash et al. (2020) found that KC causally improves infant attention and regulation, lowers scores on lethargy, and reduces nonoptimal reflexes. Moreover, a 20-year follow-up of an RCT conducted by Charpak et al. (2017) showed that KC significantly reduced the risk of neurodevelopmental disabilities among young adults who were born preterm. The study found sustained positive effects of KC on cognitive and motor outcomes compared to the traditional incubator care group. Correlational work also suggests that KC is associated with better emotional bonding between the parent and the infant, which may have an indirect positive impact on infants' cognitive development (Feldman et al., 2002), reduces infant stress hormone levels, lowers maternal postpartum depression (Cristóbal Cañadas et al., 2022), improves cognitive control, and enhances autonomic nervous system functioning (Feldman et al., 2014). Finally, a meta-analysis by Conde-Agudelo and colleagues (2011) found a positive impact of KC on developmental outcomes, and a recent systematic review by Boundy et al. (2016) suggested reduced incidence of sepsis, improved weight gain, and enhanced neurodevelopmental outcomes in preterm infants who received KC. The strength of evidence for KC's impact on reducing IDD risk is substantial, making it a cornerstone intervention in the NICU.

Attachment-Based Interventions One such early parenting program is the modified Attachment and Biobehavioral Catch-up (mABC), which was developed for low-income mothers who are dependent on opioids (Labella et al., 2021). Intervention sessions begin during the third trimester when parent coaches support mothers in preparing for anticipated challenges, such as caring for their infants who may be challenging to soothe while remaining sensitive and nurturing to the infant. mABC parenting program sessions continue throughout the first 6–12 months of the infant's life. Recent evidence from an RCT suggests that ABC causally enhances opioid-exposed infants' autonomic nervous system functioning (Tabachnick et al., 2022) and high-risk children's long-term executive functioning (Korom et al., 2021; Lind et al., 2017).

In addition to ABC, a recent systematic review by McAndrew and colleagues (2022) reviewed 19 NICU-based studies, all involving family-focused interventions that actively engaged families in some aspect of patient care in intensive care units. Although no physiological differences were observed between control and intervention groups, those receiving family-based interventions showed higher weight, shorter NICU stay, and increased alertness and total wakefulness time compared to the control group. Parents in the intervention group also endorsed reduced stress and lower respiratory rates when interacting with their vulnerable infants than parents in the control group. Although family-based interventions are a promising avenue for mitigating the risk for IDD, additional investigation into the effectiveness of interventions that involve family participation is needed.

Postnatal Follow-Up Care

Interventions that are currently being implemented through early intervention in the United States encompass many developmental areas, such as language, motor, and cognition. While the origins and principles guiding these interventions are developmentally sound, it is important to remember the developmental considerations for each individual with IDD and other co-occurring conditions. The most common interventions and services provided are occupational (adaptive, sensory, fine motor), physical (gross motor, limbic), and speech-language therapy, as well as special instruction and applied behavioral analysis, which uses behavioral psychology to teach skills acquisition through natural reinforcement. The field has shifted to implement Naturalistic Developmental Behavioral Interventions (NDBI), which are implemented in the individual's known environment (often the home), "involve shared control between child and therapist, utilize natural contingencies, and use a variety of behavioral strategies to teach developmentally appropriate and prerequisite skills" (Schreibman et al., 2015). Other important strategies during early childhood include parenting interventions which have the potential to alleviate stress in the household, empower parents, and teach parents how to play with infants in cognitively stimulating ways. These strategies may alter the course of symptoms and improve the well-being of the family members caring for a child with an IDD (Vanegas et al., 2022).

Integrative Summary, Implications for Policy, and Future Directions

This section will provide an integrative summary of our review, with the goal of highlighting key directions for research, prevention, practice, and policy. First, as reviewed in part one of our chapter, IDD may result from a broad number of environmental, experiential, and nongenetic risk factors during the perinatal period of human brain development. Adverse exposures can occur antenatally, influencing fetal brain development, and separately or in combination with risk that are encountered perinatally, related to severe premature birth, anoxia or hypoxia, and be correlated with low APGAR scores and low birth weight or poor growth. Effects of perinatal exposures can be compounded by social and caregiving stressors experienced while infants are cared for in the NICU and throughout later infancy. For this reason, perinatal risk factors may be better understood as increasing initial risk or perturbation in fetal and infant brain development that may set the infant on a trajectory of risk that is exacerbated by additional challenges encountered following birth. Timing of exposures and severity of both risk and protective factors are therefore essential for understanding the likelihood of risk for IDD.

Part two of this chapter focused on defining the phenotype associated with IDD risk. IDD involves a constellation of delays across various cognitive, memory, and learning domains and is often accompanied by additional syndromes and disorders, including autism, and ADHD. Much of the existing work has attempted to connect various perinatal exposures with late emerging latent or categorically defined IDD risk. As heterogeneity and comorbidity are the norm rather than the exception, progress in delineating etiological pathways may stem from more carefully considering developmental trajectories of specific neurocognitive processes that are measurable in newborns and infants (such as including development of processing speed, visual attention, working memory, attention, inhibition, cognitive flexibility, and social or emotional processing, attention to faces) and, over time, contribute to the development of more complex syndromes, such as IDD and related disorders later in life. In this sense, our field may need to place more focus on the developmental aspect of IDD in order to fully understand how to support individual pathways toward resiliency. As covered in our review, work must also consider the role of caregiving experiences, cultural factors, and additional environmental or systemic hazards will be essential for mitigating ongoing risk.

Considering the timing of exposures is also key for future work, given emerging evidence that the effects of some exposures are timing dependent. Advances in fetal and infant neuroimaging may offer new insight into how the impact of any given exposures may depend on the timing or extent to which it disrupts the dominant processes at that stage of central nervous system development (ranging from disruptions in neurogenesis to neural migration, to neuronal differentiation, to axonogenesis, to synaptogenesis, followed by normative apoptosis and pruning of excessive synapses through the first years of life; Linderkamp et al. (2009)). For example, certain exposures (brain insults) may have formative and severe influences if they occur early and disrupt the formation of the CNS, potentially leading to profound IDD risk. However, interruptions in synaptic remodeling at later stages of fetal and infant brain development may be more likely to impair neuronal networks that support ongoing memory, attention, and learning. With supportive caregiving over the subsequent periods of postnatal development, impacts of the perinatal exposures on risk for IDD may be more attenuated or mild.

It has long been appreciated that perinatal risk factors are not randomly distributed, but instead are stratified across the population, with the greatest burden of risk for more marginalized or impoverished families. Large-scale neuroimaging studies have drawn connections between perinatal risk and longer-term brain and cognitive delays and have revealed that effects may be exacerbated by lower socioeconomic status (Alnaes et al., 2020), pointing to the need for more progress in improved screening, detection, and prevention for marginalized populations. For example, infants with multiple significant pre- and perinatal risk factors can be offered additional follow-up or services even if they do not display overt clinical symptoms. Education to parents on early warning signs of IDD or other neurodevelopmental disorders can be offered prior to hospital discharge and at ongoing infant wellness checks. Preventative home visits that support cognitive stimulation and caregiver responsiveness can be offered to support families and reduce risk for compounding adversities that enhance IDD risk in toddlerhood and preschool years. As we reviewed in part 3, there are a multitude of existing intervention and prevention programs that aim to reduce risk for IDD. Training hospital professionals to integrate ongoing educational and prevention strategies onto existing medical or preventative treatments may maximally support feasibility and implementation.

It is important to note that social-environmental factors are complex and multidimensional and include education, occupation, household income, and material resources but also factors such as parental psychosocial stress, parenting style, and a cognitively stimulating home environment which may all impact child development. To develop effective interventions and public policies that more equitably reduce risk for IDD, further studies are needed to understand the specific mechanisms through which social-environmental factors impact neurodevelopment. The effects of specific social adversities and early life stress on brain development and neurocognition can be studied in animal models; these are difficult to control in human studies given the complex intersections between these factors (Hackman et al., 2010). For example, postnatally, chronic stress during pregnancy in rodents may result in altered mother-infant interactions including a decrease in the frequency of licking and grooming behaviors of rat pups. These rat pups who are exposed to fewer licking and grooming behaviors have decreased NMDA receptor levels and expression of growth factors in the hippocampus which results in decreased synaptic formation and cognitive performance (Liu et al., 2000). This may provide some insight into the neural basis of parenting interventions, which have been shown to be effective in modifying neurodevelopmental outcomes in children including those with brain injury and IDD. In rodent models, environmental enrichment (Rosenzweig et al., 1978) results in increased neurogenesis, gliogenesis, and synapse formation in the hippocampus and cortex as well as improved memory and learning (Kempermann et al., 1997; Rampon et al., 2000; Van Praag et al., 2000). These findings complement those of human studies observing the importance of a cognitively stimulating environment on neurodevelopmental outcomes, providing a mechanism through which interventional programs reduce IDD risk and enhance long-term educational success and employability. Human clinical trials of social and parenting interventions involving randomized trials are the gold standard approach for establishing causation in humans. Results from such trials should be leveraged to guide policy changes, with the strongest effects for reducing IDD and comorbidities in at-risk families. Interestingly, a recent randomized control trial is testing the impact of improved financial support to impoverished families (based on the provision of monthly lower or higher levels of unconditional cash transfers) can improve early life brain function and psychosocial outcomes (Noble et al., 2021). They found that infants in the high-cash group showed more power in high-frequency bands on EEG, which has been previously associated with higher language and cognitive scores in other studies (Benasich et al., 2008; Williams et al., 2012). Longitudinal follow-up of this cohort is ongoing and will provide further insight into whether these alterations in the brain persist over time (Noble et al., 2021; Troller-Renfree et al., 2022).

Closing Summary

IDD can arise from genetic or nongenetic causes and encompass a range of neurodevelopmental disorders. The relatively high prevalence of IDD motivates the search for potential mechanisms and exposures during the perinatal period that can cause an IDD. Perinatal exposures known to increase risk for IDD in later childhood include medical exposures (perinatal brain injury, infection, and prematurity), environmental exposures (chemical exposure, environmental hazards known to undermine brain development, such as lead exposure, cigarette smoke, polluted air, and pesticides), and social adversities (maternal-infant separation in the NICU, and maternal mental health affecting bonding and responsiveness). As these adversities often co-occur, disentangling their individual versus interactive developmental effects is difficult. Postnatal neural plasticity and environmental supports have the potential to buffer relations between environmental insults and offspring brain development, potentially reducing later risk for IDD.

The extant literature reviewed here provides an overall understanding of the causes of IDD. Yet, more research is needed to understand its mechanisms and improve treatment or intervention options, including for co-occurring conditions. Progress on this front will inform expanded continuity of care and help close the gap on discontinuity, by implementing interdisciplinary approaches across all medical and clinical disciplines to support both the parent and child. Early intervention and services are critical to deliver positive outcomes in later child development. Yet, we need to consider the development trajectory of the child and the potential need for lifelong services, which currently arguably lack in resources given the structural barriers. Thus, future research should not only investigate the neural pathways and systems implicated in IDD but also critically examine ways to maximize the systematic implementation of relevant intervention programs in a variety of settings, thereby informing policy and improving access.

Acknowledgments This work was supported in part by the National Center for Complementary and Integrative Health (NCCIH) of the National Institutes of Health (NIH) through a postdoctoral fellowship awarded to T.M.R. (NIH-NCCIH T32 AT002688). Its contents are solely the responsibility of the authors and do not necessarily represent the official views of the NCCIH or NIH.

References

Abella, R., Varrica, A., Satriano, A., Tettamanti, G., Pelissero, G., Gavilanes, A. D. W., Zimmermann, L. J., Vles, H. J., Strozzi, M. C., Pluchinotta, F. R., & Gazzolo, D. (2015). Biochemical markers for brain injury monitoring in children with or without congenital heart diseases. *CNS & Neurological Disorders Drug Targets, 14*, 12–23. https://doi.org/10.2174/1871527314666150116114648

Albert, M., & McCaig, L. F. (2015). Emergency department visits related to schizophrenia among adults aged 18-64: United States, 2009-2011. *NCHS Data Brief, 215*, 1–8.

Alhusen, J. L., Bower, K. M., Epstein, E., & Sharps, P. (2016). Racial discrimination and adverse birth outcomes: An integrative review. *Journal of Midwifery & Women's Health, 61*(6), 707–720. https://doi.org/10.1111/jmwh.12490

Allen, L., Leon-Attia, O., Shaham, M., Shefer, S., & Gabis, L. V. (2020). Autism risk linked to prematurity is more accentuated in girls. *PLoS One, 15*, e0236994. https://doi.org/10.1371/journal.pone.0236994

Alnæs, D., Kaufmann, T., Marquand, A. F., Smith, S. M., & Westlye, L. T. (2020). Patterns of sociocognitive stratification and perinatal risk in the child brain. *Proceedings of the National Academy of Sciences, 117*, 12419–12427. https://doi.org/10.1073/pnas.2001517117

American Psychiatric Association. (2013). Diagnostic and statistical manual of mental disorders (5th ed.). https://doi.org/10.1176/appi.books.9780890425596

American Psychiatric Association. (2022). *Diagnostic and statistical manual of mental disorders* (5th ed., text rev.). https://doi.org/10.1176/appi.books.9780890425787

Anand, K. J., & Hickey, P. R. (1987). Pain and its effects in the human neonate and fetus. *The New England Journal of Medicine, 317*, 1321–1329. https://doi.org/10.1056/NEJM198711193172105

Anderson, L. L., Humphries, K., McDermott, S., Marks, B., Sisirak, J., & Larson, S. (2013). The state of the science of health and wellness for adults with intellectual and developmental disabilities. *Intellectual and Developmental Disabilities, 51*, 385–398. https://doi.org/10.1352/1934-9556-51.5.385

Andescavage, N. N., du Plessis, A., McCarter, R., Serag, A., Evangelou, I., Vezina, G., Robertson, R., & Limperopoulos, C. (2017). Complex trajectories of brain development in the healthy human fetus. *Cerebral Cortex, 27*, 5274–5283. https://doi.org/10.1093/cercor/bhw306

Aucott, S., Donohue, P. K., Atkins, E., & Allen, M. C. (2002). Neurodevelopmental care in the NICU. *Mental Retardation and Developmental Disabilities Research Reviews, 8*(4), 298–308. https://doi.org/10.1002/mrdd.10040

Ball, G., Boardman, J. P., Rueckert, D., Aljabar, P., Arichi, T., Merchant, N., Gousias, I. S., Edwards, A. D., & Counsell, S. J. (2012). The effect of preterm birth on thalamic and cortical development. *Cerebral Cortex, 22*, 1016–1024. https://doi.org/10.1093/cercor/bhr176

Ball, G., Boardman, J. P., Aljabar, P., Pandit, A., Arichi, T., Merchant, N., Rueckert, D., Edwards, A. D., & Counsell, S. J. (2013). The influence of preterm birth on the developing thalamocortical connectome. *Cortex, 49*, 1711–1721. https://doi.org/10.1016/j.cortex.2012.07.006

Ball, G., Aljabar, P., Zebari, S., Tusor, N., Arichi, T., Merchant, N., Robinson, E. C., Ogundipe, E., Rueckert, D., Edwards, A. D., & Counsell, S. J. (2014). Rich-club organization of the newborn human brain. *Proceedings of the National Academy of Sciences of the United States of America, 111*, 7456–7461. https://doi.org/10.1073/pnas.1324118111

Barker, D. J. (1990). The fetal and infant origins of adult disease. *BMJ: British Medical Journal, 301*(6761), 1111. https://doi.org/10.1136/bmj.301.6761.1111

Barker, D. J. P. (2004). The developmental origins of adult disease. *Journal of the American College of Nutrition, 23*(sup6), 588S–595S. https://doi.org/10.1080/07315724.2004.10719428

Barre, N., Morgan, A., Doyle, L. W., & Anderson, P. J. (2011). Language abilities in children who were very preterm and/or very low birth weight: A meta-analysis. *The Journal of Pediatrics, 158*, 766–774.e1. https://doi.org/10.1016/j.jpeds.2010.10.032

Bayet, L., & Nelson, C. A. (2019). The perception of facial emotion in typical and atypical development. In *Handbook of emotional development* (pp. 105–138).

Bell, E. F., Hintz, S. R., Hansen, N. I., Bann, C. M., Wyckoff, M. H., DeMauro, S. B., Walsh, M. C., Vohr, B. R., Stoll, B. J., Carlo, W. A., Van Meurs, K. P., Rysavy, M. A., Patel, R. M., Merhar, S. L., Sánchez, P. J., Laptook, A. R., Hibbs, A. M., Cotten, C. M., D'Angio, C. T., Winter, S., et al. (2022). Mortality, in-hospital morbidity, care practices, and 2-year Outcomes for extremely preterm infants in the US, 2013–2018. *JAMA, 327*, 248–263. https://doi.org/10.1001/jama.2021.23580

Benasich, A. A., Gou, Z., Choudhury, N., & Harris, K. D. (2008). Early cognitive and language skills are linked to resting frontal gamma power across the first 3 years. *Behavioural Brain Research, 195*(2), 215–222. https://doi-org.ezproxy.cul.columbia.edu/10.1016/j.bbr.2008.08.049

Bhutta, A. T., Cleves, M. A., Casey, P. H., Cradock, M. M., & Anand, K. J. (2002). Cognitive and behavioral outcomes of school-aged children who were born preterm: A meta-analysis. *JAMA, 288*, 728–737. https://doi.org/10.1001/jama.288.6.728

Biotteau, M., Albaret, J. M., & Chaix, Y. (2020). Developmental coordination disorder. *Handbook of Clinical Neurology, 174*, 3–20. https://doi.org/10.1016/B978-0-444-64148-9.00001-6

Blencowe, H., Cousens, S., Oestergaard, M. Z., Chou, D., Moller, A. B., Narwal, R., Adler, A., Vera Garcia, C., Rohde, S., Say, L., & Lawn, J. E. (2012). National, regional, and worldwide estimates of preterm birth rates in the year 2010 with time trends since 1990 for selected countries: A systematic analysis and implications. *Lancet, 379*, 2162–2172. https://doi.org/10.1016/S0140-6736(12)60820-4

Bohlken, M. M., Brouwer, R. M., Mandl, R. C., van Haren, N. E., Brans, R. G., van Baal, G. C. M., et al. (2014). Genes contributing to subcortical volumes and intellectual ability implicate the thalamus. *Human Brain Mapping, 35*, 2632–2642. https://doi.org/10.1002/hbm.22356

Boulet, S. L., Alexander, G. R., Salihu, H. M., & Pass, M. (2003). Macrosomic births in the United States: Determinants, outcomes, and proposed grades of risk. *American Journal of Obstetrics and Gynecology, 188*, 1372–1378. https://doi.org/10.1067/mob.2003.302

Boundy, E. O., Dastjerdi, R., Spiegelman, D., Fawzi, W. W., Missmer, S. A., Lieberman, E., & Kajeepeta, S. (2016). Kangaroo mother care and neonatal outcomes: A meta-analysis. *Pediatrics, 137*(1), e20152238. https://doi.org/10.1542/peds.2015-2238

Bower, K. M., Geller, R. J., Perrin, N. A., & Alhusen, J. (2018). Experiences of racism and preterm birth: Findings from a pregnancy risk assessment monitoring system, 2004 through 2012. *Women's Health Issues, 28*(6), 495–501. https://doi.org/10.1016/j.whi.2018.06.002

Bower, K. M., Geller, R. J., Jeffers, N., McDonald, M., & Alhusen, J. (2023). Experiences of racism and perinatal depression: Findings from the Pregnancy Risk Assessment Monitoring System, 2018. *Journal of Advanced Nursing, 79*(5), 1982–1993. https://doi.org/10.1111/jan.15519

Boyle, J., Gillham, B., & Smith, N. (1996). Screening for early language delay in the 18–36 month age-range: The predictive validity of tests of production, and implications for practice. *Child Language Teaching and Therapy, 12*, 113–127. https://doi.org/10.1177/026565909601200202

Braat, S., & Kooy, R. F. (2015). The GABAA receptor as a therapeutic target for neurodevelopmental disorders. *Neuron, 86*, 1119–1130. https://doi.org/10.1016/j.neuron.2015.03.042

Brummelte, S., Grunau, R. E., Chau, V., Poskitt, K. J., Brant, R., Vinall, J., et al. (2012). Procedural pain and brain development in premature newborns. *Annals of Neurology, 71*(3), 385–396.

Cabral, T. I., da Silva, L. G., Martinez, C. M., & Tudella, E. (2016). Analysis of sensory processing in preterm infants. *Early Human Development, 103*, 77–81. https://doi.org/10.1016/j.earlhumdev.2016.06.010

Carbajal, R., Rousset, A., Danan, C., Coquery, S., Nolent, P., Ducrocq, S., Saizou, C., Lapillonne, A., Granier, M., Durand, P., Lenclen, R., Coursol, A., Hubert, P., de Saint Blanquat, L., Boëlle, P.-Y., Annequin, D., Cimerman, P., Anand, K. J. S., & Bréart, G. (2008). Epidemiology and treatment of painful procedures in neonates in intensive care units. *JAMA, 300*, 60–70. https://doi.org/10.1001/jama.300.1.60

Charpak, N., Tessier, R., Ruiz, J. G., Hernandez, J. T., Uriza, F., Villegas, J., Nadeau, L., Mercier, C., Maheu, F., Marin, J., Cortes, D., Gallego, J. M., & Maldonado, D. (2017). Twenty-year follow-up of kangaroo mother care versus traditional care. *Pediatrics, 139*, e20162063. https://doi.org/10.1542/peds.2016-2063

Chau, V., Synnes, A., Grunau, R. E., Poskitt, K. J., Brant, R., & Miller, S. P. (2013). Abnormal brain maturation in preterm neonates associated with adverse developmental outcomes. *Neurology, 81*, 2082–2089. https://doi.org/10.1212/01.wnl.0000437298.43688.b9

Chen, S. W., Zhong, X. S., Jiang, L. N., Zheng, X. Y., Xiong, Y. Q., Ma, S. J., Qiu, M., Huo, S. T., Ge, J., & Chen, Q. (2016). Maternal autoimmune diseases and the risk of autism spectrum disorders in offspring: A systematic review and meta-analysis. *Behavioural Brain Research, 296*, 61–69. https://doi.org/10.1016/j.bbr.2015.08.035

Cherubini, E., & Ben-Ari, Y. (2023). GABA signaling: Therapeutic targets for neurodegenerative and neurodevelopmental disorders. *Brain Sciences, 13*, 1240. https://doi.org/10.3390/brainsci13091240

Cicchetti, D., & Rogosch, F. A. (1996). Equifinality and multifinality in developmental psychopathology. *Development and Psychopathology, 8*, 597–600. https://doi.org/10.1017/S0954579400007318

Cieplinski, J. A. M., & Bhutani, V. K. (1996). Lactational and neonatal morbidities associated with operative vaginal deliveries. 1191. *Pediatric Research, 39*, 201–201. https://doi.org/10.1203/00006450-199604001-01213

Cioni, G., Inguaggiato, E., & Sgandurra, G. (2016). Early intervention in neurodevelopmental disorders: Underlying neural mechanisms. *Developmental Medicine and Child Neurology, 58*(Suppl 4), 61–66. https://doi.org/10.1111/dmcn.13050

Cogswell, M. E., Coil, E., Tian, L. H., Tinker, S. C., Ryerson, A. B., Maenner, M. J., Rice, C. E., & Peacock, G. (2022). Health needs and use of services among children with developmental disabilities - United States, 2014–2018. *MMWR. Morbidity and Mortality Weekly Report, 71*, 453–458. https://doi.org/10.15585/mmwr.mm7112a3

Conde-Agudelo, A., Belizán, J. M., & Diaz-Rossello, J. (2011). Kangaroo mother care to reduce morbidity and mortality in low birthweight infants. *The Cochrane Database of Systematic Reviews, 3*, CD002771. https://doi.org/10.1002/14651858.CD002771.pub2

Cong, X., Ludington-Hoe, S. M., & Walsh, S. (2011). Randomized crossover trial of kangaroo care to reduce biobehavioral pain responses in preterm infants: A pilot study. *Biological Research for Nursing, 13*, 204–216. https://doi.org/10.1177/1099800410385839

Cong, X., Wu, J., Vittner, D., Xu, W., Hussain, N., Galvin, S., Fitzsimons, M., McGrath, J. M., & Henderson, W. A. (2017). The impact of cumulative pain/stress on neurobehavioral development of preterm infants in the NICU. *Early Human Development, 108*, 9–16. https://doi.org/10.1016/j.earlhumdev.2017.03.003

Cortiella, C., & Horowitz, S. H. (2014). *The state of learning disabilities: Facts, trends and emerging issues* (Vol. 25, pp. 2–45). National Center for Learning Disabilities.

Cristóbal Cañadas, D., Parrón Carreño, T., Sánchez Borja, C., & Bonillo Perales, A. (2022). Benefits of kangaroo mother care on the physiological stress parameters of preterm infants and mothers in neonatal intensive care. *International Journal of Environmental Research and Public Health, 19*, 7183. https://doi.org/10.3390/ijerph19127183

Crump, C., Sundquist, J., & Sundquist, K. (2021). Preterm or early term birth and risk of autism. *Pediatrics, 148*, e2020032300. https://doi.org/10.1542/peds.2020-032300

Deidda, G., Bozarth, I. F., & Cancedda, L. (2014). Modulation of GABAergic transmission in development and neurodevelopmental disorders: Investigating physiology and pathology to gain therapeutic perspectives. *Frontiers in Cellular Neuroscience, 8*, 119. https://doi.org/10.3389/fncel.2014.00119

Doria, V., Beckmann, C. F., Arichi, T., Merchant, N., Groppo, M., Turkheimer, F. E., Counsell, S. J., Murgasova, M., Aljabar, P., Nunes, R. G., Larkman, D. J., Rees, G., & Edwards, A. D. (2010). Emergence of resting state networks in the preterm human brain. *Proceedings of the National Academy of Sciences of the United States of America, 107*, 20015–20020. https://doi.org/10.1073/pnas.1007921107

El-Farrash, R. A., Shinkar, D. M., Ragab, D. A., Salem, R. M., Saad, W. E., Farag, A. S., Salama, D. H., & Sakr, M. F. (2020). Longer duration of kangaroo care improves neurobehavioral performance and feeding in preterm infants: A randomized controlled trial. *Pediatric Research, 87*, 683–688. https://doi.org/10.1038/s41390-019-0558-6

Emerson, E. (2007). Poverty and people with intellectual disabilities. *Mental Retardation and Developmental Disabilities Research Reviews, 13*, 107–113. https://doi.org/10.1002/mrdd.20144

Eyre, M., Fitzgibbon, S. P., Ciarrusta, J., Cordero-Grande, L., Price, A. N., Poppe, T., Schuh, A., Hughes, E., O'Keeffe, C., Brandon, J., Cromb, D., Vecchiato, K., Andersson, J., Duff, E. P., Counsell, S. J., Smith, S. M., Rueckert, D., Hajnal, J. V., Arichi, T., O'Muircheartaigh, J., et al. (2021). The developing human connectome project: Typical and disrupted perinatal functional connectivity. *Brain, 144*, 2199–2213. https://doi.org/10.1093/brain/awab118

Eysenck, H. J. (1971). Relation between intelligence and personality. *Perceptual and Motor Skills, 32*, 637–638. https://doi.org/10.2466/pms.1971.32.2.637

Fa, F., Laup, L., Mandelbrot, L., Sibiude, J., & Picone, O. (2020). Fetal and neonatal abnormalities due to congenital herpes simplex virus infection: A literature review. *Prenatal Diagnosis, 40*(4), 408–414. https://doi.org/10.1002/pd.5587

Fatemi, S. H., & Folsom, T. D. (2009). The neurodevelopmental hypothesis of schizophrenia, revisited. *Schizophrenia Bulletin, 35*(3), 528–548. https://doi.org/10.1093/schbul/sbn187

Feldman, R., Eidelman, A. I., Sirota, L., & Weller, A. (2002). Comparison of skin-to-skin (kangaroo) and traditional care: Parenting outcomes and preterm infant development. *Pediatrics, 110*, 16–26. https://doi.org/10.1542/peds.110.1.16

Feldman, R., Rosenthal, Z., & Eidelman, A. I. (2014). Maternal-preterm skin-to-skin contact enhances child physiologic organization and cognitive control across the first 10 years of life. *Biological Psychiatry, 75*, 56–64. https://doi.org/10.1016/j.biopsych.2013.08.012

Fitzgerald, M. (1991). The development of descending brainstem control of spinal cord sensory processing. In M. Hanson (Ed.), *Foetal and neonatal brainstem: Development and clinical issues* (pp. 127–136).

Fitzgerald, M. (1993). Development of pain pathways and mechanisms. *Pain in Neonates, 5*, 19–37.

Gage, N. A., Katsiyannis, A., Rose, C., et al. (2021). Disproportionate bullying victimization and perpetration by disability status, race, and gender: A National Analysis. *Advances in Neurodevelopmental Disorders, 5*, 256–268. https://doi.org/10.1007/s41252-021-00200-2

Ganuthula, V. R. R., & Sinha, S. (2019). The looking glass for intelligence quotient tests: The interplay of motivation, cognitive functioning, and affect. *Frontiers in Psychology, 10*, 2857. https://doi.org/10.3389/fpsyg.2019.02857

Gill, P. J., Chanchlani, N., & Mahant, S. (2022). Bronchiolitis. *CMAJ: Canadian Medical Association Journal = journal de l'Association medicale canadienne, 194*, E216. https://doi.org/10.1503/cmaj.211810

Glaser, M. A., Hughes, L. M., Jnah, A., & Newberry, D. (2021). Neonatal sepsis: A review of pathophysiology and current management strategies. *Advances in Neonatal Care: Official Journal of the National Association of Neonatal Nurses, 21*, 49–60. https://doi.org/10.1097/ANC.0000000000000769

Graham, H. (2005). Intellectual disabilities and socioeconomic inequalities in health: An overview of research. *Journal of Applied Research in Intellectual Disabilities, 18*(2), 101–111. https://doi.org/10.1111/j.1468-3148.2005.00239.x

Graham, E. M., Ruis, K. A., Hartman, A. L., Northington, F. J., & Fox, H. E. (2008). A systematic review of the role of intrapartum hypoxia-ischemia in the causation of neonatal encephalopathy. *American Journal of Obstetrics and Gynecology, 199*, 587–595. https://doi.org/10.1016/j.ajog.2008.06.094

Grotheer, M., Bloom, D., Kruper, J., Richie-Halford, A., Zika, S., Aguilera González, V. A., et al. (2023). Human white matter myelinates faster in utero than ex utero. *Proceedings of the National Academy of Sciences, 120*(33), e2303491120.

Gudmundsdottir, A., Broström, L., Skiöld, B., Källén, K., Serenius, F., Norman, M., Aden, U., & Bonamy, A.-K. E. (2021). The type and timing of patent ductus arteriosus treatment was associated with neurodevelopment when extremely preterm infants reached 6.5 years. *Acta Paediatrica, 110*, 510–520. https://doi.org/10.1111/apa.15452

Gupta, R., & Cabacungan, E. T. (2021). Neonatal birth trauma: Analysis of yearly trends, risk factors, and outcomes. *The Journal of Pediatrics, 238*, 174–180. https://doi.org/10.1016/j.jpeds.2021.06.080

Gyamfi-Bannerman, C., & Ananth, C. V. (2014). Trends in spontaneous and indicated preterm delivery among singleton gestations in the United States, 2005-2012. *Obstetrics and Gynecology, 124*, 1069–1074. https://doi.org/10.1097/AOG.0000000000000546

Hack, M., & Taylor, H. G. (2000). Perinatal brain injury in preterm infants and later neurobehavioral function. *JAMA, 284*, 1973–1974. https://doi.org/10.1001/jama.284.15.1973

Hackman, D. A., Farah, M. J., & Meaney, M. J. (2010). Socioeconomic status and the brain: Mechanistic insights from human and animal research. *Nature Reviews. Neuroscience, 11*, 651–659. https://doi.org/10.1038/nrn2897

Hadders-Algra, M. (2021). Early diagnostics and early intervention in neurodevelopmental disorders-age-dependent challenges and opportunities. *Journal of Clinical Medicine, 10*, 861. https://doi.org/10.3390/jcm10040861

Hee Chung, E., Chou, J., & Brown, K. A. (2020). Neurodevelopmental outcomes of preterm infants: A recent literature review. *Translational Pediatrics, 9*(Suppl 1), S3–S8. https://doi.org/10.21037/tp.2019.09.10

Hertzman, C., & Boyce, T. (2010). How experience gets under the skin to create gradients in developmental health. *Annual Review of Public Health, 31*, 329–347. https://doi.org/10.1146/annurev.publhealth.012809.103538

Ho, J. J., Zakarija-Grkovic, I., Lok, J. W., Lim, E., Subramaniam, P., & Leong, J. J. (2023). Continuous positive airway pressure (CPAP) for apnoea of prematurity. *The Cochrane Database of Systematic Reviews, 7*(7), CD013660. https://doi.org/10.1002/14651858.CD013660.pub2

Howell, H. B., Zaccario, M., Kazmi, S. H., Desai, P., Sklamberg, F. E., & Mally, P. (2019). Neurodevelopmental outcomes of children with congenital heart disease: A review. *Current Problems in Pediatric and Adolescent Health Care, 49*(10), 100685. https://doi.org/10.1016/j.cppeds.2019.100685

Huang, J., Zhu, T., Qu, Y., & Mu, D. (2016). Prenatal, perinatal and neonatal risk factors for intellectual disability: A systemic review and meta-analysis. *PLoS One, 11*, e0153655. https://doi.org/10.1371/journal.pone.0153655

Humberg, A., Fortmann, I., Siller, B., Kopp, M. V., Herting, E., Göpel, W., Härtel, C., & German Neonatal Network, German Center for Lung Research and Priming Immunity at the beginning of life (PRIMAL) Consortium. (2020). Preterm birth and sustained inflammation: Consequences for the neonate. *Seminars in Immunopathology, 42*, 451–468. https://doi.org/10.1007/s00281-020-00803-2

Inder, T. E., Volpe, J. J., & Anderson, P. J. (2023). Defining the neurologic consequences of preterm birth. *The New England Journal of Medicine, 389*, 441–453. https://doi.org/10.1056/NEJMra2303347

Jaimes, C., Machado-Rivas, F., Afacan, O., Khan, S., Marami, B., Ortinau, C. M., Rollins, C. K., Velasco-Annis, C., Warfield, S. K., & Gholipour, A. (2020). In vivo characterization of emerging white matter microstructure in the fetal brain in the third trimester. *Human Brain Mapping, 41*, 3177–3185. https://doi.org/10.1002/hbm.25006

Joseph, R. M., Korzeniewski, S. J., Allred, E. N., O'Shea, T. M., Heeren, T., Frazier, J. A., Ware, J., Hirtz, D., Leviton, A., Kuban, K., & ELGAN Study Investigators. (2017). Extremely low gestational age and very low birthweight for gestational age are risk factors for autism spectrum disorder in a large cohort study of 10-year-old children born at 23–27 weeks' gestation. *American Journal of Obstetrics and Gynecology, 216*, 304.e1–304.e16. https://doi.org/10.1016/j.ajog.2016.11.1009

Kaufman, L., Ayub, M., & Vincent, J. B. (2010). The genetic basis of non-syndromic intellectual disability: A review. *Journal of Neurodevelopmental Disorders, 2*, 182–209. https://doi.org/10.1007/s11689-010-9055-2

Keepers, G. A., Fochtmann, L. J., Anzia, J. M., Benjamin, S., Lyness, J. M., Mojtabai, R., Servis, M., Walaszek, A., Buckley, P., Lenzenweger, M. F., Young, A. S., Degenhardt, A., & Hong, S.-H. (2020). The American Psychiatric Association practice guideline for the treatment of patients with schizophrenia. *American Journal of Psychiatry, 177*, 868–872. https://doi.org/10.1176/appi.ajp.2020.177901

Kempermann, G., Kuhn, H. G., & Gage, F. H. (1997). More hippocampal neurons in adult mice living in an enriched environment. *Nature, 386*, 493–495. https://doi.org/10.1038/386493a0

Kerr-Wilson, C. O., Mackay, D. F., Smith, G. C., & Pell, J. P. (2012). Meta-analysis of the association between preterm delivery and intelligence. *Journal of Public Health, 34*, 209–216. https://doi.org/10.1093/pubmed/fdr024

Keunen, K., Kersbergen, K. J., Groenendaal, F., Isgum, I., de Vries, L. S., & Benders, M. J. (2012). Brain tissue volumes in preterm infants: Prematurity, perinatal risk factors and neurodevelopmental outcome: A systematic review. *The Journal of Maternal-Fetal & Neonatal Medicine: The Official Journal of the European Association of Perinatal Medicine, the Federation of Asia and Oceania Perinatal Societies, the International Society of Perinatal Obstetricians, 25*(Suppl 1), 89–100. https://doi.org/10.3109/14767058.2012.664343

Korom, M., Goldstein, A., Tabachnick, A. R., Palmwood, E. N., Simons, R. F., & Dozier, M. (2021). Early parenting intervention accelerates inhibitory control development among CPS-involved children in middle childhood: A randomized clinical trial. *Developmental Science, 24*, e13054. https://doi.org/10.1111/desc.13054

Ku, L. C., Boggess, K. A., & Cohen-Wolkowiez, M. (2015). Bacterial meningitis in infants. *Clinics in Perinatology, 42*, 29–viii. https://doi.org/10.1016/j.clp.2014.10.004

Kuhl, P. K. (2010). Brain mechanisms in early language acquisition. *Neuron, 67*(5), 713–727. https://doi.org/10.1016/j.neuron.2010.08.038

Kurinczuk, J. J., White-Koning, M., & Badawi, N. (2010). Epidemiology of neonatal encephalopathy and hypoxic-ischaemic encephalopathy. *Early Human Development, 86*, 329–338. https://doi.org/10.1016/j.earlhumdev.2010.05.010

Kwon, S. H., Scheinost, D., Lacadie, C., Sze, G., Schneider, K. C., Dai, F., Constable, R. T., & Ment, L. R. (2015). Adaptive mechanisms of developing brain: Cerebral lateralization in the prematurely-born. *NeuroImage, 108*, 144–150. https://doi.org/10.1016/j.neuroimage.2014.12.032

Labella, M. H., Eiden, R. D., Roben, C. K. P., & Dozier, M. (2021). Adapting an evidence-based home visiting intervention for mothers with opioid dependence: Modified Attachment and Biobehavioral Catch-up. *Frontiers in Psychology, 12*, 675866. https://doi.org/10.3389/fpsyg.2021.675866

Lagae, L. (2008). Learning disabilities: Definitions, epidemiology, diagnosis, and intervention strategies. *Pediatric Clinics of North America, 55*, 1259–1vii. https://doi.org/10.1016/j.pcl.2008.08.001

Lampi, K. M., Lehtonen, L., Tran, P. L., Suominen, A., Lehti, V., Banerjee, P. N., Gissler, M., Brown, A. S., & Sourander, A. (2012). Risk of autism spectrum disorders in low birth weight and small for gestational age infants. *The Journal of Pediatrics, 161*, 830–836. https://doi.org/10.1016/j.jpeds.2012.04.058

Langley-Evans, S. C. (2006). Developmental programming of health and disease. *Proceedings of the Nutrition Society, 65*(1), 97–105. https://doi.org/10.1079/PNS2005478

Langley-Evans, S. C. (2015). Nutrition in early life and the programming of adult disease: A review. *Journal of Human Nutrition and Dietetics, 28*, 1–14. https://doi.org/10.1111/jhn.12212

Larroque, B., Ancel, P.-Y., Marret, S., Marchand, L., André, M., Arnaud, C., Pierrat, V., Rozé, J.-C., Messer, J., Thiriez, G., Burguet, A., Picaud, J.-C., Bréart, G., Kaminski, M., & EPIPAGE Study Group. (2008). Neurodevelopmental disabilities and special care of 5-year-old children born before 33 weeks of gestation (the EPIPAGE study): A longitudinal cohort study. *The Lancet, 371*, 813–820. https://doi.org/10.1016/S0140-6736(08)60380-3

Larsson, H. J., Eaton, W. W., Madsen, K. M., Vestergaard, M., Olesen, A. V., Agerbo, E., et al. (2005). Risk factors for autism: Perinatal factors, parental psychiatric history, and socioeconomic status. *American Journal of Epidemiology, 161*, 916–925. https://doi.org/10.1093/aje/kwi123

Laugesaar, R., Kolk, A., Tomberg, T., Metsvaht, T., Lintrop, M., Varendi, H., & Talvik, T. (2007). Acutely and retrospectively diagnosed perinatal stroke: A population-based study. *Stroke; a Journal of Cerebral Circulation, 38*, 2234–2240. https://doi.org/10.1161/STROKEAHA.107.483743

Lind, T., Lee Raby, K., Caron, E., Roben, C., & Dozier, M. (2017). Enhancing executive functioning among toddlers in foster care with an attachment-based intervention. *Development and Psychopathology, 29*, 575–586. https://doi.org/10.1017/S0954579417000190

Linder, N., Linder, I., Fridman, E., Kouadio, F., Lubin, D., Merlob, P., Yogev, Y., & Melamed, N. (2013). Birth trauma—Risk factors and short-term neonatal outcome. *The Journal of Maternal-fetal & Neonatal Medicine : The Official Journal of the European Association of Perinatal Medicine, the Federation of Asia and Oceania Perinatal Societies, the International Society of Perinatal Obstetricians, 26*, 1491–1495. https://doi.org/10.3109/14767058.2013.789850

Linderkamp, O., Janus, L., Linder, R., & Skoruppa, D. B. (2009). Time table of normal foetal brain development. *International Journal of Prenatal and Perinatal Psychology and Medicine, 21*, 4–16. https://www.researchgate.net/publication/312452357

Liu, D., Diorio, J., Day, J. C., Francis, D. D., & Meaney, M. J. (2000). Maternal care, hippocampal synaptogenesis and cognitive development in rats. *Nature Neuroscience, 3*, 799–806. https://doi.org/10.1038/77702

Lucas, B. R., Elliott, E. J., Coggan, S., Pinto, R. Z., Jirikowic, T., McCoy, S. W., & Latimer, J. (2016). Interventions to improve gross motor performance in children with neurodevelopmental disorders: A meta-analysis. *BMC Pediatrics, 16*, 193. https://doi.org/10.1186/s12887-016-0731-6

Luyster, R., Lopez, K., & Lord, C. (2007). Characterizing communicative development in children referred for autism spectrum disorders using the MacArthur-Bates Communicative Development Inventory (CDI). *Journal of Child Language, 34*, 623–654. https://doi.org/10.1017/s0305000907008094

Maenner, M. J., Warren, Z., Williams, A. R., Amoakohene, E., Bakian, A. V., Bilder, D. A., Durkin, M. S., Fitzgerald, R. T., Furnier, S. M., Hughes, M. M., Ladd-Acosta, C. M., McArthur, D., Pas, E. T., Salinas, A., Vehorn, A., Williams, S., Esler, A., Grzybowski, A., Hall-Lande, J., Nguyen, R. H. N., et al. (2023). Prevalence and characteristics of autism spectrum disorder among children aged 8 years – Autism and Developmental Disabilities Monitoring Network, 11 sites, United States, 2020. *Morbidity and Mortality Weekly Report. Surveillance Summaries, 72*, 1–14. https://doi.org/10.15585/mmwr.ss7202a1

Marrus, N., & Hall, L. (2017). Intellectual disability and language disorder. *Child and Adolescent Psychiatric Clinics of North America, 26*, 539–554. https://doi.org/10.1016/j.chc.2017.03.001

Maulik, P. K., Mascarenhas, M. N., Mathers, C. D., Dua, T., & Saxena, S. (2011). Prevalence of intellectual disability: A meta-analysis of population-based studies. *Research in Developmental Disabilities, 32*, 419–436. https://doi.org/10.1016/j.ridd.2010.12.018

McAndrew, N. S., Jerofke-Owen, T., Fortney, C. A., et al. (2022). Systematic review of family engagement interventions in neonatal, pediatric, and adult ICUs. *Nursing in Critical Care, 27*, 296–325. https://doi.org/10.1111/nicc.12564

Meltzer, A., & Van de Water, J. (2017). The role of the immune system in autism spectrum disorder. *Neuropsychopharmacology, 42*(1), 284–298. https://doi.org/10.1038/npp.2016.158

Mitha, A., Foix-L'Hélias, L., Arnaud, C., Marret, S., Vieux, R., Aujard, Y., Thiriez, G., Larroque, B., Cambonie, G., Burguet, A., Boileau, P., Rozé, J. C., Kaminski, M., Truffert, P., Ancel, P. Y., & EPIPAGE Study Group. (2013). Neonatal infection and 5-year neurodevelopmental outcome of very preterm infants. *Pediatrics, 132*, e372–e380. https://doi.org/10.1542/peds.2012-3979

Mlakar, J., Korva, M., Tul, N., Popović, M., Poljšak-Prijatelj, M., Mraz, J., Kolenc, M., Resman Rus, K., Vesnaver Vipotnik, T., Fabjan Vodušek, V., Vizjak, A., Pižem, J., Petrovec, M., & Avšič Županc, T. (2016). Zika virus associated with microcephaly. *New England Journal of Medicine, 374*(10), 951–958. https://doi.org/10.1056/nejmoa1600651

Modabbernia, A., Sandin, S., Gross, R., Leonard, H., Gissler, M., Parner, E. T., Francis, R., Carter, K., Bresnahan, M., Schendel, D., Hornig, M., & Reichenberg, A. (2019). Apgar score and risk of autism. *European Journal of Epidemiology, 34*, 105–114. https://doi.org/10.1007/s10654-018-0445-1

Moran, M. M., Gunn-Charlton, J. K., Walsh, J. M., Cheong, J. L. Y., Anderson, P. J., Doyle, L. W., Greaves, S., & Hunt, R. W. (2019). Associations of neonatal noncardiac surgery with brain structure and neurodevelopment: A prospective case-control study. *The Journal of Pediatrics, 212*, 93–101.e2. https://doi.org/10.1016/j.jpeds.2019.05.050

Murray, G. K., Jones, P. B., Kuh, D., & Richards, M. (2007). Infant developmental milestones and subsequent cognitive function. *Annals of Neurology, 62*, 128–136. https://doi.org/10.1002/ana.21120

National Center for Health Statistics. (2023). *Health, United States, 2020–2021: Annual perspective*. National Center for Health Statistics. https://doi.org/10.15620/cdc:122044

Nissen, M. D. (2007). Congenital and neonatal pneumonia. *Paediatric Respiratory Reviews, 8*, 195–203. https://doi.org/10.1016/j.prrv.2007.07.001

Noble, K. G., Magnuson, K., Gennetian, L. A., Duncan, G. J., Yoshikawa, H., Fox, N. A., & Halpern-Meekin, S. (2021). Baby's first years: Design of a randomized controlled trial of poverty reduction in the United States. *Pediatrics, 148*(4), e2020049702. https://doi.org/10.1542/peds.2020-049702

Novak, I., Morgan, C., Adde, L., Blackman, J., Boyd, R. N., Brunstrom-Hernandez, J., et al. (2017). Early, accurate diagnosis and early intervention in cerebral palsy: Advances in diagnosis and treatment. *JAMA Pediatrics, 171*, 897–907. https://doi.org/10.1001/jamapediatrics.2017.1689

Novak, C. M., Ozen, M., & Burd, I. (2018). Perinatal brain injury: Mechanisms, prevention, and Outcomes. *Clinics in Perinatology, 45*, 357–375. https://doi.org/10.1016/j.clp.2018.01.015

Oddy, W. H., Li, J., Whitehouse, A. J., Zubrick, S. R., & Malacova, E. (2010). Breastfeeding duration and academic achievement at 10 years. *Pediatrics, 126*(6), e1517–e1525. https://doi.org/10.1542/peds.2009-3489

Okoli, M. L., Ogbu, C. E., Enyi, C. O., Okoli, I. C., Wilson, R. E., & Kirby, R. S. (2022). Sociodemographic and socioeconomic correlates of learning disability in preterm children in the United States. *BMC Public Health, 22*, 1–9. https://doi.org/10.1186/s12889-022-12592-4

Olin, A., Henckel, E., Chen, Y., Lakshmikanth, T., Pou, C., Mikes, J., Gustafsson, A., Bernhardsson, A. K., Zhang, C., Bohlin, K., & Brodin, P. (2018). Stereotypic immune system development in newborn children. *Cell, 174*, 1277–1292.e14. https://doi.org/10.1016/j.cell.2018.06.045

Omidvarnia, A., Metsäranta, M., Lano, A., & Vanhatalo, S. (2015). Structural damage in early preterm brain changes the electric resting state networks. *NeuroImage, 120*, 266–273. https://doi.org/10.1016/j.neuroimage.2015.06.091

Osterman, M. J. K., Hamilton, B. E., Martin, J. A., Driscoll, A. K., & Valenzuela, C. P. (2023). Births: Final data for 2021. *National Vital Statistics Reports: From the Centers for Disease Control and Prevention, National Center for Health Statistics, National Vital Statistics System, 72*(1), 1–53.

Packard, A. E., Egan, A. E., & Ulrich-Lai, Y. M. (2016). HPA Axis interactions with behavioral systems. *Comprehensive Physiology, 6*, 1897–1934. https://doi.org/10.1002/cphy.c150042

Perlman, J. M. (2001). Neurobehavioral deficits in premature graduates of intensive care--potential medical and neonatal environmental risk factors. *Pediatrics, 108*, 1339–1348. https://doi.org/10.1542/peds.108.6.1339

Persson, M., Fadl, H., Hanson, U., & Pasupathy, D. (2013). Disproportionate body composition and neonatal outcome in offspring of mothers with and without gestational diabetes mellitus. *Diabetes Care, 36*, 3543–3548. https://doi.org/10.2337/dc13-0899

Peyre, H., Charkaluk, M. L., Forhan, A., Heude, B., Ramus, F., & EDEN Mother–Child Cohort Study Group. (2017). Do developmental milestones at 4, 8, 12 and 24 months predict IQ at 5-6 years old? Results of the EDEN mother-child cohort. *European Journal of Paediatric Neurology: EJPN: Official Journal of the European Paediatric Neurology Society, 21*, 272–279. https://doi.org/10.1016/j.ejpn.2016.11.001

Pironkova, R. P., Giamelli, J., Seiden, H., Parnell, V. A., Gruber, D., Sison, C. P., Kowal, C., & Ojamaa, K. (2017). Brain injury with systemic inflammation in newborns with congenital heart disease undergoing heart surgery. *Experimental and Therapeutic Medicine, 14*, 228–238. https://doi.org/10.3892/etm.2017.4493

Polglase, G. R., Miller, S. L., Barton, S. K., Kluckow, M., Gill, A. W., Hooper, S. B., & Tolcos, M. (2014). Respiratory support for premature neonates in the delivery room: Effects on cardiovascular function and the development of brain injury. *Pediatric Research, 75*(6), 682–688. https://doi.org/10.1038/pr.2014.40

Polin, R. A., & Sahni, R. (2017). Newborn lung disease and care. In M. A. Gleason & S. F. Devaskar (Eds.), *Avery's diseases of the newborn* (10th ed., pp. 536–547).

Prado, E. L., & Dewey, K. G. (2014). Nutrition and brain development in early life. *Nutrition Reviews, 72*, 267–284. https://doi.org/10.1111/nure.12102

Queirós, F. C., Wehby, G. L., & Halpern, C. T. (2015). Developmental disabilities and socioeconomic outcomes in young adulthood. *Public Health Reports, 130*, 213–221. https://doi.org/10.1177/003335491513000308

Rampon, C., Tang, Y. P., Goodhouse, J., Shimizu, E., Kyin, M., & Tsien, J. Z. (2000). Enrichment induces structural changes and recovery from nonspatial memory deficits in CA1 NMDAR1-knockout mice. *Nature Neuroscience, 3*, 238–244. https://doi.org/10.1038/72945

Rand, K. M., Austin, N. C., Inder, T. E., Bora, S., & Woodward, L. J. (2016). Neonatal infection and later neurodevelopmental risk in the very preterm infant. *The Journal of Pediatrics, 170*, 97–104. https://doi.org/10.1016/j.jpeds.2015.11.017

Rathbone, R., Counsell, S. J., Kapellou, O., Dyet, L., Kennea, N., Hajnal, J., Allsop, J. M., Cowan, F., & Edwards, A. D. (2011). Perinatal cortical growth and childhood neurocognitive abilities. *Neurology, 77*, 1510–1517. https://doi.org/10.1212/WNL.0b013e318233b215

Rauch, S. L., Shin, L. M., & Phelps, E. A. (2006). Neurocircuitry models of posttraumatic stress disorder and extinction: Human neuroimaging research—Past, present, and future. *Biological Psychiatry, 60*(4), 376–382. https://doi.org/10.1016/j.biopsych.2006.06.004

Ream, M. A., & Lehwald, L. (2018). Neurologic consequences of preterm birth. *Current Neurology and Neuroscience Reports, 18*, 48. https://doi.org/10.1007/s11910-018-0862-2

Roid, G. H., & Pomplun, M. (2012). The Stanford-Binet intelligence scales, fifth edition. In D. P. Flanagan & P. L. Harrison (Eds.), *Contemporary intellectual assessment: Theories, tests, and issues* (pp. 249–268). The Guilford Press.

Rose-Jacobs, R., Black, M. M., Casey, P. H., Cook, J. T., Cutts, D. B., Chilton, M., Heeren, T., Levenson, S. M., Meyers, A. F., & Frank, D. A. (2008). Household food insecurity: Associations with at-risk infant and toddler development. *Pediatrics, 121*, 65–72. https://doi.org/10.1542/peds.2006-3717

Rosenzweig, M., Bennett, E., Hebert, M., & Morimoto, H. (1978). Social grouping cannot account for cerebral effects of enriched environments. *Brain Research, 153*(3), 563–576.

Rozé, J., Darmaun, D., Boquien, C., et al. (2012). The apparent breastfeeding paradox in very preterm infants: Relationship between breastfeeding, early weight gain, and neurodevelopment based on results from two cohorts, EPIPAGE and LIFT. *BMJ Open, 2*, e000834. https://doi.org/10.1136/bmjopen-2012-000834

Sanders, R. D., Hassell, J., Davidson, A. J., Robertson, N. J., & Ma, D. (2013). Impact of anesthetics and surgery on neurodevelopment: An update. *British Journal of Anaesthesia, 110*(Suppl 1), i53–i72.

Sandin, S., Schendel, D., Magnusson, P., Hultman, C., Surén, P., Susser, E., Grønborg, T., Gissler, M., Gunnes, N., Gross, R., Henning, M., Bresnahan, M., Sourander, A., Hornig, M., Carter, K., Francis, R., Parner, E., Leonard, H., Rosanoff, M., Stoltenberg, C., et al. (2016). Autism risk associated with parental age and with increasing difference in age between the parents. *Molecular Psychiatry, 21*(5), 693–700. https://doi.org/10.1038/mp.2015.70

Sarda, S. P., Sarri, G., & Siffel, C. (2021). Global prevalence of long-term neurodevelopmental impairment following extremely preterm birth: A systematic literature review. *Journal of International Medical Research, 49*(7), 03000605211028026.

Schalock, R. L., Borthwick-Duffy, S. A., Bradley, V. J., Buntinx, W. H., Coulter, D. L., Craig, E. M., et al. (2010). *Intellectual disability: Definition, classification, and systems of supports*. American Association on Intellectual and Developmental Disabilities.

Scheinost, D., Kwon, S. H., Lacadie, C., Vohr, B. R., Schneider, K. C., Papademetris, X., Constable, R. T., & Ment, L. R. (2017). Alterations in anatomical covariance in the prematurely born. *Cerebral Cortex, 27*, 534–543. https://doi.org/10.1093/cercor/bhv248

Schmidt, B., & Davis, P. (2016). Caffeine for apnea of prematurity: Effects on the developing brain. *Neonatology, 109*(4), 241–243. https://doi.org/10.1016/j.neuro.2016.11.012

Schmidt, B., Asztalos, E. V., Roberts, R. S., Robertson, C. M. T., Sauve, R. S., Whitfield, M. F., & Trial of Indomethacin Prophylaxis in Preterms (TIPP) Investigators. (2003). Impact of bronchopulmonary dysplasia, brain injury, and severe retinopathy on the outcome of extremely low-birth-weight infants at 18 months: Results from the trial of indomethacin prophylaxis in preterms. *JAMA: The Journal of the American Medical Association, 289*(9), 1124–1129. https://doi.org/10.1001/jama.289.9.1124

Schreibman, L., Dawson, G., Stahmer, A. C., Landa, R., Rogers, S. J., McGee, G. G., Kasari, C., Ingersoll, B., Kaiser, A. P., Bruinsma, Y., McNerney, E., Wetherby, A., & Halladay, A. (2015). Naturalistic developmental behavioral interventions: Empirically validated treatments for autism spectrum disorder. *Journal of Autism and Developmental Disorders, 45*, 2411–2428. https://doi.org/10.1007/s10803-015-2407-8

Seckl, J. R., & Holmes, M. C. (2007). Mechanisms of disease: Glucocorticoids, their placental metabolism and fetal 'programming' of adult pathophysiology. *Nature Clinical Practice. Endocrinology & Metabolism, 3*(6), 479–488. https://doi.org/10.1038/ncpendmet0515

Sgadò, P., Dunleavy, M., Genovesi, S., Provenzano, G., & Bozzi, Y. (2011). The role of GABAergic system in neurodevelopmental disorders: A focus on autism and epilepsy. *International Journal of Physiology, Pathophysiology and Pharmacology, 3*, 223–235. https://www.frontiersin.org/articles/10.3389/fncel.2014.00119/full

Shaywitz, S. E., Shaywitz, B. A., Pugh, K. R., Fulbright, R. K., Constable, R. T., Mencl, W. E., Shankweiler, D. P., Liberman, A. M., Skudlarski, P., Fletcher, J. M., Katz, L., Marchione, K. E., Lacadie, C., Gatenby, C., & Gore, J. C. (1998). Functional disruption in the organization of the brain for reading in dyslexia. *Proceedings of the National Academy of Sciences of the United States of America, 95*, 2636–2641. https://doi.org/10.1073/pnas.95.5.2636

Shimony, J. S., Smyser, C. D., Wideman, G., Alexopoulos, D., Hill, J., Harwell, J., Dierker, D., Van Essen, D. C., Inder, T. E., & Neil, J. J. (2016). Comparison of cortical folding measures for evaluation of developing human brain. *NeuroImage, 125*, 780–790. https://doi.org/10.1016/j.neuroimage.2015.11.001

Shuffrey, L. C., Morales, S., Jacobson, M. H., Bosquet Enlow, M., Ghassabian, A., Margolis, A. E., Lucchini, M., Carroll, K. N., Crum, R. M., Dabelea, D., Deutsch, A., Fifer, W. P., Goldson, B., Hockett, C. W., Mason, W. A., Jacobson, L. T., O'Connor, T. G., Pini, N., Rayport, Y., et al. (2023). Association of gestational diabetes mellitus and perinatal maternal depression with early childhood behavioral problems: An environmental influences on child health outcomes (ECHO) study. *Child Development, 00*, 1–15. https://doi.org/10.1111/cdev.13938

Singer, J. K., Lichtenberger, E. O., Kaufman, J. C., Kaufman, A. S., & Kaufman, N. L. (2012). The Kaufman assessment battery for children—Second edition and the Kaufman test of educational achievement—Second edition. In D. P. Flanagan & P. L. Harrison (Eds.), *Contemporary intellectual assessment: Theories, tests, and issues* (pp. 269–296). The Guilford Press.

Slevin, M., Farrington, N., Duffy, G., Daly, L., & Murphy, J. F. (2000). Altering the NICU and measuring infants' responses. *Acta Paediatrica (Oslo, Norway: 1992), 89*(5), 577–581. https://doi.org/10.1080/080352500750027899

Smith, G. C., Gutovich, J., Smyser, C., Pineda, R., Newnham, C., Tjoeng, T. H., Vavasseur, C., Wallendorf, M., Neil, J., & Inder, T. (2011). Neonatal intensive care unit stress is associated with brain development in preterm infants. *Annals of Neurology, 70*(4), 541–549. https://doi.org/10.1002/ana.22545

Smyser, C. D., Inder, T. E., Shimony, J. S., Hill, J. E., Degnan, A. J., Snyder, A. Z., & Neil, J. J. (2010). Longitudinal analysis of neural network development in preterm infants. *Cerebral Cortex, 20*, 2852–2862. https://doi.org/10.1093/cercor/bhq035

Smyser, C. D., Snyder, A. Z., Shimony, J. S., Blazey, T. M., Inder, T. E., & Neil, J. J. (2013). Effects of white matter injury on resting state fMRI measures in prematurely born infants. *PLoS One, 8*, e68098. https://doi.org/10.1371/journal.pone.0068098

Smyser, C. D., Snyder, A. Z., Shimony, J. S., Mitra, A., Inder, T. E., & Neil, J. J. (2016). Resting-state network complexity and magnitude are reduced in prematurely born infants. *Cerebral Cortex, 26*, 322–333. https://doi.org/10.1093/cercor/bhu251

Snowling, M. J., Bishop, D. V., Stothard, S. E., Chipchase, B., & Kaplan, C. (2006). Psychosocial outcomes at 15 years of children with a preschool history of speech-language impairment. *Journal of Child Psychology and Psychiatry, and Allied Disciplines, 47*, 759–765. https://doi.org/10.1111/j.1469-7610.2006.01631.x

Soleimani, F., Azari, N., Ghiasvand, H., Shahrokhi, A., Rahmani, N., & Fatollahierad, S. (2020). Do NICU developmental care improve cognitive and motor outcomes for preterm infants? A systematic review and meta-analysis. *BMC Pediatrics, 20*, 67. https://doi.org/10.1186/s12887-020-1953-1

Spittle, A., Orton, J., Anderson, P. J., Boyd, R., & Doyle, L. W. (2015). Early developmental intervention programmes provided post hospital discharge to prevent motor and cognitive impairment in preterm infants. *The Cochrane Database of Systematic Reviews, 2015*(11), CD005495. https://doi.org/10.1002/14651858.CD005495.pub4

St Clair, M. C., Pickles, A., Durkin, K., & Conti-Ramsden, G. (2011). A longitudinal study of behavioral, emotional and social difficulties in individuals with a history of specific language impairment (SLI). *Journal of Communication Disorders, 44*, 186–199. https://doi.org/10.1016/j.jcomdis.2010.09.004

Stevens, B., Yamada, J., Ohlsson, A., Haliburton, S., & Shorkey, A. (2016). Sucrose for analgesia in newborn infants undergoing painful procedures. *The Cochrane Database of Systematic Reviews, 7*(7), CD001069.

Sucksdorff, M., Lehtonen, L., Chudal, R., Suominen, A., Joelsson, P., Gissler, M., & Sourander, A. (2015). Preterm birth and poor fetal growth as risk factors of attention-deficit/ hyperactivity disorder. *Pediatrics, 136*, e599–e608. https://doi.org/10.1542/peds.2015-1043

Tabachnick, A. R., Eiden, R. D., Labella, M. H., & Dozier, M. (2022). Effects of an attachment-based intervention on autonomic regulation among opioid-exposed infants. *Developmental Psychobiology, 64*, e22286. https://doi.org/10.1002/dev.22286

Talmi, Z., Mankuta, D., & Raz, R. (2020). Birth weight and autism spectrum disorder: A population-based nested case-control study. Autism Research: Official Journal of the International Society for Autism Research, 13, 655–665. https://doi.org/10.1002/aur.2260

Tang, X., Jaenisch, R., & Sur, M. (2021). The role of GABAergic signaling in neurodevelopmental disorders. *Nature Reviews. Neuroscience, 22*, 290–307. https://doi.org/10.1038/s41583-021-00443-x

Thompson, D. K., Warfield, S. K., Carlin, J. B., Pavlovic, M., Wang, H. X., Bear, M., Kean, M. J., Doyle, L. W., Egan, G. F., & Inder, T. E. (2007). Perinatal risk factors altering regional brain structure in the preterm infant. *Brain: A Journal of Neurology, 130*(Pt 3), 667–677. https://doi.org/10.1093/brain/awl277

Toulmin, H., Beckmann, C. F., O'Muircheartaigh, J., Ball, G., Nongena, P., Makropoulos, A., Ederies, A., Counsell, S. J., Kennea, N., Arichi, T., Tusor, N., Rutherford, M. A., Azzopardi, D., Gonzalez-Cinca, N., Hajnal, J. V., & Edwards, A. D. (2015). Specialization and integration of functional thalamocortical connectivity in the human infant. *Proceedings of the National Academy of Sciences of the United States of America, 112*, 6485–6490. https://doi.org/10.1073/pnas.1422638112

Troller-Renfree, S. V., Costanzo, M. A., Duncan, G. J., Magnuson, K., Gennetian, L. A., Yoshikawa, H., Halpern-Meekin, S., Fox, N. A., & Noble, K. G. (2022). The impact of a poverty reduction intervention on infant brain activity. *Proceedings of the National Academy of Sciences of the United States of America, 119*, e2115649119. https://doi.org/10.1073/pnas.2115649119

Turk, E., van den Heuvel, M. I., Benders, M. J., de Heus, R., Franx, A., Manning, J. H., Hect, J. L., Hernandez-Andrade, E., Hassan, S. S., Romero, R., Kahn, R. S., Thomason, M. E., & van den Heuvel, M. P. (2019). Functional connectome of the fetal brain. *The Journal of Neuroscience: The Official Journal of the Society for Neuroscience, 39*, 9716–9724. https://doi.org/10.1523/JNEUROSCI.2891-18.2019

Valeri, B. O., Holsti, L., & Linhares, M. B. M. (2015). Neonatal pain and developmental outcomes in children born preterm: A systematic review. *The Clinical Journal of Pain, 31*(4), 355–362. https://doi.org/10.1097/AJP.0000000000000114

van Baar, A. L., van Wassenaer, A. G., Briët, J. M., Dekker, F. W., & Kok, J. H. (2005). Very preterm birth is associated with disabilities in multiple developmental domains. *Journal of Pediatric Psychology, 30*(3), 247–255. https://doi.org/10.1093/jpepsy/jsi035

van den Heuvel, M. I., & Thomason, M. E. (2016). Functional connectivity of the human brain in utero. *Trends in Cognitive Sciences, 20*, 931–939. https://doi.org/10.1016/j.tics.2016.10.001

van Praag, H., Kempermann, G., & Gage, F. H. (2000). Neural consequences of environmental enrichment. *Nature Reviews. Neuroscience, 1*(3), 191–198. https://doi.org/10.1038/35044558

Vanegas, S. B., Dueñas, A. D., Kunze, M., & Xu, Y. (2022). Adapting parent-focused interventions for diverse caregivers of children with intellectual and developmental disabilities: Lessons learned during global crises. *Journal of Policy and Practice in Intellectual Disabilities, na*, 1–13.

Vinall, J., Miller, S. P., Bjornson, B. H., Fitzpatrick, K. P. V., Poskitt, K. J., Brant, R., Synnes, A. R., Cepeda, I. L., & Grunau, R. E. (2014). Invasive procedures in preterm children: Brain and cognitive development at school age. *Pediatrics, 133*, 412–421. https://doi.org/10.1542/peds.2013-1863

Volpe, J. J. (2009). Brain injury in premature infants: A complex amalgam of destructive and developmental disturbances. *The Lancet. Neurology, 8*, 110–124. https://doi.org/10.1016/S1474-4422(08)70294-1

Walani, S. R. (2020). Global burden of preterm birth. *International Journal of Gynaecology and Obstetrics: The Official Organ of the International Federation of Gynaecology and Obstetrics, 150*, 31–33. https://doi.org/10.1002/ijgo.13195

Wechsler, D. (2014). *WISC-V: Technical and interpretive manual*. Pearson.

White, R. D. (2020). Right lighting the NICU. *Acta Paediatrica (Oslo, Norway: 1992), 109*(7), 1288–1289. https://doi.org/10.1111/apa.15193

Wiggs, K., Elmore, A. L., Nigg, J. T., & Nikolas, M. A. (2016). Pre- and perinatal risk for attention-deficit hyperactivity disorder: Does neuropsychological weakness explain the link? *Journal of Abnormal Child Psychology, 44*(8), 1473–1485. https://doi.org/10.1007/s10802-016-0142-z

Williams, A. L., van Drongelen, W., & Lasky, R. E. (2007). Noise in contemporary neonatal intensive care. *The Journal of the Acoustical Society of America, 121*(5 Pt1), 2681–2690. https://doi.org/10.1121/1.2717500

Williams, I. A., Tarullo, A. R., Grieve, P. G., Wilpers, A., Vignola, E. F., Myers, M. M., & Fifer, W. P. (2012). Fetal cerebrovascular resistance and neonatal EEG predict 18-month neurodevelopmental outcome in infants with congenital heart disease. *Ultrasound in Obstetrics & Gynecology: The Official Journal of the International Society of Ultrasound in Obstetrics and Gynecology, 40*(3), 304–309. https://doi-org.ezproxy.cul.columbia.edu/10.1002/uog.11144

Woodward, L. J., Moor, S., Hood, K. M., Champion, P. R., Foster-Cohen, S., Inder, T. E., & Austin, N. C. (2009). Very preterm children show impairments across multiple neurodevelopmental domains by age 4 years. *Archives of Disease in Childhood. Fetal and Neonatal Edition, 94*(5), F339–F344. https://doi.org/10.1136/adc.2008.146282

Xiang, X., Yang, T., Chen, J., Chen, L., Dai, Y., Zhang, J., et al. (2023). Association of feeding patterns in infancy with later autism symptoms and neurodevelopment: A national multicentre survey. *BMC Psychiatry, 23*, 1–12. https://doi.org/10.1186/s12888-023-04667-2

Xie, S., Karlsson, H., Dalman, C., Widman, L., Rai, D., Gardner, R. M., Magnusson, C., Schendel, D. E., Newschaffer, C. J., & Lee, B. K. (2019). Family history of mental and neurological disorders and risk of autism. *JAMA Network Open, 2*, e190154. https://doi.org/10.1001/jamanetworkopen.2019.0154

Xiong, F., & Zhang, L. (2013). Role of the hypothalamic-pituitary-adrenal axis in developmental programming of health and disease. *Frontiers in Neuroendocrinology, 34*, 27–46. https://doi.org/10.1016/j.yfrne.2012.11.002

Yates, N., Gunn, A. J., Bennet, L., Dhillon, S. K., & Davidson, J. O. (2021). Preventing brain injury in the preterm infant—Current controversies and potential therapies. *International Journal of Molecular Sciences, 22*(4), 1671. https://doi.org/10.3390/ijms22041671

Ygberg, S., & Nilsson, A. (2012). The developing immune system – From foetus to toddler. *Acta Paediatrica (Oslo, Norway: 1992), 101*(2), 120–127. https://doi.org/10.1111/j.1651-2227.2011.02494.x

Younge, N., Goldstein, R. F., Bann, C. M., Hintz, S. R., Patel, R. M., Smith, P. B., Bell, E. F., Rysavy, M. A., Duncan, A. F., Vohr, B. R., Das, A., Goldberg, R. N., Higgins, R. D., Cotten, C. M., & Eunice Kennedy Shriver National Institute of Child Health and Human Development Neonatal Research Network. (2017). Survival and neurodevelopmental outcomes among periviable infants. *The New England Journal of Medicine, 376*(7), 617–628. https://doi.org/10.1056/NEJMoa1605566

Zablotsky, B., Black, L. I., Maenner, M. J., Schieve, L. A., Danielson, M. L., Bitsko, R. H., et al. (2023). Prevalence and trends of developmental disabilities among children in the United States: 2009–2017. *Pediatrics, 144*(4), e20190811. https://doi.org/10.1542/peds.2019-0811

Challenging Behavior in Intellectual and Developmental Disabilities

Marcella Hangen, Jennifer R. Weyman, and Maria G. Valdovinos

Abstract Many individuals diagnosed with intellectual and developmental disabilities engage in challenging behaviors such as aggression and self-injury. These behaviors may interfere with the quality of life of the individual and their families. Although biological factors may contribute to the development of challenging behavior, research has shown that the environment plays a significant role in developing all behaviors, including challenging behavior. Thus, the current chapter highlights the importance of understanding how consequences impact the development of challenging behavior and provides a discussion of the functional behavior assessment process used to isolate potential environmental factors contributing to the challenging behavior. This includes a review of indirect assessments, descriptive assessments, and functional analysis methods followed by a review of common treatments for reducing challenging behavior, including antecedent manipulations, extinction procedures, and differential reinforcement.

Keywords Challenging behavior · Function-based treatments · Functional behavior assessment · Differential reinforcement

M. Hangen (✉)
Department of Behavior Analysis, School of Sciences and Health Professions, Simmons University, Boston, MA, USA
e-mail: marcella.hangen@simmons.edu

J. R. Weyman
Department of Special Education and Counseling, California State University, Los Angeles, Los Angeles, CA, USA

M. G. Valdovinos
Department of Psychology and Neuroscience, Drake University, Des Moines, IA, USA

M. G. Valdovinos (ed.), *Intellectual and Developmental Disabilities*, Contemporary Clinical Neuroscience,
https://doi.org/10.1007/978-3-031-66932-3_10

Challenging Behavior in Intellectual Disabilities

Engagement in challenging behavior by individuals with intellectual disabilities and other neurodevelopmental disorders is not uncommon. This chapter presents the various ways these behaviors are displayed and the risk factors that contribute to challenging behavior development. Factors that lead to the development of challenging behavior are also discussed from a biological and environmental perspective. Finally, the assessment of and treatment options for challenging behavior are described.

Definitions and Prevalence of Challenging Behavior

Challenging behavior can vary in presentation. The most commonly reported forms are aggression and self-injurious behavior (Lowe et al., 2007; Lloyd & Kennedy, 2014). Aggression can involve acts toward others, such as hitting, kicking, and hair-pulling, or toward property or objects within one's surroundings, such as property destruction (Crotty et al., 2014). Self-injurious behavior (SIB) can be conceptualized as aggressive acts directed toward oneself (Huisman et al., 2018). These behaviors may involve targeting different parts of the body, such as the head (e.g., head hitting/banging) or the extremities, such as arms or hands (e.g., self-biting and skin-picking), and can range in severity. These behaviors can lead to significant injuries including tissue damage, concussion, and infection (Summers et al., 2017).

The presentation of challenging behavior varies by type, and its prevalence also varies according to diagnosis and risk factors. The prevalence of challenging behavior has been estimated to vary widely in both adults and children, with anywhere from 4% to 80% of adults (Bowring et al., 2019) and from 25% to 82% of children (Nicholls et al., 2023) diagnosed with intellectual disabilities. When examining prevalence rates according to the topography of those with intellectual disabilities who engage in challenging behavior, between 14% and 56% of adults (c.f. Crotty et al., 2014) and between 10% and 32% of children (Rojahn & Meier, 2009; Ruddick et al., 2015) with intellectual disabilities engage in aggression and between 10% and 90% of adults (c.f. Crotty et al., 2014) and between 5% and 21% of children (Rojahn & Meier, 2009; Ruddick et al., 2015) are reported to engage in SIB. Prevalence rates of challenging behavior also vary in different neurodevelopmental disorders such as autism spectrum disorder (ASD), fragile X syndrome (FXS), and Down syndrome. For example, the prevalence of SIB has been reported to occur in nearly 30% to 50% of children with ASD (Dimian et al., 2017; Soke et al., 2016), whereas the prevalence of SIB ranges from 10% to 81% in individuals with FXS (c.f. Hardiman & McGill, 2018). The prevalence of challenging behavior in individuals with Down syndrome is not well established; however, disruptive behavior disorder was prevalent in approximately 8% of a sample of children with Down syndrome (Esbensen et al., 2022).

Challenging behavior also appears to show developmental changes. For example, challenging behavior can emerge as early as 6 months but is usually first observed between two and three years of age (Mayo-Ortega et al., 2012). Moreover, frequent engagement in stereotypy (repetitive patterns of behavior such as hand flapping) in infancy and toddlerhood is predictive of engagement in SIB later in early childhood (Rojahn et al., 2016). The severity and frequency of challenging behavior often increases throughout childhood and adolescence (Allen, 2000), and younger adults engage in challenging behavior more often than older adults (Smith et al., 2022). However, these behaviors have been found to be relatively persistent throughout the lifespan if left untreated (Totsika et al., 2008).

Contributors to Challenging Behavior

There are characteristics that may be associated with the development of challenging behavior, which include neuropsychological and biological factors (Bowring et al., 2019; Mayo-Ortega et al., 2012). For example, there is some debate regarding which sex is more likely to engage in challenging behavior, with some research finding that males are more likely to engage in aggression than females, others suggesting females are more likely, and others suggesting there is no difference between the sexes (e.g., Bowring et al., 2019; Cooper et al., 2009; Tsiouris et al., 2011). Additionally, environmental factors can interact with or individually contribute to challenging behavior.

Neuropsychological and Biological Contributors

Research has found that lower intellectual ability and living in congregate settings are factors more often associated with aggressive behavior (Cooper et al., 2009). Further, the type of aggression may differ according to ability, with individuals with higher intellectual abilities more likely to engage in verbal aggression, whereas those with lower intellectual abilities are more likely to engage in physical aggression and SIB (Tsiouris et al., 2011). Conditions such as visual impairment (de Winter et al., 2011), pain (Courtemanche et al., 2016), and sleep deprivation (Schreck & Richdale, 2020; Symons et al., 2000) are all associated with engagement in SIB. Research also suggests a correlation between challenging behavior and seizure activity; however, the evidence supporting the strength of this relationship is contradictory (Brodie et al., 2016; Deb et al., 2020).

Another basis for challenging behavior often explored is neurobiology (Bowring et al., 2019). Researchers have hypothesized that aggression and SIB have neurobiological contributors to aggression, such as neurotransmitter (e.g., serotonin, dopamine, and γ-aminobutyric acid (GABA)) dysregulation and genetic disorders (e.g., Lesch-Nyhan syndrome, Cornelia de Lange syndrome, Smith-Magenis syndrome, and FXS) that alter gene expression and/or availability of neurotransmitters (e.g., Furniss & Biswas, 2020; Rosell & Siever, 2015; Willner, 2015).

Given hypothesized neurotransmitter involvement, psychotropic medication is often prescribed to treat challenging behavior. Although these medications are widely used, they are generally ineffective at addressing challenging behavior (Adler et al., 2015; Deb et al., 2008). Another increasing trend is the use of multiple psychotropic medications (either multiple classes of medication or multiple medications from the same class: polypharmacy) for which there is limited evidence to support its use in the treatment of challenging behavior (McCarthy & Chaplin, 2022). Additionally, some of these medications, such as benzodiazepines, may present side effects that adversely impact challenging behavior (e.g., worsen behavior; Antochi et al., 2003; Elbing et al., 2022; Kalachnik et al., 2002) and research has found that individuals with intellectual disabilities often respond differently to medication than those in the typical population making them more vulnerable to the negative side effects of medication (Charlot et al., 2020).

Environmental Contributors

Although biological factors may be associated with challenging behavior, research suggests that challenging behavior is often acquired and maintained by its consequences (Schlinger & Normand, 2013). In other words, many challenging behaviors are learned behaviors impacted by the events preceding and following them. Although challenging behaviors can take various forms (e.g., aggression, SIB, and property destruction), their form or *topography* provides little helpful information on the function of these behaviors or the reason why these behaviors are occurring. To address these behavior problems, it is vital to identify the function of the behavior. Function refers to the environmental events that account for that behavior. Challenging behaviors may occur for various reasons. To understand why challenging behavior occurs, a discussion on how consequences impact the future probability of behavior is necessary.

First, it is important to define the term "consequence" as it is used differently than in everyday language. A consequence is defined as an event that follows a behavior. Consequences, especially those that occur immediately after a behavior, have a major impact on behavior (Michael, 2004). These consequences influence behavior in two ways. First, some consequences increase the likelihood of the behavior occurring again. This process, in which a consequence follows a behavior and *increases* the future probability of that behavior under similar circumstances, is called reinforcement. The events that follow behavior can include accessing preferred items (positive reinforcement) or avoiding undesirable events (negative reinforcement). An example of positive reinforcement would be a child requesting "more" during dinner. The child says, "More," and the parent immediately delivers another serving of food. The consequence of receiving food for saying more results in the child requesting more in the future. An example of negative reinforcement would be when the same child is sitting in their highchair, but they are finished eating. The child then says, "All done, please." The parent then lifts the child from the highchair and puts them on the ground. The behavior of saying "all done" led to the consequence of being removed from the highchair. The child now says "all done" when they want to get out of the highchair. This is an example of negative reinforcement because the behavior (saying all done) was followed by the removal of the constricting highchair and resulted in an increase in the child saying all done when they are in uncomfortable situations.

Second, some consequences decrease the likelihood of the behavior occurring again. This process, in which a consequence follows a behavior and *decreases* the future probability of that behavior under similar circumstances, is called punishment. These events that follow behavior can include the addition of a stimulus (positive punishment) or the removal of preferred stimuli (negative punishment). An example of positive punishment would be a child who reaches out and pets a dog (behavior) without permission and gets bitten by the dog (consequence). This child no longer reaches out to pet unfamiliar dogs. This would be an example of positive punishment as the behavior (petting unfamiliar dogs) was followed by the addition of an undesirable consequence (a bite), and the behavior decreased (e.g., the child no longer pets dogs). An example of negative punishment would be a teenager cursing in front of their father and the father taking away their cell phone for the week. This teenager never curses in front of their father again. This is an example of negative punishment as the behavior (cursing) was followed by the removal of cell phone privileges (consequence) and resulted in a decrease in cursing. It is vital to note that to classify a consequence as a reinforcer or punisher, it must be known how that consequence impacts the future probability of behavior.

These processes of reinforcement and punishment are the foundations of operant behavior. Operant behavior is any behavior that is controlled by its consequences. Thinking of behavior and consequences as an if-then relationship may be helpful. If the behavior occurs, then a consequence occurs. These consequences then influence the likelihood of the behavior in the future. These relationships determine and maintain learned behavior. Research has repeatedly shown that these interactions between behavior and the environment are responsible for desirable and undesirable behaviors (Schlinger & Normand, 2013). Many challenging behaviors occur because they produce consequences that increase the likelihood of those behaviors occurring again; essentially, challenging behavior may get the individual something they want or help them avoid something they do not want.

Some individuals engage in challenging behavior to gain access to attention. That is, if a caregiver is busy and not giving attention to their child, the child may engage in behaviors that result in attention (e.g., others may make statements of concern or provide reprimands). Behavior that is reinforced by attention is labeled as social positive reinforcement because the consequence that contributes to the behavior involves another person (social), and a stimulus (i.e., attention) is added following the behavior, making it "positive."

Some challenging behaviors are reinforced by access to preferred items. These behaviors may be more likely to occur when preferred activities or items are in view but not accessible, when an individual requests them but their request is not granted, or when preferred items or activities are removed. When the behavior occurs, caregivers may provide access to the desired item or activity (e.g., an iPad is given when the challenging behavior occurs), reinforcing (and increasing) the behavior. Behavior that is reinforced by access to preferred items is also labeled social positive reinforcement because the consequence that contributes to the behavior involves another person (social), and a stimulus (i.e., preferred item) is added following the behavior, making it "positive."

Challenging behavior may also occur when non-preferred activities (e.g., academic demands, activities of daily living, and social interactions) occur. During these situations, children may engage in challenging behavior that results in the termination or postponement of the non-preferred activity. The removal or avoidance of these events could reinforce these challenging behaviors. For example, a child may hit a teacher during a difficult math session, and the teacher may send the child to the hallway or the principal's office due to the behavior. The child then avoids or delays completing the math homework. This may function as a reinforcer and result in the child being more likely to hit during non-preferred work times. Behavior that is reinforced by escape is labeled social negative reinforcement because the consequence that contributes to the behavior involves another person (social), and the stimulus or event (i.e., non-preferred activity) is removed or avoided following the behavior, making it "negative."

Additionally, some behaviors may occur regardless of how other people respond to the behavior. These behaviors continue to occur when the individual is alone. These behaviors directly produce the reinforcer. These behaviors are often referred to as sensory behaviors. If the behavior produces stimulation, it is maintained by automatic positive reinforcement. For example, individuals may flap their hands in front of a light because it results in visual stimulation.

Challenging behavior may also occur because it automatically (without the interaction of another person) terminates or avoids aversive stimulation. For example, an individual takes a pain reliever when they have a headache. The behavior of taking the medication results in the termination of the painful stimulus and results in the individual being more likely to consume pain relievers the next time they have a headache. It is hypothesized that some children may engage in challenging behavior to eliminate or distract from other painful conditions. For example, individuals may hit their head when they have an earache. Head hitting results in a decrease in ear pain (DeLissovoy, 1963).

Functional Behavior Assessment

Behavior is determined by its interaction with the environment (Schlinger & Normand, 2013). Therefore, we must change the environment to change behavior. To treat challenging behavior, it is essential to first identify the function of the challenging behavior (i.e., why problem behavior occurs) and then to develop a function-based treatment. Research has shown that interventions based on function are more likely to be effective at changing behavior (Ervin et al., 2001; Iwata et al., 1994a). A functional behavior assessment (FBA) is a series of methods used to identify why challenging behavior occurs. These assessments are used to identify the sources of reinforcement. They are used to figure out which reinforcers, social positive (e.g., attention and tangible), social negative (e.g., escape or avoidance of non-preferred activities), automatic positive (e.g., sensory stimulation), or automatic negative (e.g., pain termination), are responsible for the challenging behavior. If the reinforcers for the challenging behavior can be identified, an intervention can be designed to directly address why the behavior occurs. These interventions can then be designed to alter the environment before the challenging behavior happens, withhold the reinforcer for the challenging behavior when it does occur (the behavior no longer produces that reinforcer), and, most importantly, deliver the maintaining reinforcer for the replacement behavior (Cooper et al., 2020). In other words, an alternative behavior, such as a request, can be taught that produces the *same* consequence as the challenging behavior, providing the individual a better way to get what they want.

Ervin et al. (2001) reviewed the literature on functional assessment-based interventions in school settings and found that the FBA process resulted in positive improvements in student behavior. These methods were so successful that in 1997, the Individuals with Disabilities Education Act (IDEA) mandated using the FBA process when a child with a disability has their current placement changed for disciplinary reasons. However, the IDEA also recommends conducting an FBA when developing an Individualized Education Plan (IEP) if a student's behavior interferes with their learning.

FBA is a term used to describe various methods designed to pinpoint the environmental events that influence challenging behavior. These methods can be categorized into three types: (1) indirect assessments, (2) descriptive assessments, and (3) functional analyses. Each of these methods seeks to identify and describe the relationships between the environment and the occurrence of a specific behavior. The general FBA process consists of four steps, including gathering information via the indirect and descriptive assessments described below, interpreting the information gathered and developing a hypothesis about the function of the behavior, testing that hypothesis with a functional analysis (described below), and then developing intervention options based on the function of the challenging behavior.

Indirect Assessments

Indirect assessments involve gathering information regarding challenging behavior without directly observing the behavior. This information is gathered through verbal reports from an informant. The informant is someone who is familiar with the target individual (e.g., teacher and parent) or can be the individual themselves. This information is obtained through interviews, rating scales, or questionnaires. Because all the information is gathered without direct observation, the indirect assessment is generally not recommended in isolation. Although limited, the indirect assessment can help inform hypotheses regarding function, help build rapport, and can be completed quickly with little resources and training. However, it is important to note that research has repeatedly demonstrated that indirect assessments are unreliable and do not accurately identify the function of behavior (e.g., Iwata et al., 2013; Saini et al., 2019; Zarcone et al., 1991).

There are two main categories of indirect assessment, including open-ended and close-ended indirect assessments. Close-ended indirect assessments include rating scales and questionnaires that require the informant to answer questions by selecting from pre-specified responses, usually in the form of a Likert scale (e.g., 0 never to 3 always) or answering yes-no questions. The questions included in these assessments are designed to obtain information regarding common behavioral functions (i.e., social positive reinforcement, social negative reinforcement, automatic positive reinforcement, and automatic negative reinforcement). The survey results are then scored by categorizing the questions targeting each behavioral function and then analyzed by identifying which function has elevated scores indicating a possible function. Several of these instruments have been developed, including the Questions About Behavioral Function (QABF; Matson & Vollmer, 1995; Paclawskyj et al., 2000) and the Functional Analysis Screening Tool (FAST; Iwata et al., 2013) to name a couple. Close-ended assessments are quick and easy to administer, and they are designed to produce results that pinpoint specific functions of behavior, making interpretation easier and decreasing irrelevant information that may be obtained by using an open-ended indirect assessment. Although designing the assessment in this fashion guarantees attention to those functions of behavior, it also severely limits the quality and specificity of the information gathered (Fryling & Baires, 2016) that may help identify specific factors that might influence the likelihood of the target behavior (i.e., may miss crucial contextual information).

Open-ended indirect assessments are comprised of questions that do not have pre-specified responses for the informant to select. These assessments typically consist of structured interviews where the assessor poses questions regarding the challenging behavior, and the informant is free to answer the question openly (e.g., "Under what conditions is the challenging behavior most likely to occur?"). The questions in these assessments are designed to gather contextually rich information but are still designed to focus on obtaining information that may suggest specific functions. That is, questions gather information about specific events that precede and follow challenging behavior that would indicate the function of the challenging behavior. Like the close-ended indirect assessment, a number of these structured interviews have been created, including the Functional Assessment Interview Form (FAIF; O'Neil et al., 1997) and the Open-Ended Functional Assessment Interview (Hanley, 2012) to name a couple. The most significant advantage of the open-ended indirect assessment is the range and quality of context-specific information that can be obtained. However, they are more time-intensive, require more skills to conduct and analyze the results, and can also result in irrelevant information. Fryling and

Baires (2016) provide an excellent review and comparison of the close-ended and open-ended assessments.

Descriptive Assessments

The second method within an FBA is the descriptive assessment. A descriptive assessment involves directly observing the behavior of interest in the natural environment (e.g., observing the behavior during the situations in which it is reported to occur) and collecting data (Cooper et al., 2020; Thompson & Borrero, 2021). Like the indirect assessment, the purpose of the descriptive assessment within the FBA process is to develop a hypothesis about the function of a target behavior by gathering information about potential behavior-environment relationships. Unlike the indirect assessment, the descriptive assessment involves direct behavior observation, providing more objective information regarding behavior-environment relationships. Research has demonstrated that challenging behavior is directly related to environmental events; specifically, there is a cause-effect relationship between behavior and its consequences (Carr, 1977; Iwata et al., 1982/1994). Those potential relationships may be identified during a descriptive assessment (Vollmer et al., 2001).

There are several descriptive assessment methods. One common descriptive assessment is antecedent-behavior-consequence (ABC) recording (e.g., Bijou et al., 1968; Lalli et al., 1993; Lerman et al., 2009). ABC recording involves directly observing the individual, usually in the natural environment (e.g., home and school), and collecting information on the events that occur immediately before the behavior (antecedents), the behavior, and the events following the behavior (consequences). This chapter, covers two ABC recording procedures: narrative ABC recording and structured ABC recording.

Narrative ABC recording involves a written description of the observed events, including the antecedents and consequences for each instance of behavior (Thompson et al., 2000). This method allows the observer to collect qualitative information regarding the events surrounding the challenging behavior. The information obtained during these descriptive assessments can be used to identify what specific environmental conditions are associated with challenging behavior and what specific consequences are delivered in the natural environment. Then, those conditions and consequences can be arranged in the functional analysis. For example, a descriptive assessment can be used to identify what potential demand situations might be associated with challenging behavior (e.g., activities of daily living and academic demands) or how exactly caregivers respond when challenging behavior occurs (e.g., providing statements of concern, rubbing back, providing reprimands, and providing breaks). Additionally, narrative ABC recording requires few resources and is relatively easy to implement. Although narrative ABC recording can result in rich qualitative information and requires little training, it is at risk of gathering irrelevant information as untrained individuals may describe information that might not be useful in hypothesizing function (Lennox & Miltenberger,

1989). For example, one might record an antecedent as "he was angry," but being angry does not provide any helpful information about what environmental event is associated with challenging behavior and does not point to a specific potential function of behavior.

The second type of ABC recording is structured ABC recording, also known as the ABC checklist (Thompson & Borrero, 2021). Like narrative ABC recording, structured ABC recording gathers information on the events that precede and follow a behavior of interest, but instead of a narrative description, observers select the antecedent and consequence from a pre-defined list of common antecedents and consequences (e.g., Mace & Lalli, 1991). For example, some antecedent events included in the menu of choices might include low attention (caregiver attention is diverted to another person or activity or is otherwise unavailable), instruction presentation (the target individual is given a task to complete), or restricted access (a preferred item is removed or a request for an item or activity is denied). Some consequences may include attention (caregiver provides verbal or physical attention in the form of statements of concern or reprimands or may hold the individual, or rub their back), a break (the task is removed or postponed), or the delivery of items (providing access to any item). This method has the benefits of bringing attention to the common functions of behavior, providing more quantitative data, and it may produce more accurate data collection compared to the narrative ABC recording method (Lerman et al., 2009). However, it does not result in rich qualitative information that may be useful in identifying context-specific environmental events associated with challenging behavior.

It is essential to consider that events that follow challenging behavior do not necessarily mean they are functionally related. Many events follow behavior and do not influence behavior. Specifically, attention commonly follows behavior but is not always functionally related (St. Peter et al., 2005). For example, saying "Bless you" has a high probability of following a sneeze. However, sneezing is not likely to be maintained by attention. Additionally, it is possible that the individuals who are being observed may react to the observation and respond differently (Mercatoris & Craighead, 1974; White, 1977). This phenomenon may occur with the individual of interest or the others around them (those providing the consequence for behavior). Moreover, the consequence that ultimately maintains challenging behavior does not follow every instance of behavior, which then influences the analysis of potential, influential variables (Lerman & Iwata, 1993; Marion et al., 2003). Finally, situations that may evoke challenging behavior might not be observed during these observations because caregivers have learned to avoid situations that evoke challenging behavior. This may skew the results of the descriptive assessment because the observer would miss out on observing specific situations that are related to or evoke challenging behavior (Anderson & Long, 2002; Lerman & Iwata, 1993). For example, caregivers might avoid providing math instruction because it often triggers challenging behavior. If this situation is avoided during the observation, the observer would not have the opportunity to identify this problematic situation, missing out on a potentially relevant event.

In addition, it is critical to highlight that the relationships identified in a descriptive assessment are only correlational. Research has repeatedly shown that there is poor correspondence between descriptive assessments and the third method of the FBA, functional analyses (e.g., Camp et al., 2009; Contreras et al., 2023; Hall, 2005; Mace & Lalli, 1991; Martens et al., 2019; Pence et al., 2009; Piazza et al., 2003; Thompson & Iwata, 2007), suggesting that descriptive assessments do not accurately identify the function of behavior. Due to this poor correspondence, the descriptive assessment should not be used as a sole method to identify the function of behavior. Instead, these assessment results can be used to identify hypothesized functions of behavior for which further analysis can be tested and can be used to help design functional analysis conditions to increase the ecological validity of the functional analysis process (Thompson & Borrero, 2021).

Functional Analysis

The final assessment method included in the FBA process is the functional analysis. As initially described by Iwata et al. (1982/1994), functional analysis is a well-established procedure for determining the function of challenging behavior (Iwata & Dozier, 2008). Functional analyses assess the functional relationship between behavior and environmental events by directly manipulating antecedents and consequences and measuring their effects on challenging behavior (Iwata et al., 1982/1994). A functional analysis consists of a control condition and one or more test conditions. During functional analysis test conditions, various antecedent events (e.g., low attention and instruction delivery) and consequences (e.g., attention and breaks) suspected of maintaining the challenging behavior are arranged. These conditions are compared to a control condition in which the environment is arranged, such that antecedent events that may evoke challenging behavior are reduced or eliminated, and no programmed consequences follow the challenging behavior. Higher levels of challenging behavior during a test condition compared to the control condition suggest that the tested consequence serves as a reinforcer (evidenced by the increase in behavior) for that challenging behavior. The functional analysis is akin to an allergy test in which an individual is exposed to specific allergens and observed to see whether they react so that the specific allergen can be identified and appropriate recommendations delivered (e.g., do not consume any tree nuts).

Iwata et al. (1982/1994) was the first study to conduct this analysis and is the foundation on which various methods of functional analysis are based. Using this method, Iwata et al. found that specific conditions resulted in higher levels of challenging behavior. Further, the conditions associated with these higher levels of challenging behavior differed across participants, providing evidence that challenging behavior has specific and individualized functions.

Since the publication of Iwata et al. (1982/1994), over 400 studies have been published using functional analysis methodology (Melanson & Fahmie, 2023). As the literature has grown, various methodologies of functional analyses have been designed and used to adapt to different situations (e.g., high-intensity behavior and environmental constraints). Moreover, additional test conditions have been assessed in the literature that were not included in the Iwata et al. study (e.g., tangible, mand compliance, and access to ritualistic arrangements). These modifications have extended and refined the functional analysis methodology. It should be noted that there are no "standard" functional analysis conditions, as functional analyses should be flexible and tailored to the individual. That is, the test conditions should be designed to test specific hypotheses (Cooper et al., 2020). Although there are no "standard" functional analysis conditions, Table 1 describes common functional analysis conditions reported in the literature and the associated antecedent events and arranged consequences. In general, the test conditions are arranged to assess behavior under specific antecedent conditions and provide specific consequences that are related to that antecedent condition contingent on the challenging behavior. For example, if the antecedent condition is diverted attention, the consequence of challenging behavior would be the delivery of attention.

Functional analysis is often considered the "gold standard" assessment and has been suggested to be one of the most important advances in applied behavior analysis (Saini et al., 2021). Before the widespread adoption of the functional analysis approach to assessing and treating challenging behavior, reinforcement-based treatments for challenging behavior were not consistently effective. Research has suggested that functional analyses generally lead to more effective treatments for challenging behavior (Didden et al., 1997; Iwata et al., 1994b; Kahng et al., 2002b). Additionally, punishment-based treatments were used more frequently due to the ineffectiveness of reinforcement-based procedures (Iwata et al., 1982/1994). With the widespread adoption of the functional analysis methodology, there has been a decrease in the use of punishment-based procedures for treating SIB (Axelrod, 1987; Kahng et al., 2002a, b; Shawler et al., 2019).

Despite the significant benefits, practitioners report several barriers to using functional analyses (Oliver et al., 2015). First, there is an inherent risk to the individual and those conducting the assessment as it is designed to increase behavior. Kahng et al. (2015) found that the rate of challenging behavior was higher during a functional analysis than outside of those sessions. However, it is important to note that there were fewer injuries during the functional analysis than outside. Second, there may be practical constraints as the assessment requires time and resources (e.g., appropriate setting and staff; Oliver et al., 2015). Third, this assessment method requires expertise to design, conduct, and analyze, especially given the risk of potential harm to the individual (Oliver et al., 2015). However, alternative functional analysis methods have been designed and evaluated in the literature to address these barriers, increasing the feasibility of functional analysis implementation.

Due to these barriers, extensive training and supervision are required before implementing the functional analysis. Furthermore, a risk assessment (e.g., Deochand et al., 2020) should be conducted before deciding to conduct a functional analysis. Standard functional analysis practice must also include safeguards such as termination criteria for sessions and protective equipment to minimize participant risk (Weeden et al., 2010). Moreover, consent must be obtained from relevant

Table 1 Common functional analysis conditions

Condition	Antecedent arrangement	Consequence arrangement
Control (play)	Social attention provided, access to preferred activities/items, no demands	No programmed consequences, challenging behavior is typically ignored
Contingent attention	Attention is diverted or withheld. The therapist is busy completing other tasks or interacting with another individual	Attention is provided based on the type of attention reported or observed to follow challenging behavior in the natural environment (e.g., reprimand, statement of concern, rubbing back, and consoling)
Contingent escape	Task demands are delivered continuously (e.g., academic, chores, and activities of daily living) based on the types of demands reported to be correlated with the challenging behavior	A short break from the task is provided, task materials are removed, and prompts are suspended for a specified duration of time (e.g., 30 s)
Contingent tangible	After engaging with a preferred item (e.g., toy and activity), the item is removed (e.g., "it's time to put the iPad away")	The preferred item is returned for a brief period (e.g., 30 s)
Alone/ignore	The individual is observed in a room without any materials either alone or if necessary, another person may be in the room for safety reasons	No programmed consequences, challenging behavior is typically ignored, if warranted, challenging behavior may be blocked with minimal attention
Contingent social avoidance	Attention is delivered continuously (e.g., conversations) based on the type of attention reported to be correlated with challenging behavior	Attention is removed for a brief period (e.g., 30 s)
Contingent mand compliance	After complying with every instruction (or request) delivered by the target individual, the therapist no longer follows the instructions/requests	All requests and instructions are completed by the therapist for a brief period (e.g., 30 s)
Contingent ritual arrangement	The therapist blocks the individual from completing a ritualistic behavior (e.g., lining up items) or rearranges the environment (e.g., reordering the items the individual had manipulated)	The individual is allowed to complete the ritualistic behavior

parties, including guardians and other stakeholders such as administrators, before conducting this assessment (BACB, 2020). For more information regarding functional analyses, including an in-depth review of functional analysis methodology and best practices regarding functional analyses, see Dozier et al. (2022).

Treatments for Challenging Behavior

As mentioned previously, the general FBA process consists of four steps. Thus far, this chapter described the first three steps of the FBA process. The fourth step focuses on developing a function-based intervention. This section briefly reviews some common interventions demonstrated to be effective in reducing challenging behavior. There are three general categories of function-based interventions: antecedent manipulations, extinction, and differential reinforcement. These different treatment components can be combined to develop an effective treatment package.

Antecedent Manipulations

The purpose of antecedent manipulations is to modify the environment before challenging behavior ever occurs in an attempt to avoid challenging behavior altogether. There are three categories of antecedent manipulations: stimulus control manipulations, response effort manipulations, and establishing operation manipulations.

Stimulus control manipulations There are several types of stimulus control manipulations to prevent the onset of challenging behavior. One method is to remove the discriminative stimulus for challenging behavior. A discriminative stimulus is a stimulus that is present when a behavior is reinforced and is absent when the behavior is not reinforced. In other words, it is a stimulus that signals the availability of reinforcement. For example, a child may engage in inappropriate vocalizations (e.g., yelling profanities) in the presence of their best friend who sits next to them in class. The best friend may serve as a discriminative stimulus for challenging behavior. That is, their presence signals that they will laugh at inappropriate vocalizations, which would reinforce challenging behavior. To avoid challenging behavior, the teacher could modify the seating arrangement to change the student's proximity to their friends, removing the discriminative stimulus for challenging behavior. Another stimulus control manipulation could be to add a discriminative delta for challenging behavior. A discriminative delta is a stimulus that is present when a behavior is not reinforced and is absent when the behavior is reinforced. In other words, it is a stimulus that signals that a behavior will not be reinforced. For example, if the teacher stands close to the child that is engaging in inappropriate vocalizations, this may result in the best friend not providing attention (e.g., laughter). The close presence of the teacher may serve as a discriminative delta for challenging behavior.

Response effort manipulations Another antecedent manipulation to prevent challenging behavior is to increase the effort to engage in challenging behavior and decrease the effort to engage in an alternative appropriate behavior (e.g., Piazza et al., 2002; Wilder et al., 2021). For example, if the challenging behavior is property destruction (e.g., throwing objects) and it is maintained by social positive reinforcement in the form of attention, the practitioner could place all objects out of reach (i.e., high response effort for challenging behavior) and provide attention (e.g., speaking with the child) after an appropriate alternative behavior such as raising their hand (i.e., low response effort for the alternative appropriate behavior). It is best to have the response effort to engage in the alternative appropriate behavior be lower than the response effort to engage in the challenging behavior. In this example, raising a hand may be less effortful than getting out of a chair and walking around to find an object to throw.

A common response effort manipulation for challenging behavior maintained by social negative reinforcement in the form of escape from academic demands involves revising curricular and instructional expectations (Geiger et al., 2010). Altering curricular targets or instructional procedures might attenuate or decrease the aversive properties involved in the instructional setting. For example, a child might engage in challenging behavior during an advanced math class to avoid this aversive demand situation. The teacher could alter the assignment to teach the prerequisite skills making the task more appropriate for the child's current skill set.

Establishing operation manipulations Another antecedent manipulation is to remove the establishing operation or motivation to engage in challenging behavior. An establishing operation is an event that increases the potency or value of a reinforcer and increases the likelihood an individual will engage in behavior to gain access to that reinforcer at that time. For example, if an individual does not eat for two days, then food is more valuable, and they are more likely to engage in behavior to gain access to food. When it comes to challenging behavior, the establishing operation can be removed by implementing noncontingent reinforcement. Noncontingent reinforcement is when the reinforcer is delivered frequently regardless of challenging behavior. To reduce challenging behavior, the therapist could deliver the maintaining reinforcer on a time-based schedule regardless of the challenging behavior (e.g., Phillips et al., 2017). For example, if aggression (e.g., kicking and hitting others) is maintained by social negative reinforcement in the form of escape from demands, the therapist may deliver breaks every 5 min to reduce the motivation to engage in challenging behavior.

Noncontingent reinforcement should be linked to the function of (i.e., the reason why the individual is engaging in the behavior) the challenging behavior (Vollmer et al., 1993). If the challenging behavior is maintained by social positive reinforcement in the form of attention, the therapist should provide attention frequently regardless of challenging behavior (e.g., deliver attention every 3 min). If challenging behavior is maintained by social positive reinforcement in the form of access to tangible items, the therapist should provide access to the tangible item frequently regardless of challenging behavior (e.g., deliver tangible item every 5 min). If the challenging behavior is maintained by social negative reinforcement in the form of escape from demands, the therapist should deliver breaks from work frequently regardless of challenging behavior (e.g., provide a break every 10 min).

Because it is not possible to deliver automatic reinforcement in the same way as attention, tangibles, or escape, a competing item can be delivered instead. A competing item is something that the individual can engage with that competes with the challenging behavior (i.e., instead of engaging in the challenging behavior, they engage with the competing item). A competing item can be identified by conducting a competing stimulus assessment (Piazza et al., 1996, 2000). The purpose of a competing stimulus assessment is to identify items associated with a high level of engagement and a low level of challenging behavior (i.e., a competing item). After identifying the competing item, the therapist provides free access to the item, which might keep the individual occupied and decrease the likelihood of them engaging in these automatically maintained behaviors. For example, if the challenging behavior is eye poking (i.e., pressing fingers to eyes), it can be hypothesized that eye poking results in visual stimulation and that visual stimulation is reinforcing eye poking. A competing stimulus assessment can be conducted with a variety of items that are associated with visual stimulation (e.g., light-up wand and disco light). Once an item associated with a low level of challenging behavior and a high level of engagement is identified, the individual could have continuous access to the competing item to reduce the likelihood of engaging in challenging behavior.

Extinction

Extinction is defined as no longer reinforcing a previously reinforced behavior, and as a result, the behavior stops occurring. Extinction should also be linked to the function of challenging behavior (Iwata et al., 1994a). If the behavior is maintained by social positive reinforcement in the form of attention, the therapist should no longer provide attention contingent on challenging behavior. If the behavior is maintained by social positive reinforcement in the form of access to tangible items, the therapist should no longer provide access to the tangible item contingent on the challenging behavior. If the behavior is maintained by social negative reinforcement in the form of escape from demands, the therapist should no longer provide a break contingent on challenging behavior. In the case of challenging behavior maintained by automatic reinforcement, extinction may be more difficult to implement.

Automatically maintained behavior is behavior that produces its own reinforcement. For example, an individual may poke their eyes because the blurry sensation they get is reinforcing. In the context of extinction, it would be difficult to disconnect the relationship between the behavior of eye poking and the consequence of the visual stimulation. An alternative to extinction in treating automatically maintained challenging behavior could be to use response blocking in conjunction with protective equipment (Hagopian et al., 2015). To implement response blocking for eye poking, the therapist would physically block the individual from touching their eyes so they cannot get the sensory feeling. One way to use protective equipment for eye poking would be to place goggles on the individual so that they are not able to gain the sensory feeling from eye poking.

It should be noted that there are risks of adverse side effects when using extinction (Lerman & Iwata, 1996). Therefore, this intervention should never be used in isolation, and extinction may not be an appropriate treatment component for severe challenging behavior.

Differential Reinforcement

A common treatment for challenging behavior is differential reinforcement of alternative behavior, which involves reinforcing an alternative appropriate behavior and placing the challenging behavior on extinction (Petscher et al., 2009). One type of differential reinforcement is called functional communication training (Carr & Durand, 1985). In functional communication training, the therapist reinforces a functional communication response (e.g., handing over a picture card, raising a hand, or using sign language) and the challenging behavior is placed on extinction. The topography (i.e., vocal speech, picture-exchange communication, and sign language) of the communication response is based on the current communication repertoire of the individual.

If the challenging behavior is maintained by social positive reinforcement in the form of access to attention, the therapist delivers attention contingent on the functional communication response (e.g., "Play with me," handing over a picture card of their caregiver) and withholds attention contingent on the challenging behavior. If the challenging behavior is maintained by social positive reinforcement in the form of access to tangible items, the therapist delivers toys contingent on the functional communication response (e.g., "Toys, please," handing over a picture card of a toy bin) and withholds access to toys contingent on challenging behavior. If the challenging behavior is maintained by social negative reinforcement in the form of escape from demands, the therapist delivers a break from work contingent on the functional communication response (e.g., "Break please," handing over a picture card of a textbook with a slash through it) and withholds breaks contingent on challenging behavior.

Functional communication training may result in the individual engaging in the functional communication response at too high of a rate or engaging in the functional communication response at inappropriate times. Therefore, it is important to teach the individual when their requests will be granted and when their requests will not be granted following functional communication training. One method to do this is to use discriminative stimuli and discriminative deltas to signal when the reinforcer is available (e.g., Hanley et al., 2001). For example, a therapist could use a green piece of construction paper as the discriminative stimulus and a red piece of construction paper as a discriminative delta. These stimuli are used to teach the individual when the reinforcer is available and when it is not available. That is, all functional communication responses in the presence of the discriminative stimulus or green paper would be reinforced, and all functional communication responses in the presence of the discriminative delta or red paper would be placed on extinction (i.e., do not provide the reinforcer after the functional communication response). Initially, the duration of the discriminative delta is short (e.g., 15 s), but over time, the therapist would gradually increase the duration of the discriminative delta to more practical durations (e.g., 30 min).

When it comes to challenging behavior maintained by automatic reinforcement, differential reinforcement of other behavior may be used (e.g., Heffernan & Lyons, 2016). Differential reinforcement of other behavior involves delivering a reinforcer if the individual does not engage in the challenging behavior for a specified amount of time. For example, an hour-long session may be broken into 5-min intervals, and if the child does not engage in any challenging behavior across the entire 5-min interval, they would receive a reinforcer. This procedure is then repeated throughout the hour.

Summary

Challenging behavior, such as aggression and SIB, can severely interfere with the lives of children with intellectual and developmental disabilities and their families. Although psychotropic medications are often prescribed to treat challenging behavior, they have generally been shown to be ineffective (Adler et al., 2015; Deb et al., 2008). Most behaviors are influenced by environmental contributors; therefore, to treat challenging behavior, it is best practice to conduct a FBA to identify why an individual is engaging in the challenging behavior and then to develop a function-based intervention specifically linked to the reason why the individual is engaging in the challenging behavior.

References

Adler, B. A., Wink, L. K., Early, M., Shaffer, R., Minshawi, N., McDougle, C. J., & Erickson, C. A. (2015). Drug-refractory aggression, self-injurious behavior, and severe tantrums in autism spectrum disorders: A chart review study. *Autism: The International Journal of Research and Practice, 19*(1), 102–106. https://doi.org/10.1177/1362361314524641

Allen, D. (2000). Recent research on physical aggression in persons with intellectual disability: An overview. *Journal of Intellectual and Developmental Disability, 25*(1), 41–57. https://doi.org/10.1080/132697800112776

Anderson, C. M., & Long, E. S. (2002). Use of a structured descriptive assessment methodology to identify variables affecting problem behavior. *Journal of Applied Behavior Analysis, 35*(2), 137–154. https://doi.org/10.1901/jaba.2002.35-137

Antochi, R., Stavrakaki, C., & Emery, P. C. (2003). Psychopharmacological treatments in persons with dual diagnosis of psychiatric disorders and developmental disabilities. *Postgraduate Medical Journal, 79*(929), 139–146. https://doi.org/10.1136/pmj.79.929.139

Axelrod, S. (1987). Functional and structural analyses of behavior: Approaches leading to reduced use of punishment procedures? *Research in Developmental Disabilities, 8*(2), 165–178. https://doi.org/10.1016/0891-4222(87)90001-1

Behavior Analyst Certification Board. (2020). *Ethics code for behavior analysts*. Author.

Bijou, S. W., Peterson, R. F., & Ault, M. H. (1968). A method to integrate descriptive and experimental field studies at the level of data and empirical concepts. *Journal of Applied Behavior Analysis, 1*(2), 175–191. https://doi.org/10.1901/jaba.1968.1-175

Bowring, D. L., Painter, J., & Hastings, R. P. (2019). Prevalence of challenging behaviour in adults with intellectual disabilities, correlates, and association with mental health. *Current Developmental Disorders Reports, 6*, 173–181. https://doi.org/10.1007/s40474-019-00175-9

Brodie, M. J., Besag, F., Ettinger, A. B., Mula, M., Gobbi, G., Comai, S., Aldenkamp, A. P., & Steinhoff, B. J. (2016). Epilepsy, antiepileptic drugs, and aggression: An evidence-based review. *Pharmacological Reviews, 68*(3), 563–602. https://doi.org/10.1124/pr.115.012021

Camp, E. M., Iwata, B. A., Hammond, J. L., & Bloom, S. E. (2009). Antecedent versus consequent events as predictors of problem behavior. *Journal of Applied Behavior Analysis, 42*(2), 469–483. https://doi.org/10.1901/jaba.2009.42-469

Carr, E. G. (1977). The motivation of self-injurious behavior: A review of some hypotheses. *Psychological Bulletin, 84*(4), 800–816.

Carr, E. G., & Durand, V. M. (1985). Reducing behavior problems through functional communication training. *Journal of Applied Behavior Analysis, 18*(2), 111–126. https://doi.org/10.1901/jaba.1985.18-111

Charlot, L. R., Doerfler, L. A., & McLaren, J. L. (2020). Psychotropic medications use and side effects of individuals with intellectual and developmental disabilities. *Journal of Intellectual Disability Research, 64*(11), 852–863. https://doi.org/10.1111/jir.12777

Contreras, B. P., Tate, S. A., Morris, S. L., & Kahng, S. (2023). A systematic review of the correspondence between descriptive assessment and functional analysis. *Journal of Applied Behavior Analysis, 56*(1), 146–165. https://doi.org/10.1002/jaba.958

Cooper, S. A., Smiley, E., Jackson, A., Finlayson, J., Allan, L., Mantry, D., & Morrison, J. (2009). Adults with intellectual disabilities: Prevalence, incidence and remission of aggressive behaviour and related factors. *Journal of Intellectual Disability Research, 53*(3), 217–232. https://doi.org/10.1111/j.1365-2788.2008.01127.x

Cooper, J. O., Heron, T. E., & Heward, W. L. (2020). *Applied behavior analysis* (3rd ed.). Pearson Education.

Courtemanche, A. B., Black, W. R., & Reese, R. M. (2016). The relationship between pain, self-injury, and other problem behaviors in young children with autism and other developmental disabilities. *American Journal on Intellectual and Developmental Disabilities, 121*(3), 194–203. https://doi.org/10.1352/1944-7558-121.3.194

Crotty, G., Doody, O., & Lyons, R. (2014). Aggressive behaviour and its prevalence within five typologies. *Journal of Intellectual Disabilities, 18*(1), 76–89. https://doi.org/10.1177/1744629513511356

de Winter, C. F., Jansen, A. A., & Evenhuis, H. M. (2011). Physical conditions and challenging behaviour in people with intellectual disability: A systematic review. *Journal of Intellectual Disability Research, 55*(7), 675–698. https://doi.org/10.1111/j.1365-2788.2011.01390.x

Deb, S., Chaplin, R., Sohanpal, S., Unwin, G., Soni, R., & Lenotre, L. (2008). The effectiveness of mood stabilizers and antiepileptic medication for the management of behaviour problems in adults with intellectual disability: A systematic review. *Journal of Intellectual Disability Research, 52*, 107–113. https://doi.org/10.1111/j.1365-2788.2007.00965.x

Deb, S., Akrout Brizard, B., & Limbu, B. (2020). Association between epilepsy and challenging behaviour in adults with intellectual disabilities: Systematic review and meta-analysis. *BJPsych Open, 6*(5), e114. https://doi.org/10.1192/bjo.2020.96

DeLissovoy, V. (1963). Head banging in early childhood: A suggested cause. *The Journal of Genetic Psychology: Research and Theory on Human Development, 102*(1), 109–114. https://doi.org/10.1080/00221325.1963.10532730

Deochand, N., Eldridge, R. R., & Peterson, S. M. (2020). Toward the development of a functional analysis risk assessment decision tool. *Behavior Analysis in Practice, 13*(4), 978–990. https://doi.org/10.1007/s40617-020-00433-y

Didden, R., Duker, P. C., & Korzilius, H. (1997). Meta-analytic study on treatment effectiveness for problem behaviors with individuals who have mental retardation. *American Journal of Mental Retardation, 101*(4), 387–399.

Dimian, A. F., Botteron, K. N., Dager, S. R., Elison, J. T., Estes, A. M., Pruett, J. R., Jr., Schultz, R. T., Zwaigenbaum, L., Piven, J., Wolff, J. J., & IBIS Network. (2017). Potential risk factors for the development of self-injurious behavior among infants at risk for autism spectrum disorder. *Journal of Autism and Developmental Disorders, 47*(5), 1403–1415. https://doi.org/10.1007/s10803-017-3057-9

Dozier, C. L., Briggs, A. M., Holehan, K. M., Kanaman, N. A., & Juanico, J. F. (2022). Functional analysis methodology: Best practices and considerations. In J. B. Leaf, J. H. Cihon, J. L. Ferguson, & M. J. Weiss (Eds.), *Handbook of applied behavior analysis interventions for autism: Integrating research into practice* (pp. 417–442). Springer.

Elbing, U., Appelbaum, S., & Ostermann, T. (2022). Adverse events and contradictory effects of benzodiazepine in a case with intellectual disability and challenging behaviour. *Journal of Intellectual Disabilities, 28*, 216. https://doi.org/10.1177/17446295221134420

Ervin, R., Radford, P., Bertsch, K., Piper, A., Ehrhardt, K., & Poling, A. (2001). A descriptive analysis and critique of the empirical literature on school-based functional assessment. *School Psychology Review, 30*(2), 193–210. https://doi.org/10.1080/02796015.2001.12086109

Esbensen, A. J., Vincent, L. B., Epstein, J. N., Kamimura-Nishimura, K., Wiley, S., Angkustsiri, K., Abbeduto, L., Fidler, D., Anixt, J. S., & Froehlich, T. E. (2022). Co-occurring medical and behavioural conditions in children with Down syndrome with or without ADHD symptom presentation. *Journal of Intellectual Disability Research, 66*(3), 282–296. https://doi.org/10.1111/jir.12911

Fryling, M. J., & Baires, N. A. (2016). The practical importance of the distinction between open and closed-ended indirect assessments. *Behavior Analysis Practice, 9*(2), 146–151. https://doi.org/10.1007/s40617-016-0115-2

Furniss, F., & Biswas, A. B. (2020). Neurobiology of self-injurious behavior. In *Self-injurious behavior in individuals with neurodevelopmental conditions* (pp. 51–110). Springer Nature.

Geiger, K. B., Carr, J. E., & Leblanc, L. A. (2010). Function-based treatments for escape-maintained problem behavior: A treatment-selection model for practicing behavior analysts. *Behavior Analysis in Practice, 3*(1), 22–32. https://doi.org/10.1007/BF03391755

Hagopian, L. P., Rooker, G. W., & Zarcone, J. R. (2015). Delineating subtypes of self-injurious behavior maintained by automatic reinforcement. *Journal of Applied Behavior Analysis, 48*(3), 523–543. https://doi.org/10.1002/jaba.236

Hall, S. S. (2005). Comparing descriptive, experimental and informant-based assessments of problem behaviors. *Research in Developmental Disabilities, 26*(6), 514–526. https://doi.org/10.1016/j.ridd.2004.11.004

Hanley, G. P. (2012). Functional assessment of problem behavior: Dispelling myths, overcoming implementation obstacles, and developing new lore. *Behavior Analysis in Practice, 5*(1), 54–72. https://doi.org/10.1007/BF03391818

Hanley, G. P., Iwata, B. A., & Thompson, R. H. (2001). Reinforcement schedule thinning following treatment with functional communication training. *Journal of Applied Behavior Analysis, 34*(1), 17–38. https://doi.org/10.1901/jaba.2001.34-17

Hardiman, R. L., & McGill, P. (2018). How common are challenging behaviors amongst individuals with Fragile X Syndrome? A systematic review. *Research in Developmental Disabilities, 76*, 99–109. https://doi.org/10.1016/j.ridd.2018.02.020

Heffernan, L., & Lyons, D. (2016). Differential reinforcement of other behaviour for the reduction of severe nail biting. *Behavior Analysis in Practice, 9*(3), 253–256. https://doi.org/10.1007/s40617-016-0106-3

Huisman, S., Mulder, P., Kuijk, J., Kerstholt, M., van Eeghen, A., Leenders, A., van Balkom, I., Oliver, C., Piening, S., & Hennekam, R. (2018). Self-injurious behavior. *Neuroscience and Biobehavioral Reviews, 84*, 483–491. https://doi.org/10.1016/j.neubiorev.2017.02.027

Individuals with Disabilities Education Act, 20 U.S.C. § 1415. (1997).

Iwata, B. A., & Dozier, C. L. (2008). Clinical application of functional analysis methodology. *Behavior Analysis in Practice, 1*(1), 3–9. https://doi.org/10.1007/BF03391714

Iwata, B. A., Dorsey, M. F., Slifer, K. J., Bauman, K. E., & Richman, G. S. (1982/1994). Toward a functional analysis of self-injury. *Journal of Applied Behavior Analysis, 27*(2), 197–209. https://doi.org/10.1901/jaba.1994.27-197 Reprinted from *Analysis and Intervention in Developmental Disabilities, 1982, 2,* 3–20.

Iwata, B. A., Pace, G. M., Cowdery, G. E., & Miltenberger, R. G. (1994a). What makes extinction work: An analysis of procedural form and function. *Journal of Applied Behavior Analysis, 27*(1), 131–144. https://doi.org/10.1901/jaba.1994.27-131

Iwata, B. A., Pace, G. M., Dorsey, M. F., Zarcone, J. R., Vollmer, T. R., Smith, R. G., & Willis, K. D. (1994b). The functions of self- injurious behavior: An experimental-epidemiological analysis. *Journal of Applied Behavior Analysis, 27*(2), 215–240. https://doi.org/10.1901/jaba.1994.27-215

Iwata, B. A., DeLeon, I. G., & Roscoe, E. M. (2013). Reliability and validity of the functional analysis screening tool (FAST). *Journal of Applied Behavior Analysis, 46*(1), 271–284. https://doi.org/10.1002/jaba.31

Kahng, S., Iwata, B. A., & Lewin, A. B. (2002a). Behavioral treatment of self-injury, 1964 to 2000. *American Journal of Mental Retardation, 107*(3), 212–221. https://doi.org/10.1352/0895-8017(2002)107<0212:BTOSIT>2.0.CO;2

Kahng, S., Iwata, B. A., & Lewin, A. B. (2002b). The impact of functional assessment on the treatment of self-injurious behavior. In S. R. Schroeder, M. L. Oster-Granite, & T. Thompson (Eds.), *Self-injurious behavior: Gene-behavior relationships* (pp. 119–131). American Psychological Association. https://doi.org/10.1037/10457-007

Kahng, S., Hausman, N. L., Fisher, A. B., Donaldson, J. M., Cox, J. R., Lugo, M., & Wiskow, K. M. (2015). The safety of functional analyses of self-injurious behavior. *Journal of Applied Behavior Analysis, 48*(1), 107–114. https://doi.org/10.1002/jaba.168

Kalachnik, J. E., Hanzel, T. E., Sevenich, R., & Harder, S. R. (2002). Benzodiazepine behavioral side effects: Review and implications for individuals with mental retardation. *American Journal of Mental Retardation, 107*(5), 376–410. https://doi.org/10.1352/0895-8017(2002)107<0376:BBSERA>2.0.CO;2

Lalli, J. S., Browder, D. M., Mace, F. C., & Brown, D. K. (1993). Teacher use of descriptive analysis data to implement interventions to decrease students' problem behaviors. *Journal of Applied Behavior Analysis, 26*(2), 227–238. https://doi.org/10.1901/jaba.1993.26-227

Lennox, D. B., & Miltenberger, R. G. (1989). Conducting a functional assessment of problem behavior in applied settings. *Journal of the Association for Persons with Severe Handicaps, 14*(4), 304–311. https://doi.org/10.1177/154079698901400409

Lerman, D. C., & Iwata, B. A. (1993). Descriptive and experimental analyses of variables maintaining self-injurious behavior. *Journal of Applied Behavior Analysis, 26*(3), 293–319. https://doi.org/10.1901/jaba.1993.26-293

Lerman, D. C., & Iwata, B. A. (1996). Developing a technology for the use of operant extinction in clinical settings: An examination of basic and applied research. *Journal of Applied Behavior Analysis, 29*(3), 345–382. https://doi.org/10.1901/jaba.1996.29-345

Lerman, D., Hovanetz, A., Strobel, M., & Tetreault, A. (2009). Accuracy of teacher-collected descriptive analysis data: A comparison of narrative and structured recording formats. *Journal of Behavioral Education, 18*, 157–172. https://doi.org/10.1007/s10864-009-9084-7

Lloyd, B. P., & Kennedy, C. H. (2014). Assessment and treatment of challenging behaviour for individuals with intellectual disability: A research review. *Journal of Applied Research in Intellectual Disabilities, 27*(3), 187–199. https://doi.org/10.1111/jar.12089

Lowe, K., Allen, D., Jones, E., Brophy, S., Moore, K., & James, W. (2007). Challenging behaviours: Prevalence and topographies. *Journal of Intellectual Disability Research, 51*(8), 625–636. https://doi.org/10.1111/j.1365-2788.2006.00948.x

Mace, F. C., & Lalli, J. S. (1991). Linking descriptive and experimental analyses in the treatment of bizarre speech. *Journal of Applied Behavior Analysis, 24*(3), 553–562. https://doi.org/10.1901/jaba.1991.24-553

Marion, S. D., Touchette, P. E., & Sandman, C. A. (2003). Sequential analysis reveals a unique structure for self-injurious behavior. *American Journal of Mental Retardation, 108*(5), 301–313. https://doi.org/10.1352/0895-8017(2003)108<301:SARAUS>2.0.CO;2

Martens, B. K., Baxter, E. L., McComas, J. J., Sallade, S. J., Kester, J. S., Caamano, M., Dimian, A., Simacek, J., & Pennington, B. (2019). Agreement between structured descriptive assessments and functional analyses conducted over a telehealth system. *Behavior Analysis: Research and Practice, 19*(4), 343–356. https://doi.org/10.1037/bar0000153

Matson, J. L., & Vollmer, T. R. (1995). *User's guide: Questions about behavioral function (QABF)*. Scientific Publishers.

Mayo-Ortega, L., Oyama-Ganiko, R., Leblanc, J., Schroeder, S. R., Brady, N., Butler, M. G., Reese, R. M., Richman, D. M., Peacock, G., Foster, J., & Marquis, J. (2012). Mass screening for severe problem behavior among infants and toddlers in Peru. *Journal of Mental Health Research in Intellectual Disabilities, 5*(3–4), 246–259. https://doi.org/10.1080/19315864.2011.590626

McCarthy, J. M., & Chaplin, E. (2022). Adults with intellectual disability and autism spectrum disorder: What is the evidence around the use of polypharmacy. *International Journal of Environmental Research and Public Health, 19*(23), 15974. https://doi.org/10.3390/ijerph192315974

Melanson, I. J., & Fahmie, T. A. (2023). Functional analysis of problem behavior: A 40-year review. *Journal of Applied Behavior Analysis, 56*(2), 262–281. https://doi.org/10.1002/jaba.983

Mercatoris, M., & Craighead, W. E. (1974). Effects of nonparticipant observation on teacher and pupil classroom behavior. *Journal of Educational Psychology, 66*(4), 512–519. https://doi.org/10.1037/h0036755

Michael, J. (2004). *Concepts and principles of behavior analysis* (rev. ed.). Western Michigan University, Association for Behavior Analysis International. https://scholarworks.wmich.edu/books/353

Nicholls, G., Bailey, T., Grindle, C. F., & Hastings, R. P. (2023). Challenging behaviour and its risk factors in children and young people in a special school setting: A four wave longitudinal study. *Journal of Applied Research in Intellectual Disabilities, 36*(2), 366–373. https://doi.org/10.1111/jar.13066

O'Neill, R. E., Horner, R. H., Albin, R. W., Sprague, J. R., Storey, K., & Newton, J. S. (1997). *Functional assessment and program development for problem behavior*. Brooks/Cole Publishing.

Oliver, A. C., Pratt, L. A., & Normand, M. P. (2015). A survey of functional behavior assessment methods used by behavior analysts in practice. *Journal of Applied Behavior Analysis, 48*(4), 817–829. https://doi.org/10.1002/jaba.256

Paclawskyj, T. R., Matson, J. L., Rush, K. S., Smalls, Y., & Vollmer, T. R. (2000). Questions About Behavior Function (QABF): A behavioral checklist for functional assessment of aberrant behavior. *Research in Developmental Disabilities, 21*(3), 223–229. https://doi.org/10.1016/s0891-4222(00)00036-6

Pence, S. T., Roscoe, E. M., Bourret, J. C., & Ahearn, W. H. (2009). Relative contributions of three descriptive methods: Implications for behavioral assessment. *Journal of Applied Behavior Analysis, 42*(2), 425–446. https://doi.org/10.1901/jaba.2009.42-425

Petscher, E. S., Rey, C., & Bailey, J. S. (2009). A review of empirical support for differential reinforcement of alternative behavior. *Research in Developmental Disabilities, 30*(3), 409–425. https://doi.org/10.1016/j.ridd.2008.08.008

Phillips, C. L., Iannaccone, J. A., Rooker, G. W., & Hagopian, L. P. (2017). Noncontingent reinforcement for the treatment of severe problem behavior: An analysis of 27 consecutive applications. *Journal of Applied Behavior Analysis, 50*(2), 357–376. https://doi.org/10.1002/jaba.376

Piazza, C. C., Fisher, W. W., Hanley, G. P., Hilker, K., & Derby, K. M. (1996). A preliminary procedure for predicting the positive and negative effect of reinforcement-based procedures. *Journal of Applied Behavior Analysis, 29*(2), 137–152. https://doi.org/10.1901/jaba.1996.29-137

Piazza, C. C., Adelinis, J. D., Hanley, G. P., Goh, H. L., & Delia, M. D. (2000). An evaluation of the effects of matched stimulation on behaviors maintained by automatic reinforcement. *Journal of Applied Behavior Analysis, 33*(1), 13–27. https://doi.org/10.1901/jaba.2000.33-13

Piazza, C. C., Roane, H. S., Keeney, K. M., Boney, B. R., & Abt, K. A. (2002). Varying response effort in the treatment of pica maintained by automatic reinforcement. *Journal of Applied Behavior Analysis, 35*(3), 233–246. https://doi.org/10.1901/jaba.2002.35-233

Piazza, C. C., Fisher, W. W., Brown, K. A., Shore, B. A., Patel, M. R., Katz, R. M., Sevin, B. M., Gulotta, C. S., & Blakely-Smith, A. (2003). Functional analysis of inappropriate mealtime behaviors. *Journal of Applied Behavior Analysis, 36*(2), 187–204. https://doi.org/10.1901/jaba.2003.36-187

Rojahn, J., & Meier, L. J. (2009). Epidemiology of mental illness and maladaptive behavior in intellectual disabilities. *International Review of Research in Mental Retardation, 38*, 239–287. https://doi.org/10.1016/S0074-7750(08)38009-4

Rojahn, J., Barnard-Brak, L., Medeiros, K., & Schroeder, S. R. (2016). Stereotyped behaviours as precursors of self-injurious behaviours: A longitudinal study with infants and toddlers at risk for developmental delay. *Journal of Intellectual Disability Research, 60*(2), 156–166. https://doi.org/10.1111/jir.12224

Rosell, D. R., & Siever, L. J. (2015). The neurobiology of aggression and violence. *CNS Spectrums, 20*(3), 254–279. https://doi.org/10.1017/S109285291500019X

Ruddick, L., Davies, L., Bacarese-Hamilton, M., & Oliver, C. (2015). Self-injurious, aggressive and destructive behaviour in children with severe intellectual disability: Prevalence, service need and service receipt in the UK. *Research in Developmental Disabilities, 45-46*, 307–315. https://doi.org/10.1016/j.ridd.2015.07.019

Saini, V., Ubdegrove, K., Biran, S., & Duncan, R. (2019). A preliminary evaluation of interrater reliability and concurrent validity of open-ended indirect assessment. *Behavior Analysis in Practice, 13*, 114–125. https://doi.org/10.1007/s40617-019-00364-3

Saini, V., Fisher, W. W., Betz, A. M., & Piazza, C. C. (2021). Functional analysis: History and methods. In W. W. Fisher, C. C. Piazza, & H. S. Roane (Eds.), *Handbook of applied behavior analysis* (2nd ed., pp. 214–233). The Guilford Press.

Schlinger, H. D., Jr., & Normand, M. P. (2013). On the origin and functions of the term functional analysis. *Journal of Applied Behavior Analysis, 46*(1), 285–288. https://doi.org/10.1002/jaba.6

Schreck, K. A., & Richdale, A. L. (2020). Sleep problems, behavior, and psychopathology in autism: Inter-relationships across the lifespan. *Current Opinion in Psychology, 34*, 105–111. https://doi.org/10.1016/j.copsyc.2019.12.003

Shawler, L. A., Russo, S. R., Hilton, J. L., Kahng, S., Davis, C. J., & Dorsey, M. F. (2019). Behavioral treatment of self-injury: 2001 to 2016. *American Journal on Intellectual and Developmental Disabilities, 124*(5), 450–469. https://doi.org/10.1352/1944-7558-124.5.450

Smith, J., Baksh, R. A., Hassiotis, A., Sheehan, R., Ke, C., Wong, T. L. B., Strydom, A., & Investigators, P. E. T. A. L. (2022). Aggressive challenging behavior in adults with intellectual disability: An electronic register-based cohort study of clinical outcome and service use. *European Psychiatry: The Journal of the Association of European Psychiatrists, 65*(1), e74. https://doi.org/10.1192/j.eurpsy.2022.2336

Soke, G. N., Rosenberg, S. A., Hamman, R. F., Fingerlin, T., Robinson, C., Carpenter, L., Giarelli, E., Lee, L. C., Wiggins, L. D., Durkin, M. S., & DiGuiseppi, C. (2016). Brief report: Prevalence of self-injurious behaviors among children with autism spectrum disorder-a population-based study. *Journal of Autism and Developmental Disorders, 46*(11), 3607–3614. https://doi.org/10.1007/s10803-016-2879-1

St Peter, C. C., Vollmer, T. R., Bourret, J. C., Borrero, C. S., Sloman, K. N., & Rapp, J. T. (2005). On the role of attention in naturally occurring matching relations. *Journal of Applied Behavior Analysis, 38*(4), 429–443. https://doi.org/10.1901/jaba.2005.172-04

Summers, J., Shahrami, A., Cali, S., D'Mello, C., Kako, M., Palikucin-Reljin, A., Savage, M., Shaw, O., & Lunsky, Y. (2017). Self-Injury in autism spectrum disorder and intellectual disability: Exploring the role of reactivity to pain and sensory input. *Brain Sciences, 7*(11), 140. https://doi.org/10.3390/brainsci7110140

Symons, F. J., Davis, M. L., & Thompson, T. (2000). Self-injurious behavior and sleep disturbance in adults with developmental disabilities. *Research in Developmental Disabilities, 21*(2), 115–123. https://doi.org/10.1016/s0891-4222(00)00028-7

Thompson, R. H., & Borrero, J. C. (2021). Direct observation. In W. W. Fisher, C. C. Piazza, & H. S. Roane (Eds.), *Handbook of applied behavior analysis* (pp. 202–213). The Guilford Press.

Thompson, R. H., & Iwata, B. A. (2007). A comparison of outcomes from descriptive and functional analyses of problem behavior. *Journal of Applied Behavior Analysis, 40*(2), 333–338. https://doi.org/10.1901/jaba.2007.56-06

Thompson, T., Symons, F. J., & Felce, D. (2000). Principles of behavioral observation: Assumptions and strategies. In T. Thompson, D. Felce, & F. J. Symons (Eds.), *Behavioral observation: Technology and applications in developmental disabilities* (pp. 3–16) Brookes.

Totsika, V., Toogood, S., Hastings, R. P., & Lewis, S. (2008). Persistence of challenging behaviours in adults with intellectual disability over a period of 11 years. *Journal of Intellectual Disability Research, 52*(Pt 5), 446–457. https://doi.org/10.1111/j.1365-2788.2008.01046.x

Tsiouris, J. A., Kim, S. Y., Brown, W. T., & Cohen, I. L. (2011). Association of aggressive behaviours with psychiatric disorders, age, sex and degree of intellectual disability: A large-scale survey. *Journal of Intellectual Disability Research, 55*(7), 636–649. https://doi.org/10.1111/j.1365-2788.2011.01418.x

Vollmer, T. R., Iwata, B. A., Zarcone, J. R., Smith, R. G., & Mazaleski, J. L. (1993). The role of attention in the treatment of attention-maintained self-injurious behavior: Noncontingent reinforcement and differential reinforcement of other behavior. *Journal of Applied Behavior Analysis, 26*(1), 9–21. https://doi.org/10.1901/jaba.1993.26-9

Vollmer, T. R., Borrero, J. C., Wright, C. S., Van Camp, C., & Lalli, J. S. (2001). Identifying possible contingencies during descriptive analyses of severe behavior disorders. *Journal of Applied Behavior Analysis, 34*(3), 269–287. https://doi.org/10.1901/jaba.2001.34-269

Weeden, M., Mahoney, A., & Poling, A. (2010). Self-injurious behavior and functional analysis: Where are the descriptions of participant protections? *Research in Developmental Disabilities, 31*(2), 299–303. https://doi.org/10.1016/j.ridd.2009.09.016

White, G. D. (1977). The effects of observer presence on the activity level of families. *Journal of Applied Behavior Analysis, 10*(4), 734. https://doi.org/10.1901/jaba.1977.10-734

Wilder, D. A., Ertel, H. M., & Cymbal, D. J. (2021). A review of recent research on the manipulation of response effort in applied behavior analysis. *Behavior Modification, 45*(5), 740–768. https://doi.org/10.1177/0145445520908509

Willner, P. (2015). The neurobiology of aggression: Implications for the pharmacotherapy of aggressive challenging behaviour by people with intellectual disabilities. *Journal of Intellectual Disability Research, 59*(1), 82–92. https://doi.org/10.1111/jir.12120

Zarcone, J. R., Rodgers, T. A., Iwata, B. A., Rourke, D. A., & Dorsey, M. F. (1991). Reliability analysis of the motivation assessment scale: A failure to replicate. *Research in Developmental Disabilities, 12*(4), 349–360. https://doi.org/10.1016/0891-4222(91)90031-m

Mental Health in People with Intellectual and Developmental Disabilities

Marianne Berg Halvorsen, Arvid Nikolai Kildahl, and Sissel Berge Helverschou

Abstract Mental health (MH) disorders are common in people with intellectual and developmental disabilities (IDDs). However, these conditions are not necessarily adequately addressed in clinical practice. The current chapter discusses challenges in MH assessment and treatment among autistic people and people with intellectual disability (ID), including the applicability of general-purpose instruments, and evidence-based principles in MH assessment. Moreover, we discuss the adaptation of interventions from general MH care to the specific individual's strengths and difficulties and how contextual factors are likely to influence MH in people with IDDs. In sum, the assessment and treatment of MH disorders constitute an essential component of the care of autistic people and people with ID.

Keywords Autism spectrum disorder · Intellectual disability · Mental disorder · Mental health

Mental health (MH) can be defined as the absence of MH illness or a state of well-being that enables an individual to cope with life stressors, realize their abilities, and contribute within the community (World Health Organization [WHO], 2022).

M. B. Halvorsen (✉)
Department of Pediatric Rehabilitation, University Hospital of North Norway, Tromsø, Norway
e-mail: marianne.berg.halvorsen@unn.no

A. N. Kildahl
NevSom Norwegian Centre of Expertise for Neurodevelopmental Disorders and Hypersomnias, Oslo University Hospital, Oslo, Norway

Regional Section Mental Health, Intellectual Disabilities/Autism, Oslo University Hospital, Oslo, Norway

S. B. Helverschou
NevSom Norwegian Centre of Expertise for Neurodevelopmental Disorders and Hypersomnias, Oslo University Hospital, Oslo, Norway

M. G. Valdovinos (ed.), *Intellectual and Developmental Disabilities*, Contemporary Clinical Neuroscience,
https://doi.org/10.1007/978-3-031-66932-3_11

Feelings of anxiety, restlessness, and low mood are MH symptoms that everyone can feel from time to time and that can often be related to events and experiences. In this chapter, we will discuss MH problems that are either diagnosed or at diagnosable levels, which means that the symptom burden is considerable, lasts over time, and typically leads to functional impairment in everyday activities. We focus on MH assessment among people with intellectual disability (ID) and/or autistic people with or without co-occurring ID. ID, autism spectrum disorder (autism), and attention-deficit-hyperactivity disorder (ADHD) are considered neurodevelopmental disorders, not MH disorders, in the Diagnostic and Statistical Manual of Mental Disorders (5th Edition; DSM-5; American Psychiatric Association [APA], 2013). While they may all co-occur, the classification of neurodevelopmental disorders is beyond the scope of this chapter.

As co-occurring MH disorders are associated with reduced quality of life for the affected person and their family (Sáez-Suanes & Álvarez-Couto, 2022), careful assessment and treatment of MH is considered an essential component of care for all autistic people and people with ID (Halvorsen et al., 2019, 2023; Lord et al., 2022; Siegel et al., 2020).

Phenomenology and Recognition

Approximately 30–40% of people with ID or autism will present with a diagnosable MH disorder, a rate that is at least double that of the general population (Howlin, 2021; Mazza et al., 2020; Totsika et al., 2022). The causes of the increased risk appear to be multifactorial, including genetic vulnerabilities (e.g., Srivastava et al., 2023) and environmental factors, such as socioeconomic status and adverse childhood experiences (Munir, 2016; Vervoort-Schel et al., 2021). Some genetic syndromes are associated with an increased risk of specific MH disorders (Hodapp et al., 2016), for example, bipolar disorder and catatonia in Phelan Mc-Dermid syndrome/22q13.3 deletion syndrome (Srivastava et al., 2023; Verhoeven et al., 2020) and psychotic disorder in 22q.11 deletion syndrome (Fiksinski et al., 2021). In addition, social factors, such as the experiences of autistic people being misunderstood or misperceived by their surroundings, involve a risk of stress, isolation (Mitchell et al., 2021), and stigma (Botha & Frost, 2020; Pelleboer-Gunnink et al., 2021).

Among the most frequent MH disorders in people with ID or autism are anxiety disorders, disruptive/impulse-control/conduct disorders, and depressive disorders (Buckley et al., 2020; Lai et al., 2019; Mazza et al., 2020). Recent meta-analytical evidence finds a similar prevalence of MH problems among children and adolescents with ID, irrespective of adaptive functioning levels and communication capacity (Buckley et al., 2020). Moreover, the relationship among age, gender, and MH problems in people with ID remains unclear (Buckley et al., 2020). Among autistic people, evidence from a recent meta-analysis revealed that older age was associated with a higher prevalence of depressive, bipolar, and schizophrenia spectrum disorders (Lai et al., 2019). Additionally, a higher prevalence of depressive disorders was found in studies with a high proportion of female participants than in those with a low proportion of female participants (Lai et al., 2019). In relation to ID, studies with a high proportion of autistic people with co-occurring ID reported a higher prevalence of schizophrenia spectrum disorders than those with a low proportion of participants with ID (Lai et al., 2019).

Although an increased risk of MH conditions in autistic people or ID has been widely reported in research, this increased risk is not necessarily adequately addressed in clinical practice (Halvorsen et al., 2024; Lord et al., 2020). Identifying MH disorders in autistic people and people with ID can be challenging (Halvorsen et al., 2024; Kildahl et al., 2023b). Autistic people and people with ID are heterogeneous groups; both groups have considerable interindividual variation in, for instance, intellectual, adaptive, and verbal functioning (Burack et al., 2021; Lord et al., 2018). Limited verbal skills and severe ID complicate MH assessment, as they render self-reporting difficult. People with ID may express their MH problems differently due to their difficulties in intellectual functioning (e.g., delusions in psychosis may be less complex in people with lower IQs) as well as contextual and environmental factors (e.g., avoidance in posttraumatic stress disorder may be perceived as noncompliance in individuals with extensive care needs). Furthermore, so-called challenging behaviors have been found to be associated with MH symptoms (Bowring et al., 2019; Kildahl et al., 2023a) but are not specific to MH problems and may have other explanations (Bowring et al., 2019; Wolkorte et al., 2019).

It can be difficult to distinguish between autism/ID and MH disorders due to considerable symptom overlap (Helverschou et al., 2011). Thus, symptoms of MH disorders may be overlooked/misinterpreted or misattributed to autism or ID (Jopp & Keys, 2001). This is often referred to as "diagnostic overshadowing" (Reiss et al., 1982). Apparent symptoms overlap for autism and MH disorders make it particularly challenging to distinguish, for example, autism from schizophrenia and autism from obsessive-compulsive disorder (OCD). The ritualistic and repetitive behaviors that are one of the core characteristics of autism may be difficult to distinguish from OCD. Social withdrawal may be a feature of autism, as well as a symptom of schizophrenia. Moreover, stress, anxiety, and trauma-related disorders appear to be very common in autistic people (Helverschou et al., 2011; Kerns et al., 2021; Rumball et al., 2021), and it has been suggested that there may be a risk of overlooking these symptoms because they are so common in autistic people (Kildahl & Helverschou, 2023). Symptoms of depressive disorder may be overlooked in individuals with ID and autistic people because mood changes may be challenging to recognize if the person cannot convey these verbally and because social withdrawal, limited facial expression and apparently flattened affect may be indicators of both depression and ASD (Helverschou, 2022).

Accordingly, adapted diagnostic manuals and practice guidelines that account for how the observable manifestations of MH symptoms may differ among people with different levels of ID have been developed to assist clinicians (e.g., Diagnostic Manual –Intellectual Disability 2; Fletcher et al., 2021; Siegel et al., 2020). While these guidelines have been developed by experienced clinicians and researchers, they have not been empirically tested. There are currently no such adaptations of diagnostic criteria available for autistic people.

Mental Health Assessment

General-Purpose Instruments

The use of standardized MH instruments is often an integral part of MH assessment, along with other methods to which we will return. We use the term "broadband" to refer to instruments designed to capture multiple conditions or symptom clusters (Halvorsen et al., 2024). Broadband instruments are particularly important because they can facilitate differential diagnostic assessment in a more systematic way than single-disorder instruments and, importantly, can aid clinicians in exploring symptoms of less common MH disorders (Halvorsen et al., 2024).

We can distinguish between two main groups of instruments: instruments developed or adapted for autistic people or people with intellectual and developmental disabilities (IDDs) (referred to as ASD/IDD-specific instruments) and less specific instruments developed for the general child or adult population (referred to as conventional instruments) (Halvorsen et al., 2023, 2024). Furthermore, some instruments have been empirically developed based on specific descriptors of autism/IDD child or adult functioning via factor analyses (e.g., Aberrant Behavior Checklist: Aman & Singh, 1986, 2017; Einfeld & Tonge, 1992; Gray et al., 2018), while other instruments have been developed based on an established diagnostic MH taxonomy (Schedule for Affective Disorders and Schizophrenia for School-Age Children-Present and Lifetime version: Kaufman et al., 1997; i.e., DSM; e.g., Child Behavior Checklist: Achenbach & Rescorla, 2001). When selecting an instrument, it is recommended to base the decision on the objective of the assessment and the psychometric properties of the instrument (reliability and validity) in the specific patient population to be examined (Halvorsen et al., 2024; Siegel et al., 2020).

In two recent systematic reviews, we found consistently better documentation of the reliability and validity of autism/IDD-specific instruments compared to conventional instruments among children and adolescents with ID (Halvorsen et al., 2023) and among autistic people of all age groups (Halvorsen et al., 2024). Based on recent evidence from these and other reviews (Flynn et al., 2017; Halvorsen et al., 2023, 2024), some informant-based general-purpose scales with broad application can be recommended (see Table 1 for a detailed description of the instruments). First, the Developmental Behavior Checklist (DBC; Einfeld & Tonge, 1992; Gray et al., 2018) and the Aberrant Behavior Checklist (ABC; Aman & Singh, 1986, 2017) can be employed as screening measures to be used in an initial assessment of MH problems among people with ID (Flynn et al., 2017; Halvorsen et al., 2023) and in intellectually heterogeneous ASD samples in all age groups (Halvorsen et al., 2024). Second, the Achenbach System of Empirically Based Assessment Child Behavior Checklist (ASEBA-CBCL) can be applied among children with intellectual and verbal abilities within the normal range. If applied among intellectually heterogeneous autism samples, the use of the item-level data of the CBCL rather

Table 1 General-purpose instruments for MH assessment

Instrument	Purpose/composition	Administration and scoring	Framework
Aberrant Behavior Checklist-Community version (*ABC*; Aman & Singh, 1986, 2017)	Population: People with IDDs Age: Children–adults 58 items, 5 subscales: Irritability, social withdrawal, stereotypic behavior, hyperactivity, noncompliance, inappropriate speech	Proxy 4-point scale (0–3)	Empirically derived
Child Behavior Checklist/Youth Self-Report (*CBCL*; Achenbach & Rescorla, 2001)	Population: General Age versions: 1.5–5 years and 6–18 years 120 items, 8 syndrome scales: Withdrawn, Somatic complaints, anxious/depressed, social problems, thought problems, attention problems, delinquent behavior, aggressive behavior 2 broad-band scales: Internalizing and externalizing. A total problem score The 2001 version includes 6 DSM-oriented scales: Affective problems, anxiety problems, somatic problems, attention-deficit/hyperactivity problems, oppositional defiant problems, conduct problems	Youth Self-Report Proxy: Parent report (CBCL) and Teacher's Report Form report.3-point scale (0–2)	Empirically developed DSM-oriented scales (latest version)
Developmental Behavior Checklist (*DBC*; Einfeld & Tonge, 1992; Gray et al., 2018)	Population: People with IDD Age: 4–18 years 96 items, 6 subscales: Disruptive, self-absorbed, communication disturbance, anxiety, social relating, antisocial. Total behavior problem score	Proxy: Primary carer and teacher versions 3-point scale (0–2)	Empirically derived

(continued)

Table 1 (continued)

Instrument	Purpose/composition	Administration and scoring	Framework
Hospital Anxiety and Depression Scale (*HADS;* Zigmond & Snaith, 1983)	Population: General Age: Older youths and adults 14 items; 2 subscales: Anxiety, depression	Self-report 4-point scale (0–3)	Developed to measure anxiety and depressive symptoms (nonphysical symptoms) in a general medical population of patients
Psychopathology in Autism Checklist (*PAC;* Helverschou et al., 2009)	Population: People with autism and IDD Age: Older youth and adults 42 items, 5 subscales: General adjustment, psychosis, depression, anxiety, OCD	Proxy 4-point scale (1–4)	DSM-IV/ICD-10 and expert consensus. Empirically developed

Note: *IDDs* intellectual and developmental disabilities, *MH* mental health

than the subscales is recommended (Dovgan et al., 2019). Third, the Psychopathology in Autism Checklist (PAC; Helverschou et al., 2009) is a potentially interesting instrument; however, more research is warranted. This is a screening instrument based on a diagnostic framework (i.e., DSM-IV; ICD-10); preliminary evidence among adolescents and adults with co-occurring autism and ID indicates that it can differentiate between those with an MH disorder who need further assessment and those without an MH disorder (Helverschou et al., 2009, 2021).

Self-report instruments for MH assessment in people with ID or autism are few, frequently unreliable, and appear to lack validity (Flynn et al., 2017; Halvorsen et al., 2023, 2024; Kaiser & Halvorsen, 2022). Accordingly, interpretation of scores from self-report measures (e.g., ASEBA-Youth Self-Report; Achenbach & Rescorla, 2001; Hospital Anxiety and Depression Scale; Zigmond & Snaith, 1983) should be made with caution and always substantiated with other information. However, self-reporting is important because it may give access to vital information that family members or service providers may not be aware of (Havercamp et al., 2022). Moreover, information from caregivers is not interchangeable with self-reports regarding the subjective experience of MH disorders (Kildahl et al., 2023b). With adequate support and adaptation, many people with ID can provide highly relevant information (Havercamp et al., 2022), even if the evidence for structured instruments using self-reporting remains limited.

Evidence-Based Principles and Potential Bias in MH Assessment

Because of the challenges in conducting MH assessments in people with ID or autism, current recommendations (e.g., Bertelli & Moss, 2022; Deb et al., 2022; Kildahl et al., 2023b; Siegel et al., 2020) suggest that information from multiple informants and multiple assessment tools, as well as a comprehensive developmental and medical history, are necessary. Interdisciplinary approaches are important to rule out alternative explanations, particularly pain (e.g., migraines/headaches) and other somatic health issues (e.g., epilepsy) (Pan et al., 2021). Self-reports by the individual concerned or direct observations by the MH professional in settings in their daily lives (Kildahl et al., 2023b; Valdovinos et al., 2019) may be helpful or provide an important corrective. Information from these various sources then needs to be triangulated and integrated, accounting for the various strengths and limitations of each of these approaches to assessment.

Regarding the use of assessment tools, it may be helpful to use at least one conventional broad-band measure and combine this with more autism/IDD-specific tools. Autism/IDD tools do not always have clear implications for diagnosis, and many are primarily designed as screening tools. It may be helpful to use multiple informants, particularly for the screening checklists, as reports may vary with the relationship between the participant and the proband and with the contexts in which the informant has observed or interacted with the person in question (e.g., Kildahl & Jørstad, 2022). In the adaptation of conventional assessment tools, particular attention should be given to the context and living conditions of the individual in question. For instance, individuals with extensive care needs may be at risk of trauma experiences that are uncommon or unusual in the general population and therefore not included in conventional checklists (e.g., being scalded when taking a shower due to negligence; Kildahl & Jørstad, 2022). Whether the individual has access to sufficiently adapted services, including staff communication and behavior, is another important contextual factor that should be explored (e.g., Wolkorte et al., 2019).

Finally, to avoid diagnostic overshadowing, clinicians and others need to reflect critically on their own diagnostic practices (Kildahl et al., 2023b). Diagnostic overshadowing may also occur for clinicians who specialize in MH assessments in people with ID or autism (Kildahl et al., 2020), possibly due to the complexities of these assessments, leaving them more vulnerable to the influence of clinician bias (Kildahl et al., 2023b). For instance, "search satisficing" is a common bias in human decision-making (Bate et al., 2012; Kahneman, 2011), involving risk in assessment that when one diagnosis has been identified, assessment is discontinued (Bate et al., 2012; Kildahl et al., 2023b).

Treatment of MH Disorders

Research on treatment and intervention for MH issues in autistic people and people with ID remains sparse. Thus, the evidence base is limited, particularly for autistic individuals with co-occurring ID and limited verbal skills (Bakken et al., 2016; Helverschou, 2022; Helverschou et al., 2011; Rosen et al., 2018). However, the current literature and guidelines generally recommend adapting interventions from general MH care to the specific individual's strengths and difficulties (e.g., National Institute for Health and Care Excellence, 2016), indicating that these populations require access to the same range of intervention and treatment as others with similar MH issues (e.g., Korb & Hassiotis, 2022). However, autistic people and people with ID often struggle to gain access to MH services (Camm-Crosbie et al., 2019; Whittle et al., 2018), and research suggests that adaptation is often insufficient (Hughes et al., 2023; Lunsky et al., 2009; Maddox et al., 2020; Weise & Trollor, 2018). Moreover, some findings indicate that individuals with the most complex and severe difficulties (i.e., those presumably with the most severe need for services) have the greatest difficulties in accessing these very services (Cashin et al., 2018; Walsh et al., 2020).

In general, treatment strategies for MH disorders include psychotherapy, psychopharmacological treatment, nursing/support, and more contextually focused interventions. Pharmacological approaches constitute an important aspect of treatment for MH disorders, as in the general population, particularly for more severe disorders, such as schizophrenia and bipolar disorder (National Institute for Health and Care Excellence, 2016). Ideally, the prescription of psychopharmacological treatment should be based on an MH diagnosis following a comprehensive assessment, including somatic examinations and a pharmacological review, and should follow recommended guidelines for the specific MH disorder (Faculty of Psychiatry of Intellectual Disability, 2016; National Institute for Health and Care Excellence, 2016). A pharmacological review included critical evaluation, including potential discontinuation or optimization, of current medication (Shankar et al., 2019) and screening for atypical drug metabolism by a blood test (cytochrome P450; CYP) (Bose-Brill et al., 2017; Perera et al., 2022). While the current evidence for psychopharmacological treatment is limited, several studies indicate that targeted psychopharmacological treatment can have a positive impact on the treatment of MH disorders in autistic people and people with ID (e.g., Molina-Ruiz et al., 2017; Rysstad et al., 2022). However, studies have also found that drugs intended for the treatment of MH disorders are overprescribed in people with ID, including the use of antipsychotics for reducing or eliminating problematic behaviors (e.g., De Kuijper et al., 2010; Sheehan et al., 2015). Other studies have indicated a potential shift in prescription patterns in recent years, where the use of these drugs has increasingly targeted the treatment of MH disorders rather than the control or reduction of problematic behaviors (Tsiouris et al., 2013). Autistic people and people with ID may be particularly sensitive to the side effects of psychopharmacological treatment, and monitoring potential side effects is important with any prescription

(Scheifes et al., 2016; Sheehan et al., 2017; Srivastava et al., 2023). Moreover, discerning whether pharmacological treatment has a positive effect may be challenging unless systematic approaches to treatment evaluation are used, particularly in people with limited verbal language skills (Kildahl et al., 2023b). Studies of discontinuation of pharmacological intervention indicate mixed effects (Adams et al., 2023). While a reduction in the dosage or discontinuation of antipsychotics, for example, appears to be associated with improved physical health outcomes, such a reduction or discontinuation may increase the risk of relapse of MH symptoms or problematic behaviors (Adams et al., 2023). In summary, according to recent recommendations, psychopharmacological treatment of MH disorders in people with ID and autism should be individualized, accounting for risks as well as potential benefits, and should be systematically and critically evaluated at regular intervals, preferably using a multidisciplinary approach (Adams et al., 2023; Kildahl et al., 2023b; Molina-Ruiz et al., 2017; Shankar et al., 2019; Trollor et al., 2016).

In regard to psychotherapy, current research has mainly focused on cognitive behavioral therapy (CBT), and although high-quality evidence remains sparse, findings concerning its potential effectiveness are promising for autistic people (Kerns et al., 2016; Vasa et al., 2014; Wood et al., 2020), as well as people with ID (Dagnan et al., 2018; Fynn et al., 2023). Although recommended adaptations have been published (e.g., Spain & Happé, 2020), it has consistently been highlighted that adapting psychotherapy to the needs of autistic people or people with ID is a highly individualized process (Moore et al., 2023; Spain & Happé, 2020). While these findings have received less attention, they are promising for other types of psychotherapy, for instance, specific trauma therapies such as eye movement desensitization and reprocessing therapy (Karatzias et al., 2019). Moreover, most of the studies on psychological treatment in these populations involve younger autistic participants without ID (Kannabiran & McCarthy, 2009; Weston et al., 2016; White et al., 2013), research is particularly lacking for autistic adults with co-occurring ID, and evidence for the use of CBT in autistic people as a treatment for depressive disorders is currently more limited than for anxiety disorders (Lord et al., 2020). Overall, even though high-quality evidence concerning specific psychotherapies is lacking for autistic people and people with ID, research to date indicates a positive effect of these interventions (Tapp et al., 2023). Moreover, preliminary evidence indicates that strategies from psychotherapy can be adapted to people with severe ID and limited verbal language skills (Apanasionok et al., 2023).

Because co-occurring conditions and multimorbidity are common (e.g., Halvorsen et al., 2019; Olsen et al., 2021) and because autistic people and people with ID may depend on the support of caregivers and families in their daily lives, more comprehensive treatment plans are often necessary (Dunn et al., 2020; Mohiuddin et al., 2011). Organizing treatment as a collaborative effort involving not only the person themselves and the clinician but also family members and service providers may aid the treatment process by drawing on family members' often unique knowledge and expertise concerning the specific individual (Chester et al., 2020; Rysstad et al., 2022). Good cooperation with other services, such as primary health care and social/educational services, in addition to sufficient information about the individual's diagnoses and conditions, is associated with caregiver satisfaction (Kjærandsen et al., 2021). Moreover, ensuring that the individual is adequately supported in their daily lives and that care and support strategies are in accordance with the individual's MH needs are further important aspects of adaptation (Donner & Gustin, 2021; Sommerstad et al., 2021; Wolkorte et al., 2019). Due to the high prevalence of trauma in autistic people and people with ID, trauma-informed approaches have been held as an essential component of care (Truesdale et al., 2019). In ensuring that the individual is adequately supported, environmental factors such as sensory stimulation and the physical environment, general predictability, social demands, ambiguity, physical activity, and positive/joyful experiences may all need to be adapted to facilitate the individual's well-being (Fuentes et al., 2021).

In sum, treatment of MH disorder in autistic people and people with ID is recommended to utilize the same treatment strategies and interventions used to treat MH disorders in the general population, but these need to be delivered in a format and an organizational framework adapted to the specific individual's underlying strengths, needs, and difficulties.

Life-Long Perspectives and Outcomes

Longitudinal studies following individuals who were diagnosed with ASD in childhood throughout their development, in adolescence and into adulthood, allow us to understand more about autistic people's development over time, including their long-term outcomes. Research to date indicates that autistic individuals with good verbal skills and intellectual abilities within the typical range are more likely to do well as adults (Howlin & Magiati, 2017). However, although these studies show that many do well and improve various skills throughout development, these studies also show that many autistic people continue to have major difficulties in daily life, as well as disadvantageous life conditions as adults (Howlin & Magiati, 2017).

A recently conducted meta-analysis (Mason et al., 2021) identified 17 studies involving a total of 1076 autistic participants. Approximately 18% of the participants were assessed to have good outcomes regarding work, social contact, and independence in life, while 28% were assessed to have fair outcomes. The latter assessment referred to some dependence on others, some degree of supported work, and some social activities outside the home. As many as 51% of the participants were assessed as having poor functioning and poor living conditions as adults, which is consistent with findings from other studies. Moreover, it is worrying that comparisons between more recent longitudinal studies and earlier studies show few differences in results (Howlin, 2021), despite the fact that more recent studies probably include participants who received significantly more structured and systematic interventions as children. Among the factors that seem to be related to poorer living conditions and quality of life as adults appear to be MH problems and/or a higher incidence of autism-related characteristics.

Another pattern that has emerged in a large proportion of autistic people is that they can show reduced autism-related difficulties and function well during their school years, academically and socially. However, a change appears to occur in mid-adolescence to early adulthood for many, involving increased MH problems, such as anxiety and depression. For these individuals, MH problems often have a greater impact on their well-being and level of functioning than their autism-related difficulties. Another group of autistic people are those who are only diagnosed with ASD in adulthood after referral to MH services following the development of co-occurring MH difficulties. Thus, for a significant proportion of autistic people, indicators suggesting poor quality of life in adulthood may be at least partially explained by co-occurring MH difficulties rather than autism (Helverschou, 2022; Howlin, 2021). These findings underline the importance and urgency of providing appropriate MH services for autistic people, people with ID, and other groups with an increased prevalence of MH disorder who appear to have poorer access to services.

Mental Health in Context

Research on autistic adults constitutes a very low proportion compared to research on autistic children and adolescents (Howlin & Magiati, 2017). In particular, there has been little focus on support and services for adults (Bishop-Fitzpatrick et al., 2014; Lord et al., 2020; Shattuck et al., 2020), even though it is particularly during puberty and adulthood that MH difficulties are expressed and result in reduced quality of life. Autistic people and their families highlight access to interventions and services for somatic and MH care as important research areas to be addressed (Roche et al., 2021).

Research indicates that autistic people and people with ID are at increased risk of adverse childhood experiences (Hoover & Kaufman, 2018; McDonnell et al., 2019), including experiences involving violence and sexual abuse (Kildahl & Helverschou, 2023; Weiss & Fardella, 2018), and appear to be more vulnerable to these experiences (Kerns et al., 2015; Mevissen et al., 2016; Rumball, 2019). Autistic people and people with ID are more likely to have multimorbidity (Olsen et al., 2021), are more likely to have difficulties accessing health care (Ali et al., 2013; Camm-Crosbie et al., 2019; Whittle et al., 2018), are more likely to be unemployed (Chen et al., 2015; Howlin & Magiati, 2017; Lysaght et al., 2012), are less likely to live independently (Howlin & Magiati, 2017; Ryan et al., 2019), are less likely to have a social support network (Howlin & Magiati, 2017; Lippold & Burns, 2009), and have lower socioeconomic status (Emerson et al., 2006; Howlin & Magiati, 2017) than the general population. These are all risk or vulnerability factors associated with poor MH in the general population. Thus, the increased rates of MH difficulties in these populations are not surprising and cannot be considered an issue only associated with individual difficulties and vulnerabilities. Mitchell et al. (2021) showed how contextual influences and social experiences, including not being understood, lacking a feeling of social belonging, having difficulties decoding social cues that others take for granted, and feeling that one does not "fit in," may all influence social and psychological development and coping strategies from an early age. Inadequate support for individuals diagnosed with ASD may contribute to exacerbating these negative influences. Moreover, some autistic people are not diagnosed until adolescence or early adulthood (Huang et al., 2020), particularly autistic girls and women, and feeling "different" from their peers while not having any explanation for it may affect self-esteem, feelings of self-worth, and other aspects of psychological functioning throughout these individuals' lives (Huang et al., 2020; Mitchell et al., 2021).

Accordingly, in addition to the individual perspective, fully addressing MH issues in autistic people and people with ID will require societal adaptations and changes in how these populations are viewed and treated throughout their lives. The neurodiversity movement (e.g., Rosqvist et al., 2020) grew from the social model of disability, focusing on societal barriers, derogatory attitudes, and social exclusion faced by people who function in an atypical way. In this way, the neurodiversity movement constitutes an important and alternative viewpoint to the traditional medical model and may help us identify structural and societal contributors to the increased prevalence of MH disorder in autistic people and people with ID.

Clinical Implications and Concluding Remarks

As co-occurring MH disorders are common and impair quality of life, a focus on MH and the assessment of MH conditions need to be integral components of care for all people with IDDs. However, for some autistic people and people with ID, assessment and treatment may be challenging. Current recommendations suggest the use of information from multiple informants and multiple assessment tools and the triangulation and integration of information from different sources. It is often helpful to use at least one broad-band measure in the initial assessment, preferably autism/IDD-specific tools, and combine these with disorder-specific tools for further depth. For example, if results from one or more broad-band assessment tools indicate a possible anxiety disorder, a specific measure of anxiety disorders can be used for further exploration of these symptoms. Measures developed especially for autistic people or people with ID are available for some disorders, particularly anxiety and depressive disorders. Where no autism/IDD-specific tool exists, for instance, regarding psychotic disorders, conventional tools can be used if their results are interpreted with caution and in the context of the individual's underlying difficulties and other symptoms.

Research on treatment and interventions for MH issues in people with IDDs is in an early phase, but CBT studies in particular show promising results. However, it has consistently been highlighted that the adaptation of psychotherapy to the needs of autistic people or people with ID is a highly individualized process. Although much work remains to be done regarding research on psychotherapy for autistic people and people with ID, current knowledge suggests that these interventions can have a positive impact on the individual concerned and their difficulties, as long as the clinician is able to adapt the intervention to the specific individual. Many people with autism and ID have experiences of being misunderstood and/or experiences of ill-suited environments and approaches in, for instance, education, living arrangements, and leisure activities. Accordingly, to prevent and reduce the risk of MH problems in these groups, changes in attitudes toward people with IDDs across contexts, arenas, and services throughout these people's lives are warranted. This includes the opportunity for meaningful participation in society, as well as equitable access to respectful and competent health services.

Acknowledgments We thank MD/psychiatrist Inger Breistein Haugen at the Regional Section Mental Health, Intellectual Disabilities/Autism, Oslo University Hospital, for her helpful comments and suggestions for the paragraph describing psychopharmacological treatment.

Competing Interests Funding was supported by Helse Nord RHF (Grant No. HNF1622-22 awarded to Marianne Berg Halvorsen). Sissel Berge Helverschou has been involved in the development of the Psychopathology in Autism Checklist (PAC) but has no financial or legal interest in its use or dissemination. Arvid Nikolai Kildahl is a part of Helverschou' research group. Marianne Berg Halvorsen handled data and all aspects of the presentation related to the PAC.

References

Achenbach, T. M., & Rescorla, L. (2001). *Manual for the ASEBA school-age forms & profiles: An integrated system of multi-informant assessment*. ASEBA.

Adams, D., Hastings, R. P., Maidment, I., Shah, C., & Langdon, P. E. (2023). Deprescribing psychotropic medicines for behaviours that challenge in people with intellectual disabilities: A systematic review. *BMC Psychiatry, 23*(1), 202. https://doi.org/10.1186/s12888-022-04479-w

Ali, A., Scior, K., Ratti, V., Strydom, A., King, M., & Hassiotis, A. (2013). Discrimination and other barriers to accessing health care: Perspectives of patients with mild and moderate intellectual disability and their carers. *PLoS One, 8*(8), e70855. https://doi.org/10.1371/journal.pone.0070855

Aman, M. G., & Singh, N. N. (1986). *Aberrant behavior checklist manual*. Slosson Educational Publishers.

Aman, M. G., & Singh, N. N. (2017). *Aberrant behavior checklist manual*. Slosson Educational Publishers.

American Psychiatric Association. (2013). *Diagnostic and statistical manual of mental disorders (DSM-5)*. American Psychiatric Publishing.

Apanasionok, M., Scripps, E., Filipczuk, M, Gray, K. M., Hastings, R. P., Rai, D., McNamara, R., Jahoda, A., Bunning, K., Gillespie, D., & Langdon, P. E. (2023, September 21–23). Behavioral interventions to treat anxiety in adults with autism and moderate to severe intellectual disabilities (BEAMS-ID) [Paper presentation]. European Association for mental Health in Intellectual Disability 14th Congress, Helsinki, Finland. https://doi.org/10.1111/jir.13085

Bakken, T. L., Helvershou, S. B., Høidal, S. H., & Martinsen, H. (2016). Mental illness with intellectual disabilities and autism spectrum disorders. In C. Hemmings & N. Bouras (Eds.), *Psychiatric and behavioral disorders in intellectual and developmental disabilities* (3rd ed., pp. 119–128). Cambridge University Press.

Bate, L., Hutchinson, A., Underhill, J., & Maskrey, N. (2012). How clinical decisions are made. *British Journal of Clinical Pharmacology, 74*(4), 614–620. https://doi.org/10.1111/j.1365-2125.2012.04366.x

Bertelli, M. O., & Moss, S. (2022). Diagnostic issues. In M. O. Bertelli, S. Deb, K. Munir, A. Hassiotis, & L. Salvador-Carulla (Eds.), *Textbook of psychiatry for intellectual disability and autism spectrum disorder* (pp. 107–122). Springer International Publishing.

Bishop-Fitzpatrick, L., Minshew, N. J., & Eack, S. M. (2014). A systematic review of psychosocial interventions for adults with autism spectrum disorders. In F. R. Volkmar, B. Reichow, & J. C. McPartland (Eds.), *Adolescents and adults with autism spectrum disorders* (pp. 315–327). Springer.

Bose-Brill, S., Xing, J., Barnette, D. J., & Hanks, C. (2017). Pharmacogenomic testing: Aiding in the management of psychotropic therapy for adolescents with autism spectrum disorders. *Pharmacogenomics and Personalized Medicine, 10*, 247–252.

Botha, M., & Frost, D. M. (2020). Extending the minority stress model to understand mental health problems experienced by the autistic population. *Society & Mental Health, 10*(1), 20–34. https://doi.org/10.1177/2156869318804297

Bowring, D. L., Painter, J., & Hastings, R. P. (2019). Prevalence of challenging behaviour in adults with intellectual disabilities, correlates, and association with mental health. *Current Developmental Disorders Reports, 6*(4), 173–181. https://doi.org/10.1007/s40474-019-00175-9

Buckley, N., Glasson, E. J., Chen, W., Epstein, A., Leonard, H., Skoss, R., Jacoby, P., Blackmore, A. M., Srinivasjois, R., Bourke, J., Sanders, R. J., & Downs, J. (2020). Prevalence estimates of mental health problems in children and adolescents with intellectual disability: A systematic review and meta-analysis. *Australian and New Zealand Journal of Psychiatry, 54*(10), 970–984. https://doi.org/10.1177/0004867420924101

Burack, J. A., Evans, D. W., Russo, N., Napoleon, J. S., Goldman, K. J., & Iarocci, G. (2021). Developmental perspectives on the study of persons with intellectual disability. *Annual Review of Clinical Psychology, 17*, 339–363. https://doi.org/10.1146/annurev-clinpsy-081219-090532

Camm-Crosbie, L., Bradley, L., Shaw, R., Baron-Cohen, S., & Cassidy, S. (2019). 'People like me don't get support': Autistic adults' experiences of support and treatment for mental health difficulties, self-injury and suicidality. *Autism, 23*(6), 1431–1441. https://doi.org/10.1177/1362361318816053

Cashin, A., Buckley, T., Trollor, J. N., & Lennox, N. (2018). A scoping review of what is known of the physical health of adults with autism spectrum disorder. *Journal of Intellectual Disabilities, 22*(1), 96–108. https://doi.org/10.1177/1744629516665242

Chen, J. L., Leader, G., Sung, C., & Leahy, M. (2015). Trends in employment for individuals with autism spectrum disorder: A review of the research literature. *Review Journal of Autism and Developmental Disorders, 2*(2), 115–127. https://doi.org/10.1007/s40489-014-0041-6

Chester, V., James, N., Rogers, I., Grace, J., & Alexander, R. (2020). Family experiences of psychiatric services for their relative with intellectual and developmental disabilities. In S. Bhaumik & R. Alexander (Eds.), *Oxford textbook of the psychiatry of intellectual disability* (pp. 265–274). Oxford University Press.

Dagnan, D., Jackson, I., & Eastlake, L. (2018). A systematic review of cognitive behavioural therapy for anxiety in adults with intellectual disabilities. *Journal of Intellectual Disability Research, 62*(11), 974–991. https://doi.org/10.1111/jir.12548

De Kuijper, G., Hoekstra, P., Visser, F., Scholte, F. A., Penning, C., & Evenhuis, H. (2010). Use of antipsychotic drugs in individuals with intellectual disability (ID) in The Netherlands: Prevalence and reasons for prescription. *Journal of Intellectual Disability Research, 54*(7), 659–667. https://doi.org/10.1111/j.1365-2788.2010.01275.x

Deb, S., Perera, B., Krysta, K., Ozer, M., Bertelli, M., Novell, R., Wieland, J., & Sappok, T. (2022). The European guideline on the assessment and diagnosis of psychiatric disorders in adults with intellectual disabilities. *The European Journal of Psychiatry, 36*(1), 11–25. https://doi.org/10.1016/j.ejpsy.2021.10.002

Donner, L., & Gustin, L. W. (2021). Navigating between compassion and uncertainty—psychiatric nurses' lived experiences of communication with patients who rarely speak. *Issues in Mental Health Nursing, 42*(4), 307–316. https://doi.org/10.1080/01612840.2020.1793246

Dovgan, K., Mazurek, M. O., & Hansen, J. (2019). Measurement invariance of the child behavior checklist in children with autism spectrum disorder with and without intellectual disability: Follow-up study. *Research in Autism Spectrum Disorders, 58*, 19–29. https://doi.org/10.1016/j.rasd.2018.11.009

Dunn, K., Rydzewska, E., Fleming, M., & Cooper, S. A. (2020). Prevalence of mental health conditions, sensory impairments and physical disability in people with co-occurring intellectual disabilities and autism compared with other people: A cross-sectional total population study in Scotland. *BMJ Open, 10*(4), e035280. https://doi.org/10.1136/bmjopen-2019-035280

Einfeld, S. L., & Tonge, B. J. (1992). *Manual for the developmental behaviour checklist*. Monash University for Developmental Psychiatry and School of Psychiatry, University of New South.

Emerson, E., Graham, H., & Hatton, C. (2006). The measurement of poverty and socioeconomic position in research involving people with intellectual disability. In *International review of research in mental retardation* (pp. 77–108). Academic Press.

Faculty of Psychiatry of Intellectual Disability. (2016). *Psychotropic drug pre-scribing for people with intellectual disability, mental health problems and /or behaviours that challenge: Practice guidelines*. Retrieved February 12, 2024, from https://www.rcpsych.ac.uk/docs/default-source/members/faculties/intellectual-disability/id-faculty-report-id-09.pdf?sfvrsn=55b66f2c_6

Fiksinski, A. M., Breetvelt, E. J., Duijff, S. N., Bassett, A. S., Kahn, R. S., & Vorstman, J. A. S. (2021). Autism spectrum and psychosis risk in the 22q11.2 deletion syndrome. Findings from a prospective longitudinal study. *Schizophrenia Research, 188*, 59–62. https://doi.org/10.1016/j.schres.2017.01.032

Fletcher, R., Barnhill, J., & Cooper, S. A. (2021). *Diagnostic manual-intellectual disability 2: A textbook of diagnosis of mental disorders in persons with intellectual disability*. NADD Press.

Flynn, S., Vereenooghe, L., Hastings, R. P., Adams, D., Cooper, S. A., Gore, N., Hatton, C., Hood, K., Jahoda, A., Langdon, P. E., McNamara, R., Oliver, C., Roy, A., Totsika, V., & Waite, J. (2017). Measurement tools for mental health problems and mental well-being in people with severe or profound intellectual disabilities: A systematic review. *Clinical Psychology Review, 57*, 32–44. https://doi.org/10.1016/j.cpr.2017.08.006

Fuentes, J., Hervás, A., & Howlin, P. (2021). ESCAP practice guidance for autism: A summary of evidence-based recommendations for diagnosis and treatment. *European Child & Adolescent Psychiatry, 30*(6), 961–984. https://doi.org/10.1007/s00787-020-01587-4

Fynn, G., Porter, M., Borchard, T., Kazzi, C., Zhong, Q., & Campbell, L. (2023). The effectiveness of cognitive behavioural therapy for individuals with an intellectual disability and anxiety: A systematic review. *Journal of Intellectual Disability Research, 67*(9), 816–841. https://doi.org/10.1111/jir.13046

Gray, K., Tonge, B., Einfeld, S., Gruber, C., & Klein, A. (2018). *Developmental behavior checklist 2*. Western Psychological Services.

Halvorsen, M., Mathiassen, B., Myrbakk, E., Brøndbo, P. H., Sætrum, A., Steinsvik, O. O., & Martinussen, M. (2019). Neurodevelopmental correlates of behavioural and emotional problems in a neuropaediatric sample. *Research in Developmental Disabilities, 85*, 217–228. https://doi.org/10.1016/j.ridd.2018.11.005

Halvorsen, M. B., Helverschou, S. B., Axelsdottir, B., Brøndbo, P. H., & Martinussen, M. (2023). General measurement tools for assessing mental health problems among children and adolescents with an intellectual disability: A systematic review. *Journal of Autism and Developmental Disorders, 53*(1), 132–204. https://doi.org/10.1007/s10803-021-05419-5

Halvorsen, M. B., Kildahl, A. N., Kaiser, S., Axelsdottir, B., Aman, M. G., & Helverschou, S. B. (2024). Applicability and psychometric properties of general mental health assessment tools in autistic people: A systematic review. *Journal of Autism and Developmental Disorders*. https://doi.org/10.1007/s10803-024-06324-3

Havercamp, S. M., Barnhill, L. J., Bonardi, A., Chapman, R. A., Cobranchi, C., Fletcher, R. J., Rabidoux, P., Seeley, J. R., & Tassé, M. J. (2022). Straight from the horse's mouth: Increasing self-report in mental health assessment in individuals with intellectual disability. *Journal of Applied Research in Intellectual Disabilities, 35*(2), 471–479. https://doi.org/10.1111/jar.12952

Helverschou, S. B. (2022). *Autisme og mental helse [autism and mental health]*. Gyldendal Norsk Forlag.

Helverschou, S. B., Bakken, T. L., & Martinsen, H. (2009). The psychopathology in autism checklist (PAC): A pilot study. *Research in Autism Spectrum Disorders, 3*(1), 179–195. https://doi.org/10.1016/j.rasd.2008.05.004

Helverschou, S. B., Bakken, T. L., & Martinsen, H. (2011). Psychiatric disorders in people with autism spectrum disorders: Phenomenology and recognition. In J. L. Matson & P. Sturmey (Eds.), *International handbook of autism and pervasive developmental disorders* (pp. 53–74). Springer.

Helverschou, S. B., Ludvigsen, L. B., Hove, O., & Kildahl, A. N. (2021). Psychometric properties of the psychopathology in autism checklist (PAC). *International Journal of Developmental Disabilities, 67*(5), 318–326. https://doi.org/10.1080/20473869.2021.1910779

Hodapp, R. M., Dankner, N. A., & Dykens, E. (2016). Behavioral phenotypes / genetic syndromes. In C. Hemmings & N. Bouras (Eds.), *In psychiatric and behavioral disorders in intellectual disabilities* (pp. 196–206). Cambridge University Press.

Hoover, D. W., & Kaufman, J. (2018). Adverse childhood experiences in children with autism spectrum disorder. *Current Opinion in Psychiatry, 31*(2), 128–132. https://doi.org/10.1097/yco.0000000000000390

Howlin, P. (2021). Adults with autism: Changes in understanding since DSM-111. *Journal of Autism and Developmental Disorders, 51*(12), 4291–4308. https://doi.org/10.1007/s10803-020-04847-z

Howlin, P., & Magiati, I. (2017). Autism spectrum disorder: Outcomes in adulthood. *Current Opinion in Psychiatry, 30*(2), 69–76. https://doi.org/10.1097/yco.0000000000000308

Huang, Y., Arnold, S. R., Foley, K. R., & Trollor, J. N. (2020). Diagnosis of autism in adulthood: A scoping review. *Autism, 24*(6), 1311–1327. https://doi.org/10.1177/1362361320903128

Hughes, J., Roberts, R., Tarver, J., Warters-Louth, C., Zhang, B., Southward, E., Shaw, R., Edwards, G., Waite, J., & Pearson, E. (2023). 'It wasn't the strategies on their own': Exploring caregivers' experiences of accessing services in the development of interventions for autistic people with intellectual disability. *Autism, 28*, 1231. https://doi.org/10.1177/13623613231196084

Jopp, D. A., & Keys, C. B. (2001). Diagnostic overshadowing reviewed and reconsidered. *American Journal on Mental Retardation, 106*(5), 416–433. https://doi.org/10.1352/0895-8017(2001)106<0416:Dorar>2.0.Co;2

Kahneman, D. (2011). *Thinking, fast and slow*. Penguin Books.

Kaiser, S., & Halvorsen, M. B. (2022). The strengths and difficulties questionnaire self-report-, parent-, and teacher version in children with intellectual and developmental disabilities. *Research in Developmental Disabilities, 123*, 104194. https://doi.org/10.1016/j.ridd.2022.104194

Kannabiran, M., & McCarthy, J. (2009). The mental health needs of people with autism spectrum disorders. *Psychiatry, 8*(10), 398–401. https://doi.org/10.1016/j.mppsy.2009.06.016

Karatzias, T., Brown, M., Taggart, L., Truesdale, M., Sirisena, C., Walley, R., Mason-Roberts, S., Bradley, A., & Paterson, D. (2019). A mixed-methods, randomized controlled feasibility trial of eye movement desensitization and reprocessing (EMDR) plus standard care (SC) versus SC alone for DSM-5 posttraumatic stress disorder (PTSD) in adults with intellectual disabilities. *Journal of Applied Research in Intellectual Disabilities, 32*(4), 806–818. https://doi.org/10.1111/jar.12570

Kaufman, J., Birmaher, B., Brent, D., Rao, U., Flynn, C., Moreci, P., Williamson, D., & Ryan, N. (1997). Schedule for affective disorders and schizophrenia for school-age children-present and lifetime version (K-SADS-PL): Initial reliability and validity data. *Journal of the American Academy of Child and Adolescent Psychiatry, 36*(7), 980–988. https://doi.org/10.1097/00004583-199707000-00021

Kerns, C. M., Newschaffer, C. J., & Berkowitz, S. J. (2015). Traumatic childhood events and autism spectrum disorder. *Journal of Autism and Developmental Disorders, 45*(11), 3475–3486. https://doi.org/10.1007/s10803-015-2392-y

Kerns, C. M., Wood, J. J., Kendall, P. C., Renno, P., Crawford, E. A., Mercado, R. J., Fujii, C., Collier, A., Hoff, A., Kagan, E. R., Small, B. J., Lewin, A. B., & Storch, E. A. (2016). The treatment of anxiety in autism spectrum disorder (TAASD) study: Rationale, design and methods. *Journal of Child and Family Studies, 25*(6), 1889–1902. https://doi.org/10.1007/s10826-016-0372-2

Kerns, C. M., Winder-Patel, B., Iosif, A. M., Nordahl, C. W., Heath, B., Solomon, M., & Amaral, D. G. (2021). Clinically significant anxiety in children with autism spectrum disorder and varied intellectual functioning. *Journal of Clinical Child & Adolescent Psychology, 50*(6), 780–795. https://doi.org/10.1080/15374416.2019.1703712

Kildahl, A. N., & Helverschou, S. B. (2023). Post-traumatic stress disorder and experiences involving violence or sexual abuse in a clinical sample of autistic adults with intellectual disabilities: Prevalence and clinical correlates. *Autism, 13623613231190948*, 1075. https://doi.org/10.1177/13623613231190948

Kildahl, A. N., & Jørstad, I. (2022). Post-traumatic stress disorder symptom manifestations in an autistic man with severe intellectual disability following coercion and scalding. *Journal of Intellectual & Developmental Disability, 47*(2), 190–194. https://doi.org/10.3109/13668250.2021.1995930

Kildahl, A. N., Helverschou, S. B., & Oddli, H. W. (2020). Clinicians' retrospective perceptions of failure to detect sexual abuse in a young man with autism and mild intellectual disability. *Journal of Intellectual & Developmental Disability, 45*(2), 194–202. https://doi.org/10.3109/13668250.2019.1680821

Kildahl, A. N., Ludvigsen, L. B., Hove, O., & Helverschou, S. B. (2023a). Exploring the relationship between challenging behaviour and mental health disorder in autistic individuals with intellectual disabilities. *Research in Autism Spectrum Disorders, 104*, 102147. https://doi.org/10.1016/j.rasd.2023.102147

Kildahl, A. N., Oddli, H. W., & Helverschou, S. B. (2023b). Bias in assessment of co-occurring mental disorder in individuals with intellectual disabilities: Theoretical perspectives and implications for clinical practice. *Journal of Intellectual Disabilities, 17446295231154119*, 393. https://doi.org/10.1177/17446295231154119

Kjærandsen, K. S., Brøndbo, P. H., & Halvorsen, M. B. (2021). Determinants of caregiver satisfaction with child neurodevelopmental assessment in neuropaediatric clinics. *BMC Health Services Research, 21*(1), 139. https://doi.org/10.1186/s12913-021-06153-5

Korb, L., & Hassiotis, A. (2022). Psychotic disorders. In M. O. Bertelli, S. Deb, K. Munir, A. Hassiotis, & L. Salvador-Carulla (Eds.), *Textbook of psychiatry for intellectual disability and autism spectrum disorder* (pp. 537–555). Springer International Publishing.

Lai, M. C., Kassee, C., Besney, R., Bonato, S., Hull, L., Mandy, W., Szatmari, P., & Ameis, S. H. (2019). Prevalence of co-occurring mental health diagnoses in the autism population: A systematic review and meta-analysis. *The Lancet Psychiatry, 6*(10), 819–829. https://doi.org/10.1016/s2215-0366(19)30289-5

Lippold, T., & Burns, J. (2009). Social support and intellectual disabilities: A comparison between social networks of adults with intellectual disability and those with physical disability. *Journal of Intellectual Disability Research, 53*(5), 463–473. https://doi.org/10.1111/j.1365-2788.2009.01170.x

Lord, C., Elsabbagh, M., Baird, G., & Veenstra-Vanderweele, J. (2018). Autism spectrum disorder. *Lancet, 392*(10146), 508–520. https://doi.org/10.1016/s0140-6736(18)31129-2

Lord, C., Brugha, T. S., Charman, T., Cusack, J., Dumas, G., Frazier, T., Jones, E. J. H., Jones, R. M., Pickles, A., State, M. W., Taylor, J. L., & Veenstra-VanderWeele, J. (2020). Autism spectrum disorder. *Nature Reviews Disease Primers, 6*(1), 5. https://doi.org/10.1038/s41572-019-0138-4

Lord, C., Charman, T., Havdahl, A., Carbone, P., Anagnostou, E., Boyd, B., Carr, T., De Vries, P. J., Dissanayake, C., Divan, G., Freitag, C. M., Gotelli, M. M., Kasari, C., Knapp, M.,

Mundy, P., Plank, A., Scahill, L., Servili, C., Shattuck, P., Simonoff, E., Singer, A. T., Slonims, V., Wang, P. P., Ysrraelit, M. C., Jellett, R., Pickles, A., Cusack, J., Howlin, P., Szatmari, P., Holbrook, A., Toolan, C., & McCauley, J. B. (2022). The lancet commission on the future of care and clinical research in autism. *Lancet, 399*(10321), 271–334. https://doi.org/10.1016/s0140-6736(21)01541-5

Lunsky, Y., Gracey, C., & Bradley, E. (2009). Adults with autism spectrum disorders using psychiatric hospitals in Ontario: Clinical profile and service needs. *Research in Autism Spectrum Disorders, 3*(4), 1006–1013. https://doi.org/10.1016/j.rasd.2009.06.005

Lysaght, R., Ouellette-Kuntz, H., & Lin, C. J. (2012). Untapped potential: Perspectives on the employment of people with intellectual disability. *Work, 41*(4), 409–422. https://doi.org/10.3233/wor-2012-1318

Maddox, B. B., Crabbe, S., Beidas, R. S., Brookman-Frazee, L., Cannuscio, C. C., Miller, J. S., Nicolaidis, C., & Mandell, D. S. (2020). "I wouldn't know where to start": Perspectives from clinicians, agency leaders, and autistic adults on improving community mental health services for autistic adults. *Autism, 24*(4), 919–930. https://doi.org/10.1177/1362361319882227

Mason, D., Capp, S. J., Stewart, G. R., Kempton, M. J., Glaser, K., Howlin, P., & Happé, F. (2021). A meta-analysis of outcome studies of autistic adults: Quantifying effect size, quality, and meta-regression. *Journal of Autism and Developmental Disorders, 51*(9), 3165–3179. https://doi.org/10.1007/s10803-020-04763-2

Mazza, M. G., Rossetti, A., Crespi, G., & Clerici, M. (2020). Prevalence of co-occurring psychiatric disorders in adults and adolescents with intellectual disability: A systematic review and meta-analysis. *Journal of Applied Research in Intellectual Disabilities, 33*(2), 126–138. https://doi.org/10.1111/jar.12654

McDonnell, C. G., Boan, A. D., Bradley, C. C., Seay, K. D., Charles, J. M., & Carpenter, L. A. (2019). Child maltreatment in autism spectrum disorder and intellectual disability: Results from a population-based sample. *Journal of Child Psychology and Psychiatry, 60*(5), 576–584. https://doi.org/10.1111/jcpp.12993

Mevissen, L., Didden, R., & De Jongh, A. (2016). Assessment and treatment of PTSD in people with intellectual disabilities. In C. R. Martin, V. R. Preedy, & V. B. Patel (Eds.), *Comprehensive guide to post-traumatic stress disorder* (pp. 1–15). Springer International Publishing.

Mitchell, P., Sheppard, E., & Cassidy, S. (2021). Autism and the double empathy problem: Implications for development and mental health. *British Journal of Developmental Psychology, 39*(1), 1–18. https://doi.org/10.1111/bjdp.12350

Mohiuddin, S., Bobak, S., Gih, D., & Ghaziuddin, M. (2011). Comorbid psychopathology and treatment. In J. L. Matson & P. Sturmey (Eds.), *International handbook of autism and pervasive developmental disorders* (pp. 453–462). Springer.

Molina-Ruiz, R. M., Martín-Carballeda, J., Asensio-Moreno, I., & Montañés-Rada, F. (2017). A guide to psychopharmacological treatment of patients with intellectual disability in psychiatry. *The International Journal of Psychiatry in Medicine, 52*(2), 176–189. https://doi.org/10.1177/0091217417720896

Moore, L., Larkin, F., & Foley, S. (2023). Mental health professionals' experiences of adapting mental health interventions for autistic adults: A systematic review and thematic synthesis. *Journal of Autism and Developmental Disorders*. https://doi.org/10.1007/s10803-023-06006-6

Munir, K. M. (2016). The co-occurrence of mental disorders in children and adolescents with intellectual disability/intellectual developmental disorder. *Current Opinion in Psychiatry, 29*(2), 95–102. https://doi.org/10.1097/yco.0000000000000236

National Institute for Health and Care Excellence. (2016). *Mental health problems in people with learning disabilities: Prevention, assessment and management*. Retrieved November 1, 2023, from https://www.nice.org.uk/guidance/ng54

Olsen, M. I., Halvorsen, M. B., Søndenaa, E., Langballe, E. M., Bautz-Holter, E., Stensland, E., Tessem, S., & Anke, A. (2021). How do multimorbidity and lifestyle factors impact the perceived health of adults with intellectual disabilities? *Journal of Intellectual Disability Research, 65*(8), 772–783. https://doi.org/10.1111/jir.12845

Pan, P. Y., Bölte, S., Kaur, P., Jamil, S., & Jonsson, U. (2021). Neurological disorders in autism: A systematic review and meta-analysis. *Autism, 25*(3), 812–830. https://doi.org/10.1177/1362361320951370

Pelleboer-Gunnink, H. A., Van Weeghel, J., & Embregts, P. (2021). Public stigmatisation of people with intellectual disabilities: A mixed-method population survey into stereotypes and their relationship with familiarity and discrimination. *Disability and Rehabilitation, 43*(4), 489–497. https://doi.org/10.1080/09638288.2019.1630678

Perera, B., Steward, C., Courtenay, K., Andrews, T., & Shankar, R. (2022). Pharmacogenomics: An opportunity for personalised psychotropic prescribing in adults with intellectual disabilities. *BJPsych Open, 8*(5), e157. https://doi.org/10.1192/bjo.2022.554

Reiss, S., Levitan, G. W., & Szyszko, J. (1982). Emotional disturbance and mental retardation: Diagnostic overshadowing. *American Journal of Mental Deficiency, 86*(6), 567–574.

Roche, L., Adams, D., & Clark, M. (2021). Research priorities of the autism community: A systematic review of key stakeholder perspectives. *Autism, 25*(2), 336–348. https://doi.org/10.1177/1362361320967790

Rosen, T. E., Mazefsky, C. A., Vasa, R. A., & Lerner, M. D. (2018). Co-occurring psychiatric conditions in autism spectrum disorder. *International Review of Psychiatry, 30*(1), 40–61. https://doi.org/10.1080/09540261.2018.1450229

Rosqvist, H. B., Chown, N., & Stenning, A. (2020). *Neurodiversity studies: A new critical paradigm*. Routledge.

Rumball, F. (2019). A systematic review of the assessment and treatment of posttraumatic stress disorder in individuals with autism spectrum disorders. *Review Journal of Autism and Developmental Disorders, 6*(3), 294–324. https://doi.org/10.1007/s40489-018-0133-9

Rumball, F., Brook, L., Happé, F., & Karl, A. (2021). Heightened risk of posttraumatic stress disorder in adults with autism spectrum disorder: The role of cumulative trauma and memory deficits. *Research in Developmental Disabilities, 110*, 103848. https://doi.org/10.1016/j.ridd.2020.103848

Ryan, J. B., Randall, K. N., Walters, E., & Morash-MacNeil, V. (2019). Employment and independent living outcomes of a mixed model post-secondary education program for young adults with intellectual disabilities. *Journal of Vocational Rehabilitation, 50*(1), 61–72. https://doi.org/10.3233/JVR-180988

Rysstad, A. L., Kildahl, A. N., Skavhaug, J. O., Dønnum, M. S., & Helverschou, S. B. (2022). Case study: Organizing outpatient pharmacological treatment of bipolar disorder in autism, intellectual disability and Phelan-McDermid syndrome (22q13.3 deletion syndrome). *International Journal of Developmental Disabilities, 68*(3), 378–387. https://doi.org/10.1080/20473869.2020.1756113

Sáez-Suanes, G. P., & Álvarez-Couto, M. (2022). Factors associated with quality of life in adults with autism spectrum disorder: A systematic review. *Review Journal of Autism and Developmental Disorders, 9*(3), 307–319. https://doi.org/10.1007/s40489-021-00254-x

Scheifes, A., Walraven, S., Stolker, J. J., Nijman, H. L., Egberts, T. C., & Heerdink, E. R. (2016). Adverse events and the relation with quality of life in adults with intellectual disability and challenging behaviour using psychotropic drugs. *Research in Developmental Disabilities, 49-50*, 13–21. https://doi.org/10.1016/j.ridd.2015.11.017

Shankar, R., Wilcock, M., Oak, K., McGowan, P., & Sheehan, R. (2019). Stopping, rationalising or optimising antipsychotic drug treatment in people with intellectual disability and/or autism. *Drug and Therapeutics Bulletin, 57*(1), 10–13. https://doi.org/10.1136/dtb.2018.000009

Shattuck, P. T., Garfield, T., Roux, A. M., Rast, J. E., Anderson, K., Hassrick, E. M., & Kuo, A. (2020). Services for adults with autism spectrum disorder: A systems perspective. *Current Psychiatry Reports, 22*(3), 13. https://doi.org/10.1007/s11920-020-1136-7

Sheehan, R., Hassiotis, A., Walters, K., Osborn, D., Strydom, A., & Horsfall, L. (2015). Mental illness, challenging behaviour, and psychotropic drug prescribing in people with intellectual disability: UK population based cohort study. *BMJ, 351*, h4326. https://doi.org/10.1136/bmj.h4326

Sheehan, R., Horsfall, L., Strydom, A., Osborn, D., Walters, K., & Hassiotis, A. (2017). Movement side effects of antipsychotic drugs in adults with and without intellectual disability: UK population-based cohort study. *BMJ Open, 7*(8), e017406. https://doi.org/10.1136/bmjopen-2017-017406

Siegel, M., McGuire, K., Veenstra-VanderWeele, J., Stratigos, K., King, B., Bellonci, C., Hayek, M., Keable, H., Rockhill, C., Bukstein, O. G., & Walter, H. J. (2020). Practice parameter for the assessment and treatment of psychiatric disorders in children and adolescents with intellectual disability (intellectual developmental disorder). *Journal of the American Academy of Child and Adolescent Psychiatry, 59*(4), 468–496. https://doi.org/10.1016/j.jaac.2019.11.018

Sommerstad, H. S., Kildahl, A. N., Munkhaugen, E. K., Karlsen, K., & Bakken, T. L. (2021). Experiences of ward atmosphere in inpatients with intellectual disability and mental illness: Clinical implications for mental health nursing. *International Journal of Developmental Disabilities, 67*(5), 339–348. https://doi.org/10.1080/20473869.2021.1900507

Spain, D., & Happé, F. (2020). How to optimise cognitive behaviour therapy (CBT) for people with autism spectrum disorders (ASD): A delphi study. *Journal of Rational-Emotive & Cognitive-Behavior Therapy, 38*(2), 184–208. https://doi.org/10.1007/s10942-019-00335-1

Srivastava, S., Sahin, M., Buxbaum, J. D., Berry-Kravis, E., Soorya, L. V., Thurm, A., Bernstein, J. A., Asante-Otoo, A., Bennett, W. E., Betancur, C., Brickhouse, T. H., Bueno, M. R. P., Chopra, M., Christensen, C. K., Cully, J. L., Dies, K., Friedman, K., Gummere, B., Holder, J. L., Jr., Jimenez-Gomez, A., Kerins, C. A., Khan, O., Kohlenberg, T., Lacro, R. V., Levi, L. A., Levy, T., Linnehan, D., Eva, L., Moshiree, B., Neumeyer, A., Paul, S. M., Phelan, K., Persico, A., Rapaport, R., Rogers, C., Saland, J., Sethuram, S., Shapiro, J., Tarr, P. I., White, K. M., Wickstrom, J., Williams, K. M., Winrow, D., Wishart, B., & Kolevzon, A. (2023). Updated consensus guidelines on the management of Phelan-McDermid syndrome. *American Journal of Medical Genetics Part A, 191*(8), 2015–2044. https://doi.org/10.1002/ajmg.a.63312

Tapp, K., Vereenooghe, L., Hewitt, O., Scripps, E., Gray, K. M., & Langdon, P. E. (2023). Psychological therapies for people with intellectual disabilities: An updated systematic review and meta-analysis. *Comprehensive Psychiatry, 122*, 152372. https://doi.org/10.1016/j.comppsych.2023.152372

Totsika, V., Liew, A., Absoud, M., Adnams, C., & Emerson, E. (2022). Mental health problems in children with intellectual disability. *The Lancet Child & Adolescent Health, 6*(6), 432–444. https://doi.org/10.1016/s2352-4642(22)00067-0

Trollor, J. N., Salomon, C., & Franklin, C. (2016). Prescribing psychotropic drugs to adults with an intellectual disability. *Australian Prescriber, 39*(4), 126–130. https://doi.org/10.18773/austprescr.2016.048

Truesdale, M., Brown, M., Taggart, L., Bradley, A., Paterson, D., Sirisena, C., Walley, R., & Karatzias, T. (2019). Trauma-informed care: A qualitative study exploring the views and experiences of professionals in specialist health services for adults with intellectual disabilities. *Journal of Applied Research in Intellectual Disabilities, 32*(6), 1437–1445. https://doi.org/10.1111/jar.12634

Tsiouris, J. A., Kim, S. Y., Brown, W. T., Pettinger, J., & Cohen, I. L. (2013). Prevalence of psychotropic drug use in adults with intellectual disability: Positive and negative findings from a large scale study. *Journal of Autism and Developmental Disorders, 43*(3), 719–731. https://doi.org/10.1007/s10803-012-1617-6

Valdovinos, M. G., Seibert, H. N., Piersma, D., Wilkinson, A., Griffith, G., Hillring, S., & Boyle, B. (2019). Characterizing mood states in individuals diagnosed with bipolar disorder and intellectual disability. *Journal of Mental Health Research in Intellectual Disabilities, 12*(1–2), 26–44. https://doi.org/10.1080/19315864.2018.1561770

Vasa, R. A., Carroll, L. M., Nozzolillo, A. A., Mahajan, R., Mazurek, M. O., Bennett, A. E., Wink, L. K., & Bernal, M. P. (2014). A systematic review of treatments for anxiety in youth with autism spectrum disorders. *Journal of Autism and Developmental Disorders, 44*(12), 3215–3229. https://doi.org/10.1007/s10803-014-2184-9

Verhoeven, W. M. A., Egger, J. I. M., & De Leeuw, N. (2020). A longitudinal perspective on the pharmacotherapy of 24 adult patients with Phelan McDermid syndrome. *European Journal of Medical Genetics, 63*(3), 103751. https://doi.org/10.1016/j.ejmg.2019.103751

Vervoort-Schel, J., Mercera, G., Wissink, I., Van Der Helm, P., Lindauer, R., & Moonen, X. (2021). Prevalence of and relationship between adverse childhood experiences and family context risk factors among children with intellectual disabilities and borderline intellectual functioning. *Research in Developmental Disabilities, 113*, 103935. https://doi.org/10.1016/j.ridd.2021.103935

Walsh, C., Lydon, S., Hehir, A., & O'Connor, P. (2020). Development and evaluation of a novel caregiver-report tool to assess barriers to physical healthcare for people on the autism spectrum. *Research in Autism Spectrum Disorders, 79*, 101680. https://doi.org/10.1016/j.rasd.2020.101680

Weise, J., & Trollor, J. N. (2018). Preparedness and training needs of an Australian public mental health workforce in intellectual disability mental health. *Journal of Intellectual & Developmental Disability, 43*(4), 431–440. https://doi.org/10.3109/13668250.2017.1310825

Weiss, J. A., & Fardella, M. A. (2018). Victimization and perpetration experiences of adults with autism. *Frontiers in Psychiatry, 9*, 203. https://doi.org/10.3389/fpsyt.2018.00203

Weston, L., Hodgekins, J., & Langdon, P. E. (2016). Effectiveness of cognitive behavioural therapy with people who have autistic spectrum disorders: A systematic review and meta-analysis. *Clinical Psychology Review, 49*, 41–54. https://doi.org/10.1016/j.cpr.2016.08.001

White, S. W., Ollendick, T., Albano, A. M., Oswald, D., Johnson, C., Southam-Gerow, M. A., Kim, I., & Scahill, L. (2013). Randomized controlled trial: Multimodal anxiety and social skill intervention for adolescents with autism spectrum disorder. *Journal of Autism and Developmental Disorders, 43*(2), 382–394. https://doi.org/10.1007/s10803-012-1577-x

Whittle, E. L., Fisher, K. R., Reppermund, S., Lenroot, R., & Trollor, J. (2018). Barriers and enablers to accessing mental health services for people with intellectual disability: A scoping review. *Journal of Mental Health Research in Intellectual Disabilities, 11*(1), 69–102. https://doi.org/10.1080/19315864.2017.1408724

Wolkorte, R., Van Houwelingen, I., & Kroezen, M. (2019). Challenging behaviours: Views and preferences of people with intellectual disabilities. *Journal of Applied Research in Intellectual Disabilities, 32*(6), 1421–1427. https://doi.org/10.1111/jar.12631

Wood, J. J., Kendall, P. C., Wood, K. S., Kerns, C. M., Seltzer, M., Small, B. J., Lewin, A. B., & Storch, E. A. (2020). Cognitive behavioral treatments for anxiety in children with autism spectrum disorder: A randomized clinical trial. *JAMA Psychiatry, 77*(5), 474–483. https://doi.org/10.1001/jamapsychiatry.2019.4160

World Health Organization. (2022). *Mental health*. Retrieved November 1, 2023, https://www.who.int/news-room/fact-sheets/detail/mental-health-strengthening-our-response

Zigmond, A. S., & Snaith, R. P. (1983). The hospital anxiety and depression scale. *Acta Psychiatrica Scandinavica, 67*(6), 361–370. https://doi.org/10.1111/j.1600-0447.1983.tb09716.x

Women's Health and Sexuality

Hilary K. Brown and Lesley Tarasoff

Abstract Sexual and reproductive health (SRH) in women with intellectual and developmental disabilities (IDD) has long been neglected in health care, driven by historical eugenic practices related to institutionalization and involuntary sterilization of women with IDD. Growing evidence from multiple regions shows women with IDD experience important disparities in sexual health, violence, cancer screening and outcomes, contraception, pregnancy outcomes and care, and mental health across the reproductive life course. There is an urgent need to address these SRH disparities through changes to the content and delivery of health care for women with IDD, training and resources for healthcare providers, and data to monitor SRH outcomes and evaluate interventions developed to address them. Such changes are needed to better support and promote the SRH of women with IDD globally.

Keywords Intellectual disability · Developmental disabilities · Sexual health · Reproductive health

Intellectual and developmental disabilities (IDD) are common, occurring in one to three percent of the population (Anderson et al., 2019). Research has long shown significant health and healthcare disparities among adults with IDD (Havercamp & Scott, 2015). However, less attention has been paid to the sexual and reproductive health (SRH) of women with IDD (Richards et al., 2008). Throughout the twentieth

H. K. Brown (✉)
Department of Health & Society, University of Toronto Scarborough, Toronto, ON, Canada

Dalla Lana School of Public Health, University of Toronto, Toronto, ON, Canada
e-mail: hk.brown@utoronto.ca

L. Tarasoff
Department of Health & Society, University of Toronto Scarborough, Toronto, ON, Canada

M. G. Valdovinos (ed.), *Intellectual and Developmental Disabilities*,
Contemporary Clinical Neuroscience,
https://doi.org/10.1007/978-3-031-66932-3_12

century, women with IDD frequently experienced institutional segregation and involuntary sterilization, driven by eugenic notions about the need to prevent procreation in "feebleminded" individuals (Tilley et al., 2012). In 2006, the United Nations Convention on the Rights of Persons with Disabilities (CRPD) was enacted to protect the human rights of people with disabilities, including their SRH rights (United Nations, 2006). Yet, examples of "new eugenics" persist, including genome editing, prenatal screening for disabilities, and continued stigma related to pregnancy and parenting among women with IDD (McConnell & Phelan, 2022). The impacts of these historical and contemporary injustices are seen in the SRH outcomes and care experiences of women with IDD, with growing evidence showing disparities in sexual health, violence, cancer screening and outcomes, contraception, pregnancy outcomes and care, and mental health across the reproductive life course.

Historical and Contemporary Context

Women with IDD endured harsh eugenic practices for much of the twentieth century. In many regions, these practices were supported and promoted by medical professionals (Roy et al., 2012). In the 1920s, many countries in North America, Europe, and elsewhere introduced sexual sterilization acts that permitted involuntary sterilization of people with IDD and protected surgeons performing sterilizations from civil action (Tilley et al., 2012). Most widely known is the case of Carrie Buck, the plaintiff in the United States Supreme Court Case, *Buck v. Bell.* Buck was the first person to be sterilized under a 1924 Virginia sterilization law that allowed anyone described as "feeble-minded, an imbecile, or epileptic" to undergo compulsory sterilization (Holmes and The Supreme Court of the United States, 1926). The Supreme Court's move to uphold the decision in *Buck v. Bell* was seen as validating sterilization and led to rising rates of sterilization throughout the United States, with an estimated 65,000 Americans with IDD and psychiatric disabilities sterilized between the 1920s and the 1970s (Roy et al., 2012). In Canada, Germany, Sweden, and many other countries, there were similarly high numbers of sterilizations of people with IDD during this time (Roy et al., 2012).

There has been significant progress in the protection of the SRH rights of people with IDD. Most regions repealed their sterilization acts in the 1970s and subsequently created guidelines for obtaining informed consent for sterilization procedures (Roy et al., 2012). In 2006, the United Nations adopted the CRPD. As the first international treaty on disability, the CRPD promotes the idea that people with disabilities are active members of society, have the same human rights as people without disabilities, and are able to exercise these rights and freely make their own decisions. In relation to SRH, the CRPD specifically includes statements about the rights of people with disabilities to retain their fertility, marry and have a family, determine the number and spacing of their children, and obtain family planning services and quality, affordable SRH care (United Nations, 2006). Over 180 countries have now ratified the CRPD, legally requiring them to ensure that their policies and services reflect the CRPD's values. A 2009 guidance note from the World Health Organization and the United Nations Population Fund provides specific recommendations for the implementation of the CRPD in relation to SRH care delivery, including the requirement to create laws, policies, services, and budgets that address the SRH needs of people with disabilities and to enhance accessibility within the existing healthcare system rather than creating separate or parallel services for people with disabilities (World Health Organization/United Nations Population Fund, 2009).

However, negative societal perceptions of disability, sexuality, pregnancy, and parenting persist despite these milestones. For instance, some argue that genome editing and prenatal screening for disabilities like Down syndrome perpetuate the idea of disability as an undesirable characteristic to be prevented (McConnell & Phelan, 2022). Many pregnant women with IDD report that healthcare providers and families pressure them to terminate their pregnancies (Björnsdóttir et al., 2017). Parents with IDD also disproportionally experience child protective services involvement, which more often results in child custody loss than for parents without IDD, even when there is no evidence of neglect or abuse (Tøssebro et al., 2017). Thus, stigma and discrimination related to SRH continue for many women with IDD.

Most women with IDD receive health care in general practice, not IDD-specific, settings (Sullivan et al., 2018). However, family physicians, gynecologists, and obstetricians receive little to no training on the SRH care needs of women with IDD in their formal or continuing education (Amir et al., 2022; Tarasoff et al., 2023). There is also a lack of clinical practice guidelines and other tools (e.g., written materials in an accessible format) to support delivery of SRH care to this population (Amir et al., 2022). This lack of healthcare provider training perpetuates the stigma and discrimination around IDD and SRH via poor provider knowledge and confidence, as well as negative attitudes. These issues contribute to SRH disparities for women with IDD.

Sexual Health

Sexuality is a complex phenomenon of the human experience involving an individual's emotions, beliefs, and biology and is shaped strongly by culture and context (Parchomiuk, 2022; Whittle & Butler, 2018). Studies show many women with IDD report having the same desires for sexual relationships as their peers without IDD (Ignani et al., 2015; Whittle & Butler, 2018). However, paradoxical societal assumptions of women with IDD as being either asexual or hypersexual have led to sexual health care for this population being designed around protecting them from sex, rather than taking on a positive sexuality model that emphasizes pleasure and freedom (McCarthy, 2014; Parchomiuk, 2022). This has resulted in significant disparities for women with IDD in the areas of sexual health knowledge, autonomy in sexual relationships, and rates of adverse outcomes such as sexually transmitted infections (STIs).

Disparities in sexual health knowledge for women with IDD begin in childhood and adolescence. For example, adolescents with IDD are less likely than their peers without IDD to receive sexual health education in school (Grove et al., 2018). Further, much of their sexual health education is disparities focused (i.e., focused on negative sexual health outcomes or sexual violence prevention) and heteronormative (i.e., with little acknowledgment that people with IDD may have same-sex attraction or identify as lesbian, gay, bisexual, queer, or another sexual minority identity) (Duke, 2011; Tarasoff, 2021; Smith et al., 2022). Parents report being less likely to initiate conversations about puberty, relationships, and pregnancy with their adolescents with IDD (Crehan et al., 2023). Similarly, girls and women with IDD are less likely to have health care for SRH, including for family planning (Greenwood & Wilkinson, 2013; Kopac et al., 1998; Marquis et al., 2023). They are therefore more likely to try to obtain information on sexuality and SRH from often unreliable sources such as television and the Internet (Greenwood & Wilkinson, 2013). This lack of education and resources results in them having lower knowledge than their peers about SRH topics such as menstruation, contraceptive methods, sexual behaviors, sexual identity, consent and sexual violence, and the processes of sexual intercourse and pregnancy (Matin et al., 2021).

Women with IDD also report having little control over their sexual lives, including limited opportunities to form romantic relationships and to make decisions related to sexual activity, fertility, and pregnancy (Bernet, 2011; Matin et al., 2021; McCarthy, 2014). They report having to hide their sexual experiences and consensual sexual contact, particularly when they reside in group homes and other institutional settings, because of negative service provider attitudes as well as prohibitive rules around sex and relationships (Chou et al., 2015; Matin et al., 2021). Women with IDD who desire an intimate relationship also report having limited support from service providers, family, and friends in navigating various relationship issues, such as finding a romantic partner, initiating sexual intimacy, negotiating safe sexual practices such as condom use, and even ending relationships (Ignani et al., 2015; Matin et al., 2021). All of these barriers to sexual and romantic relationships may be heightened for sexual minority women with IDD; some sexual minority women with IDD have reported that others assume they only engage in same-sex relationships because they cannot find a male partner, and others have reported being forced into heterosexual social situations (Tarasoff, 2016).

This lack of knowledge and autonomy has significant negative repercussions in terms of the sexual health outcomes of women with IDD. For example, sexually active adolescents and adults with IDD are more likely than those without IDD to contract STIs, such as chlamydia, gonorrhea, herpes, and syphilis (Mandell et al., 2008; Parekh et al., 2023). Erroneous assumptions by healthcare providers that women with IDD are not sexually active or are all heterosexual can lead to significant disparities in access to sexual health services for those who need them. For example, although adults with and without IDD have similar rates of HIV (Lunsky et al., 2017), adults with IDD are less likely to be tested for HIV (Neri et al., 2007). There is clearly a need to equip service providers, and women with IDD, with appropriate education and support related to sexual health to address these disparities.

Violence

Violence against women is a major public health and human rights issue and includes intentional use of physical force and other forms of power against an individual by an intimate partner, family member, or other member of the community, including service providers (World Health Organization, 2013). Research on violence against women with IDD is difficult because, compared to their peers without IDD, they are less likely to report abuse (Horner-Johnson & Drum, 2006). This under-reporting is thought to be due to a combination of a lack of access to information about violence and personal rights, complex power dynamics with the perpetrators of violence including intimate partners and service providers, and negative reactions from others upon reporting abuse (Carmody, 1991; Horner-Johnson & Drum, 2006; Matin et al., 2021). Service providers, such as emergency department physicians and nurses as well as shelter workers, lack training on the needs of women with IDD, and frequently used violence screening tools do not usually include questions relevant to the types of violence experienced by many women with IDD, including refusal to provide disability-related and practical supports (Curry et al., 2004). Finally, social stereotypes of IDD lead women with IDD to be seen as less credible when disclosing violence (McGilloway et al., 2018).

Nevertheless, studies suggest at least half of women with IDD experience violence in their lifetime (Amborski et al., 2021; Bowen & Swift, 2017; Chirwa et al., 2020; Fanslow et al., 2021; Garcia-Cuellar et al., 2023; Hughes et al., 2012; Matin et al., 2021). Intimate partner violence is especially common, experienced by women with IDD 40% more often than women without IDD (McCarthy et al., 2018). The types of intimate partner violence that women with IDD experience include physical assault, sometimes involving weapons and life-threatening injury; sexual violence, including rape; financial abuse; emotional and psychological distress; and coercive control, including intimidation and social isolation (McCarthy et al., 2018). Women with IDD are also vulnerable to abuse from their caregivers, including formal service providers and family members, with the most common types of abuse including theft, verbal aggression, physical assault, and neglect (Codina et al., 2020). It is also important to note that with an increasing use of the Internet by women with IDD to explore friendships and intimate relationships, technology has emerged as a means of violence against women with IDD, including through cyberbullying, stalking, and harassment (Woodlock & Harris, 2022).

There are a number of social and structural factors that explain why women with IDD are at elevated risk of experiencing violence. Due to their disability-related and socioeconomic needs, women with IDD experience greater reliance on intimate partners, family, and service providers, making it more difficult for them to leave violent situations (Bowen & Swift, 2017; Chirwa et al., 2020; Garcia-Cuellar et al., 2023; Hughes et al., 2012). Women with IDD are less likely to be offered accessible information about violence, resulting in a lack of knowledge about personal rights and the ability to recognize abusive situations (Bowen & Swift, 2017; Chirwa et al., 2020; Garcia-Cuellar et al., 2023; Hughes et al., 2012). Moreover, women with IDD typically learn about sexual violence only in the context of heterosexual relationships and interactions. As such, they have more difficulty recognizing sexual violence within same-sex relationships, including both intimate and caregiving relationships (Tarasoff, 2021). Women with IDD also experience an elevated prevalence of other risk factors for violence, including mental illness and poverty (Brucker & Nord, 2016; Cooper et al., 2006; Roebuck, 2021). These risk factors, along with the barriers to reporting previously described, make them particularly vulnerable to abuse.

Cervical and Breast Cancer Screening and Outcomes

Cervical and breast cancer are major causes of morbidity and death in women worldwide (Sun et al., 2021). As such, reproductive cancer screening has become a critical aspect of preventive health care for women, resulting in important reductions in the occurrence of and death associated with these cancers (World Health Organization, 2006). While specific cancer screening guidelines differ across jurisdictions, breast cancer screening at regular intervals using mammography is usually recommended for average-risk women approximately 50–70 years of age (Canadian Task Force on Preventive Health Care, 2011; Oeffinger et al., 2015), and regular cervical cancer screening using Papanicolaou (Pap) tests is recommended for average-risk women approximately 20–70 years of age (Canadian Task Force on Preventive Health Care, 2013; Fontham et al., 2020). Disparities in the occurrence and frequency of these screening tests are therefore an important marker of access to SRH care for all women.

Historically, studies suggested that rates of cervical and breast cancer were lower in women with IDD compared to those without IDD, leading to assumptions that screening for these cancers was not required in this population (Cuypers et al., 2022). However, emerging evidence suggests that the lower cancer rates initially observed might have been due to a shorter overall life expectancy in women with IDD, and not a lower risk of cancer per se (Cuypers et al., 2022). Indeed, emerging research suggests that as life expectancy for women with IDD has increased, cancer has become at least as common in women with IDD as in those without IDD (Batchelor, 2017). Yet, numerous studies from diverse healthcare systems in the United States, Canada, the United Kingdom, and elsewhere show that women with IDD are less likely than their peers without IDD to receive breast and cervical cancer screening (Cobigo et al., 2013; Horsbøl et al., 2023; Osborn et al., 2012; Reidy et al., 2014; Shin et al., 2018; Xu et al., 2017), with the magnitude of the reported disparities ranging from 4% to 64%, depending on the use of clinical data, surveys, or population-based data (Horsbøl et al., 2023).

These disparities are driven by multiple factors. System-level factors include the lack of accessible information and education about reproductive cancer screening (Greenwood & Wilkinson, 2013). Provider-level factors include negative healthcare provider attitudes or assumptions that women with IDD do not require screening, as well as concerns about patient trauma due to the invasive nature of the tests (Chan et al., 2022; Swaine et al., 2013). These issues ultimately result in individual-level barriers, including women's inadequate knowledge of SRH, including risk factors for cancer and the importance of screening, and embarrassment and worry about screening procedures; sometimes, there are also concerns about individuals' ability to consent to screening (Chan et al., 2022; Greenwood & Wilkinson, 2013; Swaine et al., 2013). Cervical cancer is an area in which disparities in screening rates have been particularly stark. Since most cases of cervical cancer are thought to be sexually transmitted via the human papillomavirus (Fontham et al., 2020), some healthcare providers have argued that screening is not needed for women with IDD, under the assumption that they are less likely than women without IDD to be sexually active and are therefore not at risk (Greenwood & Wilkinson, 2013). However, studies show that even in women with a prior pregnancy, those with IDD are still 60% less likely than women without IDD to receive a Pap test (Brown et al., 2016), demonstrating that disparities exist even in women who are clearly eligible for screening.

Notably, there is emerging evidence of the negative repercussions of these disparities in reproductive cancer screening. Studies suggest women with IDD are more likely than their peers to receive cancer diagnoses at later stages (Batchelor, 2017; Heslop et al., 2022; Mahar et al., 2023; Satgé et al., 2014), possibly because of missed opportunities to detect cancers earlier with screening. However, although a later stage of cancer diagnosis has implications for subsequent morbidity and mortality (Sun et al., 2021), few studies have examined longer-term cancer outcomes in women with IDD. One recent study found that cancer is the cause of death 1.5 times more often in people with IDD compared to those without IDD (Cuypers et al., 2022), suggesting important disparities in cancer-related mortality. There is therefore clearly a need for more research to fully understand the health outcomes of women with IDD with breast and cervical cancer, in the context of ongoing disparities in screening access.

Contraception

Contraception use is a critical marker of SRH care access among reproductive-aged, sexually active women. It is important for preventing unintended pregnancy (Chor et al., 2011), with better planning of the timing and spacing of pregnancies leading to opportunities to benefit from preconception health promotion efforts to address risk factors, such as nutrition deficits, smoking, and chronic physical and mental health conditions, associated with reproductive and pregnancy complications (American College of Obstetricians & Gynecologists, 2005). However, due to barriers to sexual health education (Grove et al., 2018) and family planning services (Greenwood & Wilkinson, 2013; Kopac et al., 1998; Marquis et al., 2023), women with IDD have lower contraception knowledge compared to their peers and utilize a narrower range of contraception methods (Horner-Johnson et al., 2019b).

For example, studies have shown that women with IDD are more likely than those without IDD to have a tubal ligation and hysterectomy, which are invasive and permanent procedures (Brown et al., 2018a; Mosher et al., 2018; Schuengel et al., 2022). They are also more likely to use the progestin shot (i.e., Depo-Provera; Brown et al., 2018a, b; Schuengel et al., 2022; Wu et al., 2018), which can have negative impacts on mood, bone density, and weight (Bigrigg et al., 1999). On the other hand, women with IDD are less likely to use oral contraception and highly effective long-acting reversible contraception options such as copper or hormonal intrauterine devices (Schuengel et al., 2022; Wu et al., 2018).

Some of these contraception use patterns may relate to health considerations for women with IDD. For example, higher rates of menstrual suppression may be due to patient or provider concerns about menstrual difficulties such as dysmenorrhea and adverse cyclic changes in mood and behavior (Tracy et al., 2016; Wilkinson & Cerreto, 2008). Similarly, oral contraception may be contraindicated for some women with IDD with co-occurring conditions such as epilepsy or cardiovascular disease, because of concerns about medications interfering with the effectiveness of oral contraception, or elevated risk of side effects such as blood clots (Quint, 2016). However, there is also evidence that healthcare providers hold paternalistic assumptions about the ability of women with IDD to reliably use certain forms of contraception, such as oral contraceptive pills (Taouk et al., 2018). Moreover, some studies have suggested that women with IDD are not given accessible information to understand the side effects associated with different contraceptive options, or the full implications of their contraception choices for their future reproductive life plans, despite the documented need for informed consent for invasive procedures such as hysterectomy (Rowlands & Amy, 2019). There are therefore significant concerns that women with IDD are not afforded the same freedom and autonomy in contraception decision-making as their peers.

There is also other evidence of unmet contraception needs among women with IDD. For example, surveys and population-based data show that women with IDD are at increased risk of unintended pregnancy (Horner-Johnson et al., 2020; Schuengel et al., 2022). They are also more likely than women without IDD to have an induced abortion (Goldacre et al., 2015; Schuengel et al., 2022). Finally, women with IDD have elevated rates of rapid repeat pregnancy (Brown et al., 2018a, b), defined as a pregnancy conceived within 12 months of a prior delivery, which increases the risk of subsequent adverse pregnancy outcomes such as stillbirth and preterm birth (Wendt et al., 2012). These data point to a need for access to family planning services for women with IDD, to help them plan the timing and spacing of their pregnancies.

Pregnancy Outcomes and Care

Indicators of maternal and newborn health, such as maternal mortality and preterm birth, are important barometers of pregnancy care access and quality (Aoyama et al., 2019; Creanga et al., 2014; Richter et al., 2019). Historical eugenic practices have led to a dearth of data on pregnancy in women with IDD, until relatively recently, based on the assumption that pregnancy is uncommon in this population. However, an increasing number of population-based studies provide evidence to refute this assumption. For example, Medicaid data from the United States suggest relatively similar fertility rates in women with and without IDD (62.1 vs. 77.1 live births per 1000 women) (Rubenstein et al., 2022). Likewise, national survey data in the United States report only slightly lower rates of past-year pregnancy in women with IDD and other cognitive disabilities (9.2%) compared to those without disabilities (12.3%) (Horner-Johnson et al., 2016). Studies from other regions suggest lower overall fertility rates in women with versus without IDD, but elevated rates of adolescent pregnancy in young women with IDD (Brown et al., 2016). Taken together, these data show that pregnancy is an important aspect of SRH for women with IDD that requires attention with respect to health outcomes and care.

Indeed, a growing number of studies now show women with IDD have elevated rates of common complications of pregnancy, with meta-analyses reporting higher risk of gestational diabetes and cesarean delivery compared to women without IDD (Tarasoff et al., 2020c). Fewer studies have examined rare, but serious, outcomes; however, new population-based studies from the United States and Canada suggest women with IDD have double the risk of severe maternal morbidity and mortality (Brown et al., 2021; Gleason et al., 2021; Horner-Johnson et al., 2022; Mitra et al., 2021). Women with IDD also experience disparities in fetal and newborn complications. For example, studies suggest that they have higher rates of miscarriage and stillbirth than women without IDD (Schuengel et al., 2022). Meta-analyses also show that newborns of women with IDD have a 50% increased risk of preterm birth before 37 weeks' gestation and low birth weight of less than 2500 grams compared to newborns of women without IDD (Tarasoff et al., 2020b). Smaller numbers of studies report elevated risks of rare but severe neonatal outcomes such as neonatal morbidity resulting in intensive care unit admission as well as neonatal mortality (Brown et al., 2022; Gleason et al., 2023).

There are several possible reasons for these disparities in pregnancy outcomes. For example, women with IDD experience a greater burden of preconception health risk factors compared to those without IDD, including chronic physical conditions such as diabetes mellitus, chronic hypertension, and cardiovascular disease, mental illness and substance use, potentially teratogenic medication use, smoking, and physical inactivity (Havercamp & Scott, 2015; Krahn & Fox, 2013; Krahn et al., 2006; Tarasoff et al., 2020a). Each of these factors carries known risks for adverse pregnancy outcomes and is the target of preconception health interventions in the general population (World Health Organization, 2012). However, women with IDD are less likely than those without IDD to receive preconception health care, since they are not viewed as a population in which reproductive life plans need to be considered (Tarasoff et al., 2020a). This lack of focus on preconception health in the context of health care for women with IDD makes them more likely to enter pregnancy with these risk factors.

Barriers to accessing prenatal care may also be an important determinant of these disparities. Several population-based studies have quantified the timing and frequency of prenatal care received by women with IDD. For example, studies in the United States (Horner-Johnson et al., 2019a) and Canada (Nishat et al., 2022) show women with IDD are at elevated risk for delayed entry into prenatal care and receive fewer than the recommended number of visits. At a healthcare system level, this inadequate receipt of prenatal care is also reflected in elevated rates of emergency department visits and hospital admissions among women with IDD during pregnancy, and in the postpartum period as well, for obstetric, other medical, and psychiatric reasons (Brown et al., 2017; Mitra et al., 2018a, b, 2019). This elevated use of acute care among women with IDD across the perinatal period suggests the presence of complications that could have been managed in outpatient settings.

In addition to the quantity of prenatal care, studies show important gaps in the quality of prenatal care for women with IDD. For example, surveys from the United Kingdom show women with IDD are more likely than those without IDD to report being unsatisfied with their prenatal, delivery, and postpartum care (Redshaw et al., 2013). In qualitative studies, they frequently report negative healthcare provider attitudes about their pregnancies and their ability to parent; barriers to communication, including healthcare provider use of complex medical terminology; a lack of autonomy in decisions around labor and delivery, including pain management and delivery mode; lack of coordination across health and social service providers; and fear of child protective service involvement and intrusive surveillance of their parenting activities by service providers (Höglund & Larsson, 2013, 2014; Khan et al., 2021; Khanna et al., 2022; Potvin et al., 2016; Rosenthal et al., 2022; Saeed et al., 2022; Xie & Gemmill, 2018). These issues in the quality of pregnancy care for women with IDD have important repercussions on pregnancy outcomes; for example, reviews of maternal deaths in the United Kingdom have shown how issues in the coordination and quality of prenatal care led to severe maternal complications, and subsequently, death for a woman with IDD (Cantwell et al., 2011). Together, these studies show an urgent need to improve the pregnancy care of women with IDD.

Mental Health Across the Reproductive Life Course

It is well-established that women with IDD are at elevated risk of mental illness, with higher rates of depression and anxiety compared to both women without IDD as well as men with IDD (Cooper et al., 2006; Havercamp & Scott, 2015; Krahn & Fox, 2013; Taggart et al., 2008). However, mental health in this population has infrequently been examined in relation to key time points across the reproductive life course, including the menstrual cycle, pregnancy and the postpartum period, and menopause. The lack of consideration of mental health in relation to SRH has led to important gaps in mental health care for women with IDD.

Most women experience mild menstrual cycle-related changes in mood, or premenstrual syndrome, driven by neuroendocrine factors (Bigg & Demuth, 2011). Premenstrual dysphoric disorder is a more severe condition that includes significant worsening of an individual's mood, impacting their quality of life, and is experienced by 3% to 8% of women (Bigg & Demuth, 2011). Research suggests that women with IDD are more likely than their peers without IDD to experience both premenstrual syndrome and premenstrual dysphoric disorder (Silka & Hurley, 2005). However, they experience difficulty describing their experiences, and their symptoms may be behavioral in nature (Kyrkou, 2005). This often leads to misdiagnosis and use of inappropriate treatments. For example, behavioral changes are often treated with behavioral management strategies rather than addressing underlying issues related to mood or physical symptoms of menstruation such as pain (Tracy et al., 2016). More severe symptoms related to premenstrual dysphoric disorder may also be misdiagnosed as bipolar disorder due to their cyclic nature (Silka & Hurley, 2005). There is therefore a need for better recognition of the mental health of women with IDD in relation to their menstrual cycle.

Another key time point for considering the mental health needs of women with IDD is the perinatal period. In the general population, nearly 20% of women experience mental illness in pregnancy or the postpartum period (O'Hara & Wisner, 2014). Perinatal mental illness is associated with difficulty coping, problems with relationships, and elevated risk of chronic mental illness, maternal suicide, birth complications, and offspring developmental delays (Deave et al., 2008; Meltzer-Brody & Stuebe, 2014). The strongest risk factors for perinatal mental illness are a pre-pregnancy history of mental illness, low social support, and stress (Robertson et al., 2004). However, while women with IDD have elevated rates of these risk factors, few studies have examined their mental health in pregnancy and postpartum. A handful of studies have shown elevated risk of anxiety and distress in women with IDD in pregnancy (McConnell et al., 2009; Schuengel et al., 2022). There is also evidence that they experience severe postpartum mental health outcomes, with rates of hospital admissions for postpartum mental illness 10 times higher than in women without IDD (Brown et al., 2017). It is important to note that numerous studies have shown that women with IDD experience unique, acute stressors in the perinatal period, including negative reactions of service providers and family members toward their pregnancies, questioning of their parenting abilities, and involvement of child protective services (Björnsdóttir et al., 2017; Tøssebro et al., 2017). Unfortunately, these factors may also make it more difficult for women with IDD to seek help when they are experiencing perinatal mental illness, due to fear of the repercussions of disclosing mental health challenges (Brown et al., 2017). There is a need for more research to better understand the perinatal mental health of women with IDD and the support they need.

On the other end of the reproductive life course, menopause may be particularly difficult for women with IDD in terms of their mental health. Menopause is a critical transition in women's lives marked by biological, psychological, behavioral, and social change. Yet, little research has focused on menopause in women with IDD, possibly due to the historically lower life expectancy among women with IDD and the corresponding de-prioritization of aging-related research and clinical care (Willis, 2008). Preliminary research suggests that women with IDD may be more likely than women without IDD to experience menopause-related mood problems, including increased depression, irritability, and anger (Langer-Shapland et al., 2023). However, women with IDD may also have more difficulty understanding whether these mood swings are related to menopause or other daily stressors, making identification and appropriate management difficult (Willis et al., 2010). Service providers may also under-recognize menopause-related symptoms in women with IDD (Langer-Shapland et al., 2023), showing a need for more research to support the mental health care of women with IDD in this critical life stage.

Recommendations

Growing evidence from multiple countries shows women with IDD experience important disparities in sexual health, violence, cancer screening and outcomes, contraception, pregnancy outcomes and care, and mental health across the reproductive life course. These disparities show an urgent need to improve SRH care for women with IDD. Such improvements require changes related to the content and delivery of care, training and resources for service providers, and data to support the monitoring and evaluation of health outcomes and care experiences.

In the general population, primary care providers play a critical role in SRH promotion (Marcell et al., 2017). However, health care for women with IDD tends to focus on the management of IDD and co-occurring conditions, rather than addressing their needs as women (Greenwood & Wilkinson, 2013). When SRH care is offered, it tends to focus on menstruation management and provision of contraception in the context of concerns about sexual abuse and pregnancy prevention, rather than supporting positive sexuality and fertility (McCarthy, 2014; Parchomiuk, 2022). This has resulted in serious gaps in the SRH care of women with IDD, including those identified in the prior sections of this chapter. There needs to be an increased focus on SRH promotion in the primary care of women with IDD, supported by the view that women with IDD have the same SRH rights as those without IDD (United Nations, 2006). The content of primary care for girls and women with IDD should include accessible sexual health education, including information on positive sexuality, starting in adolescence. Across the reproductive life course, care should include information on critical SRH topics such as breast and cervical cancer screening procedures and their importance, violence prevention and identification, and contraception and family planning (Greenwood & Wilkinson, 2013). For women who could become pregnant, preconception health promotion is also important, including discussions of reproductive life plans related to the timing and spacing of pregnancies, management of chronic conditions and medications, and addressing health behaviors such as nutrition and smoking (Tarasoff et al., 2020a). Mental health also needs to be addressed, including the unique ways that menstruation, pregnancy, and menopause might impact the mental health of women with IDD. While these topics are routine in primary care for women without IDD, they need to also be regularly addressed for girls and women with IDD.

Coverage of these topics may require changes to the ways in which health care for women with IDD is delivered. In many health systems, healthcare visits are organized around infrequent appointments that are restricted in time (e.g., 10–15 minutes). However, women with IDD often experience difficulties with communication, memory, and organizational skills that impact their learning. Many SRH topics are complex (e.g., violence, pregnancy) and may require longer and more frequent healthcare appointments to ensure sufficient time to build trust, provide accessible explanations, answer questions, and ensure understanding (Brown et al., 2019). SRH care delivery for this group may also require creativity beyond routine healthcare appointments. For example, providing "easy read" brochures and videos on SRH topics may be useful for women with IDD to take home and review with family members or social service providers to ensure longer-term understanding and uptake of advice; however, few such resources exist (Hollins et al., 1992), pointing to a need for the development of resources. Likewise, SRH care delivery for women with IDD may benefit from better communication and coordination between healthcare and social service providers who interact with women with IDD more regularly and may be able to better support their learning and ongoing SRH promotion between healthcare visits. As such, delivery of SRH care for women with IDD may require more flexibility and collaboration than is typical in the healthcare system.

Such efforts require better training of healthcare providers on the unique needs and experiences of women with IDD. Many senior healthcare providers who are now in positions of leadership were trained at a time when institutionalization and involuntary sterilization of women with IDD was either common, or recently so. However, with closure of institutions and greater recognition of the SRH rights of women with IDD, education of healthcare providers has not yet "caught up to" the current needs of the populations with IDD that they serve. For example, most physicians, midwives, and nurses receive little or no training on IDD, and when IDD is taught, it is conceptualized as a health outcome to be prevented (e.g., prevention of prenatal risk factors for IDD) or as a childhood condition to be managed (e.g., screening and interventions for autistic children) (Amir et al., 2022; Höglund et al., 2013; Smith et al., 2023; Tarasoff et al., 2023). However, while girls with IDD grow up to be women with IDD with their own SRH needs and desires, the lack of exposure of healthcare providers to IDD and SRH in their training perpetuates misinformed stereotypes of such topics as being unimportant and results in negative attitudes around disability, sexuality, and pregnancy (Boyd et al., 2008). There needs to be a curriculum on the SRH of women with IDD embedded into healthcare provider education, including broad topics early in individuals' training (e.g., general disability competency training in undergraduate education; Havercamp et al., 2021) and more specific topics in specialty training and continuing professional development (e.g., information on IDD and pregnancy for obstetric care providers). Such training is consistent with the growing recognition of the need for equity, diversity, and inclusion education in health professional training (Janz, 2019) and is needed to equip healthcare providers to support SRH in women with IDD.

There is also a need for the development of resources to support healthcare provider delivery of evidence-based SRH care to women with IDD. For example, healthcare providers report a lack of clinical care guidelines on the SRH care of women with IDD (Greenwood & Wilkinson, 2013; Wilkinson & Cerreto, 2008). Broader primary care guidelines for adults with IDD (e.g., Sullivan et al., 2018) only briefly discuss SRH, or not at all. Clinical care guidelines related to IDD and SRH tend to focus on menstruation and contraception (e.g., Quint et al., 2016), neglecting other important SRH topics such as pregnancy. There have been some important advancements in the creation of clinical care guidelines related to disability and pregnancy in other populations, with recent Canadian obstetric care guidelines created for women with physical disabilities (Berndl et al., 2021). However, to our knowledge, no formal clinical care guidelines have been created for pregnancy care for women with IDD. Given the volume of evidence on the pregnancy and other SRH outcomes and experiences of women with IDD, there is a need to formalize the resulting care recommendations in clinical care guidelines promoted by health professional organizations.

Finally, to support these activities, there is a need for data to monitor the SRH disparities experienced by women with IDD and to evaluate the effectiveness of interventions developed to address them. In the general population, national SRH surveys and population-based health administrative data are used to track SRH outcomes such as reproductive cancer screening rates and maternal mortality rates. However, to date, there have been significant barriers to the use of such data sources to understand SRH in women with IDD (Brown & Mitra, 2022; Tarasoff et al., 2020a, b, c). General population health surveys and SRH surveys often exclude people who experience communication barriers or live in specific settings (e.g., group homes) and frequently do not include questions about disability. On the other hand, surveys that target people with disabilities usually do not address SRH topics. There is thus a need for inclusion of disability questions in SRH surveys, and of SRH questions in disability surveys, with supports provided to facilitate participation of people with a range of communication and other needs. Likewise, population-based health administrative data sources often do not include adequate identification of disability status (Lin et al., 2013). For example, birth records used to study pregnancy outcomes such as maternal mortality typically only include diagnoses recorded during the birth hospital stay. If IDD status is unknown at that time, or not perceived to be relevant to that episode of care, it may not be recorded, leading to misclassification of IDD status (Brown & Mitra, 2022). There is a need to leverage record linkages over time as well as with other sources of information on disability (e.g., disability-related income receipt) to better identify IDD status in health administrative data (Lin et al., 2013). These efforts also require recognition of women with IDD as a priority population within broader health and SRH surveillance efforts. Furthermore, these health and SRH surveillance efforts must acknowledge the intersections of IDD with other important identities, providing information on the SRH of sexual minority women with IDD and the SRH of Black and other racialized women with IDD, to further understand how disability, gender and sexual identity, race, and other identities and structural forms of oppression (e.g., ableism, racism) intersect to shape SRH care experiences and outcomes in women with IDD.

Conclusion

SRH has long been an area neglected, and stigmatized, for women with IDD. Women with IDD have the same SRH rights as those without IDD but experience important disparities in many aspects of SRH, including sexual health, violence, cancer screening and outcomes, contraception, pregnancy outcomes and care, and mental health across the reproductive life course. The evidence of these disparities from multiple jurisdictions can no longer be ignored; urgent changes are needed in the content and delivery of care, training and resources for healthcare providers, and data to support these activities, in order to promote the SRH of women with IDD globally.

References

Amborski, A. M., Bussieres, E. L., Vaillancourt-Morel, M. P., & Joyal, C. C. (2021). Sexual violence against persons with disabilities: A meta-analysis. *Trauma, Violence, & Abuse, 23*(4), 1330–1343. https://doi.org/10.1177/1524838021995975

American College of Obstetricians & Gynecologists. (2005). The importance of preconception care in the continuum of women's healthcare: Committee Opinion No. 313. *Obstetrics & Gynecology, 106*(3), 665–666. https://doi.org/10.1097/00006250-200509000-00052

Amir, N., Smith, L. D., Valentine, A. M., Mitra, M., Parish, S. L., & Moore Sima, T. A. (2022). Clinician perspectives on the need for training on caring for pregnant women with intellectual and developmental disabilities. *Disability and Health Journal, 150*(2), 101262. https://doi.org/10.1016/j.dhjo.2021.101262

Anderson, L. L., Larson, S. A., MapelLentz, S., & Hall-Lande, J. (2019). A systematic review of U.S. studies on the prevalence of intellectual or developmental disabilities since 2000. *Intellectual and Developmental Disabilities, 57*(5), 421–438. https://doi.org/10.1352/1934-9556-57.5.421

Aoyama, K., Ray, J. G., Pinto, R., Hil, L. A., Scales, D. C., & Lapinsky, S. E. (2019). Temporal variations in incidence and outcomes of critical illness among pregnant and postpartum women in Canada: A population-based observational study. *Journal of Obstetrics & Gynaecology Canada, 41*(5), 631–640. https://doi.org/10.1016/j.jogc.2018.07.021

Batchelor, N. C. (2017). *Breast cancer incidence and stage at diagnosis in Ontarians with and without intellectual disabilities*. University of Ontario Institute of Technology.

Berndl, A., Ladhani, N., Wilson, D., Basso, M., Jung, E., Tarasoff, L. A., et al. (2021). Guideline No. 416: Labour, delivery, and postpartum care for people with physical disabilities. *Journal of Obstetrics and Gynaecology Canada, 43*(6), 769–780. https://doi.org/10.1016/j.jogc.2021.02.111

Bernet, D. J. (2011). Sexuality and disability in the lives of women with intellectual disabilities. *Sexuality and Disability, 29*, 129–141. https://doi.org/10.1007/s11195-010-9190-4

Bigg, W. S., & Demuth, R. H. (2011). Premenstrual syndrome and premenstrual dysphoric disorder. *American Family Physician, 84*(8), 918–924. https://doi.org/10.1007/s11195-010-9190-4

Bigrigg, A., Evans, M., Gbolade, B., Newton, J., Pollard, L., Szarewski, A., et al. (1999). Depo Provera: Position paper on clinical use, effectiveness and side effects. *British Journal of Family Planning, 25*(2), 69–76.

Björnsdóttir, K., Stefánsdóttir, Á., & Stefánsdóttir, G. V. (2017). People with intellectual disabilities negotiate autonomy, gender and sexuality. *Sexuality and Disability, 35*(3), 295–311. https://doi.org/10.1007/s11195-017-9492-x

Bowen, E., & Swift, C. (2017). The prevalence and correlates of partner violence used and experienced by adults with intellectual disabilities: A systematic review and call to action. *Trauma, Violence & Abuse, 20*(5), 693–705. https://doi.org/10.1177/1524838017728707

Boyd, S., Sanders, C. L., Kleinert, H. L., Huff, M. B., Lock, S., Johnson, S., et al. (2008). Virtual patient training to improve reproductive health care for women with intellectual disabilities. *Journal of Midwifery & Women's Health, 53*(5), 453–460. https://doi.org/10.1016/j.jmwh.2008.04.017

Brown, H. K., & Mitra, M. (2022). Improved obstetric care for people with disabilities. An urgent call for accessibility and inclusion. *Journal of Women's Health, 31*(1), 4–6. https://doi.org/10.1089/jwh.2021.0510

Brown, H. K., Plourde, N., Ouellette-Kuntz, H., Vigod, S., & Cobigo, V. (2015). Brief report: Cervical cancer screening in women with intellectual and developmental disabilities who have had a pregnancy. *Journal of Intellectual Disability Research, 60*(1), 22–27. https://doi.org/10.1111/jir.12225

Brown, H. K., Lunsky, Y., Wilton, A. S., Cobigo, V., & Vigod, S. N. (2016). Pregnancy in women with intellectual and developmental disabilities. *Journal of Obstetrics & Gynaecology Canada, 38*(1), 9–16. https://doi.org/10.1016/j.jogc.2015.10.004

Brown, H. K., Cobigo, V., Lunsky, Y., & Vigod, S. (2017). Postpartum acute care utilization among women with intellectual and developmental disabilities. *Journal of Women's Health, 26*(4), 329–337. https://doi.org/10.1089/jwh.2016.5979

Brown, H. K., Kirkham, Y., Lunsky, Y., Cobigo, V., & Vigod, S. N. (2018a). Contraceptive provision to postpartum women with intellectual and developmental disabilities: A population-based cohort study. *Perspectives in Sexual and Reproductive Health, 50*(3), 93–99. https://doi.org/10.1363/psrh.12060

Brown, H. K., Ray, J. G., Liu, N., Lunsky, Y., & Vigod, S. N. (2018b). Rapid repeat pregnancy among women with intellectual and developmental disabilities: A population-based cohort study. *Canadian Medical Association Journal, 190*(32), E949–E956. https://doi.org/10.1503/cmaj.170932

Brown, H. K., Cobigo, V., Lunsky Y., & Vigod, S. (2019). Reproductive health in women with intellectual and developmental disabilities in Ontario: Implications for policy and practice. *Healthcare Quarterly, 21*(4), 6–9. https://doi.org/10.12927/hcq.2019.25748

Brown, H. K., Ray, J. G., Chen, S., Guttmann, A., Havercamp, S. M., Parish, S. L., et al. (2021). Association of preexisting disability with severe maternal morbidity or mortality in Ontario. *Canada. JAMA Network Open, 4*(2), e2034993. https://doi.org/10.1001/jamanetworkopen.2020.34993

Brown, H. K., Chen, S., Guttmann, A., Havercamp, S. M., Parish, S. L., Ray, J. G., et al. (2022). Neonatal outcomes of mothers with a disability. *Pediatrics, 150*(3), e2021055318. https://doi.org/10.1542/peds.2021-055318

Brucker, D. L., & Nord, D. (2016). Food insecurity among young adults with intellectual and developmental disabilities in the United States: Evidence from the National Health Interview Survey. *American Journal of Intellectual and Developmental Disabilities, 121*(6), 520–532. https://doi.org/10.1352/1944-7558-121.6.520

Canadian Task Force on Preventive Health Care. (2011). Recommendations on screening for breast cancer in average-risk women aged 40-74 years. *Canadian Medical Association Journal, 183*(17), 1991–2001. https://doi.org/10.1503/cmaj.110334

Canadian Task Force on Preventive Health Care. (2013). Recommendations on screening for cervical cancer. *Canadian Medical Association Journal, 185*(1), 35–45. https://doi.org/10.1503/cmaj.121505

Cantwell, R., Clutton-Brock, T., Cooper, G., Dawson, A., Drife, J., Garrod, D., et al. (2011). Saving Mothers' Lives: Reviewing maternal deaths to make motherhood safer: 2006-2008. The Eighth Report of the Confidential Enquiries into Maternal Deaths in the United Kingdom. *BJOG: An International Journal of Obstetrics & Gynacology, 118*(S1), 1–203. https://doi.org/10.1111/j.1471-0528.2010.02847.x

Carmody, M. (1991). Invisible victims: Sexual assault of people with an intellectual disability. *Australia and New Zealand Journal of Developmental Disabilities, 17*(2), 229–236.

Chan, D. N. S., Law, B. M. H., Au, D. W. H., So, W. K. W., & Fan, N. (2022). A systematic review of the barriers and facilitators influencing the cancer screening behavior among people with intellectual disabilities. *Cancer Epidemiology, 76*, 102084. https://doi.org/10.1016/j.canep.2021.102084

Chirwa, E., Jewkes, R., Van Der Heijden, I., & Dunkle, K. (2020). Intimate partner violence among women with and without disabilities: A pooled analysis of baseline data from seven violence-prevention programmes. *BMJ Global Health, 5*, e002156. https://doi.org/10.1136/bmjgh-2019-002156

Chor, J., Rankin, K., Harwood, B., & Handler, A. (2011). Unintended pregnancy and postpartum contraceptive use in women with and without chronic medical disease who experienced a live birth. *Contraception, 84*(1), 57–63. https://doi.org/10.1016/j.contraception.2010.11.018

Chou, Y.-C., Lu, Z.-Y. J., & Pu, C.-Y. (2015). Attitudes toward male and female sexuality among men and women with intellectual disabilities. *Women & Health, 55*(6), 663–678. https://doi.org/10.1080/03630242.2015.1039183

Cobigo, V., Ouellette-Kuntz, H., Balogh, R., Leung, F., Lin, E., & Lunsky, Y. (2013). Are cervical and breast cancer screening programmes equitable? The case of women with intellectual and developmental disabilities. *Journal of Intellectual Disability Research, 57*(5), 478–488. https://doi.org/10.1111/jir.12035

Codina, M., Pereda, N., & Guilera, G. (2020). Lifetime victimization and poly-victimization in a sample of adults with intellectual disabilities. *Journal of Interpersonal Violence, 37*(5–6), 2062–2082. https://doi.org/10.1177/0886260520936372

Cooper, S. A., Smiley, E., Morrison, J., Williamson, A., & Allan, L. (2006). Mental ill health in adults with intellectual disabilities: Prevalence and associated factors. *British Journal of Psychiatry, 190*(1), 27–35. https://doi.org/10.1192/bjp.bp.106.022483

Creanga, A. A., Berg, C. J., Ko, J. Y., Farr, S. L., & Tong, V. T. (2014). Maternal mortality and morbidity in the United States: Where are we now? *Journal of Women's Health, 23*(1), 3–9. https://doi.org/10.1089/jwh.2013.4617

Crehan, E. T., Schwartz, A. E., & Schmidt, E. K. (2023). Who is delivering sexual health education content to young adults with intellectual or developmental disability? A survey of US-based school-based professionals and parents. *Sexuality and Disability, 41*, 189–200. https://doi.org/10.1007/s11195-023-09790-2

Curry, M. A., Powers, L. E., & Oschwald, M. (2004). Development of an abuse screening tool for women with disabilities. *Journal of Aggression, Maltreatment & Trauma, 8*(4), 123–141. https://doi.org/10.1300/J146v08n04_06

Cuypers, M., Schalk, B. W. M., Boonman, A. J. N., Naaldenberg, J., & Leusink, G. L. (2022). Cancer-related mortality among people with intellectual disabilities: A nationwide population-based cohort study. *Cancer, 128*(6), 1267–1274. https://doi.org/10.1002/cncr.34030

Deave, T., Heron, J., Evans, J., & Emond, A. (2008). The impact of maternal depression in pregnancy on early child development. *British Journal of Obstetrics & Gynaecology, 115*(8), 1043–1051. https://doi.org/10.1111/j.1471-0528.2008.01752.x

Duke, T. S. (2011). Lesbian, gay, bisexual, and transgender youth with disabilities: A meta-synthesis. *Journal of LGBT Youth, 8*(1), 1–52. https://doi.org/10.1080/19361653.2011.519181

Fanslow, J. L., Malihi, Z. A., Hasemi, L., Gulliver, P. J., & McIntosh, T. K. D. (2021). Lifetime prevalence of intimate partner violence and disability: Results from a population-based study in New Zealand. *American Journal of Preventive Medicine, 61*(3), 320–328. https://doi.org/10.1016/j.amepre.2021.02.022

Fontham, E. T. H., Wolf, A. M. D., Church, T. R., Etzioni, R., Flowers, C. R., Herzig, A., et al. (2020). Cervical cancer screening for individuals at average risk: 2020 guideline update from the American Cancer Society. *CA: a Cancer Journal for Clinicians, 70*(5), 321–346. https://doi.org/10.3322/caac.21628

García-Cuéllar, M. M., Pastor-Moreno, G., Ruiz-Pérez, I., & Henares-Montiel, J. (2023). The prevalence of intimate partner violence against women with disabilities: A systematic review of the literature. *Disability and Rehabilitation, 45*(1), 1–8. https://doi.org/10.1080/09638288.2022.2025927

Gleason, J. L., Grewal, J., Chen, Z., Cernich, A. N., & Grantz, K. L. (2021). Risk of adverse maternal outcomes in pregnant women with disabilities. *JAMA Network Open, 4*(12), 2138414. https://doi.org/10.1001/jamanetworkopen.2021.38414

Gleason, J. L., Grewal, J., Chen, Z., Cernich, A. N., & Grantz, K. L. (2023). Risk of adverse neonatal outcomes among pregnant women with disabilities. *International Journal of Epidemiology, 52*(1), 203–213. https://doi.org/10.1093/ije/dyac183

Goldacre, A. D., Gray, R., & Goldacre, M. J. (2015). Childbirth in women with intellectual disability: Characteristics of their pregnancies and outcomes in an archived epidemiological dataset. *Journal of Intellectual Disability Research, 59*(7), 653–663. https://doi.org/10.1111/jir.12169

Greenwood, N. W., & Wilkinson, J. (2013). Sexual and reproductive health care for women with intellectual disabilities: A primary care perspective. *International Journal of Family Medicine*, 642472. https://doi.org/10.1155/2013/642472

Grove, L., Morrison-Beedy, D., Kirby, R., & Hess, J. (2018). The birds, bees, and special needs: Making evidence-based sex education accessible for adolescents with intellectual disabilities. *Sexuality and Disability, 36*, 313–329. https://doi.org/10.1007/s11195-018-9547-7

Havercamp, S. M., & Scott, H. M. (2015). National health surveillance of adults with disabilities, adults with intellectual and developmental disabilities, and adults with no disabilities. *Disability and Health Journal, 8*(2), 165–172. https://doi.org/10.1016/j.dhjo.2014.11.002

Havercamp, S. M., Barnhart, W. R., Robinson, A. C., & Whalen Smith, C. N. (2021). What should we teach about disability? National consensus on disability competencies for health care education. *Disability and Health Journal, 14*(2), 100989. https://doi.org/10.1016/j.dhjo.2020.100989

Heslop, P., Cook, A., Sullivan, B., Calkin, R., Pollard, J., & Byrne, V. (2022). Cancer in deceased adults with intellectual disabilities: English population-based study using linked data from three sources. *BMJ Open, 12*(3), e056974. https://doi.org/10.1136/bmjopen-2021-056974

Höglund, B., & Larsson, M. (2013). Struggling for motherhood with an intellectual disability—A qualitative study of women's experiences in Sweden. *Midwifery, 29*(6), 698–704. https://doi.org/10.1016/j.midw.2012.06.014

Höglund, B., & Larsson, M. (2014). Professional and social support enhances maternal well-being in women with intellectual disability—A Swedish interview study. *Midwifery, 30*(11), 1118–1123. https://doi.org/10.1016/j.midw.2014.03.018

Höglund, B., Lindgren, P., & Larsson, M. (2013). Midwives' knowledge of, attitudes towards and experiences of caring for women with intellectual disability during pregnancy and childbirth: A cross-sectional study in Sweden. *Midwifery, 29*(8), 950–955. https://doi.org/10.1016/j.midw.2012.12.002

Hollins, S., Hutchinson, D., & Sinason, V. (1992). Books beyond words. *British Journal of Psychiatry, 171*(1), 1–2.

Holmes, O. W. & Supreme Court of the United States. (1926). *U.S. Reports: Buck v. Bell*, 274.

Horner-Johnson, W., & Drum, C. E. (2006). Prevalence of maltreatment of people with intellectual disabilities: A review of recently published research. *Mental Retardation and Developmental Disabilities Research Reviews, 12*(1), 57–69. https://doi.org/10.1002/mrdd.20097

Horner-Johnson, W., Darney, B. G., Kulkarni-Rajasekhara, S., Quigley, B., & Caughey, A. B. (2016). Pregnancy among US women: Differences by presence, type, and complexity of disability. *American Journal of Obstetrics & Gynecology, 214*(4), e1–e9. https://doi.org/10.1016/j.ajog.2015.10.929

Horner-Johnson, W., Biel, F. M., Caughey, A. B., & Darney, B. G. (2019a). Differences in prenatal care by presence and type of maternal disability. *American Journal of Preventive Medicine, 56*(3), 376–382. https://doi.org/10.1016/j.amepre.2018.10.021

Horner-Johnson, W., Moe, E. L., Stoner, R. C., Klein, K. A., Edelman, A. B., Eden, K. B., et al. (2019b). Contraception knowledge and use among women with intellectual, physical, or sensory disabilities: A systematic review. *Disability and Health Journal, 12*(2), 139–154. https://doi.org/10.1016/j.dhjo.2018.11.006

Horner-Johnson, W., Dissanayake, M., Wu, J. P., Caughey, A. B., & Darney, B. G. (2020). Pregnancy intendedness by maternal disability status and type in the United States. *Perspectives on Sexual and Reproductive Health, 52*(1), 31–38. https://doi.org/10.1363/psrh.12130

Horner-Johnson, W., Garb, B., Darney, B. G., Biel, F. M., & Caughey, A. B. (2022). Severe maternal morbidity and other perinatal complications among women with physical, sensory, or intellectual and developmental disabilities. *Paediatric and Perinatal Epidemiology, 36*(5), 759–768. https://doi.org/10.1111/ppe.12873

Horsbøl, T. A., Michelsen, S. I., Lassen, T. H., Juel, K., Bigaard, J., Hoei-Hansen, C., et al. (2023). Breast cancer screening among women with intellectual disability in Denmark. *JAMA Network Open, 6*(1), e2248980. https://doi.org/10.1001/jamanetworkopen.2022.48980

Hughes, K., Bellis, M. A., Jones, L., Wood, S., Bates, G., Eckley, L., et al. (2012). Prevalence and risk of violence against adults with disabilities: A systematic review and meta-analysis of observational studies. *The Lancet, 379*(9826), 1621–1629. https://doi.org/10.1016/S0140-6736(11)61851-5

Ignani, E., Fudge Shormans, A., Liddiard, K., & Runskwick-Cole, K. (2015). "Some people are not allowed to love": Intimate citizenship in the lives of people labeled with intellectual disabilities. *Disability & Society, 31*(1), 131–135. https://doi.org/10.1080/09687599.2015.1136148

Janz, H. L. (2019). Ableism: The undiagnosed malady afflicting medicine. *Canadian Medical Association Journal, 19*(17), e478–e479. https://doi.org/10.1503/cmaj.180903

Khan, M., Brown, H. K., Lunsky, Y., Welsh, K., Havercamp, S. M., Proulx, L., et al. (2021). A socio-ecological approach to understanding the perinatal care experiences of people with intellectual and/or developmental disabilities in Ontario, Canada. *Women's Health Issues, 31*(6), 550–559. https://doi.org/10.1016/j.whi.2021.08.002

Khanna, A., Smith, L. D., Parish, S. L., & Mitra, M. (2022). Pregnancy recommendations from women with intellectual and developmental disabilities to their peers. *Disability and Health Journal, 15*(4), 101343. https://doi.org/10.1016/j.dhjo.2022.101343

Kopac, C. A., Fritz, J., & Holt, R. A. (1998). Gynecologic and reproductive services for women with developmental disabilities. *Clinical Excellence in Nursing Practice, 2*(2), 88–95.

Krahn, G. L., & Fox, M. H. (2013). Health disparities of adults with intellectual disabilities: What do we know? What do we do? *Journal of Applied Research in Intellectual Disabilities, 27*(5), 431–446. https://doi.org/10.1111/jar.12067

Krahn, G. L., Hammond, L., & Turner, A. (2006). A cascade of disparities: Health and health care for people with intellectual disabilities. *Mental Retardation and Developmental Disabilities Research Reviews, 12*(1), 70–82. https://doi.org/10.1002/mrdd.20098

Kyrkou, M. (2005). Health issues and quality of life in women with intellectual disability. *Journal of Intellectual Disability Research, 49*(10), 770–772. https://doi.org/10.1111/j.1365-2788.2005.00749.x

Langer-Shapland, K., Minton, S. J., & Richards, N. (2023). "It should be more outspoken and not hushed away, not like put in a dark box": An interpretive phenomenological analysis of experiences of menopause voiced by women with learning disabilities. *British Journal of Learning Disabilities, 51*(4), 509–521. https://doi.org/10.1111/bld.12527

Lin, E., Balogh, R., Cobigo, V., Ouellette-Kuntz, H., Wilton, A. S., & Lunsky, Y. (2013). Using administrative health data to identify individuals with intellectual and developmental disabilities: A comparison of algorithms. *Journal of Intellectual Disability Research, 57*(5), 462–477. https://doi.org/10.1111/jir.12002

Lunsky, Y., Durbin, A., Brown, H. K., Bansal, S., Heifetz, M., & Antoniou, T. (2017). Health profiles and associated service using among adults with HIV and intellectual and developmental disabilities. *AIDS, 31*, 697–705. https://doi.org/10.1097/QAD.0000000000001361

Mahar, A., Biggs, K., Hansford, R. L., Derksen, S., Griffiths, R., Enns, J. E., et al. (2023). Stage IV breast, colorectal, and lung cancer at diagnosis in adults living with intellectual or developmental disabilities: A population-based cross-sectional study. *Cancer, 130*, 740. https://doi.org/10.1002/cncr.35068

Mandell, D. S., Eleey, C. C., Cederbaum, J. A., Noll, E., Hutchinson, M. K., Jemmott, L. S., et al. (2008). Sexually transmitted infection among adolescents receiving special education services. *Journal of School Health, 78*(7), 382–388. https://doi.org/10.1111/j.1746-1561.2008.00318.x

Marcell, A. V., Burstein, G. R., & the Committee on Adolescence. (2017). Sexual and reproductive health care services in the pediatric setting. *Pediatrics, 140*(5), e20172858. https://doi.org/10.1542/peds.2017-2858

Marquis, S., Lunsky, Y., McGrail, K. M., & Baumbusch, J. (2023). Population-level use of gynecological health services by female youth with intellectual/developmental disabilities in British Columbia Canada. *Disability and Health Journal, 16*(3), 101478. https://doi.org/10.1016/j.dhjo.2023.101478

Matin, B. K., Ballan, M., Darabi, F., Karyani, A. K., Soofi, M., & Soltani, S. (2021). Sexual health concerns in women with intellectual disabilities: A systematic review in qualitative studies. *BMC Public Health, 21*, 1965. https://doi.org/10.1186/s12889-021-12027-6

McCarthy, M. (2014). Women with intellectual disability: Their sexual lives in the 21st century. *Journal of Intellectual & Developmental Disability, 39*, 124–131. https://doi.org/10.3109/13668250.2014.894963

McCarthy, M., Bates, C., Triantafyllopoulou, P., Hunt, S., & Milne Skillman, K. (2018). "Put bluntly, they are targeted by the worst creeps society has to offer": Police and professionals' views and actions relating to domestic violence and women with intellectual disabilities. *Journal of Applied Research in Intellectual Disabilities, 32*(1), 71–81. https://doi.org/10.1111/jar.12503

McConnell, D., & Phelan, S. (2022). The devolution of eugenic practices: Sexual and reproductive health and oppression of people with intellectual disability. *Social Science & Medicine, 298*, 114877. https://doi.org/10.1016/j.socscimed.2022.114877

McConnell, D., Mayes, R., & Llewellyn, G. (2009). Pre-partum distress in women with intellectual disabilities. *Journal of Intellectual & Developmental Disability, 33*(2), 117–183. https://doi.org/10.1080/13668250802007903

McGilloway, C., Smith, D., & Galvin, R. (2018). Barriers faced by adults with intellectual disabilities who experience sexual assault: A systematic review and meta-synthesis. *Journal of Applied Research in Intellectual Disabilities, 33*(1), 51–66. https://doi.org/10.1111/jar.12445

Meltzer-Brody, S., & Stuebe, A. (2014). The long-term psychiatric and medical prognosis of perinatal mental illness. *Best Practices in Research in Clinical Obstetrics & Gynaecology, 28*(1), 49–60. https://doi.org/10.1016/j.bpobgyn.2013.08.009

Mitra, M., Parish, S. L., Clements, K. M., Zhang, J., & Moore Simas, T. A. (2018a). Antenatal hospitalizations among U.S. women with intellectual and developmental disabilities: A retrospective cohort study. *American Journal of Intellectual and Developmental Disabilities, 123*(5), 399–411. https://doi.org/10.1352/1944-7558-123.5.399

Mitra, M., Parish, S. L., Akobirshoev, I., Rosenthal, E., & Moore Simas, T. A. (2018b). Postpartum hospital utilization among Massachusetts women with intellectual and developmental disabilities: A retrospective cohort study. *Maternal and Child Health Journal, 22*, 1492–1501. https://doi.org/10.1007/s10995-018-2546-6

Mitra, M., Akobirshoev, I., Parish, S. L., Valentine, A., Clements, K. M., & Moore Simas, T. A. (2019). Postpartum emergency department use among women with intellectual and developmental disabilities: A retrospective cohort study. *Journal of Epidemiology and Community Health, 73*(6), 557–563. https://doi.org/10.1136/jech-2018-211589

Mitra, M., Akobirshoev, I., Valentine, A., Brown, H. K., & Moore Simas, T. A. (2021). Severe maternal morbidity and maternal mortality in women with intellectual and developmental disabilities. *American Journal of Preventive Medicine, 61*(6), 872–881. https://doi.org/10.1016/j.amepre.2021.05.041

Mosher, W., Hughes, R. B., Bloom, T., Horton, L., Mojtabai, R., & Alhusen, J. L. (2018). Contraceptive use by disability status: New national estimates from the National Survey of family growth. *Contraception, 97*(6), 552–558. https://doi.org/10.1016/j.contraception.2018.03.031

Neri, S. V., Bradley, E. H., & Groce, N. E. (2007). Frequency of HIV testing among persons with disabilities: Results from the National Health Interview Survey, 2002. *AIDS Education and Prevention, 19*(6), 545–554. https://doi.org/10.1521/aeap.2007.19.6.545

Nishat, F., Lunsky, Y., Tarasoff, L. A., & Brown, H. K. (2022). Prenatal care adequacy among women with disabilities: A population-based study. *American Journal of Preventive Medicine, 62*(1), 39–49. https://doi.org/10.1016/j.amepre.2021.05.037

O'Hara, M. W., & Wisner, K. L. (2014). Perinatal mental illness: Definition, description and aetiology. *Best Practices in Research in Clinical Obstetrics & Gynaecology, 28*(1), 3–12. https://doi.org/10.1016/j.bpobgyn.2013.09.002

Oeffinger, K. C., Fontham, E. T. H., Etzioni, R., Herzig, A., Michaelson, J. A., Shih, Y.-C. T., et al. (2015). Breast cancer screening for women at average risk: 2015 guideline update from the American Cancer Society. *JAMA, 315*(15), 1499–1614. https://doi.org/10.1001/jama.2015.12783

Osborn, D. P. J., Horsfall, L., Hassiotis, A., Petersen, I., Walters, K., & Nazareth, I. (2012). Access to cancer screening people with learning disabilities in the UK: Cohort study in the health improvement network, a primary care research database. *PLoS One, 7*(8), e43841. https://doi.org/10.1371/journal.pone.0043841

Parchomiuk, M. (2022). Sexuality of people with intellectual disabilities: A proposal to use the positive sexuality model. *Sexuality & Culture, 26*, 418–448. https://doi.org/10.1007/s12119-021-09893-y

Parekh, T., Gimm, G., & Kitsantas, P. (2023). Sexually transmitted infections in women of reproductive age by disability type. *American Journal of Preventive Medicine, 64*(3), 393–404. https://doi.org/10.1016/j.amepre.2022.10.013

Potvin, L., Brown, H. K., & Cobigo, V. (2016). Social support received by women with intellectual and developmental disabilities during pregnancy and childbirth. *Midwifery, 37*, 57–64. https://doi.org/10.1016/j.midw.2016.04.005

Quint, E. H. (2016). Adolescents with special needs: Clinical challenges in reproductive health care. *Journal of Pediatric and Adolescent Gynecology, 29*(1), 2–6. https://doi.org/10.1016/j.jpag.2015.05.003

Quint, E. H., O'Brien, R. F., & the Committee on Adolescence. (2016). Menstrual management for adolescents with disabilities. *Pediatrics, 138*(1), e20160295. https://doi.org/10.1542/peds.2016-0295

Redshaw, M., Malouf, R., Gao, H., & Gray, R. (2013). Women with disability: the experience of maternity care during pregnancy, labour and birth and the postnatal period. *BMC Pregnancy and Childbirth, 13*, 174. https://doi.org/10.1186/1471-2393-13-174

Reidy, M., Denieffe, S., & Foran, S. (2014). Cancer screening in women with intellectual disabilities: An Irish perspective. *Journal of Intellectual Disabilities, 18*(1), 51–60. https://doi.org/10.1177/1744629514521067

Richards, D., Miodrag, N., Watson, S. L., Feldman, M., Aunos, M., Cox-Lindenbaum, D., et al. (2008). Sexuality and human rights of persons with intellectual disabilities. In F. Owen & D. Griffiths (Eds.), *Challenges to the human rights of people with intellectual disabilities* (pp. 137–155). Jessica Kingsley Publishers.

Richter, L. L., Ting, J., Muraca, G. M., Boutin, A., Wen, Q., & Lyons, J. (2019). Temporal trends in preterm birth, neonatal mortality, and neonatal morbidity following spontaneous and clinician-initiated delivery in Canada, 2009-2016. *Journal of Obstetrics & Gynaecology Canada, 41*(12), 1742–1751. https://doi.org/10.1016/j.jogc.2019.02.151

Robertson, E., Grace, S., Wallington, T., & Stewart, D. E. (2004). Antenatal risk factors for postpartum depression: A synthesis of recent literature. *General Hospital Psychiatry, 26*(4), 289–295. https://doi.org/10.1016/j.genhosppsych.2004.02.006

Roebuck, M. (2021). Housing for people with intellectual disabilities: A scoping review. *Journal on Developmental Disabilities, 26*(2), 1–25.

Rosenthal, E., Parish, S. L., Ransom, C., Smith, L. D., & Mitra, M. (2022). Formal and informal supports for women with intellectual and developmental disabilities during pregnancy.

American Journal of Intellectual and Developmental Disabilities, 60(4), 261–272. https://doi.org/10.1352/1934-9556-60.4.261

Rowlands, S., & Amy, J.-J. (2019). Sterilization of those with intellectual disability: Evolution from non-consensual interventions to strict safeguards. *Journal of Intellectual Disabilities, 23*(2), 233–249. https://doi.org/10.1177/1744629517747162

Roy, A., Roy, A., & Roy, M. (2012). The human rights of women with intellectual disability. *Journal of the Royal Society of Medicine, 105*, 384–389. https://doi.org/10.1258/jrsm.2012.110303

Rubenstein, E., Ehrenthal, D. B., Nobles, J., Mallinson, D. C., Bishop, L., Jenkins, M. C., et al. (2022). Fertility rates in women with intellectual and developmental disabilities in Wisconsin Medicaid. *Disability and Health Journal, 15*(3), 101321. https://doi.org/10.1016/j.dhjo.2022.101321

Saeed, G., Brown, H. K., Lunsky, Y., Welsh, K., Proulx, L., Havercamp, S. M., et al. (2022). Barriers to and facilitators of effective communication in perinatal care: A qualitative study of the experiences of birthing people with sensory, intellectual, and/or developmental disabilities. *BMC Pregnancy and Childbirth, 22*, 364. https://doi.org/10.1186/s12884-022-04691-2

Satgé, D., Sauleau, E.-A., Jacot, W., Raffi, F., Azéma, B., Bouyat, J.-C., et al. (2014). Age and stage at diagnosis: A hospital series of 11 women with intellectual disability and breast carcinoma. *BMC Cancer, 14*, 150. https://doi.org/10.1186/1471-2407-14-150

Schuengel, C., Cuypers, M., Bakkum, L., & Leusink, G. L. (2022). Reproductive health of women with intellectual disability: Antenatal care, pregnancies and outcomes in the Dutch population. *Journal of Intellectual Disability Research, 67*(12), 1306–1316. https://doi.org/10.1111/jir.12982

Shin, D. W., Lee, J.-W., Jung, J. H., Han, K., Kim, S. Y., Choi, K. S., et al. (2018). Disparities in cervical cancer screening among women with disabilities: A national database study in South Korea. *Journal of Clinical Oncology, 36*(27), 2778–2786. https://doi.org/10.1200/JCO.2018.77.7912

Silka, V. R., & Hurley, A. D. (2005). Premenstrual syndrome and premenstrual dysphoric disorder in individuals with intellectual disability. *Mental Health Aspects of Developmental Disabilities, 8*(1), 29–30. https://doi.org/10.1016/j.cppeds.2022.101187

Smith, E., Zirnsak, T. M., Power, J., Lyons, A., & Bigby, C. (2022). Social inclusion of LGBTQ and gender diverse adults with intellectual disability in disability services: A systematic review of the literature. *Journal of Applied Research in Intellectual Disabilities, 35*(1), 46–59. https://doi.org/10.1111/jar.12925

Smith, L., Valentine, A., Moore Simas, T. A., Parish, S. L., Levy, A., & Mitra, M. (2023). Clinician-reported barriers to providing perinatal care to women with intellectual and developmental disabilities. *Journal of Intellectual and Developmental Disabilities, 48*(1), 12–22. https://doi.org/10.3109/13668250.2022.2086110

Sullivan, W. F., Diepstra, H., Heng, J., Ally, S., Bradley, E., Casson, E., et al. (2018). Primary care of adults with intellectual and developmental disabilities: 2018 Canadian consensus guidelines. *Canadian Family Physician, 64*(4), 254–279.

Sun, H., Ferlay, J., Siege, L. R. L., Laversanne, M., Soerjomataram, I., Jemal, A., et al. (2021). Global cancer statistics 2020: GLOBOCAN estimates of incidence and mortality worldwide for 36 cancers in 185 countries. *CA: a Cancer Journal for Clinicians, 71*(3), 209–249. https://doi.org/10.3322/caac.21660

Swaine, J. G., Dababnah, S., Parish, S. L., & Luken, K. (2013). Family caregivers' perspectives on barriers and facilitators of cervical and breast cancer screening for women with intellectual disability. *Intellectual and Developmental Disabilities, 51*(1), 62–73. https://doi.org/10.1352/1934-9556-51.01.062

Taggart, L., McMillan, R., & Lawson, A. (2008). Women with and without intellectual disability and psychiatric disorders: An examination of the literature. *Journal of Intellectual Disabilities, 12*(3), 191–211. https://doi.org/10.1177/1744629508095323

Taouk, L. H., Failkow, M. F., & Schulkin, J. A. (2018). Women with disabilities: A survey of obstetrician-gynecologists' training, practices, and perceived barriers. *Health Equity, 2*(1), 207–215. https://doi.org/10.1089/heq.2018.0014

Tarasoff, L. A. (2016). "We exist": The health and Well-being of sexual minority women and trans people with disabilities. In S. E. Miles-Cohen & C. Signore (Eds.), *Eliminating inequities for women with disabilities: An agenda for health and wellness* (pp. 179–207). American Psychological Association.

Tarasoff, L. A. (2021). A call for comprehensive, disability- and LGBTQ-inclusive sexual and reproductive health education. *The Journal of Adolescent Health, 69*(2), 185–186. https://doi.org/10.1016/j.jadohealth.2021.05.013

Tarasoff, L. A., Lunsky, Y., Chen, S., Guttmann, A., Havercamp, S. M., Parish, S. L., et al. (2020a). Preconception health of women with disabilities in Ontario: A population-based cross-sectional study. *Journal of Women's Health, 29*(12), 1564–1575. https://doi.org/10.1089/jwh.2019.8273

Tarasoff, L. A., Murtaza, F., Carty, A., Salaeva, D., Hamilton, A. D., & Brown, H. K. (2020b). Health of newborns and infants born to women with disabilities: A meta-analysis. *Pediatrics, 146*(6), e20201635. https://doi.org/10.1542/peds.2020-1635

Tarasoff, L. A., Ravindran, S., Malik, H., Salaeva, D., & Brown, H. K. (2020c). Maternal disability and risk for pregnancy, delivery, and postpartum complications: A systematic review and meta-analysis. *American Journal of Obstetrics & Gynecology, 222*(1), e1–e27. https://doi.org/10.1016/j.ajog.2019.07.015

Tarasoff, L. A., Lunsky, Y., Welsh, K., Havercamp, S. M., Vigod, S. N., & Brown, H. K. (2023). The disability-related education and training experiences of perinatal care providrs in Ontario. *Journal of Obstetrics and Gynaecology Canada, 45*(8), 581–586. https://doi.org/10.1016/j.jogc.2023.05.032

Tilley, E., Walmsley, J., Earle, S., & Atkinson, D. (2012). "The silence is roaring": Sterilization, reproductive rights and women with intellectual disabilities. *Disability & Society, 27*(3), 413–426. https://doi.org/10.1080/09687599.2012.654991

Tøssebro, J., Midjo, T., Paulsen, V., & Berg, B. (2017). Prevalence, trends and custody among children of parents with intellectual disabilities in Norway. *Journal of Applied Research in Intellectual Disabilities, 30*(3), 533–542. https://doi.org/10.1111/jar.12304

Tracy, J., Grover, S., & Macgibbon, S. (2016). Menstrual issues for women with intellectual disability. *Australian Prescriber, 39*(2), 54–57. https://doi.org/10.18773/austprescr.2016.024

United Nations. (2006). *Convention on the rights of persons with disabilities. Treaty series, 2515, 3*. United Nations General Assembly.

Wendt, A., Gibbs, C. M., Peters, S., & Hogue, C. J. (2012). Impact of increasing inter-pregnancy interval on maternal and infant health. *Paediatric and Perinatal Epidemiology, 26*(S1), 239–258. https://doi.org/10.1111/j.1365-3016.2012.01285.x

Whittle, C., & Butler, C. (2018). Sexuality in the lives of people with intellectual disabilities: A meta-ethnographic synthesis of qualitative studies. *Research in Developmental Disabilities, 75*, 68–81. https://doi.org/10.1016/j.ridd.2018.02.008

Wilkinson, J. E., & Cerreto, M. C. (2008). Primary care for women with intellectual disabilities. *The Journal of the American Board of Family Medicine, 21*(3), 215–222. https://doi.org/10.3122/jabfm.2008.03.070197

Willis, D. S. (2008). A decade on: What have we learnt about supporting women with intellectual disabilities through the menopause? *Journal of Intellectual Disabilities, 12*(1), 9–23. https://doi.org/10.1177/1744629507086604

Willis, D. S., Wishart, J. G., & Muir, J. (2010). Menopausal experiences of women with intellectual disabilities. *Journal of Applied Research in Intellectual Disabilities, 24*(1), 74–85. https://doi.org/10.1111/j.1468-3148.2010.00566.x

Woodlock, D., & Harris, B. (2022). "You have to be really careful": Technology and the abuse of women with intellectual and cognitive disabilities. *Disability & Society*. https://doi.org/10.1080/09687599.2022.2114886

World Health Organization. (2006). *Comprehensive cervical cancer control: A guide to essential practice*. World Health Organization.

World Health Organization. (2012). *Meeting to develop a global consensus on preconception care to reduce maternal and childhood mortality and morbidity*. World Health Organization.

World Health Organization. (2013). *Global and regional estimates of violence against women: Prevalence and health effects of intimate partner violence and non-partner sexual violence*. World Health Organization.

World Health Organization and United Nations Population Fund. (2009). *Promoting sexual and reproductive health for persons with disabilities—WHO/UNFPA guidance note*. World Health Organization. Retrieved from: https://apps.who.int/iris/bitstream/handle/10665/44207/9789241598682_eng.pdf?sequence=1

Wu, J., Zhang, M., Mitra, M., Parish, S. L., & Minima Reddy, G. K. (2018). Provision of moderately and highly effective reversible contraception to insured women with intellectual and developmental disabilities. *Obstetrics & Gynecology, 132*(3), 565–574. https://doi.org/10.1097/aog.0000000000002777

Xie, E., & Gemmill, M. (2018). Exploring the prenatal experience of women with intellectual and developmental disabilities in a southeastern Ontario family health team. *Canadian Family Physician, 64*(S2), S70–S75.

Xu, X., McDermott, S. W., Mann, J. R., Hardin, J. W., Deroche, C. B., Carroll, D. D., et al. (2017). A longitudinal assessment of adherence to breast and cervical cancer screening recommendations among women with and without intellectual disability. *Preventive Medicine, 100*, 167–172. https://doi.org/10.1016/j.ypmed.2017.04.034

Down Syndrome

Anna J. Esbensen, Emily K. Schworer, and Sigan L. Hartley

Abstract Down syndrome is a genetic cause of intellectual disability that predisposes individuals to distinct physical and clinical profiles. While there are commonalities in physical features, slower rates of growth, and neurodevelopmental delays among individuals with Down syndrome, there is also significant heterogeneity. Individuals with Down syndrome are at higher risk of developing some co-occurring medical and mental health conditions, and at lower risk of developing others. This chapter reviews the etiology and prevalence of Down syndrome, the impact of neurobiology on cognition, language, and motor development primarily focused on childhood and adolescence, and advances in intervention and educational supports specific to individuals with Down syndrome. Common co-occurring medical and mental health conditions among individuals with Down syndrome are reviewed. Given the increased lifespan experienced by individuals with Down syndrome in recent decades, the chapter reviews factors related to adulthood in Down syndrome, specifically focused on health, employment, and enhancing healthy aging. The chapter concludes by discussing the family system in the context of having a family member with Down syndrome and the role of community engagement in the lives of individuals with Down syndrome.

A. J. Esbensen (✉)
Division of Developmental and Behavioral Pediatrics, Cincinnati Children's Hospital Medical Center, Cincinnati, OH, USA

Department of Pediatrics, University of Cincinnati College of Medicine, Cincinnati, OH, USA
e-mail: anna.esbensen@cchmc.org

E. K. Schworer
Waisman Center, University of Wisconsin-Madison, Madison, WI, USA

S. L. Hartley
Waisman Center, University of Wisconsin-Madison, Madison, WI, USA

Department of Human Development and Family Studies, University of Wisconsin-Madison, Madison, WI, USA

M. G. Valdovinos (ed.), *Intellectual and Developmental Disabilities*, Contemporary Clinical Neuroscience,
https://doi.org/10.1007/978-3-031-66932-3_13

Keywords Down syndrome · Etiology · Early development · Health · Aging

Down syndrome (DS) is the most common genetic cause of intellectual disability with a prevalence of approximately 1/700 to 1500 births per year (de Graaf et al., 2017a; Parker et al., 2010; Patterson & Costa, 2005). Individuals with DS present with a similar phenotype of behavioral characteristics and health problems that have been well-described in recent decades. Yet, there are also important differences among individuals with DS in their behaviors, level of cognition, and everyday functioning and health, making each individual unique. For many living with a disability, the environment focuses on what supports are needed or what skills are lacking, and while this is necessary for safety and intervention, it is equally important to showcase individuals' strengths and abilities. This chapter will focus on the etiology and neurobiology of DS, interventions and co-occurring health conditions, the aging and health of adults with DS, the families of individuals with DS, and on the abilities and valuable contributions from the community of persons with DS.

Persons with DS were originally grouped with others with intellectual disability until the 1860s. Langdon Down first described the physical features (e.g., flat face, wide distance between eyes, skin lacking elasticity) and several behavioral features (e.g., motor control difficulties, intelligibility of speech) classifying individual as having DS (Down, 1866). Although DS could be classified based on physical and behavioral characteristics, it was not until the 1930s that DS was suggested to be caused by a chromosomal anomaly. In the 1950s, Jérôme Lejeune reported discovery of the cause of DS, with the identification of the third copy of chromosome 21 (Brown et al., 2020). However, Marthe Gautier completed extensive experimental work and was the first to identify an extra chromosome associated with DS (Brown et al., 2020). Despite Gautier's science, Lejeune was the first to publish and is given credit for the identification of trisomy of chromosome 21 as the cause for DS. Originally, the condition was described as Mongolian Idiocy. However, in 1961, a group of 20 of the world's leading geneticists came together to recommend that trisomy 21 be designated as Langdon Down anomaly, Down's syndrome, or anomaly or congenital acromicria. Currently, groups such as the National Down Syndrome Society endorse the term Down syndrome (rather than "Down's"). There are several resources available to learn more about the history of DS including *Downs—The History of a Disability* and *The Oxford Handbook of Down Syndrome and Development* (Burack et al., 2022; Prost & Roberts, 2013).

Genetic Identification, Etiology, and Prevalence

With the help of advances in genetic research and karyotyping, we now have a better understanding of the different genetic underpinnings of DS. A third copy of chromosome 21, also called trisomy 21, occurs through nondisjunction (96% of cases) or translocation (3–4% of cases, where part of chromosome 21 attaches to chromosome 14, 21, or 22) (Bull, 2020). Mosaicism is another type of DS, in which not all cells are affected with trisomy 21 (1–2% of cases) and leads to lesser impact on cognitive difficulties and clinical presentation (Bull, 2020). The availability of prenatal testing has contributed to the earlier identification of DS. This prenatal testing originally included amniocentesis; however, in recent years, highly accurate (99%) noninvasive cell-free prenatal testing has become more widely available and affordable. Among women who receive prenatal testing that indicates DS, an estimated 43% go on to have a miscarriage, which is attributed to the fact that older mothers are more likely to have a child with DS and also are at-risk for miscarriages (Morris et al., 1999; Savva et al., 2006). Historically, some parents who receive a prenatal diagnosis of DS have elected to terminate the pregnancy. In a review of termination rates from 1995 to 2011 in the United States, the mean termination rate was 61–93% following a prenatal diagnosis of DS (Natoli et al., 2012). Rates of elective termination vary widely across countries. Postnatally, diagnoses of DS are made based on inspection of physical characteristics and a karyotype to confirm diagnosis. Physical features associated with DS include hypotonia, excessive skin at the nape of the neck, features of the head and face (small brachycephalic head, flat nasal bridge, small mouth, small ears), features of and around the eye (epicanthal folds, upward-slanting palpebral fissures, Brushfield spots), and features of the hands and feet (single transverse palmar crease, short fifth finger with clinodactyly, and wide spacing between the first and second toes) (Bull et al., 2022). Despite the high rate of pregnancy terminations, infants born with DS now are having better survival outcomes than in prior decades due to improvements in congenital heart defect and gastrointestinal tract surgeries and common risk factors associated with DS (Weijerman & De Winter, 2010).

The number of children living to adulthood and their average life expectancy have increased dramatically over the past century. In 1900, the life expectancy of people with DS was 9 years. By 1951, the average life expectancy for a child with DS had only increased to 12 years (Penrose & Smith, 1966). Largely due to medical care improvements for infants and children with DS, the life expectancy of people with DS increased to 28 years in 1984 and is now about 60 years (Bittles & Glasson, 2004; Coppus et al., 2008; Zhu et al., 2013). In recent years, the life expectancy of people with DS has been shown to vary based on race, with Hispanic White and Black adults with DS having a lower mean and median life expectancy than non-Hispanic White adults with DS (de Graaf et al., 2017b).

Neurobiology and Early Development

While there are individual profiles of strength and challenge in the development of cognition, language, and motor domains, there are also areas of commonality among those with DS. This probabilistic behavioral phenotype associated with DS begins to emerge as early as infancy or toddlerhood and consists of relative strengths in receptive language, nonverbal social skills, and visual processing and challenges with expressive language, short-term memory, goal-directed behavior, and motor skills (Chapman & Hesketh, 2000; Fidler, 2005; Schworer et al., 2020). We describe in-depth these early cognitive, language, and motor abilities in the following sections, concluding with discussion of a selection of interventions that show promise for optimizing developmental potential within this population.

Cognition

Anatomically, children with DS have smaller total brain volume and gray matter compared to typically developing peers, which impacts cognitive abilities (Becker et al., 1986; Hamner et al., 2018; Menghini et al., 2011); however, there is still a dearth of research characterizing the anatomical correlates of the cognitive profile associated with DS. Specific regions have been identified as having reduced volume including the frontal lobe, temporal lobe, cerebellum, and hippocampus (Baburamani et al., 2019; Hamner et al., 2018), and some have found that the density of brain volume in certain regions aligns with language abilities, short-term memory, and long-term memory (Lott & Dierssen, 2010; Menghini et al., 2011). Although there has been progress with understanding brain development in DS, there is still much to learn from longitudinal studies and additional work with new imaging technology (e.g., fetal MRI). It is likely that these differences in brain development are strongly connected with behavioral presentations of cognition and executive functions in DS.

Although we are still understanding the neuroanatomical connections between structure and cognitive function in DS, there is a more established understanding of the behavioral characterization of cognitive abilities. Intellectual disability (IQ < 70) is a feature of most individuals with DS. Generally, IQ in persons with DS ranges from 30 to 70 and a variety of IQ tests and functional assessments are used to characterize overall cognition and functioning throughout the lifespan (Chapman & Hesketh, 2000; Hamburg et al., 2019). Overall, ID is a well-known feature of individuals with DS, yet there continue to be studies that examine within-syndrome heterogeneity of more specific cognitive abilities (i.e., verbal versus visuomotor cognition; (Channell et al., 2021; Onnivello et al., 2022, 2023)). One recent study of verbal and nonverbal cognitive profiles in 7- to 16-year-olds with DS found three varying profiles of cognitive abilities, one with higher nonverbal than verbal cognition, one with greater verbal compared to nonverbal cognition, and one with high scores on both nonverbal and verbal cognition (Onnivello et al., 2022). Evidence of within-syndrome variability is critical to document to avoid stereotyping of cognitive abilities of those with DS and has important implications for educational supports for these individuals who do not follow the standard behavioral phenotype associated with DS.

A distinct profile of executive functioning abilities has also been well-defined in persons with DS. Most commonly reported are challenges with working memory and cognitive flexibility (Loveall et al., 2017; Tungate & Conners, 2021; Van Deusen et al., 2022). More ambiguity is noted in areas such as inhibitory control (Borella et al., 2013; Carney et al., 2013; Daunhauer et al., 2017; Tungate & Conners, 2021), although it is common for global executive function difficulties to be used to describe the cognitive abilities for this population (Daunhauer et al., 2014). These cognitive features also may impact other aspects of development, as social interactions rely on working memory and shifting in topics throughout a conversation, placing cognition abilities at the forefront of supports and interventions designed for those with DS. Although the majority of research to date on executive functioning in DS focuses on children and adolescents, there is evidence that patterns of challenges with goal-directed planning begin in infancy (Schworer et al., 2020, 2022b).

Language

Language development is another domain that is delayed compared to typically developing peers throughout the lifespan in DS, with challenges beginning in infancy (Berglund et al., 2001). Particular difficulties with expressive language domains of phonology, syntax, and grammar are observed in young children (Harris, 1983; Hilvert & Sterling, 2019; Roberts et al., 2007). Hearing loss, frequent otitis media, and mouth structure and tone are common factors that impact language learning in DS (Abbeduto et al., 2007; Roberts et al., 2007). Before language develops, infants use gestures and nonword sounds to communicate. Gestures have been identified as a relative strength in young children with DS, and this skill is often leveraged in early intervention through the use of sign language (Bird et al., 2000). Production of first words is often delayed in DS as well, with children speaking their first word at 21 months on average (Stoel-Gammon, 2001), with one large study of over 300 children with DS indicating 12% uttered their first word by 12 months and 80% by 24 months (Berglund et al., 2001). Language delays continue throughout early childhood with challenges associated with DS including difficulty with articulation and combining of words into phrases (Abbeduto et al., 2007; Thurman et al., 2022). Similarly to cognitive development, there is marked variability in language phonology, vocabulary, and grammar (Thurman et al., 2022). Most children with DS are referred for speech therapy from birth to encourage the development of early communication, speech sounds, turn-taking, and pointing skills. Speech therapies continue for children with DS to address articulation, grammar, and sentence formation, with several evidence-based treatments available. Strategies employed in speech therapy for individuals with DS often include the use of functional communication, such as picture exchange, sign language, and augmentative and alternative communication devices.

Motor Development

Gross and fine motor development are known to be common areas of relative challenge and are delayed throughout development in DS, starting in infancy (das Neves Cardoso et al., 2015; de Campos et al., 2013; Winders et al., 2019). Difficulties with motor skills in DS can be impacted by reduced size of the cerebellum (Pinter et al., 2001), hypotonia (Lott, 2012), or other pathophysiological processes (Malak et al., 2015). Recovery from surgery (e.g., cardiac repair) in infancy can also delay motor milestones for this population. Growth curves and a schedule of gross motor development have been created for children with DS (Lauteslager et al., 2020; Palisano et al., 2001; Winders et al., 2019). These studies show that children with DS generally sit unsupported between 10 and 22 months, crawl forward between 14 and 25 months, and walk between 26 and 38 months (Lauteslager et al., 2020; Palisano et al., 2001; Winders et al., 2019). Having a schedule of motor development based within DS is beneficial for clinicians and families when evaluating the motor progress of young children with DS. Fine motor skills are also impacted, and infants with DS show less fine motor behaviors when exploring objects compared to typically developing peers (de Campos et al., 2013), and as fine motor skills become more complex, there is greater heterogeneity in mastery of skills in DS (Frank & Esbensen, 2015).

Delays in motor skills can have a cascade effect and lead to continued motor delays. For example, sitting was identified as a predictor of gross motor development in preschool (Locatelli et al., 2021). However, sitting was not a predictor of school-age motor skills, indicating the importance may become less substantial over time and other factors may play a larger role in motor outcomes (Locatelli et al., 2021). Motor development also has implications for several areas of development. In infancy, delays in fine motor development can impact a child's ability to explore their environment and learn about objects (de Campos et al., 2013; Fidler et al., 2019). Gross motor skills can also affect other areas of development, such as walking acquisition, which was correlated with later cognitive and language development in a study of 156 young children with DS less than 4 years old (Yamauchi et al., 2019). Motor skills have also been correlated with adaptive skills (i.e., communication, daily living, and social skills) in children 2–7 years old with DS (Schworer et al., 2022a). Identifying the best ways to support motor development may be an important target for intervention aimed to support a variety of developmental domains within DS. Most children with DS are referred for physical or occupational therapy from birth to address goals related to gross and fine motor development. Adaptations of toys, clothing, and tricycle/bicycle are often used to support motor development of children with DS.

Intervention and Education

With the growing understanding of areas of challenge within DS from the last several decades of research, there is potential to utilize intervention to positively impact the developmental trajectory of cognitive, language, and motor skills in DS. A few interventions are highlighted here, yet many more exist and continue to be developed to improve outcomes for individuals with DS. Working memory is one cognitive domain that has been targeted in training programs for children and adolescents with DS (Conners et al., 2008; Pulina et al., 2015). There remain challenges with the generalizability of these types of training programs, although the Pulina et al.'s study did report transfer of skills to everyday working memory tasks (2015). In a review of 18 speech-language interventions in DS, improvements were found in all studies (Moraleda-Sepúlveda et al., 2022). While it is exciting to see such success with the use of language interventions in this population, there is still work to be done to identify active components of the interventions to be better able to generalize results to clinical practice. Reading interventions have also improved word reading, letter-sound knowledge, and phoneme blending skills directly taught in 20 weeks of intervention for 5- to 10-year-old children with DS (Burgoyne et al., 2012), and there are several other phonics-based reading interventions show success in this population (Baylis & Snowling, 2012; Goetz et al., 2008). Finally, the feasibility of a syndrome-informed facilitated reaching intervention has also been evaluated and shown to increase the speed and frequency of reaching and swatting in infants with DS, with the most promising improvements identified in infants 5 to 10 months old (Fidler et al., 2021). Reaching is a critical skill for early exploration, which has potential to not only impact motor domains, but also cognitive skills longitudinally. Each of these interventions requires additional steps to integrate these research-based findings into service provision and educational experiences to continue to improve upon the lives of children with DS.

Health

Physical Health

Individuals with DS experience certain medical conditions at higher rates than the general population. Because of the specific risk for certain medical conditions, health-care guidelines are available for both children and adults with DS (Bull et al., 2022; Tsou et al., 2020). Not all individuals with DS will experience these medical conditions, but it is important to screen all individuals with DS for conditions of high risk over their lifespan to ensure adequate detection and treatment of these co-occurring health conditions.

In infancy, 44% of individuals with DS are diagnosed with congenital heart defects (CHD) (Freeman et al., 2008). For comparison, only 0.8–0.9% of the general population have CHD (Liu et al., 2019; Reller et al., 2008). These defects are sometimes identified prenatally, and surgery often occurs during the first year of life to repair the incomplete development of heart chambers and valves. Prior to surgical advances for repairing CHD, many children with DS died in childhood due to compromised circulation (Hijii et al., 1997).

Children with DS are screened for obstructive sleep apnea (OSA) in preschool, as it is found in 50–76% of children (Lee et al., 2018; Maris et al., 2016). Individuals with DS are predisposed to a higher rate of OSA given their craniofacial features and hypotonia. While OSA can be treated with positive airway pressure, surgical implants, and surgery (including removal of tonsils and adenoids), ongoing screening is necessary as OSA can reoccur despite intervention (Donnelly et al., 2004).

Individuals with DS also experience high rates of visual and hearing impairments across the lifespan. In addition to near- and far-sightedness (refractive errors and astigmatism), individuals with DS also experience high rates of strabismus (36%; misalignment of the eyes), nystagmus (3–33%; involuntary eye movements), keratoconus (1–13%; thinning of cornea), and cataracts (3%; clouding of lens) (Bull et al., 2022). Conductive hearing loss is common in childhood, often the result of high rates of recurrent middle ear infections (50–70%). Without proper screening across the lifespan, these visual and hearing impairments can go undetected by families and care providers (Van Buggenhout et al., 1999).

Other conditions reported at higher rates in DS include hypothyroidism and gastrointestinal problems. Hypothyroidism is routinely screened for and treated with medication during childhood, and rates of thyroid problems increase with age. Gastrointestinal atresia (narrowing) impacts 12% of children with DS and can impact the small intestine, lower bowel, or the connections to the trachea and esophagus (Freeman et al., 2009).

A compromised autoimmune system contributes to other medical conditions reported at higher rates in individuals with DS. Celiac disease is an immune reaction to gluten and impacts 1–5% of children with DS (Szaflarska-Popławska et al., 2016). Alopecia areata, or hair loss, is an immune reaction that impacts hair follicles and impacts 6% of children with DS (Du Vivier & Munro, 1975). Grave's disease and Hashimoto's thyroiditis are autoimmune diseases that either contribute to an over- or under-production of thyroid hormones, both more common among individuals with DS than their same-age peers (Amr, 2018). While the rate of leukemia in DS is low, affecting only 1% of children with DS, the rate is 20 times that of the rate for their typically developing peers (Ross et al., 2005). Treatment of leukemia often requires lower doses of chemotherapy for children with DS.

Mental Health

In addition to screening for physical health conditions, individuals with DS also warrant screening for mental health conditions, some of which also appear at higher rates among individuals with DS compared to their typically developing peers. Two conditions that appear at higher rates are attention-deficit/hyperactivity disorder (ADHD) and autism spectrum disorder (ASD). The rate of ADHD in youth with DS (44%) is 3–5 times the rate in typically developing peers (Dekker & Koot, 2003; Ekstein et al., 2011; Froehlich et al., 2007). However, rates of medication management are a quarter of what is typical in the general population (Esbensen et al., 2022). The rate of ASD is 16–18% and continues to skew toward individuals with DS at lower levels of functioning in terms of cognition, language, and daily living skills (DiGuiseppi et al., 2010; Richards et al., 2015). However, diagnoses also tend to be made at an older age than is typical in the general population.

Other mental health conditions, such as depression and anxiety, are reported to be less frequent among individuals with DS (Mantry et al., 2008). Making accurate mental health diagnoses among individuals with DS is complicated due to their intellectual disability. There are adaptations to diagnostic criteria to consider when making clinical diagnoses among individuals with intellectual disabilities, which may contribute to diagnostic overshadowing, or the likelihood of misattributing clinical symptoms to intellectual disability (Fletcher et al., 2016).

There are additional complications when accurately diagnosing mental health conditions in individuals with DS. Untreated medical conditions can present similar to symptoms of mental health conditions. For example, lethargy and decreased interest in activities are commonly present in individuals with untreated hypothyroidism and may thus not reflect clinical symptoms of depression. Some behavioral presentations common in DS, such as self-talk, are normative based on challenges with self-monitoring and internalizing thoughts (McGuire & Chicoine, 2006). Yet, these normative behaviors are misjudged to be symptoms of hallucinations or delusions. A newer clinical presentation, called Down syndrome regression disorder, is observed among adolescents and young adults with DS. This disorder involves the loss of cognitive and/or functional skills that can occur suddenly or over a period of weeks to months, and symptoms that mirror depression, delusions, stereotypies, whispering, and catatonia. Causes of this disorder are suspected of being linked to autoimmune functioning or catatonia (Santoro et al., 2020).

Adulthood and Aging

Health and Aging

Individuals with DS continue to experience certain medical conditions at higher rates than the general population into adulthood. Adult health-care guidelines are available for the care of adults with DS, and screening for these conditions of high risk continues to be important (Tsou et al., 2020). This older population of adults with DS highlights the need for aging care for those with DS and health care and adaptive supports throughout the lifespan (Iulita et al., 2022).

Adults with DS experience a high rate of medical conditions relative to the general population (Carfì et al., 2014; Franceschi et al., 2019; Zigman, 2013). It is estimated that 63% of middle-aged adults with DS have hypothyroidism (Aversa et al., 2015; Guaraldi et al., 2017). Hyperthyroidism is also more prevalent in adults with DS than in adults in the general population (Goday-Arno et al., 2009). Adults with DS are also more likely than adults in the general population to have celiac disease (de Asua et al., 2015), diabetes (Alexander et al., 2016), and vitamin D and B12 deficiencies (Stagi et al., 2015). Gout, which is a form of arthritis, is more common in adults with DS compared to adults in the general population (Capone et al., 2018). Declines in bone mineral density have been shown to occur more often in middle-aged adults with versus without DS, and thought to contribute to the increased risk for osteoporosis and bone fractures observed in DS (Capone et al., 2018; Carfì et al., 2014). It is estimated that 8–11% of adults with DS experience atlantoaxial instability, a condition that affects the bones in the upper spine or neck under the base of the skull (de Asua et al., 2015; Tsou et al., 2020). Overweight or obesity is estimated to occur in about 80% of adults with DS (Ptomey et al., 2020). The high incidence (40–50%) of CHD in individuals with DS (Benhaourech et al., 2016) is thought to contribute to an elevated likelihood of pulmonary hypertension in adulthood (Baraona et al., 2013). Hyperlipidemia is also prevalent in DS (Adelekan et al., 2012). While individuals with DS have a lower risk of hypertension and atherosclerosis than adults in the general population (Alexander et al., 2016; Roy-Vallejo et al., 2020), hypotension is observed in adults with DS (Alexander et al., 2016), as well as an increased rate of cardiac valve pathology (Capone et al., 2018; Kennedy et al., 2018), and strokes (Sobey et al., 2015). Further, sensorineural hearing loss related to aging onsets earlier among adults with DS than in the general population.

Adults with DS also experience a high rate of OSA. Across research published to date, OSA is estimated to occur in 40–94% of adults with DS (Tietze et al., 2012; Trois et al., 2009) which is higher than the prevalence of 30–50% reported for adults in the general population in the United States (Franklin & Lindberg, 2015). However, the actual prevalence and severity of sleep-disordered breathing problems in adults with DS is difficult to determine because most studies to date are based on small-sized convenience samples and use informant-report of diagnosed OSA (Tietze et al., 2012; Trois et al., 2009). Yet, many adults with DS may not undergo regular OSA screenings because they typically require an overnight sleep lab study.

During adulthood, people with DS can experience a variety of mental health problems and are more likely to than are adults without DS (Rivelli et al., 2022). Overall, 18–23% of adults with DS are estimated to have a clinically significant mental health condition (Tassé et al., 2016; Vicari et al., 2013). The most common mental health conditions reported for adults with DS include depressive disorders, anxiety disorders (and especially obsessive-compulsive disorder), schizophrenia, anger problems, psychosis and hallucinations, personality disorders, and tic disorders (Dykens et al., 2015; Rivelli et al., 2022; Urv et al., 2010). The likelihood of having mental health problems increases from younger to middle adulthood in people with DS (Tassé et al., 2016).

Adults with DS also experience signs of accelerated aging relative to the general population, including decreased elasticity of skin and thinning, graying, and loss of hair at a younger age than adults without DS. Alopecia that involves hair loss occurs in about 6–18% of individuals with DS and is thought to be due to immunological problems that target hair follicles. The onset of menopause occurs 4–6 years earlier among women with DS than women in the general population (Schupf et al., 1997). This earlier age of menopause has been associated with increasing the risk for heart disease, osteoporosis, and depression in women with DS.

Adults with DS are also at increased risk for Alzheimer's disease and experience an early age of dementia onset relative to that seen in sporadic late-onset Alzheimer's disease (e.g., (Zigman & Lott, 2007). The average age of dementia onset for people with sporadic late-onset Alzheimer's disease (the most common form) is in the late 70s (Breijyeh & Karaman, 2020). In contrast, it is estimated that over half of adults with DS exhibit Alzheimer's dementia by 55 years of age (Rubenstein et al., 2020), and this rises to 88–100% after 65 years of age (Iulita et al., 2022; Mann & Esiri, 1989). The early hallmark features of Alzheimer's brain pathology, including the accumulation of amyloid-beta, are present in people with DS by the age of 40 years (Fortea et al., 2021; Head et al., 2012). Alzheimer's disease has been identified as the proximate cause of death in 70% of cases of adults with DS (Hithersay et al., 2019; McCarron et al., 2017). These findings have led researchers to view trisomy 21 as a genetic form of Alzheimer's disease (Fortea et al., 2021). The triplication of the amyloid precursor protein (APP) gene, which is located on chromosome 21 (e.g., Prasher et al., 1998; Wiseman et al., 2015), is thought to be the cause of the high risk for Alzheimer's disease in DS. This APP gene regulates the expression and proteolytic processing of amyloid-beta.

Large research consortiums and studies (e.g., Alzheimer Biomarker Consortium-Down syndrome, Down Alzheimer Barcelona Neuroimaging Initiative, London Down syndrome Research, and LuMIND International Down syndrome Community) have formed in recent years to understand the decades-long sequence of brain changes that cause Alzheimer's disease in people with DS and to identify feasible and valid measures for screening for early cognitive declines in this community. The goal of this research is to generate the necessary science to open the door for more clinical intervention trials aimed at delaying or preventing Alzheimer's disease in adults with DS in the coming years.

In contrast to the elevated rates of conditions reviewed above, individuals with DS are also at lower risk of other medical conditions than their typically developing peers. Adults with DS experience lower rates of cardiovascular and cerebrovascular disease, coronary heart disease, high cholesterol, and emphysema (Chicoine et al., 2021; Tsou et al., 2020). Lower rates of some of these conditions are related to lower blood pressure and resting heart rates related to suppressor genes found on chromosome 21 (Kerins et al., 2008; Prasher, 1994; Satge et al., 1998). Individuals with DS experience lower rates of solid tumors (Hasle et al., 2000). This lower risk for solid tumor cancers has implications for health screening among adults with DS, where women with DS are not required to have annual mammograms comparable to other middle-aged women without DS (Alagoz et al., 2019).

Employment

Employment activities are an important part of daily life for many adults with DS (Kumin & Schoenbrodt, 2016). These activities include competitive employment positions, meaning full- or part-time employment within the community with compensation at a competitive pay level. A job coach or support person may initially accompany the adult with DS to provide training until the adult with DS is successfully independent. It is estimated that ~16% of adults with DS are involved in competitive employment positions (Bush & Tassé, 2017). Other adults with DS engage in supported employment, which involves a full- or part-time position within the community that is paid but differs from competitive employment in that long-term or ongoing support is provided through a job coach or support person. Sheltered employment, or work center employment, involves work activities done at a center for adults with disabilities. These activities often involve subcontract work (e.g., mailings, packaging, and assembly) and are paid according to how much they can accomplish compared to workers without disabilities. Finally, some adults with DS engage in work activities through volunteer positions in which they are not paid but may receive job training. In a national survey of more than 500 adults with DS, the most common employment activities were described as involving the "five Fs," which included *food* (fast food and kitchen), *filth* (cleaning and janitorial services), *flowers* (florists landscaping teams), *factories* (light assembly line work), and *filing and office mail delivery* (Kumin & Schoenbrodt, 2016).

Other adults with DS engage in post-secondary education and job training programs. Some of these programs involve taking academic courses, gaining technical training, and/or earning certificates through universities or colleges. In some of these programs, adults with DS live in housing on campus, gain meaningful social experiences, and learn life and work skills.

Employment has been associated with several benefits for adults with disabilities, including increases in quality of life (Eggleton et al., 1999). Improvements in quality of life stemming from being employed are often attributed to the structure, social support, and income that is provided by having a job. In addition, engaging in employment activities (regardless of whether these activities are paid or non-paid) may help prevent aging-related cognitive decline in adults with DS. These benefits are thought to be related to the theory of cognitive reserve (Stern, 2006), which posits that cognitive stimulation (e.g., problem-solving, learning new information, and thinking flexibly) helps people tolerate aging-related brain pathology (e.g., decreases in hippocampal volume) because the person can draw on a broader range of preexisting cognitive strategies or use compensatory approaches to mitigate the adverse effects of this pathology. In a study by Piro-Gambetti et al. (2023), after controlling for chronological age, intellectual disability level, and hours spent in employment, adults with DS who were involved in more complex work in terms of *people* (e.g., frequent and diverse social interactions) and *things* (assembling and using objects in specific ways) evidenced less cognitive decline across 16–20 months than did adults with DS involved in no employment activities or less complex activities (Piro-Gambetti et al., 2023). Thus, social policies aimed at increasing opportunities for adults with DS to engage in work activities that allow for social interactions and new learning may be a way to promote healthy cognitive aging.

Efforts to Promote Healthy Aging in Down Syndrome

Given the increased life span and elevated rate of health- and aging-related problems in adults with DS, there has been a push to identify policies and interventions to support healthy aging in this population. For example, a growing body of work has focused on evaluating the benefits of physical activity lifestyles in adults with DS. Adults with intellectual and developmental disabilities, including those with DS, have been reported to engage in less overall physical activity as compared to the general population of adults (Cowley et al., 2010; Izquierdo-Gomez et al., 2014). Studies have also reported that, on average, individuals with DS spend more time in sedentary behavior—sitting or lying around—than do adults without DS (Gabel et al., 2016; Ramirez-Rico et al., 2014). In part, adults with DS may engage in low levels of physical activity due to problems with chronotropic incompetence, low muscular strength, and muscle hypotonia that are associated with having three copies of chromosome 21 (Phillips & Holland, 2011). However, in part, the high rate of sedentary behavior and low rate of moderate-to-vigorous physical activity by adults with DS may be due to differences in adult lives—e.g., lower rate of employment, lower income, transportation difficulties, lack of social inclusion, and reliance on programs offered in the community.

The benefits of engaging in physical activity on the health of adults with DS are clear in cross-sectional studies. Indeed, within these studies, spending more time in moderate-to-vigorous intensity physical activity was associated with a lower likelihood of having OSA, endocrine/metabolic conditions, and anxiety disorders and having a lower number of overall physical health problems in adults with DS (Fleming et al., 2022). The benefits of engaging in more physical activity appeared to be, in part, driven by having a lower body mass index (BMI) within the study. However, engaging in more physical activity and less sedentary behavior continued to have associations with health outcomes after controlling for BMI, suggesting that physical activity can also improve health by strengthening muscles, altering the distribution of fat, and regulating hormones and metabolism in beneficial ways.

Programs and social policies aimed at promoting regular physical activity could be a way to improve overall health of adults with DS. Research has shown that adults with intellectual disability who participate in Special Olympics have fewer trips to healthcare professionals and few medical claims than do their non-athlete counterparts (Al-Heizan et al., 2021). Moreover, results from a 12-week exercise intervention conducted with adolescents and young adults with DS found that participation in the intervention leads to short-term improvements in cognitive functioning, and longer-term effects are yet to be explored (Ptomey et al., 2018, 2020).

Family

The family system has an impact on the development and well-being of the individual with DS. Similarly, a diagnosis of DS has an impact on the family system. These reciprocal influences change across the lifespan.

Related to the diagnosis of DS, it is important how information is communicated and how families are supported when receiving this information. These key factors can impact how a diagnosis is received and parents' decisions regarding options for continuation, termination, or adoption that follow. Diagnoses of DS can occur during prenatal screening or postnatally. Although *when* the diagnosis is shared with families (pre- or postnatally) has minimal influence on parental well-being, *how* the diagnosis is presented to caregivers and families is reported to have a substantial influence on subsequent parental well-being and adjustment to the diagnosis (Skotko, 2005b; Skotko & Bedia, 2005). There is evidence that the communication style and sensitivity of the physician's language impact parents' coping when receiving a DS diagnosis for their child (Muggli et al., 2009; Skotko et al., 2009). Knowledge of the person delivering the diagnosis and ability to connect families with DS support groups also are critical factors for families (Skotko et al., 2009). Including information about the possibilities for children and adults with DS in relation to functioning and stories from mothers of children with DS are identified as important to include in initial conversations with families (Skotko, 2005b). By informing parents of a diagnosis and including positive aspects of raising a child with DS, the negative impact on parental well-being can be minimized (Skotko, 2005a; Skotko & Bedia, 2005). Supporting families with sufficient information to prevent fear-based decision-making is critical in initial communications of a diagnosis and can have lasting implications for coping and emotional recovery after a child is diagnosed with DS (Muggli et al., 2009).

Parental well-being is heavily influenced by behavior problems exhibited by their child with an intellectual disability, including diagnoses of DS, and this relationship is bidirectional (Neece et al., 2012). Parents report more difficulties with parenting and strain on the marital relationship when the child is exhibiting more frequent behavioral challenges (Cahill & Glidden, 1996). Parents also report more challenges in parenting a child or adolescent with DS in comparison to a typically developing child or adolescent, yet fewer challenges in comparison to other children and adolescents with intellectual disabilities (Abbeduto et al., 2004; Esbensen & Seltzer, 2011; Fidler et al., 2000; Hauser-Cram et al., 2001; Skotko et al., 2011). This pattern of better well-being among parents of children with DS than parents of children with other developmental disabilities also persists into the adulthood of the individual with the intellectual disability (Corrice & Glidden, 2009).

Community Engagement

With the growth of medical and behavioral supports, individuals with DS are able to engage in meaningful ways through school, their communities, and work, and continued advances in understanding the abilities of persons with DS will allow this population to engage and thrive in their everyday activities. Individuals with DS continue to benefit from advances in health care and therapies, educational inclusion, vocational opportunities, community inclusion, and increased research knowledge. National organizations and self-advocates are working together through NIH-led DS Consortiums and other avenues to support advances in research. Among these larger-scale research and policy efforts, there are numerous methods through which undergraduate students can contribute to improving outcomes for individuals with DS.

Undergraduate students can volunteer with their local DS organization or other programs supporting individuals with intellectual and developmental disabilities. Many universities and colleges will have a Best Buddies program, a nonprofit organization that supports friendships, employment, and inclusive living for individuals with intellectual and developmental disabilities. Many regions have Special Olympic programs and/or recreational sports programs for individuals with developmental disabilities. All states have an equivalent of a DS association assisting and creating community among families of and individuals with DS.

Students interested in government can partner with a local or national DS organization to lobby on behalf of programs or supports that promote optimal quality of life for individuals with DS. Students interested in business can learn more about the opportunities for individuals with intellectual disabilities within the workforce as valued employees to increase inclusion in vocational settings. Students could also consider specializing in the medical field (such as becoming a developmental-behavioral pediatrician, internal medicine with a focus on intellectual disabilities, genetic counselor) or within allied health fields (such as becoming a behavior analyst, psychologist, speech therapist, occupational therapist, or physical therapist focused on DS). Students interested in DS can also consider research careers in basic, behavioral, and clinical sciences. It can also include becoming a specialist in financial planning for families of individuals with DS and other types of intellectual disabilities, a special education teacher, social worker, or working with an agency supporting individuals with DS or other types of intellectual disabilities. Most importantly, all students can understand that individuals with DS have much in common with them regarding life goals of health, happiness, work, friends, and family.

References

Abbeduto, L., Seltzer, M. M., Shattuck, P., Krauss, M. W., Orsmond, G., & Murphy, M. M. (2004). Psychological well-being and coping in mothers of youths with autism, Down syndrome, or Fragile X syndrome. *American Journal on Mental Retardation, 109*(3), 237–254.

Abbeduto, L., Warren, S. F., & Conners, F. A. (2007). Language development in Down syndrome: From the prelinguistic period to the acquisition of literacy. *Mental Retardation and Developmental Disabilities Research Reviews, 13*(3), 247–261.

Adelekan, T., Magge, S., Shults, J., Stallings, V., & Stettler, N. (2012). Lipid profiles of children with Down syndrome compared with their siblings. *Pediatrics, 129*(6), e1382–e1387.

Alagoz, O., Hajjar, A., Chootipongchaivat, S., van Ravesteyn, N. T., Yeh, J. M., Ergun, M. A., de Koning, H. J., Chicoine, B., & Martin, B. (2019). Benefits and harms of mammography screening for women with Down syndrome: A collaborative modeling study. *Journal of General Internal Medicine, 34*, 2374–2381.

Alexander, M., Petri, H., Ding, Y., Wandel, C., Khwaja, O., & Foskett, N. (2016). Morbidity and medication in a large population of individuals with Down syndrome compared to the general population. *Developmental Medicine & Child Neurology, 58*(3), 246–254.

Al-Heizan, M. O., St. John, B., DuBois, L., Niemeier, B., & Ausderau, K. K. (2021). Difference in healthcare usage, pharmacy prescriptions, and healthcare cost for special Olympic athletes compared to non-athletes with intellectual disabilities. *Journal of Policy and Practice in Intellectual Disabilities, 18*(2), 141–150.

Amr, N. H. (2018). Thyroid disorders in subjects with Down syndrome: An update. *Acta Bio Medica: Atenei Parmensis, 89*(1), 132.

Aversa, T., Salerno, M., Radetti, G., Faienza, M. F., Iughetti, L., Corrias, A., Predieri, B., Mussa, A., Mirabelli, S., & De Luca, F. (2015). Peculiarities of presentation and evolution over time of Hashimoto's thyroiditis in children and adolescents with Down's syndrome. *Hormones, 14*, 410–416.

Baburamani, A. A., Patkee, P. A., Arichi, T., & Rutherford, M. A. (2019). New approaches to studying early brain development in Down syndrome. *Developmental Medicine & Child Neurology, 61*(8), 867–879.

Baraona, F., Gurvitz, M., Landzberg, M. J., & Opotowsky, A. R. (2013). Hospitalizations and mortality in the United States for adults with Down syndrome and congenital heart disease. *The American Journal of Cardiology, 111*(7), 1046–1051.

Baylis, P., & Snowling, M. J. (2012). Evaluation of a phonological reading programme for children with Down syndrome. *Child Language Teaching and Therapy, 28*(1), 39–56.

Becker, L., Armstrong, D., & Chan, F. (1986). Dendritic atrophy in children with Down's syndrome. *Annals of Neurology: Official Journal of the American Neurological Association and the Child Neurology Society, 20*(4), 520–526.

Benhaourech, S., Drighil, A., & Hammiri, A. E. (2016). Congenital heart disease and Down syndrome: Various aspects of a confirmed association. *Cardiovascular Journal of Africa, 27*(5), 287–290.

Berglund, E., Eriksson, M. R., & Johansson, I. N. (2001). *Parental reports of spoken language skills in children with Down syndrome.*

Bird, E. K.-R., Gaskell, A., Babineau, M. D., & Macdonald, S. (2000). Novel word acquisition in children with Down syndrome: Does modality make a difference? *Journal of Communication Disorders, 33*(3), 241–266.

Bittles, A., & Glasson, E. (2004). Clinical, social, and ethical implications of changing life expectancy in Down syndrome. *Developmental Medicine & Child Neurology, 46*(4), 282–286.

Borella, E., Carretti, B., & Lanfranchi, S. (2013). Inhibitory mechanisms in Down syndrome: Is there a specific or general deficit? *Research in Developmental Disabilities, 34*(1), 65–71.

Breijyeh, Z., & Karaman, R. (2020). Comprehensive review on Alzheimer's disease: Causes and treatment. *Molecules, 25*(24), 5789.

Brown, R. I., Kyrkou, M. R., Watchman, K., & Hodapp, R. M. (2020). *A brief history of Down syndrome: Key moments and reflections*.

Bull, M. J. (2020). Down syndrome. *New England Journal of Medicine, 382*(24), 2344–2352.

Bull, M. J., Trotter, T., Santoro, S. L., Christensen, C., Grout, R. W., Genetics, C., & o. (2022). Health supervision for children and adolescents with Down syndrome. *Pediatrics, 149*(5), e2022057010.

Burack, J. A., Edgin, J., & Abbeduto, L. (2022). *The Oxford handbook of Down syndrome and development*. Oxford University Press.

Burgoyne, K., Duff, F. J., Clarke, P. J., Buckley, S., Snowling, M. J., & Hulme, C. (2012). Efficacy of a reading and language intervention for children with Down syndrome: A randomized controlled trial. *Journal of Child Psychology and Psychiatry, 53*(10), 1044–1053.

Bush, K. L., & Tassé, M. J. (2017). Employment and choice-making for adults with intellectual disability, autism, and down syndrome. *Research in Developmental Disabilities, 65*, 23–34.

Cahill, B. M., & Glidden, L. M. (1996). Influence of child diagnosis on family and parental functioning: Down syndrome versus other disabilities. *American Journal of Mental Retardation: AJMR, 101*(2), 149–160.

Capone, G. T., Chicoine, B., Bulova, P., Stephens, M., Hart, S., Crissman, B., Videlefsky, A., Myers, K., Roizen, N., & Esbensen, A. (2018). Co-occurring medical conditions in adults with Down syndrome: A systematic review toward the development of health care guidelines. *American Journal of Medical Genetics Part A, 176*(1), 116–133.

Carfì, A., Antocicco, M., Brandi, V., Cipriani, C., Fiore, F., Mascia, D., Settanni, S., Vetrano, D. L., Bernabei, R., & Onder, G. (2014). Characteristics of adults with down syndrome: Prevalence of age-related conditions. *Frontiers in Medicine, 1*, 51.

Carney, D. P., Brown, J. H., & Henry, L. A. (2013). Executive function in Williams and Down syndromes. *Research in Developmental Disabilities, 34*(1), 46–55.

Channell, M. M., Mattie, L. J., Hamilton, D. R., Capone, G. T., Mahone, E. M., Sherman, S. L., Rosser, T. C., Reeves, R. H., Kalb, L. G., & Project, D. S. C. (2021). Capturing cognitive and behavioral variability among individuals with Down syndrome: A latent profile analysis. *Journal of Neurodevelopmental Disorders, 13*, 1–14.

Chapman, R. S., & Hesketh, L. J. (2000). Behavioral phenotype of individuals with Down syndrome. *Mental Retardation and Developmental Disabilities Research Reviews, 6*(2), 84–95.

Chicoine, B., Rivelli, A., Fitzpatrick, V., Chicoine, L., Jia, G., & Rzhetsky, A. (2021). Prevalence of common disease conditions in a large cohort of individuals with Down syndrome in the United States. *Journal of Patient-Centered Research and Reviews, 8*(2), 86.

Conners, F., Rosenquist, C., Arnett, L., Moore, M., & Hume, L. (2008). Improving memory span in children with Down syndrome. *Journal of Intellectual Disability Research, 52*(3), 244–255.

Coppus, A. M., Evenhuis, H. M., Verberne, G. J., Visser, F. E., Oostra, B. A., Eikelenboom, P., Van Gool, W. A., Janssens, A. C. J., & Van Duijn, C. M. (2008). Survival in elderly persons with Down syndrome. *Journal of the American Geriatrics Society, 56*(12), 2311–2316.

Corrice, A. M., & Glidden, L. M. (2009). The Down syndrome advantage: Fact or fiction? *American Journal on Intellectual and Developmental Disabilities, 114*(4), 254–268.

Cowley, P. M., Ploutz-Snyder, L. L., Baynard, T., Heffernan, K., Jae, S. Y., Hsu, S., Lee, M., Pitetti, K. H., Reiman, M. P., & Fernhall, B. (2010). Physical fitness predicts functional tasks in individuals with Down syndrome. *Medicine and Science in Sports and Exercise, 42*(2), 388–393.

das Neves Cardoso, A. C., de Campos, A. C., Dos Santos, M. M., Santos, D. C. C., & Rocha, N. A. C. F. (2015). Motor performance of children with Down syndrome and typical development at 2 to 4 and 26 months. *Pediatric Physical Therapy, 27*(2), 135–141.

Daunhauer, L. A., Fidler, D. J., Hahn, L., Will, E., Lee, N. R., & Hepburn, S. (2014). Profiles of everyday executive functioning in young children with Down syndrome. *American Journal on Intellectual and Developmental Disabilities, 119*(4), 303–318.

Daunhauer, L. A., Gerlach-McDonald, B., Will, E., & Fidler, D. (2017). Performance and ratings based measures of executive function in school-aged children with Down syndrome. *Developmental Neuropsychology, 42*(6), 351–368.

de Asua, D. R., Quero, M., Moldenhauer, F., & Suarez, C. (2015). Clinical profile and main comorbidities of Spanish adults with Down syndrome. *European Journal of Internal Medicine, 26*(6), 385–391.

de Campos, A. C., da Costa, C. S. N., Savelsbergh, G. J., & Rocha, N. A. C. F. (2013). Infants with Down syndrome and their interactions with objects: Development of exploratory actions after reaching onset. *Research in Developmental Disabilities, 34*(6), 1906–1916.

de Graaf, G., Buckley, F., Dever, J., & Skotko, B. G. (2017a). Estimation of live birth and population prevalence of Down syndrome in nine US states. *American Journal of Medical Genetics Part A, 173*(10), 2710–2719.

de Graaf, G., Buckley, F., & Skotko, B. G. (2017b). Estimation of the number of people with Down syndrome in the United States. *Genetics in Medicine, 19*(4), 439–447.

Dekker, M. C., & Koot, H. M. (2003). DSM-IV disorders in children with borderline to moderate intellectual disability. I: Prevalence and impact. *Journal of the American Academy of Child & Adolescent Psychiatry, 42*(8), 915–922.

DiGuiseppi, C., Hepburn, S., Davis, J. M., Fidler, D. J., Hartway, S., Lee, N. R., Miller, L., Ruttenber, M., & Robinson, C. (2010). Screening for autism spectrum disorders in children with Down syndrome: Population prevalence and screening test characteristics. *Journal of Developmental and Behavioral Pediatrics, 31*(3), 181.

Donnelly, L. F., Shott, S. R., LaRose, C. R., Chini, B. A., & Amin, R. S. (2004). Causes of persistent obstructive sleep apnea despite previous tonsillectomy and adenoidectomy in children with down syndrome as depicted on static and dynamic cine MRI. *American Journal of Roentgenology, 183*(1), 175–181.

Down, J. L. H. (1866). Observations on an ethnic classification of idiots. *London Hospital Reports, 3*(1866), 259–262.

Du Vivier, A., & Munro, D. D. (1975). Alopecia areata, autoimmunity, and Down's syndrome. *British Medical Journal, 1*(5951), 191–192.

Dykens, E., Shah, B., Davis, B., Baker, C., Fife, T., & Fitzpatrick, J. (2015). Psychiatric disorders in adolescents and young adults with Down syndrome and other intellectual disabilities. *Journal of Neurodevelopmental Disorders, 7*(1), 1–8.

Eggleton, I., Robertson, S., Ryan, J., & Kober, R. (1999). The impact of employment on the quality of life of people with an intellectual disability. *Journal of Vocational Rehabilitation, 13*(2), 95–107.

Ekstein, S., Glick, B., Weill, M., Kay, B., & Berger, I. (2011). Down syndrome and attention-deficit/hyperactivity disorder. *Journal of Child Neurology, 26*, 1290–1295.

Esbensen, A. J., & Seltzer, M. M. (2011). Accounting for the "Down syndrome advantage". *American Journal of Intellectual and Developmental Disabilities, 116*(1), 3–15.

Esbensen, A., Vincent, L., Epstein, J., Kamimura-Nishimura, K., Wiley, S., Angkustsiri, K., Abbeduto, L., Fidler, D., Anixt, J., & Froehlich, T. (2022). Co-occurring medical and behavioural conditions in children with Down syndrome with or without ADHD symptom presentation. *Journal of Intellectual Disability Research, 66*(3), 282–296.

Fidler, D. J. (2005). The emerging Down syndrome behavioral phenotype in early childhood: Implications for practice. *Infants & Young Children, 18*(2), 86–103.

Fidler, D., Hodapp, R., & Dykens, E. (2000). Stress in families of young children with Down syndrome, Williams syndrome, and Smith-Magenis syndrome. *Early Education and Development, 11*(4), 395–406.

Fidler, D. J., Needham, A. W., & Schworer, E. (2019). Infant foundations in Down syndrome: Early constraints on cognitive skill acquisition. *International Review of Research in Developmental Disabilities, 56*, 41–65.

Fidler, D., Schworer, E., Needham, A., Prince, M., Patel, L., Will, E., & Daunhauer, L. (2021). Feasibility of a syndrome-informed micro-intervention for infants with Down syndrome. *Journal of Intellectual Disability Research, 65*(4), 320–339.

Fleming, V., Piro-Gambetti, B., Handen, B., Christian, B. T., Cohen, A., Tudorascu, D., Plante, D. T., Okonkwo, O., & Hartley, S. L. (2022). Physical activity and physical and mental health in middle-aged adults with Down syndrome. *Journal of Policy and Practice in Intellectual Disabilities, 19*(4), 408–418.

Fletcher, R., Barnhill, J., & Cooper, S. A. (2016). *Diagnostic manual, intellectual disability (DM-ID-2): A textbook of diagnosis of mental disorders in persons with intellectual disability.* NADD Press.

Fortea, J., Zaman, S. H., Hartley, S., Rafii, M. S., Head, E., & Carmona-Iragui, M. (2021). Alzheimer's disease associated with Down syndrome: A genetic form of dementia. *The Lancet Neurology, 20*(11), 930–942.

Franceschi, C., Garagnani, P., Gensous, N., Bacalini, M. G., Conte, M., & Salvioli, S. (2019). Accelerated bio-cognitive aging in Down syndrome: State of the art and possible deceleration strategies. *Aging Cell, 18*(3), e12903.

Frank, K., & Esbensen, A. (2015). Fine motor and self-care milestones for individuals with Down syndrome using a retrospective chart review. *Journal of Intellectual Disability Research, 59*, 719–729.

Franklin, K. A., & Lindberg, E. (2015). Obstructive sleep apnea is a common disorder in the population—A review on the epidemiology of sleep apnea. *Journal of Thoracic Disease, 7*(8), 1311.

Freeman, S. B., Bean, L. H., Allen, E. G., Tinker, S. W., Locke, A. E., Druschel, C., Hobbs, C. A., Romitti, P. A., Royle, M. H., & Torfs, C. P. (2008). Ethnicity, sex, and the incidence of congenital heart defects: A report from the National Down Syndrome Project. *Genetics in Medicine, 10*(3), 173–180.

Freeman, S., Torfs, C., Romitti, P., Royle, M., Druschel, C., Hobbs, C., & Sherman, S. (2009). Congenital gastrointestinal defects in Down syndrome: A report from the Atlanta and National Down Syndrome Projects. *Clinical Genetics, 75*(2), 180–184.

Froehlich, T. E., Lanphear, B. P., Epstein, J. N., Barbaresi, W. J., Katusic, S. K., & Kahn, R. S. (2007). Prevalence, recognition, and treatment of attention-deficit/hyperactivity disorder in a national sample of US children. *Archives of Pediatrics & Adolescent Medicine, 161*(9), 857–864.

Gabel, L., Ridgers, N. D., Della Gatta, P., Arundell, L., Cerin, E., Robinson, S., Daly, R., Dunstan, D. W., & Salmon, J. (2016). Associations of sedentary time patterns and TV viewing time with inflammatory and endothelial function biomarkers in children. *Pediatric Obesity, 11*(3), 194–201.

Goday-Arno, A., Cerda-Esteva, M., Flores-Le-Roux, J. A., Chillaron-Jordan, J. J., Corretger, J. M., & Cano-Pérez, J. F. (2009). Hyperthyroidism in a population with Down syndrome (DS). *Clinical Endocrinology, 71*(1), 110–114.

Goetz, K., Hulme, C., Brigstocke, S., Carroll, J. M., Nasir, L., & Snowling, M. (2008). Training reading and phoneme awareness skills in children with Down syndrome. *Reading and Writing, 21*, 395–412.

Guaraldi, F., Giaccherino, R. R., Lanfranco, F., Motta, G., Gori, D., Arvat, E., Ghigo, E., & Giordano, R. (2017). Endocrine autoimmunity in Down's syndrome. *Endocrine Immunology, 48*, 133–146.

Hamburg, S., Lowe, B., Startin, C. M., Padilla, C., Coppus, A., Silverman, W., Fortea, J., Zaman, S., Head, E., & Handen, B. L. (2019). Assessing general cognitive and adaptive abilities in adults with Down syndrome: A systematic review. *Journal of Neurodevelopmental Disorders, 11*(1), 1–16.

Hamner, T., Udhnani, M. D., Osipowicz, K. Z., & Lee, N. R. (2018). Pediatric brain development in Down syndrome: A field in its infancy. *Journal of the International Neuropsychological Society, 24*(9), 966–976.

Harris, J. (1983). What does mean length of utterance mean? Evidence from a comparative study of normal and Down's syndrome children. *British Journal of Disorders of Communication, 18*(3), 153–169.

Hasle, H., Clemmensen, I. H., & Mikkelsen, M. (2000). Risks of leukaemia and solid tumours in individuals with Down's syndrome. *The Lancet, 355*(9199), 165–169.

Hauser-Cram, P., Warfield, M. E., Shonkoff, J. P., & Krauss, M. W. (2001). Children with disabilities: A longitudinal study of child development and parent well-being. *Monographs of the Society for Research in Child Development, 66*(3), 1–126.

Head, E., Powell, D., Gold, B. T., & Schmitt, F. A. (2012). Alzheimer's disease in Down syndrome. *European Journal of Neurodegenerative Disease, 1*(3), 353.

Hijii, T., Fukushige, J., Igarashi, H., Takahashi, N., & Ueda, K. (1997). Life expectancy and social adaptation in individuals with Down syndrome with and without surgery for congenital heart disease. *Clinical Pediatrics, 36*(6), 327–332.

Hilvert, E., & Sterling, A. (2019). Vocabulary and grammar development in children with autism spectrum disorder, fragile X syndrome, and Down syndrome. In *International review of research in developmental disabilities* (Vol. 57, pp. 119–169). Elsevier.

Hithersay, R., Startin, C. M., Hamburg, S., Mok, K. Y., Hardy, J., Fisher, E. M., Tybulewicz, V. L., Nizetic, D., & Strydom, A. (2019). Association of dementia with mortality among adults with Down syndrome older than 35 years. *JAMA Neurology, 76*(2), 152–160.

Iulita, M. F., Chavez, D. G., Christensen, M. K., Tamayo, N. V., Plana-Ripoll, O., Rasmussen, S. A., Figuls, M. R., Alcolea, D., Videla, L., & Barroeta, I. (2022). Association of Alzheimer disease with life expectancy in people with Down syndrome. *JAMA Network Open, 5*(5), e2212910–e2212910.

Izquierdo-Gomez, R., Martínez-Gómez, D., Acha, A., Veiga, O. L., Villagra, A., Diaz-Cueto, M., & UP, & DOWN study. (2014). Objective assessment of sedentary time and physical activity throughout the week in adolescents with Down syndrome. The UP&DOWN study. *Research in Developmental Disabilities, 35*(2), 482–489.

Kennedy, J., Devlin, P., Wilson, C., & McGlinchey, P. (2018). Sinoatrial node disease in adults with Down's syndrome. *The Ulster Medical Journal, 87*(1), 37.

Kerins, G., Petrovic, K., Bruder, M., & Gruman, C. (2008). Medical conditions and medication use in adults with Down syndrome: A descriptive analysis. *Down Syndrome Research and Practice, 12*(2), 141–147.

Kumin, L., & Schoenbrodt, L. (2016). Employment in adults with Down syndrome in the United States: Results from a national survey. *Journal of Applied Research in Intellectual Disabilities, 29*(4), 330–345.

Lauteslager, P. E., Lauteslager, T., Van den Heuvel, M. E., Jongerling, J., & Klugkist, I. G. (2020). Basic motor skills of children with Down syndrome: Creating a motor growth curve. *Pediatric Physical Therapy, 32*(4), 375–380.

Lee, C.-F., Lee, C.-H., Hsueh, W.-Y., Lin, M.-T., & Kang, K.-T. (2018). Prevalence of obstructive sleep apnea in children with Down syndrome: A meta-analysis. *Journal of Clinical Sleep Medicine, 14*(5), 867–875.

Liu, Y., Chen, S., Zühlke, L., Black, G. C., Choy, M.-K., Li, N., & Keavney, B. D. (2019). Global birth prevalence of congenital heart defects 1970–2017: Updated systematic review and meta-analysis of 260 studies. *International Journal of Epidemiology, 48*(2), 455–463.

Locatelli, C., Onnivello, S., Antonaros, F., Feliciello, A., Filoni, S., Rossi, S., Pulina, F., Marcolin, C., Vianello, R., & Toffalini, E. (2021). Is the age of developmental milestones a predictor for future development in down syndrome? *Brain Sciences, 11*(5), 655.

Lott, I. T. (2012). Neurological phenotypes for Down syndrome across the life span. *Progress in Brain Research, 197*, 101–121.

Lott, I. T., & Dierssen, M. (2010). Cognitive deficits and associated neurological complications in individuals with Down's syndrome. *The Lancet Neurology, 9*(6), 623–633.

Loveall, S., Conners, F., Tungate, A., Hahn, L., & Osso, T. (2017). A cross-sectional analysis of executive function in Down syndrome from 2 to 35 years. *Journal of Intellectual Disability Research, 61*(9), 877–887.

Malak, R., Kostiukow, A., Krawczyk-Wasielewska, A., Mojs, E., & Samborski, W. (2015). Delays in motor development in children with Down syndrome. *Medical Science Monitor: International Medical Journal of Experimental and Clinical Research, 21*, 1904.

Mann, D. M., & Esiri, M. M. (1989). The pattern of acquisition of plaques and tangles in the brains of patients under 50 years of age with Down's syndrome. *Journal of the Neurological Sciences, 89*(2–3), 169–179.

Mantry, D., Cooper, S. A., Smiley, E., Morrison, J., Allan, L., Williamson, A., Finlayson, J., & Jackson, A. (2008). The prevalence and incidence of mental ill-health in adults with Down syndrome. *Journal of Intellectual Disability Research, 52*(2), 141–155.

Maris, M., Verhulst, S., Wojciechowski, M., Van de Heyning, P., & Boudewyns, A. (2016). Prevalence of obstructive sleep apnea in children with Down syndrome. *Sleep, 39*(3), 699–704.

McCarron, M., McCallion, P., Reilly, E., Dunne, P., Carroll, R., & Mulryan, N. (2017). A prospective 20-year longitudinal follow-up of dementia in persons with Down syndrome. *Journal of Intellectual Disability Research, 61*(9), 843–852.

McGuire, D., & Chicoine, B. (2006). *Mental wellness in adults with Down syndrome*. Woodbine.

Menghini, D., Costanzo, F., & Vicari, S. (2011). Relationship between brain and cognitive processes in Down syndrome. *Behavior Genetics, 41*, 381–393.

Moraleda-Sepúlveda, E., López-Resa, P., Pulido-García, N., Delgado-Matute, S., & Simón-Medina, N. (2022). Language intervention in Down syndrome: A systematic literature review. *International Journal of Environmental Research and Public Health, 19*(10), 6043.

Morris, J., Wald, N., & Watt, H. (1999). Fetal loss in Down syndrome pregnancies. *Prenatal Diagnosis: Published in Affiliation with the International Society for Prenatal Diagnosis, 19*(2), 142–145.

Muggli, E. E., Collins, V. R., & Marraffa, C. (2009). Going Down a different road: First support and information needs of families with a baby with Down syndrome. *Medical Journal of Australia, 190*(2), 58–61.

Natoli, J. L., Ackerman, D. L., McDermott, S., & Edwards, J. G. (2012). Prenatal diagnosis of Down syndrome: A systematic review of termination rates (1995–2011). *Prenatal Diagnosis, 32*(2), 142–153.

Neece, C. L., Green, S. A., & Baker, B. L. (2012). Parenting stress and child behavior problems: A transactional relationship across time. *American Journal on Intellectual and Developmental Disabilities, 117*(1), 48–66.

Onnivello, S., Pulina, F., Locatelli, C., Marcolin, C., Ramacieri, G., Antonaros, F., Vione, B., Caracausi, M., & Lanfranchi, S. (2022). Cognitive profiles in children and adolescents with Down syndrome. *Scientific Reports, 12*(1), 1936.

Onnivello, S., Schworer, E., Prince, M., Daunhauer, L., & Fidler, D. (2023). Early developmental profiles among infants with Down syndrome. *Journal of Intellectual Disability Research, 67*(3), 228–238.

Palisano, R. J., Walter, S. D., Russell, D. J., Rosenbaum, P. L., Gémus, M., Galuppi, B. E., & Cunningham, L. (2001). Gross motor function of children with Down syndrome: Creation of motor growth curves. *Archives of Physical Medicine and Rehabilitation, 82*(4), 494–500.

Parker, S. E., Mai, C. T., Canfield, M. A., Richard, R., Wang, Y., Meyer, R. E., Anderson, P., Mason, C. A., Collins, J. S., Kirby, R. S., & Correa, A. (2010). Updated National Birth Prevalence estimates for selected birth defects in the United States, 2004-2006. *Birth Defects Research Part A: Clinical and Molecular Teratology, 88*, 1008–1016.

Patterson, D., & Costa, A. C. (2005). Down syndrome and genetics—A case of linked histories. *Nature Reviews Genetics, 6*(2), 137–147.

Penrose, L. S., & Smith, G. F. (1966). *Down's anomaly*.

Phillips, A. C., & Holland, A. J. (2011). Assessment of objectively measured physical activity levels in individuals with intellectual disabilities with and without Down's syndrome. *PLoS One, 6*(12), e28618.

Pinter, J. D., Eliez, S., Schmitt, J. E., Capone, G. T., & Reiss, A. L. (2001). Neuroanatomy of Down's syndrome: A high-resolution MRI study. *American Journal of Psychiatry, 158*(10), 1659–1665.

Piro-Gambetti, B., Schworer, E. K., Handen, B., Glukhovskaya, M., & Hartley, S. L. (2023). Does employment complexity promote healthy cognitive aging in Down syndrome? *Journal of Intellectual Disabilities*. https://doi.org/10.1177/17446295231169379

Prasher, V. (1994). Screening of medical problems in adults with Down syndrome. *Down Syndrome Research and Practice, 2*(2), 59–66.

Prasher, V., Farrer, M. J., Kessling, A. M., Fisher, E., West, R., Barber, P., & Butler, A. (1998). Molecular mapping of Alzheimer-type dementia in Down's syndrome. *Annals of Neurology, 43*(3), 380–383.

Prost, E., & Roberts, N. (2013). Downs: The history of a disability. *Journal of the Canadian Academy of Child and Adolescent Psychiatry, 22*(2), 180.

Ptomey, L. T., Szabo, A. N., Willis, E. A., Gorczyca, A. M., Greene, J. L., Danon, J. C., & Donnelly, J. E. (2018). Changes in cognitive function after a 12-week exercise intervention in adults with Down syndrome. *Disability and Health Journal, 11*(3), 486–490.

Ptomey, L., Walpitage, D., Mohseni, M., Dreyer Gillette, M., Davis, A. M., Forseth, B., Dean, E., & Waitman, L. (2020). Weight status and associated comorbidities in children and adults with Down syndrome, autism spectrum disorder and intellectual and developmental disabilities. *Journal of Intellectual Disability Research, 64*(9), 725–737.

Pulina, F., Carretti, B., Lanfranchi, S., & Mammarella, I. C. (2015). Improving spatial-simultaneous working memory in Down syndrome: Effect of a training program led by parents instead of an expert. *Frontiers in Psychology, 6*, 1265.

Ramirez-Rico, E., Hilland, T. A., Foweather, L., Fernandez-Garcia, E., & Fairclough, S. J. (2014). Weekday and weekend patterns of physical activity and sedentary time among Liverpool and Madrid youth. *European Journal of Sport Science, 14*(3), 287–293.

Reller, M. D., Strickland, M. J., Riehle-Colarusso, T., Mahle, W. T., & Correa, A. (2008). Prevalence of congenital heart defects in metropolitan Atlanta, 1998-2005. *The Journal of Pediatrics, 153*(6), 807–813.

Richards, C., Jones, C., Groves, L., Moss, J., & Oliver, C. (2015). Prevalence of autism spectrum disorder phenomenology in genetic disorders: A systematic review and meta-analysis. *The Lancet Psychiatry, 2*(10), 909–916.

Rivelli, A., Fitzpatrick, V., Chaudhari, S., Chicoine, L., Jia, G., Rzhetsky, A., & Chicoine, B. (2022). Prevalence of mental health conditions among 6078 individuals with Down syndrome in the United States. *Journal of Patient-Centered Research and Reviews, 9*(1), 58.

Roberts, J. E., Price, J., & Malkin, C. (2007). Language and communication development in Down syndrome. *Mental Retardation and Developmental Disabilities Research Reviews, 13*(1), 26–35.

Ross, J. A., Spector, L. G., Robison, L. L., & Olshan, A. F. (2005). Epidemiology of leukemia in children with Down syndrome. *Pediatric Blood & Cancer, 44*(1), 8–12.

Roy-Vallejo, E., Galván-Román, J. M., Moldenhauer, F., & Real de Asua, D. (2020). Adults with Down syndrome challenge another paradigm: When aging no longer entails arterial hypertension. *The Journal of Clinical Hypertension, 22*(7), 1127–1133.

Rubenstein, E., Hartley, S., & Bishop, L. (2020). Epidemiology of dementia and Alzheimer disease in individuals with Down syndrome. *JAMA Neurology, 77*(2), 262–264.

Santoro, S. L., Cannon, S., Capone, G., Franklin, C., Hart, S. J., Hobensack, V., Kishnani, P. S., Macklin, E. A., Manickam, K., & McCormick, A. (2020). Unexplained regression in Down syndrome: 35 cases from an international Down syndrome database. *Genetics in Medicine, 22*(4), 767–776.

Satge, D., Sommelet, D., Geneix, A., Nishi, M., Malet, P., & Vekemans, M. (1998). A tumor profile in Down syndrome. *American Journal of Medical Genetics, 78*(3), 207–216.

Savva, G. M., Morris, J. K., Mutton, D. E., & Alberman, E. (2006). Maternal age-specific fetal loss rates in Down syndrome pregnancies. *Prenatal Diagnosis: Published in Affiliation with the International Society for Prenatal Diagnosis, 26*(6), 499–504.

Schupf, N., Zigman, W., Kapell, D., Lee, J., Kline, J., & Levin, B. (1997). Early menopause in women with Down's syndrome. *Journal of Intellectual Disability Research, 41*(3), 264–267.

Schworer, E., Fidler, D., Kaur, M., Needham, A., Prince, M., & Daunhauer, L. (2020). Goal-directed action planning in infants with Down syndrome. *Journal of Intellectual Disability Research, 64*(9), 713–724.

Schworer, E. K., Esbensen, A. J., Nguyen, V., Bullard, L., Fidler, D. J., Daunhauer, L. A., Mervis, C. B., Becerra, A. M., Abbeduto, L., & Thurman, A. J. (2022a). Patterns and predictors of adaptive skills in 2-to 7-year-old children with Down syndrome. *Journal of Neurodevelopmental Disorders, 14*(1), 1–11.

Schworer, E. K., Fidler, D. J., Kaur, M., Needham, A., Prince, M. A., & Daunhauer, L. A. (2022b). Infant precursors of executive function in Down syndrome. *Journal of Intellectual Disability Research, (66)*, 108–120. https://doi.org/10.1111/jir.12824

Skotko, B. G. (2005a). Mothers of children with Down syndrome reflect on their postnatal support. *Pediatrics, 115*(1), 64–77.

Skotko, B. G. (2005b). Prenatally diagnosed Down syndrome: Mothers who continued their pregnancies evaluate their health care providers. *American Journal of Obstetrics and Gynecology, 192*(3), 670–677.

Skotko, B. G., & Bedia, R. C. (2005). Postnatal support for mothers of children with Down syndrome. *Mental Retardation, 43*(3), 196–212.

Skotko, B. G., Kishnani, P. S., Capone, G. T., & Group, D. S. D. S. (2009). Prenatal diagnosis of Down syndrome: How best to deliver the news. *American Journal of Medical Genetics Part A, 149*(11), 2361–2367.

Skotko, B. G., Levine, S. P., & Goldstein, R. (2011). Having a son or daughter with Down syndrome: Perspectives from mothers and fathers. *American Journal of Medical Genetics Part A, 155*(10), 2335–2347.

Sobey, C. G., Judkins, C. P., Sundararajan, V., Phan, T. G., Drummond, G. R., & Srikanth, V. K. (2015). Risk of major cardiovascular events in people with Down syndrome. *PLoS One, 10*(9), e0137093.

Stagi, S., Lapi, E., Romano, S., Bargiacchi, S., Brambilla, A., Giglio, S., Seminara, S., & de Martino, M. (2015). Determinants of vitamin d levels in children and adolescents with down syndrome. *International Journal of Endocrinology, 2015*, 896758.

Stern, Y. (2006). Cognitive reserve and Alzheimer disease. *Alzheimer Disease & Associated Disorders, 20*, S69–S74.

Stoel-Gammon, C. (2001). Down syndrome phonology: Developmental patterns and intervention strategies. *Down Syndrome Research and Practice, 7*(3), 93–100.

Szaflarska-Popławska, A., Soroczyńska-Wrzyszcz, A., Barg, E., Józefczuk, J., Korczowski, B., Grzybowska-Chlebowczyk, U., Więcek, S., & Cukrowska, B. (2016). Assessment of coeliac disease prevalence in patients with Down syndrome in Poland–a multi-centre study. *Gastroenterology Review/Przegląd Gastroenterologiczny, 11*(1), 41–46.

Tassé, M. J., Navas Macho, P., Havercamp, S. M., Benson, B. A., Allain, D. C., Manickam, K., & Davis, S. (2016). Psychiatric conditions prevalent among adults with Down syndrome. *Journal of Policy and Practice in Intellectual Disabilities, 13*(2), 173–180.

Thurman, A. J., Bullard, L., Kelly, L., Wong, C., Nguyen, V., Esbensen, A. J., Bekins, J., Schworer, E. K., Fidler, D. J., & Daunhauer, L. A. (2022). Defining expressive language benchmarks for children with Down syndrome. *Brain Sciences, 12*(6), 743.

Tietze, A.-L., Blankenburg, M., Hechler, T., Michel, E., Koh, M., Schlüter, B., & Zernikow, B. (2012). Sleep disturbances in children with multiple disabilities. *Sleep Medicine Reviews, 16*(2), 117–127.

Trois, M. S., Capone, G. T., Lutz, J. A., Melendres, M. C., Schwartz, A. R., Collop, N. A., & Marcus, C. L. (2009). Obstructive sleep apnea in adults with Down syndrome. *Journal of Clinical Sleep Medicine, 5*(4), 317–323. http://www.ncbi.nlm.nih.gov/pubmed/19968008

Tsou, A. Y., Bulova, P., Capone, G., Chicoine, B., Gelaro, B., Harville, T. O., Martin, B. A., McGuire, D. E., McKelvey, K. D., & Peterson, M. (2020). Medical care of adults with Down syndrome: A clinical guideline. *JAMA, 324*(15), 1543–1556.

Tungate, A. S., & Conners, F. A. (2021). Executive function in Down syndrome: A meta-analysis. *Research in Developmental Disabilities, 108*, 103802.

Urv, T. K., Zigman, W. B., & Silverman, W. (2010). Psychiatric symptoms in adults with Down syndrome and Alzheimer's disease. *American Journal on Intellectual and Developmental Disabilities, 115*(4), 265–276.

Van Buggenhout, G., Trommelen, J., Schoenmaker, A., De Bal, C., Verbeek, J., Smeets, D., Ropers, H., Devriendt, K., Hamel, B., & Fryns, J.-P. (1999). Down syndrome in a population of elderly mentally retarded patients: Genetic-diagnostic survey and implications for medical care. *American Journal of Medical Genetics, 85*(4), 376–384.

Van Deusen, K., Prince, M. A., Esbensen, A. J., Edgin, J. O., Schworer, E. K., Thurman, A. J., Patel, L. R., Daunhauer, L. A., & Fidler, D. J. (2022). Profiles of caregiver-reported executive function in children with Down syndrome. *Brain Sciences, 12*(10), 1333.

Vicari, S., Pontillo, M., & Armando, M. (2013). Neurodevelopmental and psychiatric issues in Down's syndrome: Assessment and intervention. *Psychiatric Genetics, 23*(3), 95–107.

Weijerman, M. E., & De Winter, J. P. (2010). Clinical practice: The care of children with Down syndrome. *European Journal of Pediatrics, 169*, 1445–1452.

Winders, P., Wolter-Warmerdam, K., & Hickey, F. (2019). A schedule of gross motor development for children with Down syndrome. *Journal of Intellectual Disability Research, 63*(4), 346–356.

Wiseman, F. K., Al-Janabi, T., Hardy, J., Karmiloff-Smith, A., Nizetic, D., Tybulewicz, V. L., Fisher, E. M., & Strydom, A. (2015). A genetic cause of Alzheimer disease: Mechanistic insights from Down syndrome. *Nature Reviews Neuroscience, 16*(9), 564–574.

Yamauchi, Y., Aoki, S., Koike, J., Hanzawa, N., & Hashimoto, K. (2019). Motor and cognitive development of children with Down syndrome: The effect of acquisition of walking skills on their cognitive and language abilities. *Brain and Development, 41*(4), 320–326.

Zhu, J. L., Hasle, H., Correa, A., Schendel, D., Friedman, J., Olsen, J., & Rasmussen, S. A. (2013). Survival among people with Down syndrome: A nationwide population-based study in Denmark. *Genetics in Medicine, 15*(1), 64–69.

Zigman, W. B. (2013). Atypical aging in Down syndrome. *Developmental Disabilities Research Reviews, 18*(1), 51–67.

Zigman, W. B., & Lott, I. T. (2007). Alzheimer's disease in Down syndrome: Neurobiology and risk. *Mental Retardation and Developmental Disabilities Research Reviews, 13*(3), 237–246.

The FMR1 Phenotype: Premutation and Fragile X Syndrome

Holley P. Arnold, Erin E. Hunt, Rachel M. Hantman, Conner J. Black, Chandler E. Knott, Danielle R. Lynch, and Jane E. Roberts

Abstract Fragile X syndrome (FXS) and the fragile X premutation (FXpm) are caused by a mutation on the fragile X messenger ribonucleoprotein 1 (FMR1) gene, which is located on the X chromosome. As these conditions are X-linked, sex differences often present. While there are some similarities between the FXS and FXpm phenotypes, there are many differences. Although the phenotype of FXpm is typically milder than that of FXS, it has become clear that impairment exists both across and within FXS and FXpm. This chapter describes the motor, cognitive, social, and clinical characteristics of individuals with FXS and FXpm across the lifespan as well as possible interventions. Being informed about FMR1 gene mutations is important to support early diagnosis and targeted treatment and intervention for FXS and FXpm. Thus, this chapter highlights the complexity of the FMR1 gene and the importance of understanding how individuals are impacted across the lifespan.

Keywords FMR1 phenotype · Fragile X syndrome · Fragile X premutation · Fragile X carriers · Lifespan development

---

H. P. Arnold · E. E. Hunt · R. M. Hantman · C. J. Black · C. E. Knott · D. R. Lynch · J. E. Roberts (✉)
Department of Psychology, Institute for Mind and Brain, University of South Carolina, Columbia, SC, USA
e-mail: jerobert@mailbox.sc.edu

M. G. Valdovinos (ed.), *Intellectual and Developmental Disabilities*, Contemporary Clinical Neuroscience,
https://doi.org/10.1007/978-3-031-66932-3_14

Example Case Study

Cathy had concerns about her son Nathan when he was less than 6 months old. Cathy noticed that Nathan was not very interested in interacting with people; she expected him to make eye contact and smile more, even as a baby. Nathan also showed some motor movement delays. Most babies in his daycare were crawling by 10 months and walking by 17 months, but Nathan did not crawl until 12 months nor walk until 20 months. When Nathan was 2 years old, he started to receive early intervention services, including occupational, speech, and physical therapy. Even though Nathan had many different toys, Nathan seemed only interested in a toy giraffe that he chewed on and banged repeatedly. Cathy often had a hard time getting Nathan's attention, even when she loudly called his name. Nathan made fewer sounds than most babies, and he did not say his first word until he was 31 months old, even though his younger sister said her first word when she was 12 months old. He did not put two words together, like "Mommy up," until he was about 4 years old. Once he started talking, he usually repeated words and phrases in a sing-song voice. Nathan also had some disruptive behaviors, like hitting his mom.

At 4 years old, he was diagnosed with autism spectrum disorder (ASD) by his pediatrician, who also recommended Cathy get Nathan genetically tested for other disorders; Nathan was then diagnosed with fragile X syndrome (FXS). After getting diagnosed with FXS and ASD, Nathan also began receiving private applied behavioral analysis (ABA) therapy to decrease his disruptive behaviors. Once Nathan began elementary school, Cathy worked with his teachers to set up an individualized education plan (IEP), which gave him academic accommodations and additional occupational, physical, and speech therapy services. Nathan's IQ has been tested many times over his life, and his scores have always been in the moderate intellectual disability range.

Nathan is now 22 years old; he continues to live with his parents. He can do many things independently, like dressing, eating, and bathing. He needs help with some activities of daily living, like buttoning small buttons, preparing and heating food, and reading. Nathan has a part-time job at a local organization that hires people with disabilities; he enjoys working with his coworkers. He continues to receive therapy to increase his independence, like telling time, using money, and going grocery shopping.

After Nathan was diagnosed with FXS, Cathy decided to get herself and her two younger daughters tested for genetic disorders related to FXS. These tests showed that Cathy and one of her daughters had the fragile X premutation (FXpm), while the other daughter did not have FXS or FXpm. Although Cathy did not know about her own genetic diagnosis before Nathan was diagnosed with FXS, when she thinks back on her life, she reflects that she has always had social difficulties. She was even diagnosed with social anxiety in high school, which she managed successfully through medication and therapy. In school, Cathy excelled in her humanities courses, but she struggled with organization and time management. Even though Cathy did not have fragile X primary ovarian insufficiency (FXPOI), which 20% of women with FXpm experience, she did start menopause in her late 30s, which is sooner than average. Cathy's parents also decided to get genetic testing after Nathan's diagnosis. These results showed that Cathy inherited the FXpm from her father, Simon, who also has FXpm. In his mid-60s, Simon began to show symptoms of tremor and unsteady movement as he walked, as well as personality changes, like more irritability. Simon's family encouraged him to talk to his doctor about these physical and personality changes. Simon's doctor and neurologist who, aware of his genetic status, diagnosed him with fragile X-associated tremor/ataxia syndrome (FXTAS).

Fragile X syndrome (FXS) and the fragile X premutation (FXpm) are caused by a mutation on the fragile X messenger ribonucleoprotein 1 (FMR1) gene, which is located on the X chromosome. This chapter describes the physical, cognitive, social, and other characteristics of individuals with FXS and FXpm, called a phenotype, across the lifespan. While there are some similarities across the FXS and FXpm phenotypes, there are many differences that we will describe. Overall, FXS is associated with more severe impairments, including moderate intellectual disability (ID), difficulties with adaptive functioning, language and social communication impairments, high rates of autism spectrum disorder (ASD) and anxiety, executive function deficits, elevated risk of seizures, and specific facial features. The phenotype of FXpm is typically milder than that of FXS. Historically, FXpm was considered a "silent carrier" of the mutated FMR1 gene and was not associated with any symptoms of a genetic disorder. Recent research has begun to shed light on a broad FXpm phenotype, which includes mild social communication difficulties, executive function deficits, elevated rates of anxiety and depression, and a neurodegenerative disorder at older ages called fragile X-associated tremor/ataxia syndrome (FXTAS). Furthermore, females with FXpm are at risk of developing fragile X primary ovarian insufficiency (FXPOI), which can cause early menopause and infertility. Thus, it has become clear that impairment exists both across and within FXS and FXpm, highlighting the complexity of the FMR1 gene and the importance of understanding how individuals are impacted across the lifespan.

Genetic Etiology

The fragile X messenger ribonucleoprotein 1 (FMR1) gene is located on the long arm of the X chromosome. FMR1 mRNA contains three DNA building blocks (nucleotides) that make up CGG trinucleotides. The CGG triplicate is repeated between 6 and 54 times in the FMR1 allele (an allele is a "version" of a gene) in individuals who do not have FXS or FXpm. In these unaffected individuals, the FMR1 gene produces a level of FMR1 protein (FMRP) that is considered typical, which is important as FMR1 protein is essential for healthy brain growth and cognitive functioning (Bassell & Warren, 2008; Darnell & Klann, 2013). The full mutation of the FMR1 gene is called FXS and is caused by CGG repeat expansions of >200 on the FMR1 gene, which reduces FMR1 protein. Specifically, by 10–12 week gestation, the gene with expanded CGG repeats becomes hypermethylated, or additional methyl markers expand beyond the typical methylation boundary found upstream to this gene and attach to the gene promoter and repeat region (Poeta et al., 2020). Gene methylation alters the amount of protein produced; in the context of FXS, hypermethylation turns off (i.e., silences) the FMR1 gene, stopping the production of FMR1 protein. FMR1 mRNA levels in those with the FXS are generally lower than those without the full genetic mutation. The FXpm is caused by CGG repeat expansions of 55–200 that can increase transcription of the gene and elevate levels of FMR1 mRNA. Higher levels of mRNA (i.e., mRNA toxicity) impacts neuronal activity, likely contributing to symptoms of FXTAS and FXPOI. The production of FMR1 protein in those with the FXpm is generally at the same level (or slightly lower) as those without the FXpm genetic mutation. Although FXS and FXpm are both caused by the mutation on the FMR1 gene, differences in genetic and molecular variability (e.g., CGG repeat lengths, levels of FMR1 mRNA, and FMR protein production) are related to FXS and FXpm phenotypes.

Prevalence, Sex Effects, Hereditability, and Diagnosis

FXS is a relatively rare syndrome with an estimated prevalence of 1:7000 for males and 1:11,000 for females (Hunter et al., 2014). The FXpm is much more common and is estimated to occur in about 1:148–300 for females and 1:290–855 for males (Hunter et al., 2014; Maenner et al., 2013). Due to the X-linked nature of FXS and FXpm, sex differences are apparent. Females have two X chromosomes, so they likely have one X chromosome that has a "healthy" (i.e., unmutated) FMR1 allele and one that is affected with the FMR1 mutation. Given that one of the X chromosomes in females is randomly inactivated, the ratio of activated genes that are affected with the FMR1 mutation (i.e., X-activation ratio) can vary. Thus, the proportion of activated genes that are not affected with the FMR1 mutation serves as a protective factor and facilitates FMR1 protein production in females, typically producing a milder, less impairing phenotype. Given that males only have one X chromosome, they generally have more impairment than females. To date, most research has primarily focused on males with FXS, and the literature on females with FXS is somewhat limited.

FXS and FXpm are inherited disorders; however, hereditability is complicated (see https://www.cdc.gov/ncbddd/fxs/inherited.html for more information). A female with FXpm or FXS will pass either the FMR1-mutated gene or the unaffected FMR1 gene to her child. If the child receives the mutated FMR1 gene, the child will have FXS or FXpm. If the child receives the unaffected gene, the child will not have FXS or FXpm. The number of CGG repeats the mother has impacts if her child will have FXS or FXpm because the CGG expansion length can increase in size when inherited (i.e., during meiosis). In contrast to females with an FMR1 mutation, who can pass the affected gene to either their sons or daughters, males with an FMR1 mutation pass their affected X chromosome onto all their daughters. Given that males pass their Y chromosome to their sons, they cannot pass the X-lined genetic mutation to their sons. Thus, all females who receive an FMR1 mutation from their fathers will have FXpm and none of their sons will have FXpm.

FXS and FXpm are diagnosed through genetic testing, which can be conducted prenatally or after birth. DNA is analyzed to detect CGG length, methylation status (i.e., if the FMR1 gene has been effectively silenced or not), X-activation ratio, and if mosaicism is present (i.e., mixture of different cells with or without the FMR1 mutation or cells with different methylation statuses). Research suggests associations between phenotypes and molecular markers, explaining some differences in the level of difficulties experienced in FXS and FXpm (Meng et al., 2021). For FXS, the average age of diagnosis is 35–37 months for males and 42 months for females, although symptoms are present much earlier (Bailey et al., 2009a). For FXpm, the diagnosis generally occurs much later. Given that individuals with FXpm generally experience mild symptoms and universal screening is not in place to detect FMR1 mutations, many individuals without a family history of FXS are unaware of their status. As a result, FXpm is often diagnosed only after a child with FXS or FXpm is born and testing of the child reveals the FMR1 mutation, which generates testing of multiple family members. While FMR1 mutations are most often detected following an affected child, the diagnosis of FXS or FXpm may also be given based on other symptoms, including infertility (i.e., FXPOI), onset of tremors (i.e., FXTAS), or a family history of autism, intellectual disability, or FXS.

Motor Profile

Both gross motor (e.g., sitting up, crawling, and walking) and fine motor (e.g., grasping, pointing, and manipulating objects) skills are important to support a child's independence and ability to explore and engage in their surroundings. Strong motor skills foster development across multiple domains, such as language and communication skills (Iverson, 2010; Will & Roberts, 2021). For example, when a child can sit independently, then the visual and social input that a child receives is greatly expanded in comparison to when the child was only able to lay on the floor; this expanded input can then increase the child's communication skills.

As with most areas of development, motor impairments emerge earlier and are more severe in individuals with FXS than in those with FXpm. Most individuals with FXS display motor skill delays as early as the first year of life. These delays typically remain and may get more severe over time. In contrast, the motor abilities of children with FXpm are generally in line with age expectations, with some children displaying mild impairment. However, while those with FXpm are often "spared" the moderate to severe motor impairments observed in FXS, some sex differences are apparent. Furthermore, those with FXpm, particularly males, are at higher risk for a severe neurodegenerative disorder called fragile X-associated tremor/ataxia syndrome (FXTAS). FXTAS emerges in late adulthood and is characterized by intention tremor and cerebellar ataxia (i.e., poor muscle control), leading to frequent falling, neuropathy, parkinsonian features, and cognitive decline (Hagerman & Hagerman, 2016).

Fragile X Syndrome

Most infants and toddlers with FXS have low muscle tone and have hyper-flexible joints (Crawford et al., 2020; Kau et al., 2002), which impacts motor development. Delays in motor skill acquisition are among the earliest signs of impairment in FXS, with delays often evident as early as 6 months of age (Hinton et al., 2013; Will & Roberts, 2021). On average, male infants and toddlers with FXS sit, crawl, and walk later than expected (Bailey et al., 2003; Roberts et al., 2001). Their development of motor skills across age also occurs at a slower rate than for females. For females, clinically significant delays are also present early in life, but they do not emerge as early as males, with impairment typically noted by the second year of life. Motor skill development in females with FXS is slower than those without FXS; however, it is not as delayed as males with FXS.

Although the research of motor skills is somewhat sparse, the majority of research has focused on early development and demonstrated that impairments in motor development persist into childhood (Will et al., 2019; Zingerevich et al., 2008). Continued difficulties in motor control, planning, and coordination (Friefeld & Macgregor, 1994; Simko et al., 1989) impact daily living skills in adolescents and adults. Although there is limited research on aging in FXS, it is important to note that FXTAS is not associated with the full mutation. However, there are case studies in which males with FXS developed FXTAS—all of which had unmethylated alleles or had mosaicisms (Lozano et al., 2014). This highlights the complexity of the FXS spectrum.

Fragile X Premutation

On average, the motor abilities of infants and toddlers with FXpm do not significantly differ from their neurotypical peers (Cordeiro et al., 2015; Wheeler et al., 2016, 2021). However, a range of abilities has been indicated, with some children evidencing mild, yet clinically significant delayed motor abilities, particularly for males. For example, significant delays in fine motor skills emerged by 3 years for males. Relative to males with FXS, the delays were mild. Significant impairments were not indicated for females with FXpm.

Aging adults with FXpm are at an elevated risk of developing FXTAS, which has parkinsonian features. No studies have evaluated if there is any association between early motor delays and FXTAS. Approximately 30–40% of males and 16% of females experience symptoms in their 50s, with the likelihood of developing FXTAS increasing with age (e.g., 75% of in males in their 80s; Jacquemont et al., 2007; Polussa et al., 2014). Mild declines in motor abilities have also been reported in males without FXTAS (Famula et al., 2022).

Cognitive and Adaptive Functioning

Cognitive functioning includes a range of abilities, including basic understanding of information, acquiring new knowledge, and reasoning skills. Adaptive functioning describes the daily living, socialization, and communication skills used to function independently; these skills are strongly predictive of independence, employment, and the amount of support an individual needs in everyday life. Both cognitive and adaptive abilities emerge early in development and develop rapidly over time. Diagnosis of an intellectual disability (ID) requires consideration of both general cognitive abilities and adaptive functioning (American Psychiatric Association [APA], 2022).

FXS is the most common inherited cause of ID. The majority of males and some females with FXS have some degree of ID (i.e., mild to severe), though level of impairment varies greatly across individuals. For FXpm, cognitive and adaptive skills generally resemble the broader neurotypical population. However, difficulties with learning and more complex cognitive skills, such as regulating attention, working memory, and planning, have been implicated in FXpm. Although more research is needed to better understand how these abilities evolve across the lifespan, cognitive decline in adulthood has been suggested for both FXS and FXpm.

Fragile X Syndrome

Developmental delays are evident within the first year of life in males with FXS (Bailey et al., 2009a, Wheeler et al., 2021). Developmental delays are less frequently seen in females with FXS and are not usually evident until 16–24 months. Developmental abilities advance at approximately half the rate observed in neurotypical infants and toddlers for males with FXS and at three-fourths the rate for females with FXS. As a result of this slowed development, impairments in cognitive and adaptive abilities extend across childhood and into adulthood in FXS (Del Hoyo Soriano et al., 2018; Glaser et al., 2003; Joga-Elvira et al., 2021).

Intellectual abilities typically range from mild to severe ID for males with FXS. Intellectual abilities are generally higher and more variable in females with FXS, ranging from severe ID to average IQ, likely resulting from their protective unaffected X chromosome and differences in their X-activation ratios. Females with FXS often experience difficulties with learning, attention, and impulse control, which can emerge during childhood (Huddleston et al., 2014). Considering cognitive strengths and weaknesses, children with FXS tend to exhibit relatively better verbal skills compared to visual-spatial cognitive skills (Quintin et al., 2016). In line with cognitive abilities, both males and females with FXS typically exhibit adaptive skills that are below age expectations (Klaiman et al., 2014), though there is some evidence of age-typical adaptive skills in school-aged females with FXS (Jordan et al., 2023). Sex differences in adaptive functioning are also evident throughout childhood and adolescence, with males demonstrating lower communication, daily living, and socialization skills (Glaser et al., 2003). The presence of co-occurring conditions also influences individual differences in cognitive and adaptive functioning; specifically, poorer cognitive and adaptive functioning is seen in children and adults with FXS with co-occurring autism spectrum disorder (FXS+ASD; Caravella & Roberts, 2017; Hernandez et al., 2009; Roberts et al., 2009b).

Uneven cognitive profiles are also evident in adults with FXS, with stronger receptive understanding of language compared to both expressive use of words and language as well as nonverbal cognitive skills being noted (Hoffmann et al., 2019). Notably, most research examining cognitive functioning in adulthood in FXS has only included women. Lower cognitive functioning has been documented in adult females with FXS compared to both neurotypical females and those with the FXpm; furthermore, even when considering IQ and education level, adult women with FXS have a neurocognitive profile characterized by deficits in executive function (e.g., planning, decision-making, and self-control; Bennetto et al., 2001). Growth in IQ and adaptive behavior skills over time lags behind neurotypical peers in adolescents and adults with FXS (Huddleston et al., 2014; Klaiman et al., 2014). In some cases, cognitive regression (i.e., loss of skills) may even occur (Hahn et al., 2015). Adaptive outcomes vary widely in adults with FXS. A large national survey indicated that approximately 44% of females and 10% of males lived independently; the remaining adults either lived with their parents or in a group home; 73% of females and 60% of males were employed at least part-time in an array of jobs. Of those employed, 27% of females and 61% of males had a job coach. Furthermore, 82% of females and 43% of males were reported by their families to need minimal to no assistance in their everyday lives (Hartley et al., 2011).

Fragile X Premutation

Infants and toddlers with FXpm demonstrate age-typical adaptive behavior and age-typical or mildly delayed developmental abilities (Myers et al., 2001; Wheeler et al., 2016, 2021), such that early trajectories and developmental profiles largely resemble neurotypical peers. There is some evidence to suggest that males with FXpm demonstrate somewhat lower cognitive functioning compared to neurotypical children in middle childhood (Wang et al., 2020); these differences were not seen for females with FXpm.

Intellectual abilities in adult males with FXpm resemble those of neurotypical males and fall in the average to high average range across various domains (e.g., sustained attention, processing speed, verbal comprehension, and visual–spatial functioning; Allen et al., 2005; Cornish et al., 2008). In middle-aged males with FXpm without ragile X-associated tremor/ataxia syndrome (FXTAS), specific deficits in long-term memory (i.e., memory for information from the past, such as previously learned concepts and procedures) and semantic memory (i.e., general knowledge) have been seen, even in individuals who have IQs in the high average range (Moore et al., 2004). Cognitive functioning in adult females with FXpm may be more variable, as findings have been inconsistent; several studies have found no differences in cognitive functioning compared to neurotypical adult females (Bennetto et al., 2001; Loesch et al., 2003; Roberts et al., 2009a). In contrast, lower verbal reasoning has been seen in females with FXpm compared to neurotypical females, as well as males with FXpm (Allen et al., 2005). Both males and females with FXpm may also demonstrate relative weaknesses in numeric reasoning, visual-spatial- processing, and working memory (Cornish et al., 2008; Grigsby et al., 2008).

Cognitive decline in FXpm has primarily been documented in the context of FXTAS (Adams et al., 2010). Cognitive impairment is usually evident after age 50, following the emergence of motor symptoms, both of which likely impact adaptive skills (Jacquemont et al., 2007). Notably, cognitive impairments are evident independent of FXTAS motor symptoms or disease progression, suggesting these deficits are not secondary to FXTAS.

Language and Social Communication

Language encompasses words and sounds that are produced (expressive language), as well as the ability to understand spoken and written words (receptive language), whereas social communication refers to using communication skills to interact with others. Social communication specifically includes the use and understanding of gestures, facial expressions, and body language, as well as knowing what to say when and how to say it (i.e., pragmatics; understanding conversational turn-taking, jokes, and sarcasm). Aspects of language and social communication are impacted to varying degrees in FXS and FXpm across the lifespan. For FXS, expressive language, receptive language, and social communication abilities are significantly below age expectations. However, there is a wide range of variability. These difficulties are particularly evident for males with FXS and those with co-occurring autism spectrum disorder (FXS+ASD). For FXpm, difficulties are less apparent, yet challenges with social communication have been reported.

Fragile X Syndrome

Infants and preschoolers with FXS have expressive and receptive delays (Hoffmann, 2022) that are clear by 6 months for males and 12 months for females (Wheeler et al., 2021). These difficulties continue, as children with FXS have delayed onset of language. For example, two studies reported that boys with FXS produced their first words between 21 and 31 months (Hinton et al., 2013; Roberts et al., 2002) and one study indicated that over half of toddlers with FXS did not produce any words spontaneously (Brady et al., 2006), whereas neurotypical children tend to produce their first words by around 12 months of age. Based on parent report of communication abilities in children ages 5 years and younger, approximately 35% of boys and 70% of girls used two- to three-word sentences, 18% of boys and 48% of girls could engage in conversation, and just over 60% of males and just under 80% of females could understand others (Bailey et al., 2009b). Most individuals with FXS will develop spoken language.

Contributing to language delays, infants with FXS also have difficulty with social communication skills, such as more limited eye contact, delayed social smiling, more limited range of facial expressions, and fewer gestures to direct others' attention and to express needs (Rague et al., 2018; Raspa et al., 2017; Roberts et al., 2002). For instance, by 24 months, toddlers with FXS produce six fewer gestures than neurotypical toddlers and four fewer gestures than toddlers with an elevated-likelihood of ASD (Rague et al., 2018). In infants with FXS+ASD, expressive language and overall language communication delays are also greater than those with FXS only (Mattie & Hamrick, 2022; Philofsky et al., 2004). For example, 72% of infants with FXS+ASD, but only 45% of infants with FXS, scored significantly below what would be expected given their age for their social communication skills (Mattie & Hamrick, 2022).

The language patterns are relatively similar in childhood and adolescence, as impairments in expressive and receptive language are common. Expressive and receptive abilities appear to be stronger in female children and adolescents than in males (Finestack & Abbeduto, 2010; Martin et al., 2018). Receptive language skills are a relative strength over expressive language skills in those with FXS and FXS+ASD (Martin et al., 2018; Thurman et al., 2017). Like in infancy, autism symptoms and co-occurring ASD appear to impact language abilities. Both receptive and expressive language skills are positively related to nonverbal IQ for children with FXS and FXS+ASD (Brady et al., 2020; McDuffie et al., 2012). Adolescents with FXS+ASD are more unintelligible than those with just FXS (Kover & Abbeduto, 2010), and language skills are more developed in those with FXS compared to those with FXS+ASD (Martin et al., 2018). Higher autism symptoms have also been linked to lower receptive language skills (McDuffie et al., 2012), and children with FXS tend to have higher receptive language abilities than children with ASD (Thurman et al., 2017). Finally, difficulties with social communication also arise, especially for those with FXS+ASD (Abbeduto et al., 2008; Martin et al., 2018; Roberts et al., 2007).

In adults, there is little research with FXS in terms of language and social communication abilities. Most individuals can functionally communicate their wants and needs, with females having better expressive language skills than males. For example, one study reported 91% of adult females used 2- or 3-word sentences and engaged in conversation, whereas 86% of adult males used 2- or 3-word sentences and 59% engaged in conversation (Bailey et al., 2009b). Speech impairments, which impacts intelligibility, have been indicated in approximately 50% of males (Bangert et al., 2022). Although overall language skills tend to be lower than expected given adults' chronological age, receptive language skills remain a relative strength (Hoffmann et al., 2019). Adults with FXS also experience difficulties with aspects of pragmatics (Diez-Itza et al., 2022), especially when compared to neurotypical females and those with FXpm (Simon et al., 2001).

Fragile X Premutation

In general, the language abilities of males and females with FXpm are similar to chronological age expectations. However, subtle differences have been reported, most notably in social communication. For infants and preschoolers with FXpm, two studies indicated that receptive and expressive abilities and rate of development were similar to the abilities of their neurotypical peers (Wheeler et al., 2016, 2021). However, differences in social communication were indicated as early as age 12 months, with approximately one-third of parents reporting fewer gestures and less eye gaze and emotion (Wheeler et al., 2016). The rate of development of social communication skills also appears to be slower relative to the neurotypical infants (Wheeler et al., 2016), suggesting that these difficulties may become more apparent throughout childhood and persist into adulthood. Very little is known about language abilities in early childhood through adolescence. An early study showed that 50% of children and adolescents (ages 5–15 years) had below average receptive language abilities, 67% had below average expressive abilities, and 50% had both (Aziz et al., 2003), but much more research is needed before conclusions can be made.

In adulthood, females with FXpm experience pragmatic language difficulties (Klusek et al., 2016, 2019; Losh et al., 2012). One study found that females with FXpm made nearly twice as many pragmatic language violations, such as interrupting, verbosity, and persistently talking about one topic, when compared to neurotypical women (Klusek et al., 2019). Women with FXpm also show challenges with speech dysfluency, or interruptions in the flow of their speech (e.g., using filler words like "um," changing the words in a sentence, or repetitions of words; Klusek et al., 2018; Movaghar et al., 2017; Sterling et al., 2013). One study found that they have 28% more dysfluencies than mothers of children with ASD, who were used as the comparison group to control for the stress of raising a child with a developmental disability (Sterling et al., 2013). It also appears that these dysfluencies get worse over time (Bredin-Oja et al., 2021; Sterling et al., 2013). There is much less research on adult males with FXpm. One early study indicated that males with and without FXpm had similar receptive, expressive, and pragmatic language abilities (Moore et al., 2004).

Related Conditions in Fragile X Syndrome and Fragile X Premutation

Individuals with FXS or FXpm are more likely to have specific mental health-related conditions, which overlap with motor, cognitive, language, and social phenotypic features. Hyperarousal, which is increased physiological responsiveness and sensitivity to sensory stimuli in the environment (e.g., lights, sounds, social cues, and textures), has been described as an underlying mechanisms related to the behavioral phenotype of FXS. For example, hyperarousal has been related to the increased severity of autism and anxiety symptoms in FXS (Hogan et al., 2021). Much less research has been conducted investigating hyperarousal in in FXpm (Klusek et al., 2017). Relative to the general population, males and females with FXS or FXpm are at an elevated-likelihood of developing autism, anxiety, depression, attention-deficit/hyperactivity disorder (ADHD), and other related impairments or disorders (Bartholomay et al., 2019; Cordeiro et al., 2011; Ezell et al., 2019; Roberts et al., 2016, 2020). These conditions are generally more common and more severe in FXS than in FXpm. Many of these co-occurring conditions emerge early in life and persist into adulthood, often causing significant impairment, which impacts quality of life (Bailey et al., 2008; Cordeiro et al., 2011; Ezell et al., 2019; Wheeler et al., 2014).

Fragile X Syndrome

Autism Spectrum Disorder Autism spectrum disorder (ASD) is characterized by persistent deficits in social communication and interaction, combined with the presence of restricted and repetitive behaviors or interests (APA, 2022). ASD is highly prevalent in FXS, with 51–75% of males and 16–45% of females with FXS also diagnosed with autism (Kaufmann et al., 2017; Klusek et al., 2014; Roberts et al., 2020).

FXS is the most common identified genetic cause of ASD, which makes it a model disorder to study. The presentation of ASD symptoms is often more profound in males (Bartholomay et al., 2019; Kaufmann et al., 2017). Early features of ASD symptoms and physiological hyperarousal can be detected in infancy, and many of these early features predict ASD features later in development (Hogan et al., 2021; Wall et al., 2019). While early signs of ASD are present in many infants and toddlers with FXS, the diagnosis of ASD in FXS often does not occur until during the preschool years or later, which prevents specific treatment to target the reduction of ASD symptom severity.

Attention-Deficit/Hyperactivity Disorder The core symptoms of Attention-Deficit/Hyperactivity Disorder (ADHD) include inattention, disorganization, and/or hyperactivity and impulsivity (APA, 2022). Attention problems are commonly indicated, with criteria for ADHD met by 42–84% of males and 45–67% of females (Bailey et al., 2008; Grefer et al., 2016; Klusek et al., 2022). ADHD generally manifests in the preschool years as hyperactivity and impulsivity, rather than as inattention (Grefer et al., 2016). While more research is needed to better understand how the symptoms of ADHD may change across maturation in FXS, inattention is more prevalent in adolescence and adulthood (Klusek et al., 2022). Though some children with FXS may not meet diagnostic criteria or receive a formal a diagnosis of ASD or ADHD, behavior problems, such as inattention, impulsivity, aggression, repetitive and self-injurious behavior, and hyperactivity, are common among children and adolescents with FXS (Creganzan-Royo et al., 2022). These behaviors are often targets for intervention.

Anxiety and Depression Anxiety is highly prevalent, with up to 70–86% of males and 56–77% of females with FXS meeting criteria for at least one anxiety disorder at some point in their lifetime (Bailey et al., 2008; Cordeiro et al., 2011; Ezell et al., 2019). Social anxiety and specific phobia are the most experienced types of anxiety; 60% of males and 55% of females meet criteria for social anxiety, and 60% of males and 51% of females meet criteria for specific phobia (Cordeiro et al., 2011). Physiological hyperarousal and behavioral indicators of anxiety (e.g., increased distress, gaze aversion, and avoidance in social contexts) observed early in development are related to the later emergence of anxiety (Black et al., 2021; Hogan et al., 2021; Smith et al., 2021; Wall et al., 2019).

Individuals with FXS also experience depression at a higher rate (12–22%) than is observed in the general population, with females likely having somewhat higher rates (Bailey et al., 2008; Lightbody et al., 2022). This is consistent with the neurotypical population, in which females are significantly more likely to have symptoms of depression than males. One recent study of females with FXS found that social avoidance and withdrawal were related to depression and higher levels of depression were reported after puberty (Lightbody et al., 2022); this has not been evaluated in males.

Fragile X Premutation

Autism Spectrum Disorder The association of autism in FXpm has not been well-studied. However, there is some evidence suggesting that the diagnosis of ASD is somewhat elevated in individuals with FXpm relative to the general population (4% males, 2% females; Maenner et al., 2023). For example, previous reports have indicated that 14–19% of boys and 1–5% of girls with FXpm meet criteria for ASD (Bailey et al., 2008; Clifford et al., 2007). Although most individuals with FXpm do not meet diagnostic criteria for ASD, many exhibit some traits related to autism such as social communication difficulties (Wheeler et al., 2017).

Attention-Deficit/Hyperactivity Disorder Increased hyperactivity and inattention have been indicated in individuals with FXpm (Aishworiya et al., 2022; Bailey et al., 2008; Hunter et al., 2012). The likelihood of having ADHD is particularly elevated in males, with as many as 42–75% having an ADHD diagnosis (Aishworiya et al., 2022; Klusek et al., 2022). Prevalence rates for females (11–27%) are much lower than males with FXpm, yet higher than the general population (Aishworiya et al., 2022; Farzin et al., 2006; Wheeler et al., 2014). Given that individuals with FXpm have subtle difficulties in executive function (which plays a role in attention, impulse control, and planning/organization), many exhibit traits of ADHD that are mild enough to not reach the threshold for a diagnosis.

Anxiety and Depression Research shows that anxiety and depression are highly prevalent in FXpm, with lifetime prevalence rates far surpassing the general population (Bailey et al., 2008; Kenna et al., 2013; Roberts et al., 2016). Approximately 40% of individuals with FXpm have an anxiety disorder (Kenna et al., 2013). These rates vary by the type of anxiety disorder: specific phobia (15%), generalized anxiety disorder (7%), and social phobia (21%; Kenna et al., 2013). Overall, there do not seem to be differences in rates of anxiety in males and females with FXpm (Flavell et al., 2023). Depression also occurs in about 41% of individuals with FXpm, with onset in adulthood (Flavell et al., 2023; Seritan et al., 2013). While males and females with FXpm tend to have similar rates of depression overall, males have a much higher prevalence (33%) of dysthymia (also known as persistent depressive disorder; APA, 2022) than females (11%). Other individual characteristics impact the occurrence of depression and anxiety. For some mothers with FXpm, the feelings of anxiety and depression may be compounded by the stress of having child with FXS. For example, challenging child behavior has been linked with higher depression and anxiety (Lowell et al., 2017; Roberts et al., 2009a).

Intervention

Across the lifespan, individually tailored interventions designed and implemented by a team of professionals knowledgeable about the cognitive, behavioral, and medical characteristics associated with FXS are critical to improving functional outcomes for individuals with FXS and FXpm. Although treatment is typically most indicated for individuals with FXS, individuals with FXpm may also benefit from intervention following level of impairment. Emphasis is placed on the early detection of FXS through prenatal screening and early identification of co-occurring conditions, such as ASD, anxiety, and ADHD, to allow for the provision of appropriate therapies and supports beginning early in life (Hagerman et al., 2009). However, screening for FXpm is also important given the elevated rates of mental health and developmental disorders (e.g., anxiety, depression, ASD, and ADHD), as well as the increased risk for fragile X primary ovarian insufficiency (FXPOI, which can have implications for family planning and fertility) in females and fragile X-associated tremor/ataxia syndrome (FXTAS) in males. Although no single therapeutic intervention exists for FXS, many treatments are available to help address the specific symptoms associated with FXS and the various conditions that frequently co-occur alongside FXS. These include non-medication treatments, educational support, and medications. Early in development, non-medication treatments for FXS, such as physical therapy (PT), occupational therapy (OT), speech-language therapy, and applied behavioral analysis (ABA), are implemented to address the motor, speech, adaptive, social communication, and sensory processing difficulties associated with the FXS phenotype (Kaufmann et al., 2017). The benefits of these therapies have been shown in children with FXS with or without co-occurring conditions, such as ASD, although individuals with FXS+ASD are more likely to benefit from intensive ABA services (Protic et al., 2022) than individuals with FXS without ASD. Other interventions, such as behavioral parent training and parent-led communication/language interventions, have also been shown to reduce behavioral problems, such as aggression, self-injury, hyperactivity, and non-compliance in children with FXS (Bullard & Abbeduto, 2021; Protic et al., 2022). In adults with FXS, interventions focused on mindfulness and stress reduction have been shown to improve functional outcomes. In addition, exposure therapy and cognitive behavioral therapy may be appropriate for reducing co-occurring symptoms of anxiety through fear extinction, habituation, and inhibitory learning in older individuals with FXS and higher cognitive functioning (Protic et al., 2022; Weiss et al., 2018).

The school setting plays a critical role in the provision of services for individuals with FXS, and some individuals with FXpm, from birth up to 21 years of age. Many individuals with FXS qualify for special education and related services in the public school system under the Individuals with Disabilities Education Act (IDEA; Nash et al., 2019). When academic impairment is present, some individuals with FXpm may also qualify for special education and related services, such as access to a special education resource room. Generally, educational supports and services for individuals with FXS are typically provided through an individualized education program (IEP), which is designed to address the educational needs of the individual based on their unique cognitive and behavioral strengths and weaknesses. Educational supports and services provided for individuals with FXS served under IDEA include but are not limited to special education, academic accommodations, individualized instruction, OT, PT, speech services, and social skill groups (Nash et al., 2019). In adolescents with FXS, independence and vocational skills are often included in an individual's IEP to teach life skills related to computer skills, money management, and social skills (Protic et al., 2022). Once adolescents and young adults transition out of the school system, there is generally a decrease in access to therapies and services (Martin et al., 2013). However, adults with FXS evidence persisting difficulties throughout their lives. Thus, it recommended that adults with

FXS continue to receive services, particularly those focused on the practical skills necessary for employment. Although little research has examined school-based supports for FXpm, given their subtle but still impacted language, cognitive, and motor development, advocating for school-based supports would likely still be beneficial.

Medications targeting co-occurring symptoms of ADHD, anxiety, hyperarousal, and aggression have been shown to be effective in approximately 50–70% of individuals with FXS (Berry-Kravis et al., 2012). Stimulant medications (e.g., Adderall and Ritalin) are frequently prescribed to address symptoms related to ADHD (Hagerman et al., 2009). Similarly, non-stimulant medications (e.g., Clonidine and Guanfacine) have been found to be effective in reducing ADHD symptoms in younger individuals with FXS, individuals with adverse reactions to stimulant medications, and individuals with co-occurring sleep difficulties (Hagerman et al., 2009). Selective serotonin reuptake inhibitors (SSRIs) are typically regarded as the front-line treatment for anxiety in FXS, whereas antipsychotic drugs (e.g., Risperidone and Aripiprazole) are helpful for targeting irritability, aggression, mood instability, and repetitive behaviors in individuals with FXS (Hagerman et al., 2009). While improvements in outcomes are generally modest (Erickson et al., 2017), clinical trials designed to develop targeted treatments to reverse the cognitive deficits caused by the lack of FMR1 protein (FMRP) in FXS suggest the potential utility of drugs targeting metabotropic glutamate receptor 5 (mGluR5) and GABA receptors (Gross et al., 2015). However, research in this area is still ongoing. More research is needed for medication-based treatments in FXpm.

In sum, non-medication treatments, educational supports, and medications work together to address common co-occurring conditions in individuals with FXS and some individuals with FXpm when impairment is present. Continuation of services from early childhood and into adulthood is critical to improve functional outcomes for individuals with FXS throughout the lifespan.

Conclusion

Knowledge regarding mutations on the FMR1 gene has rapidly advanced across the past 30 years. We now know that FMR1 mutations are present in a large proportion of individuals with up to 1 in 148 females or 1 in 290 males having FXpm (Maenner et al., 2013). As we have detailed in this chapter, FXpm is associated with mild to moderately impairing symptoms that vary across individuals, with the risk of expansion into FXS for females with FXpm who have children. In contrast to FXpm, FXS affects fewer individuals (e.g., 1 in 11,000 females and 1 in 7,000 males) and is characterized by moderate to severe symptoms in nearly all males and many females, requiring lifelong support to achieve optimal outcomes. Being informed about FMR1 gene mutations is important to support early diagnosis and targeted treatment for FXS and FXpm. However, this knowledge is also critical to advance understanding about genetic mutations more generally, as FMR1 gene mutations serve as an excellent model to understand related disorders.

Additional Resources

- Centers for Disease Control and Prevention (CDC): https://www.cdc.gov/ncbddd/fxs/index.html
- National Fragile X Foundation: https://fragilex.org/
- FRAXA Research Foundation: https://www.fraxa.org/
- National Institute of Child Health and Human Development (NICHD): https://www.nichd.nih.gov/health/topics/fragilex

References

Abbeduto, L., Murphy, M. M., Kover, S. T., Giles, N. D., Karadottir, S., Amman, A., Bruno, L., Kim, J. S., Schroeder, S., Anderson, J. A., & Nollin, K. A. (2008). Signaling noncomprehension of language: A comparison of fragile X syndrome and Down syndrome. *American Journal on Mental Retardation, 113*(3). https://doi.org/10.1352/0895-8017(2008)113[214:SNOLAC]2.0.CO;2

Adams, P. E., Adams, J. S., Nguyen, D. V., Hessl, D., Brunberg, J. A., Tassone, F., Zhang, W., Koldewyn, K., Rivera, S. M., Grigsby, J., Zhang, L., DeCarli, C., Hagerman, P. J., & Hagerman, R. J. (2010). Psychological symptoms correlate with reduced hippocampal volume in fragile X premutation carriers. *American Journal of Medical Genetics, Part B: Neuropsychiatric Genetics, 153*(3), 775–785. https://doi.org/10.1002/ajmg.b.31046

Aishworiya, R., Protic, D., Tang, S. J., Schneider, A., Tassone, F., & Hagerman, R. (2022). Fragile X-associated neuropsychiatric disorders (FXAND) in young fragile X premutation carriers. *Genes, 13*(12). https://doi.org/10.3390/genes13122399

Allen, E. G., Sherman, S., Abramowitz, A., Leslie, M., Novak, G., Rusin, M., Scott, E., & Letz, R. (2005). Examination of the effect of the polymorphic CGG repeat in the FMR1 gene on cognitive performance. *Behavior Genetics, 35*(4), 435–445. https://doi.org/10.1007/s10519-005-2792-4

American Psychiatric Association. (2022). Diagnostic and statistical manual of mental disorders (5th ed., text rev.). https://doi.org/10.1176/appi.books.9780890425787

Aziz, M., Stathopulu, E., Callias, M., Taylor, C., Turk, J., Oostra, B., Willemsen, R., & Patton, M. (2003). Clinical features of boys with fragile X premutations and intermediate alleles. *American Journal of Medical Genetics – Neuropsychiatric Genetics, 121 B*(1), 119–127. https://doi.org/10.1002/ajmg.b.20030

Bailey, D. B., Skinner, D., & Sparkman, K. L. (2003). *Discovering fragile X syndrome: Family experiences and perceptions*. http://publications.aap.org/pediatrics/article-pdf/111/2/407/1008016/pe0203000407.pdf

Bailey, D. B., Raspa, M., Olmsted, M., & Holiday, D. B. (2008). Co-occurring conditions associated with FMR1 gene variations: Findings from a national parent survey. *American Journal of Medical Genetics, Part A, 146*(16), 2060–2069. https://doi.org/10.1002/ajmg.a.32439

Bailey, D. B., Raspa, M., Bishop, E., & Holiday, D. (2009a). No change in the age of diagnosis for fragile X syndrome: Findings from a national parent survey. *Pediatrics, 124*(2), 527–533. https://doi.org/10.1542/peds.2008-2992

Bailey, D. B., Raspa, M., Holiday, D., Bishop, E., & Olmsted, M. (2009b). Functional skills of individuals with fragile X syndrome: A lifespan cross-sectional analysis. *American Journal on Intellectual and Developmental Disabilities, 114*(4), 289–303. https://doi.org/10.1352/1944-7558-114.4.289-303

Bangert, K., Scott, K. S., Adams, C., Kisenwether, J. S., Giuffre, L., Reed, J., Thurman, A. J., Abbeduto, L., & Klusek, J. (2022). Cluttering in the speech of young men with fragile X syndrome. *Journal of Speech, Language, and Hearing Research, 65*(3), 954–969. https://doi.org/10.1044/2021_JSLHR-21-00446

Bartholomay, K. L., Lee, C. H., Bruno, J. L., Lightbody, A. A., & Reiss, A. L. (2019). Closing the gender gap in fragile X syndrome: Review on females with FXS and preliminary research findings. *Brain Sciences, 9*(1). https://doi.org/10.3390/brainsci9010011

Bassell, G. J., & Warren, S. T. (2008). Fragile X syndrome: Loss of local mRNA regulation alters synaptic development and function. *Neuron, 60*, 201–214. https://doi.org/10.1016/j.neuron.2008.10.004

Bennetto, L., Pennington, B. F., Porter, D., Taylor, A. K., & Hagerman, R. J. (2001). Profile of cognitive functioning in women with the fragile X mutation. *Neuropsychology, 15*(2), 290–299. https://doi.org/10.1037/0894-4105.15.2.290

Berry-Kravis, E., Sumis, A., Hervey, C., & Mathur, S. (2012). Clinic-based retrospective analysis of psychopharmacology for behavior in fragile x syndrome. *International Journal Of Pediatrics, 2012*, 1. https://doi.org/10.1155/2012/843016

Black, C. J., Hogan, A. L., Smith, K. D., & Roberts, J. E. (2021). Early behavioral and physiological markers of social anxiety in infants with fragile X syndrome. *Journal of Neurodevelopmental Disorders, 13*(1), 11. https://doi.org/10.1186/s11689-021-09356-3

Brady, N., Skinner, D., Roberts, J., & Hennon, E. (2006). Communication in young children with fragile X syndrome: A qualitative study of mothers' perspectives. *American Journal of Speech-Language Pathology, 15*(4), 353. https://doi.org/10.1044/1058-0360(2006/033)

Brady, N. C., Fleming, K., Bredin-Oja, S. L., Fielding-Gebhardt, H., & Warren, S. F. (2020). Language development from early childhood to adolescence in youths with fragile X syndrome. *Journal of Speech, Language, and Hearing Research, 63*(11), 3727–3742. https://doi.org/10.1044/2020_JSLHR-20-00198

Bredin-Oja, S. L., Warren, S. F., Swinburne Romine, R. E., Fleming, K. K., Brady, N., & Berry-Kravis, E. (2021). Word retrieval difficulty in adult females with the FMR1 premutation: Changes over time and across contexts. *Brain and Cognition, 148*(January), 1–7. https://doi.org/10.1016/j.bandc.2021.105694

Bullard, L., & Abbeduto, L. (2021). Responsive parenting as a target for telehealth language interventions in fragile X syndrome: Implications for scalability and best practices. *Seminars in Speech and Language, 42*(4), 287–300. https://doi.org/10.1055/s-0041-1730989

Caravella, K. E., & Roberts, J. E. (2017). Adaptive skill trajectories in infants with fragile X syndrome contrasted to typical controls and infants at high risk for autism. *Research in Autism Spectrum Disorders, 40*, 1–12. https://doi.org/10.1016/j.rasd.2017.05.002

Clifford, S., Dissanayake, C., Bui, Q. M., Huggins, R., Taylor, A. K., & Loesch, D. Z. (2007). Autism spectrum phenotype in males and females with fragile X full mutation and premutation. *Journal of Autism and Developmental Disorders, 37*(4), 738–747. https://doi.org/10.1007/S10803-006-0205-Z/TABLES/2

Cordeiro, L., Ballinger, E., Hagerman, R., & Hessl, D. (2011). Clinical assessment of DSM-IV anxiety disorders in fragile X syndrome: Prevalence and characterization. *Journal of Neurodevelopmental Disorders, 3*(1), 57–67. https://doi.org/10.1007/s11689-010-9067-y

Cordeiro, L., Abucayan, F., Hagerman, R., Tassone, F., & Hessl, D. (2015). Anxiety disorders in fragile X premutation carriers: Preliminary characterization of probands and non-probands. *Intractable and Rare Diseases Research, 4*(3), 123–130. https://doi.org/10.5582/irdr.2015.01029

Cornish, K. M., Li, L., Kogan, C. S., Jacquemont, S., Turk, J., Dalton, A., Hagerman, R. J., & Hagerman, P. J. (2008). Age-dependent cognitive changes in carriers of the fragile X syndrome. *Cortex, 44*(6), 628–636. https://doi.org/10.1016/j.cortex.2006.11.002

Crawford, H., Abbeduto, L., Hall, S. S., Hardiman, R., Hessl, D., Roberts, J. E., Scerif, G., Stanfield, A. C., Turk, J., & Oliver, C. (2020). Fragile X syndrome: An overview of cause, characteristics, assessment, and management. *Paediatrics and Child Health, 30*(11), 400–403. https://doi.org/10.1016/j.paed.2020.08.007

Cregenzán-Royo, O., Brun-Gasca, C., & Fornieles-Deu, A. (2022). Behavior problems and social competence in fragile X syndrome: A systematic review. *Genes, 13*(2), 280. https://doi.org/10.3390/genes13020280

Darnell, J. C., & Klann, E. (2013). The translation of translational control by FMRP: Therapeutic targets for FXS. *Nature Neuroscience, 16*(11), 1530. https://doi.org/10.1038/nn.3379

Del Hoyo Soriano, L., Thurman, A. J., Harvey, D. J., Ted Brown, W., & Abbeduto, L. (2018). Genetic and maternal predictors of cognitive and behavioral trajectories in females with fragile X syndrome. *Journal of Neurodevelopmental Disorders, 10*(1), 22. https://doi.org/10.1186/s11689-018-9240-2

Diez-Itza, E., Viejo, A., & Fernández-Urquiza, M. (2022). Pragmatic profiles of adults with fragile X syndrome and Williams syndrome. *Brain Sciences, 12*(3). https://doi.org/10.3390/brainsci12030385

Erickson, C. A., Davenport, M. H., Schaefer, T. L., Wink, L. K., Pedapati, E. V., Sweeney, J. A., Fitzpatrick, S. E., Brown, W. T., Budimirovic, D., Hagerman, R. J., Hessl, D., Kaufmann, W. E., & Berry-Kravis, E. (2017). Fragile X targeted pharmacotherapy: Lessons learned and future directions. *Journal of Neurodevelopmental Disorders, 9*(1), 7. https://doi.org/10.1186/s11689-017-9186-9

Ezell, J., Hogan, A., Fairchild, A., Hills, K., Klusek, J., Abbeduto, L., & Roberts, J. (2019). Prevalence and predictors of anxiety disorders in adolescent and adult males with autism spectrum disorder and fragile X syndrome. *Journal of Autism and Developmental Disorders, 49*(3), 1131–1141. https://doi.org/10.1007/s10803-018-3804-6

Famula, J., Ferrer, E., Hagerman, R. J., Tassone, F., Schneider, A., Rivera, S. M., & Hessl, D. (2022). Neuropsychological changes in FMR1 premutation carriers and onset of fragile X-associated tremor/ataxia syndrome. *Journal of Neurodevelopmental Disorders, 14*(1), 23. https://doi.org/10.1186/s11689-022-09436-y

Farzin, F., Perry, H., Hessl, D., Loesch, D., Cohen, J., Bacalman, S., Louise Gane, M., Tassone, F., Hagerman, P., & Hagerman, R. (2006). Autism spectrum disorders and attention-deficit/hyperactivity disorder in boys with the fragile X premutation. *Journal of Developmental & Behavioral Pediatric, 27*(2), S137.

Finestack, L. H., & Abbeduto, L. (2010). Expressive language profiles of verbally expressive adolescents and young adults with Down syndrome or fragile X syndrome. *Journal of Speech Language and Hearing Research, 53*(5), 1334–1348. https://doi.org/10.1044/1058-0360(2011/10-0095)

Flavell, J., Franklin, C., & Nestor, P. J. (2023). A systematic review of fragile X-associated neuropsychiatric disorders. *The Journal of Neuropsychiatry and Clinical Neurosciences, 35*(2), 110–120. https://doi.org/10.1176/appi.neuropsych.21110282

Friefeld, S. J., & Macgregor, D. (1994). Sensorimotor coordination in boys with fragile X syndrome. *Occupational Therapy International, 1*(3), 174–182. https://doi.org/10.1002/OTI.6150010305

Glaser, B., Hessl, D., Dyer-Friedman, J., Johnston, C., Wisbeck, J., Taylor, A., & Reiss, A. (2003). Biological and environmental contributions to adaptive behavior in fragile X syndrome. *American Journal of Medical Genetics, 117A*(1), 21–29. https://doi.org/10.1002/ajmg.a.10549

Grefer, M., Flory, K., Cornish, K., Hatton, D., & Roberts, J. (2016). The emergence and stability of attention deficit hyperactivity disorder in boys with fragile X syndrome. *Journal of Intellectual Disability Research, 60*(2), 167–178. https://doi.org/10.1111/JIR.12226

Grigsby, J., Brega, A. G., Engle, K., Leehey, M. A., Hagerman, R. J., Tassone, F., Hessl, D., Hagerman, P. J., Cogswell, J. B., Bennett, R. E., Cook, K., Hall, D. A., Bounds, L. S., Paulich, M. J., & Reynolds, A. (2008). Cognitive profile of fragile X premutation carriers with and without fragile X-associated tremor/ataxia syndrome. *Neuropsychology, 22*(1), 48–60. https://doi.org/10.1037/0894-4105.22.1.48

Gross, C., Hoffmann, A., Bassell, G. J., & Berry-Kravis, E. M. (2015). Therapeutic strategies in fragile X syndrome: From bench to bedside and back. *Neurotherapeutics, 12*(3), 584–608. Springer New York LLC. https://doi.org/10.1007/s13311-015-0355-9

Hagerman, R. J., & Hagerman, P. (2016). Fragile X associated tremor/ataxia syndrome—features, mechanisms and management. *Nature Reviews, 12*, 403–412.

Hagerman, R. J., Berry-Kravis, E., Kaufmann, W. E., Ono, M. Y., Tartaglia, N., Lachiewicz, A., Kronk, R., Delahunty, C., Hessl, D., Visootsak, J., Picker, J., Gane, L., & Tranfaglia, M. (2009). Advances in the treatment of fragile x syndrome. *Pediatrics, 123*(1), 378–390. https://doi.org/10.1542/peds.2008-0317

Hahn, L. J., Brady, N. C., Warren, S. F., & Fleming, K. K. (2015). Do children with fragile X syndrome show declines or plateaus in adaptive behavior? *American Journal on Intellectual and Developmental Disabilities, 120*(5), 412–432. https://doi.org/10.1352/1944-7558-120.5.412

Hartley, S. L., Seltzer, M. M., Raspa, M., Olmstead, M., Bishop, E., & Bailey, D. B. (2011). Exploring the adult life of men and women with fragile X syndrome: Results from a national survey. *American Journal on Intellectual and Developmental Disabilities, 116*(1), 16–35. https://doi.org/10.1352/1944-7558-116.1.16

Hernandez, R. N., Feinberg, R. L., Vaurio, R., Passanante, N. M., Thompson, R. E., & Kaufmann, W. E. (2009). Autism spectrum disorder in fragile X syndrome: A longitudinal evaluation. *American Journal of Medical Genetics. Part A, 149A*(6), 1125. https://doi.org/10.1002/AJMG.A.32848

Hinton, R., Budimirovic, D. B., Marschik, P. B., Talisa, V. B., Einspieler, C., Gipson, T., & Johnston, M. V. (2013). Parental reports on early language and motor milestones in fragile X syndrome with and without autism spectrum disorders. *Developmental Neurorehabilitation, 16*(1), 58–66. https://doi.org/10.3109/17518423.2012.704414

Hoffmann, A. (2022). Communication in fragile X syndrome: Patterns and implications for assessment and intervention. *Frontiers in Psychology, 13*. https://doi.org/10.3389/FPSYG.2022.929379

Hoffmann, A., Krause, S. E., Wuu, J., Leurgans, S., Guter, S. J., Block, S. S., Salt, J., Cook, E., Maino, D. M., & Berry-Kravis, E. (2019). Vocabulary comprehension in adults with fragile X syndrome (FXS). *Journal of Neurodevelopmental Disorders, 11*(1), 1–9. https://doi.org/10.1186/s11689-019-9285-x

Hogan, A., Hunt, E., Smith, K., Black, C., Bangert, K., Klusek, J., & Roberts, J. (2021). Trajectories of heart activity across infancy to early childhood differentially predict autism and anxiety symptoms in fragile X syndrome. *Frontiers in Psychiatry, 12*. https://doi.org/10.3389/fpsyt.2021.727559

Huddleston, L. B., Visootsak, J., & Sherman, S. L. (2014). Cognitive aspects of fragile X syndrome. *Wiley Interdisciplinary Reviews: Cognitive Science, 5*(4), 501–508. https://doi.org/10.1002/wcs.1296

Hunter, J. E., Epstein, M. P., Tinker, S. W., Abramowitz, A., & Sherman, S. L. (2012). The FMR1 premutation and attention-deficit hyperactivity disorder (ADHD): Evidence for a complex inheritance. *Behavior Genetics, 42*(3), 415–422. https://doi.org/10.1007/S10519-011-9520-Z/TABLES/3

Hunter, J., Rivero-Arias, O., Angelov, A., Kim, E., Fotheringham, I., & Leal, J. (2014). Epidemiology of fragile X syndrome: A systematic review and meta-analysis. *American Journal of Medical Genetics Part A, 164*(7), 1648–1658. https://doi.org/10.1002/ajmg.a.36511

Iverson, J. M. (2010). Developing language in a developing body: The relationship between motor development and language development. *Journal of Child Language, 37*, 229–261. https://doi.org/10.1017/S0305000909990432

Jacquemont, S., Hagerman, R. J., Hagerman, P. J., & Leehey, M. A. (2007). Review fragile-X syndrome and fragile X-associated tremor/ataxia syndrome: Two faces of FMR1. *The Lancet Neurology, 6*(1), 45–55. http://neurology.thelancet.comVol

Joga-Elvira, L., Jacas, C., Joga, M. L., Roche-Martínez, A., & Brun-Gasca, C. (2021). Fragile X syndrome in young females: Influence of executive function on the neurocognitive profile and adaptive behavior. *Research in Developmental Disabilities, 112*(November 2020), 103912. https://doi.org/10.1016/j.ridd.2021.103912

Jordan, T. L., Bartholomay, K. L., Lee, C. H., Lightbody, A. A., & Reiss, A. L. (2023). Research in developmental disabilities cognition, academic achievement, and adaptive behavior in school-aged girls with fragile X syndrome. *Research in Developmental Disabilities, 143*(May), 1–10.

Kau, A. S. M., Meyer, W. A., & Kaufmann, W. E. (2002). Early development in males with fragile X syndrome: A review of the literature. *Microscopy Research and Technique, 57*(3), 174–178. https://doi.org/10.1002/JEMT.10069

Kaufmann, W. E., Kidd, S. A., Andrews, H. F., Budimirovic, D. B., Esler, A., Haas-Givler, B., Stackhouse, T., Riley, C., Peacock, G., Sherman, S. L., Brown, W. T., & Berry-Kravis, E. (2017). Autism spectrum disorder in fragile X syndrome: Cooccurring conditions and current treatment. *Pediatrics, 139*(Supplement_3), S194–S206. https://doi.org/10.1542/PEDS.2016-1159F

Kenna, H. A., Tartter, M., Hall, S. S., Lightbody, A. A., Nguyen, Q., de los Angeles, C. P., Reiss, A. L., & Rasgon, N. L. (2013). High rates of comorbid depressive and anxiety disorders among women with premutation of the FMR1 gene. *American Journal of Medical Genetics Part B: Neuropsychiatric Genetics, 162*(8), 872–878. https://doi.org/10.1002/AJMG.B.32196

Klaiman, C., Quintin, E. M., Jo, B., Lightbody, A. A., Hazlett, H. C., Piven, J., Hall, S. S., & Reiss, A. L. (2014). Longitudinal profiles of adaptive behavior in fragile X syndrome. *Pediatrics, 134*(2), 315–324. https://doi.org/10.1542/peds.2013-3990

Klusek, J., Martin, G. E., & Losh, M. (2014). Consistency between research and clinical diagnoses of autism among boys and girls with fragile X syndrome. *Journal of Intellectual Disability Research, 58*(10), 940–952. https://doi.org/10.1111/jir.12121

Klusek, J., McGrath, S. E., Abbeduto, L., & Roberts, J. E. (2016). Pragmatic language features of mothers with the FMR1 premutation are associated with the language outcomes of adolescents and young adults with fragile X syndrome. *Journal of Speech, Language, and Hearing Research, 59*(2), 49–61. https://doi.org/10.1044/2015

Klusek, J., LaFauci, G., Adayev, T., Brown, W. T., Tassone, F., & Roberts, J. E. (2017). Reduced vagal tone in women with the FMR1 premutation is associated with FMR1 mRNA but not depression or anxiety. *Journal of Neurodevelopmental Disorders, 9*(1), 16. https://doi.org/10.1186/s11689-017-9197-6

Klusek, J., Porter, A., Abbeduto, L., Adayev, T., Tassone, F., Mailick, M. R., Glicksman, A., Tonnsen, B. L., & Roberts, J. E. (2018). Curvilinear association between language disfluency and FMR1 CGG repeat size across the normal, intermediate, and premutation range. *Frontiers in Genetics, 9*(AUG), 1–14. https://doi.org/10.3389/fgene.2018.00344

Klusek, J., Fairchild, A. J., & Roberts, J. E. (2019). Vagal tone as a putative mechanism for pragmatic competence: An investigation of carriers of the FMR1 premutation. *Journal of Autism and Developmental Disorders, 49*(1), 197–208. https://doi.org/10.1007/s10803-018-3714-7

Klusek, J., O'Connor, S. L., Hickey, A., Hills, K. J., Abbeduto, L., & Roberts, J. E. (2022). Attention/deficit hyperactivity disorder in adolescent and young adult males with fragile X syndrome. *American Journal on Intellectual and Developmental Disabilities, 127*(3), 213–230. https://doi.org/10.1352/1944-7558-127.3.213

Kover, S. T., & Abbeduto, L. (2010). Expressive language in male adolescents with fragile X syndrome with and without comorbid autism. *Journal of Intellectual Disability Research, 54*(3), 246–265. https://doi.org/10.1111/j.1365-2788.2010.01255.x.Expressive

Lightbody, A. A., Bartholomay, K. L., Jordan, T. L., Lee, C. H., Miller, J. G., & Reiss, A. L. (2022). Anxiety, depression, and social skills in girls with fragile X syndrome: Understanding the cycle to improve outcomes. *Journal of Developmental and Behavioral Pediatrics, 43*(9), E565–E572. https://doi.org/10.1097/DBP.0000000000001128

Loesch, D. Z., Bui, Q. M., Butler, E., Huggins, R. M., Grigsby, J., Epstein, J., Taylor, A. K., & Hagerman, R. J. (2003). Effect of the fragile X status categories and the fragile X mental retardation protein levels on executive functioning in males and females with fragile X. *Neuropsychology, 17*(4), 646–657. https://doi.org/10.1037/0894-4105.17.4.646

Losh, M., Martin, G. E., Klusek, J., Hogan-Brown, A. L., & Sideris, J. (2012). Social communication and theory of mind in boys withautism and fragile X syndrome. *Frontiers in Psychology, 3*(AUG), 26952. https://doi.org/10.3389/FPSYG.2012.00266/BIBTEX

Lowell, E. P., Tonnsen, B. L., Bailey, D. B., & Roberts, J. E. (2017). The effects of optimism, religion, and hope on mood and anxiety disorders in women with the FMR1 premutation. *Journal of Intellectual Disability Research, 61*(10), 916–927. https://doi.org/10.1111/jir.12409

Lozano, R., Rosero, C. A., & Hagerman, R. J. (2014). Fragile X spectrum disorders. *Intractable and Rare Diseases Research, 3*(4), 134–146. International Advancement Center for Medicine and Health Research. https://doi.org/10.5582/irdr.2014.01022

Maenner, M. J., Baker, M. W., Broman, K. W., Tian, J., Barnes, J. K., Atkins, A., Mcpherson, E., Hong, J., Brilliant, M. H., & Mailick, M. R. (2013). *FMR1 CGG expansions: Prevalence and sex ratios* (Vol. 162B, p. 466). https://doi.org/10.1002/ajmg.b.32176

Maenner, M. J., Warren, Z., Williams, A. R., Amoakohene, E., Bakian, A. V., Bilder, D. A., Durkin, M. S., Fitzgerald, R. T., Furnier, S. M., Hughes, M. M., Ladd-Acosta, C. M., McArthur, D., Pas, E. T., Salinas, A., Vehorn, A., Williams, S., Esler, A., Grzybowski, A., Hall-Lande, J., … Shaw, K. A. (2023). Prevalence and Characteristics of Autism Spectrum Disorder Among Children Aged 8 Years—Autism and Developmental Disabilities Monitoring Network, 11 Sites, United States, 2020. MMWR Surveillance Summaries, 72(2), 1–14. https://doi.org/10.15585/mmwr.ss7202a1

Martin, G. E., Ausderau, K. K., Raspa, M., Bishop, E., Mallya, U., & Bailey, D. B. (2013). Therapy service use among individuals with fragile X syndrome: Findings from a US parent survey. *Journal of Intellectual Disability Research, 57*(9), 837–849. https://doi.org/10.1111/j.1365-2788.2012.01608.x

Martin, G. E., Bush, L., Klusek, J., Patel, S., & Losh, M. (2018). A multimethod analysis of pragmatic skills in children and adolescents with fragile X syndrome, autism spectrum disorder, and Down syndrome. *Journal of Speech, Language, and Hearing Research, 61*(12), 3023–3037. https://doi.org/10.1044/2018_JSLHR-L-18-0008

Mattie, L. J., & Hamrick, L. R. (2022). Early communication development in infants and toddlers with fragile X syndrome. *Autism and Developmental Language Impairments, 7*, 7–14. https://doi.org/10.1177/23969415221099403

McDuffie, A. S., Kover, S. T., Abbeduto, L., Lewis, P., & Ted Brown, W. (2012). Profiles of receptive and expressive language abilities in males with comorbid fragile X syndrome and autism. *American Journal on Intellectual and Developmental Disabilities, 117*(1), 18–32. https://doi.org/10.1352/1944-7558-117.1.18.Profiles

Meng, L., Kaufmann, W. E., Frye, R. E., Ong, K., Kaminski, J. W., Velinov, M., & Berry-Kravis, E. (2021). The association between mosaicism type and cognitive and behavioral functioning among males with fragile X syndrome. *American Journal of Medical Genetics Part A, 188*, 858. https://doi.org/10.1002/ajmg.a.62594

Moore, C. J., Daly, E. M., Schmitz, N., Tassone, F., Tysoe, C., Hagerman, R. J., Hagerman, P. J., Morris, R. G., Murphy, K. C., & Murphy, D. G. M. (2004). A neuropsychological investigation of male premutation carriers of fragile X syndrome. *Neuropsychologia, 42*(14), 1934–1947. https://doi.org/10.1016/j.neuropsychologia.2004.05.002

Movaghar, A., Mailick, M., Sterling, A., & Greenberg, J. (2017). Automated screening for fragile X premutation carriers based on linguistic and cognitive computational phenotypes. *Scientific Reports, 7*(1), 1–11. https://doi.org/10.1038/s41598-017-02682-4

Myers, G. F., Miche, M., Mazzocco, M. M. M., Maddalena, A., & Reiss, A. L. (2001). No widespread psychological effect of the fragile X premutation in childhood: Evidence from a preliminary controlled study original articles. *Developmental and Behavioral Pediatrics, 22*(6), 353. https://doi.org/10.1097/00004703-200112000-00001

Nash, R., Riley, C., Paramsothy, P., Gilbertson, K., Raspa, M., Wheeler, A., Dziuban, E. J., & Peacock, G. (2019). A description of the educational setting among individuals with fragile x syndrome. *American Journal on Intellectual and Developmental Disabilities, 124*(1), 57–76. https://doi.org/10.1352/1944-7558-124.1.57

Philofsky, A., Hepburn, S. L., Hayes, A., Hagerman, R., & Rogers, S. J. (2004). Linguistic and cognitive functioning and autism symptoms in young children with fragile X syndrome. *American Journal on Mental Retardation, 109*(3), 208–218. https://doi.org/10.1352/0895-8017(2004)109<208:LACFAA>2.0.CO;2

Poeta, L., Drongitis, D., Verrillo, L., & Miano, M. G. (2020). DNA Hypermethylation and unstable repeat diseases: A paradigm of transcriptional silencing to decipher the basis of pathogenic mechanisms. *Genes, 11*(6), 684. https://doi.org/10.3390/genes11060684

Polussa, J., Schneider, A., & Hagerman, R. (2014). Molecular advances leading to treatment implications for fragile X premutation carriers. *Brain Disorders & Therapy, 03*(02). https://doi.org/10.4172/2168-975x.1000119

Protic, D. D., Aishworiya, R., Salcedo-Arellano, M. J., Tang, S. J., Milisavljevic, J., Mitrovic, F., Hagerman, R. J., & Budimirovic, D. B. (2022). Fragile X syndrome: From molecular aspect to clinical treatment. *International Journal of Molecular Sciences, 23*(4), 1935. MDPI. https://doi.org/10.3390/ijms23041935

Quintin, E.-M., Jo, B., Hall, S. S., Bruno, J. L., Chromik, L., Raman, M., Lightbody, A. A., Martin, A., & Reiss, A. (2016). The cognitive developmental profile associated with fragile X syndrome: A longitudinal investigation of cognitive strengths and weaknesses through childhood and adolescence. *Development and Psychopathology, 28*(4), 1457–1469. https://doi.org/10.1017/S0954579415001200

Rague, L., Caravella, K. E., Tonnsen, B. L., Klusek, J., & Roberts, J. E. (2018). Early gesture use in fragile X syndrome. *Journal of Intellectual & Developmental Disability, 62*(7), 625–636. https://doi.org/10.1111/jir.12498.Early

Raspa, M., Wheeler, A. C., & Riley, C. (2017). Public health literature review of fragile X syndrome. *Pediatrics, 139*(Supplement_3), S153–S171. https://doi.org/10.1542/PEDS.2016-1159C

Roberts, J. E., Hatton, D. D., & Bailey, D. B. (2001). Development and behavior of male toddlers with fragile X syndrome. *Journal of Early Intervention, 24, 207.*

Roberts, J. E., Mirrett, P., Anderson, K., Burchinal, M., & Neebe, E. (2002). Early communication, symbolic behavior, and social profiles of young males with fragile X syndrome. *American Journal of Speech-Language Pathology, 11*(3), 295–304. https://doi.org/10.1044/1058-0360(2002/034)

Roberts, J., Martin, G. E., Moskowitz, L., Harris, A. A., Foreman, J., & Nelson, L. (2007). Discourse skills of boys with fragile X syndrome in comparison to boys with down syndrome. *Journal of Speech, Language, and Hearing Research, 50*(2), 475–492. https://doi.org/10.1044/1092-4388(2007/033)

Roberts, J. E., Bailey, D. B., Mankowski, J., Ford, A., Sideris, J., Weisenfeld, L. A., Heath, T. M., & Golden, R. N. (2009a). Mood and anxiety disorders in females with the FMR1 premutation. *American Journal of Medical Genetics, Part B: Neuropsychiatric Genetics, 150*(1), 130–139. https://doi.org/10.1002/ajmg.b.30786

Roberts, J. E., Mankowski, J. B., Sideris, J., Goldman, B. D., Hatton, D. D., Mirrett, P. L., Baranek, G. T., Reznick, J. S., Long, A. C. J., & Bailey, D. B. (2009b). Trajectories and predictors of the development of very young boys with fragile X syndrome. *Journal of Pediatric Psychology, 34*(8), 827–836. https://doi.org/10.1093/jpepsy/jsn129

Roberts, J. E., Tonnsen, B. L., McCary, L. M., Ford, A. L., Golden, R. N., & Bailey, D. B. (2016). Trajectory and predictors of depression and anxiety disorders in mothers with the FMR1 premutation. *Biological Psychiatry, 79*(10), 850–857. https://doi.org/10.1016/j.biopsych.2015.07.015

Roberts, J. E., Bradshaw, J., Will, E., Hogan, A. L., McQuillin, S., & Hills, K. (2020). Emergence and rate of autism in fragile X syndrome across the first years of life. *Development and Psychopathology, 32*(4), 1335–1352. https://doi.org/10.1017/S0954579420000942

Seritan, A. L., Bourgeois, J. A., Schneider, A., Mu, Y., Hagerman, R. J., & Nguyen, D. V. (2013). Ages of onset of mood and anxiety disorders in fragile X premutation carriers. *Current Psychiatry Reviews, 9*(1), 65–71. https://doi.org/10.2174/157340013805289662

Simko, A., Hornstein, L., Soukup, S., & Bagamery, N. (1989). Fragile X syndrome: Recognition in young children. *Pediatrics, 83*(4), 547. http://publications.aap.org/pediatrics/article-pdf/83/4/547/1053886/547.pdf

Simon, J. A., Keenan, J. M., Pennington, B. F., Taylor, A. K., & Hagerman, R. J. (2001). Discourse processing in women with fragile X syndrome: Evidence for a deficit establishing coherence. *Cognitive Neuropsychology, 18*(1), 1–18. https://doi.org/10.1080/02643290126042

Smith, K., Hogan, A. L., Will, E., & Roberts, J. E. (2021). Attention bias and prodromal anxiety symptoms in toddlers with fragile X syndrome and down syndrome. *American Journal on Intellectual and Developmental Disabilities, 126*(2), 167–181. https://doi.org/10.1352/1944-7558-126.2.167

Sterling, A. M., Mailick, M., Greenberg, J., Warren, S. F., & Brady, N. (2013). Language dysfluencies in females with the FMR1 premutation. *Brain and Cognition, 82*(1), 84–89. https://doi.org/10.1016/j.bandc.2013.02.009

Thurman, A. J., McDuffie, A., Hagerman, R. J., Josol, C. K., & Abbeduto, L. (2017). Language skills of males with fragile X syndrome or nonsyndromic autism spectrum disorder. *Journal of Autism and Developmental Disorders, 47*(3), 728–743. https://doi.org/10.1007/s10803-016-3003-2

Wall, C. A., Hogan, A. L., Will, E. A., McQuillin, S., Kelleher, B. L., & Roberts, J. E. (2019). Early negative affect in males and females with fragile X syndrome: Implications for anxiety and autism. *Journal of Neurodevelopmental Disorders, 11*(1), 22. https://doi.org/10.1186/s11689-019-9284-y

Wang, J. Y., Danial, M., Soleymanzadeh, C., Kim, B., Xia, Y., Kim, K., Tassone, F., Hagerman, R. J., & Rivera, S. M. (2020). Cortical gyrification and its relationships with molecular measures and cognition in children with the FMR1 premutation. *Scientific Reports, 10*(1), 1–13. https://doi.org/10.1038/s41598-020-73040-0

Weiss, J. A., Thomson, K., Burnham Riosa, P., Albaum, C., Chan, V., Maughan, A., Tablon, P., & Black, K. (2018). A randomized waitlist-controlled trial of cognitive behavior therapy to improve emotion regulation in children with autism. *Journal of Child Psychology and Psychiatry and Allied Disciplines, 59*(11), 1180–1191. https://doi.org/10.1111/jcpp.12915

Wheeler, A., Raspa, M., Bann, C., Bishop, E., Hessl, D., Sacco, P., & Bailey, D. B. (2014). Anxiety, attention problems, hyperactivity, and the aberrant behavior checklist in fragile X syndrome. *American Journal of Medical Genetics, Part A, 164*(1), 141–155. https://doi.org/10.1002/ajmg.a.36232

Wheeler, A. C., Sideris, J., Hagerman, R., Berry-Kravis, E., Tassone, F., & Bailey, D. B. (2016). Developmental profiles of infants with an FMR1 premutation. *Journal of Neurodevelopmental Disorders, 8*(1), 1–11. https://doi.org/10.1186/s11689-016-9171-8

Wheeler, A., Raspa, M., Hagerman, R., Mailick, M., & Riley, C. (2017). Implications of the FMR1 premutation for children, adolescents, adults, and their families. *Pediatrics, 139*, S172–S182. American Academy of Pediatrics. https://doi.org/10.1542/peds.2016-1159D

Wheeler, A. C., Gwaltney, A., Raspa, M., Okoniewski, K. C., Berry-Kravis, E., Botteron, K. N., Budimirovic, D., Hazlett, H. C., Hessl, D., Losh, M., Martin, G. E., Rivera, S. M., Roberts, J. E., & Bailey, D. B. (2021). Emergence of developmental delay in infants and toddlers with an FMR1 mutation. *Pediatrics, 147*(5), 1–11. https://doi.org/10.1542/peds.2020-011528

Will, E. A., & Roberts, J. E. (2021). Motor influences on communication: Comparisons between down syndrome and fragile X syndrome. *American Journal of Intellectual and Developmental Disabilities, 126*(6), 460–476. https://doi.org/10.1352/1944-7558-126.6.460

Will, E. A., Bishop, S. L., & Roberts, J. E. (2019). Developmental divergence: Motor trajectories in children with fragile X syndrome with and without co-occurring autism. *Journal of Neurodevelopmental Disorders, 11*(1), 23. https://doi.org/10.1186/s11689-019-9281-1

Zingerevich, C., Greiss-Hess, L., Lemons-Chitwood, K., Harris, S. W., Hessl, D., Cook, K., & Hagerman, R. J. (2008). Motor abilities of children diagnosed with fragile X syndrome with and without autism. *Journal of Intellectual Disability Research, 53*(1), 11. https://doi.org/10.1111/j.1365-2788.2008.01107.x

Autism Spectrum Disorder: A Lifespan Overview

Rachel A. Gordon and Anastasia Dimitropoulos

Abstract Autism spectrum disorder (ASD) is a neurodevelopmental disability associated with difficulties in social communication and interaction and patterns of repetitive and restricted behaviors, interests, and preferences. Autism is a spectrum condition, presenting diversely across individuals, often intertwined with co-occurring medical or psychological disorders, and related to differing levels of clinical impairment. ASD can be detected in early infancy and changes across the lifespan. Research thus far has identified a combination of genetic and environmental risk factors that contribute to the etiology of autism. Accurate diagnosis requires an interdisciplinary team alongside child, parent, and medical or behavioral assessments. Following diagnosis, various supports including early intervention, school-based interventions, vocational training, mental health care, and independent living options are available. Quality interventions are individualized and implemented in home, school, and community settings to support an autistic person's well-being across the lifespan. From a public health perspective, the increasing prevalence of ASD alongside difficulties in accessing autism-specific supports particularly impact marginalized and diverse groups of autistic individuals. These challenges have led to calls for policy reform to increase equitable and quality care. Overall, the tenants of neurodiversity that value differences, not deficits, and a strength-based, lifespan approach are imperative to support this community.

Keywords Autism · ASD · Social communication · Early intervention · Neurodiversity

---

R. A. Gordon (✉) · A. Dimitropoulos
Department of Psychological Sciences, Case Western Reserve University, Cleveland, OH, USA
e-mail: rxg612@case.edu

M. G. Valdovinos (ed.), *Intellectual and Developmental Disabilities*, Contemporary Clinical Neuroscience,
https://doi.org/10.1007/978-3-031-66932-3_15

Autism was first described by Leo Kanner in 1943, who noticed patterns in intelligent children who appeared to prefer sameness and being alone. Later captured by Hans Asperger in "milder" forms, autism was initially characterized by social challenges, specific interests, and intelligence. From its first introduction in the *Diagnostic and Statistical Manual of Mental Disorders* (3rd ed., rev.; DSM-III-R; American Psychiatric Association, 1987) in 1987, through its various modifications over time, autism has been defined as a developmental disorder capturing a range of heterogenous presentations in addition to core deficits of social communication and restricted, repetitive behaviors. Other terms previously used to define autism spectrum disorder (ASD) as we currently know it include autistic disorder, Asperger's disorder, and pervasive developmental disorder–not otherwise specified. In 2013, the *DSM-5* collapsed these sub-categories and established ASD as a singular diagnosis that ranges in severity. Alongside these diagnostic changes, clinical and public understandings of autism have expanded, initially focused on behaviorally driven treatments, and now encompassing considerations for mental health, diversity, and advocacy in ASD.

A Note on Language

Autism-referential language preferences are diverse across the autism community. While the medical community emphasizes the utilization of person-first language [persons with autism], it is important to recognize that some individuals prefer to identify themselves as autistic and wish to be referred to as such. Individuals within the autism community have historically disagreed on language preferences, rendering it necessary to listen to and respect nuances in viewpoints (Dwyer et al., 2022). This chapter reflects respect for individual differences in self-referential language in relation to autism or ASD. As a result, a variety of autism-referential language will be employed, including terms such as: autistic, person with autism, individual on the spectrum, and person who is autistic, as suggested by Arnhart et al., 2022. This chapter will also use both terms of autism and ASD as done in clinical practice.

Understanding the Diagnosis

Autism is a spectrum disorder with significant heterogeneity across presentations, severity level, and comorbid conditions. The primary DSM-5-TR criteria for ASD include core symptoms across two domains: social communication and interaction, and restricted, repetitive behavior patterns (see Table 1 for DSM-5-TR criteria). Recent conceptualizations have shifted to view ASD as a series of traits that deviate from normative development and present in early years due to brain-based differences. Notably, different symptoms and associated challenges are suggested to result as cascading effects of challenges due to lack of appropriate supports in the environment, creating a symptom profile that comprises autism at its core (Klin et al., 2020).

Table 1 DSM-5-TR autism spectrum disorder diagnostic criteria (American Psychiatric Association, 2022)

A. *Persistent deficits in social communication and social interaction* across multiple contexts, as manifested by the following, currently or by history.	1. Deficits in social–emotional reciprocity, ranging, for example, from abnormal social approach and failure of normal back-and-forth conversation; to reduced sharing of interests, emotions, or affect; to failure to initiate or respond to social interactions.
	2. Deficits in nonverbal communicative behaviors used for social interaction, ranging, for example, from poorly integrated verbal and nonverbal communication; to abnormalities in eye contact and body language or deficits in understanding and use of gestures; to a total lack of facial expressions and nonverbal communication.
	3. Deficits in developing, maintaining, and understanding relationships, ranging, for example, from difficulties adjusting behavior to suit various social contexts; to difficulties in sharing imaginative play or in making friends; to absence of interest in peers.
B. *Restricted, repetitive patterns of behavior, interests, or activities,* as manifested by at least two of the following, currently or by history.	1. Stereotyped or repetitive motor movements, use of objects, or speech (e.g., simple motor stereotypies, lining up toys or flipping objects, echolalia, idiosyncratic phrases).
	2. Insistence on sameness, inflexible adherence to routines, or ritualized patterns or verbal nonverbal behavior (e.g., extreme distress at small changes, difficulties with transitions, rigid thinking patterns, greeting rituals, need to take the same route or eat food every day).
	3. Highly restricted, fixated interests that are abnormal in intensity or focus (e.g., strong attachment to or preoccupation with unusual objects, excessively circumscribed or perseverative interest).
	4. Hyper- or hyporeactivity to sensory input or unusual interests in sensory aspects of the environment (e.g., apparent indifference to pain/temperature, adverse response to specific sounds or textures, excessive smelling or touching of objects, visual fascination with lights or movement).
C. Symptoms must be present in the early developmental period (but may not become fully manifest until social demands exceed limited capacities or may be masked by learned strategies in later life).	
D. Symptoms cause clinically significant impairment in social, occupational, or other important areas of current functioning.	
E. These disturbances are not better explained by intellectual disability (ID) (intellectual developmental disorder) or global developmental delay. ID and autism spectrum disorder frequently co-occur; to make comorbid diagnoses of autism spectrum disorder and ID, social communication should be below that expected for general developmental level.	

Social Communication and Interactions

Within the first few months of development, reduced social engagement is present in ASD, and is thought to lead to atypical behaviors as a result of reduced social learning and cognition. Of note, social interaction is a platform for brain development in infancy and throughout the lifespan. This platform for social growth is typically built off of innate abilities for social orienting, reciprocity, and neuroplasticity key in human learning (Johnson et al., 2005; Shultz et al., 2018). However, individuals with autism show differences in how they construct the world and are less attuned to social factors as a result of biological differences in these areas.

Nonverbal Communication Individuals with autism demonstrate impairments in social communication, evident in early non-verbal patterns, such as facial expressions, gestures, and eye contact. Individuals with ASD show reduced eye contact and attention to human faces, as eye tracking studies have demonstrated increased fixation on objects versus humans, and human mouths versus eyes (Dawson et al., 2005; Klin et al., 2002). In addition, eye contact and gestures are often not properly coordinated, as children with ASD are less likely to demonstrate facial expressions or utilize eye gaze to communicate (Landa et al., 2007; Stone et al., 1997). Unlike typical development where children utilize gestures prior to the use of vocalization to direct attention, children with ASD may vocalize before utilizing gestures, and later on, may use gestures to compensate for language delays (Özçalışkan et al., 2016; Talbott et al., 2020). Overall, research has found autistic individuals to be less likely to point and gesture to share, reference, or engage others, demonstrating gestures that are less frequent, delayed, less coordinated with eye contact and spoken language, and atypical (Mishra et al., 2021; Manwaring et al., 2018; Talbott et al., 2020). For example, children who are autistic may physically use others' hands as a "tool" to accomplish goals, such as putting their parents' hands on a jar to open it, rather than communicating their needs via gestures and language (Mishra et al., 2021). People on the spectrum may demonstrate flat affect, produce fewer and lower quality facial expressions than neurotypical peers, and are less likely to share facial expressions with others (Trevisan et al., 2018). However, this difference does not necessarily mean that these individuals experience emotions any differently; rather, this stems from a difficulty in social cognition related to demonstrating consistent, socially relevant emotional states (Trevisan et al., 2018). Facial expressions may appear awkward, forced, or unusual, impacting social interactions, while autistic individuals with higher IQ and more life experiences are often able to produce more facial expressions consistent with neurotypical "norms" (Trevisan et al., 2018). Overall, nonverbal communication deficits stem from challenges with social cognition and *joint attention*, or the ability to share focus or attention with another person, seen through reduced social orienting and attention to others, as well as reduced empathy.

Expressive and Receptive Communication While language challenges are one of the most frequent concerns in autism, language disorders are not required for a diagnosis. Expressive language is often delayed in children with autism, as language skills may appear initially typical, but then plateau or regress in the early preschool years (Rapin & Dunn, 1997; Ozonoff et al., 2010; Zwaigenbaum et al., 2013). The majority of children with ASD are late talkers or develop speech at a slower rate; however, less than 20% of those with ASD are fully nonvocal and never develop speech (Lord et al., 2004; Tager-Flusberg et al., 2005). Individuals may use supportive language tools, such as augmentative and alterative communication, pictures such as picture exchange communicative system boards, or other devices to communicate with others. By the school-aged years, a language profile typical to autism often emerges in which comprehension, semantics, and language morphology is most impacted. Specifically, language is characterized by odd utterances, sometimes including grammar or articulation challenges, and may be overly precise, descriptive, or concrete—often leading others to describe children with autism to speak as "little professors" or in a "bookish" manner (Boucher, 2012; Tager-Flusberg et al., 2005). Individuals with ASD may demonstrate flat or unusual intonation or prosody in speech patterns and may speak in a high-pitched or sing-song voice (Landa et al., 2007; Rapin & Dunn, 1997). Furthermore, individuals with autism often display immediate and delayed *echolalia*, or repetition of the last word or phrase they heard, suggesting impaired comprehension (Tager-Flusberg et al., 2005). Similarly, autistic individuals may exhibit "*scripting*" in which they repeat phrases, songs, or dialogues they learn from others, daily routines, or from media, such as TV shows or radio jingles, in a variety of settings. Those who are more verbal struggle with the social rules of conversations, showing impairments in pragmatic skills, or higher-order language use including sarcasm, idioms, or speech acts such as adding to a social conversation, commenting, initiating interactions, turn-taking, or requesting information (Rapin & Dunn 1997; Tager-Flusberg et al., 2005). Overall, autistic individuals often have discrepancies in their language skills, frequently stronger in expressive than receptive language skills, which may result in the appearance of strong language profiles alongside difficulties connecting and relating to others (Boucher, 2012). Individuals on the autism spectrum demonstrate atypical neural processing of speech and may benefit from visual supports to complement speech interactions (Patel et al., 2023; Rapin & Dunn; 1997).

Social Cognition Overall, language challenges are indicative of broader socio-emotional impairments. Impaired *social cognition*, or the understanding and accurate interpretation of social information, is central to autism. Namely, autistic individuals are known to have impaired *theory of mind*, or understandings of others' emotions, thoughts, and perspectives that differ from one's own, and show difficulties inferring others' beliefs, behaviors, and responding to the distress of others (Bacon et al., 1998; Baron-Cohen et al., 1985). This phenomenon relates to additional challenges with interpreting facial expressions, tone, body language, and gestures of others and comprehending how this information conveys meaning or mental states (Baron-Cohen et al., 1995; Dawson et al., 2005).

Social Interaction/Reciprocity Taken together, the social cognitive profile of autism creates substantial challenges in *socioemotional reciprocity*, or shared connections and understandings with others, often seen through challenges with back-and-forth conversations. Children with ASD show impaired joint attention from early infancy (Charman et al., 1997; Dawson et al., 2004). This overall reduction in social responsiveness, as seen through challenges with joint attention, imitation, lack of social interest, and difficulties in reciprocal interaction with others, is critical to the ASD diagnosis and often emerges after the first year of life (Baron-Cohen et al., 1995). Relatedly, children with autism demonstrate impairments in their play abilities. Autistic children may engage in functional play or may often replay scenes of movies or daily routines through scripted play, but demonstrate difficulties engaging in pretend or symbolic play, in which children use imagination and creativity in an "as if" way (Charman et al., 1997; Kasari et al., 2006). Children with ASD thus may lack the imaginative, perspective-taking, and organizational skills needed to engage in a pretend story, and further, may not have the social skills to appropriately play with others.

Creating and Maintaining Relationships As a result of this reduced social cognition, autistic individuals may miss key opportunities for social learning and have difficulties connecting with others, such as when building relationships (Klin et al., 2020). Thus, people on the spectrum struggle with initiating and maintaining social relationships across the lifespan. While autistic people often develop close bonds with caregivers or a few individuals close to them, peer and romantic relationships may be complex and challenging, as people with ASD more often have fewer friends, are less popular, and get bullied and teased by peers when compared to neurotypical peers (Tsai et al., 2020). For example, young children who are autistic often prefer to play alone or may avoid other children, while others may be socially motivated but lack the skills to appropriately engage with peers. When interacting with peers, children with autism lack the socioemotional competencies of their agemates, associated with difficulties in emotional understanding and emotion dysregulation (Gev et al., 2021). Although often fewer and lower in quality, children with ASD do form friendships, but they may not fully experience common benefits of friendship, such as shared emotional warmth (Mendelson et al., 2016). In adolescence, teens with ASD often exhibit either aloof social attitudes, characterized by a lack of responsivity or motivation to interact, or an active, but odd interaction style in which they seek out social interaction through unusual modes, such as standing too close in proximity or talking excessively about their specific interests (Scheeren et al., 2012). People on the spectrum also may develop normative interests in dating and sexual activity and self-identify as LGBTQ+ at a much higher rate than the general population, which may further complicate navigation of social and romantic experiences (Turner et al., 2017). Specifically, individuals with ASD or autistic traits are more commonly represented in groups identifying as transgender and nonbinary, with a higher prevalence of autistic individuals with gender dysphoria than in the general population (De Vries et al., 2010; Glidden et al., 2016; Stagg & Vincent, 2019). Consideration of multiple diverse identities and intersectionality is critical in supporting autistic individuals to navigate adulthood and relationships (Hillier et al., 2019).

Restricted/Repetitive Behaviors

Restricted and Repetitive Behaviors Individuals with autism also demonstrate patterns of *restricted and repetitive behaviors (RRBs)*, characterized by a variety of preferences, interests, and observable behavioral patterns. For example, autistic individuals often exhibit strong preferences for and dependence upon routines and rituals. This insistence on sameness was one of the earliest symptoms noted by Leo Kanner in 1943 and may include rigidity through inflexible behavior patterns, insistence on particular foods or items of clothing, and challenges adapting to changes in the environment, such as a scheduling change, new type of yogurt, or having a substitute teacher in school (Leekam et al., 2011). Although wanting things to be "just right" is relatively normative in all children ages 2–4, individuals with ASD are found to have greater frequency and severity of RRBs throughout their lifespan (Bodfish et al., 2000). In addition to repetitive and stereotyped speech patterns, such as echolalia, repetitive motor mannerisms, such as rocking, hand flapping, or pacing, and repetitive use of objects, such as lining up toys, spinning parts of objects, or having excessive attachments to objects, are common. Often known as "*stimming*," these behaviors are self-stimulating, repetitive, and stereotyped in nature. Children who are autistic may have a stereotyped quality to their object play, such as using toys in repetitive or nonconventional manners, and may demonstrate unusual posturing of their body or body parts (Baranek, 1999). Overall, repetitive behaviors vary in their severity and may be associated with related behavioral challenges common to this population, such as hyperactivity, inattention, and self-injurious behavior, such as head-banging or skin-picking (Richler et al., 2007; Tsai et al., 2020).

Sensory Differences These repetitive motor behaviors often have a strong sensory component to them, such as spinning or tapping, as atypical sensory input is also common in ASD (Leekam et al., 2011). Sensory-motor deficits in autism emerge as early as 9–12 months and are conceptualized as either *hyper-* or, more commonly, *hypo-reactivity* and under-arousal to sensory stimuli (Baranek, 1999; Rogers & Ozonoff, 2005). Specifically, autistic individuals exhibit broad differences in sensory responsiveness, orientation, and attention to environmental stimuli and bodily sensations, and may be more attentive to objects through visual staring or fixation (Baranek, 1999). Furthermore, children with autism may have *sensory aversions*, preference or dislik of certain textures, sensitivity to bright lights or loud sounds, and seek out certain sensory experiences, such as by touching their bodies to different surfaces or mouthing (i.e., licking or chewing) of non-edible objects.

Restricted Interests Finally, individuals on the spectrum demonstrate restrictive interests from an early age, often referred to as *circumscribed interests* and applicable across a range of topics. These special interests often first appear in the preschool years, with a preoccupation with letters and numbers, and are later seen through the learning and memorization of facts pertaining to topics of interest

(Klin et al., 2007). Some interests may be age-appropriate, for example, in young children who enjoy dinosaurs or garbage trucks, some may be more advanced for a child's age, such as astronomy or the stock market, and some topics may be idiosyncratic or appear odd to others, such as radio station codes, kitchen appliances, or public transit systems (Klin et al., 2007; Leekam et al., 2011). Overall, preoccupations with preferred interests may vary depending on an individual's age and verbal ability; however, high-level preoccupations and interests are common and manifest differently across the spectrum. The ways in which individuals pursue and learn the content of their preferred interests are atypical to the extent that they can interfere with their daily experiences and relationships and may strongly affect mood and behavior. For example, some individuals may bond with others over shared interests, whereas others may have challenges connecting with peers because the intensity of their preferred topics and activities is so strong. Nevertheless, there can be various benefits to special interests in ASD, as they may foster success in education, careers, and activities of interest. Special interests often allow autistic individuals to bond with or help others through their strong knowledge base.

Neurodiversity, Advocacy, and Strengths of the Autism Spectrum

ASD is importantly a *spectrum disorder*—no two autistic individuals are alike. The "spectrum" aspect of ASD indicates the range in abilities and symptoms seen across individuals—whether in verbal, cognitive, behavioral, or social abilities. However, autism is also a spectrum in its role and intersection with various diverse identities an individual may hold. It is important to acknowledge and carefully consider how autism may interact with these identities.

Neurodiversity pertains to the idea that each brain is different, celebrating differences across individuals that are thought to be natural and beneficial aspects of life, rather than forcing conformity or "curing" differences in lieu of a societal standard (Kapp, 2020). Neurodiversity is both a political idea and concept, as well as a scientific mode of thinking about functioning, norms, and impairment (Chapman, 2021). As such, autism advocacy, stemming from the disability rights movement started by individuals with intellectual and developmental disabilities, embraces the tenants of neurodiversity by fighting for equal access, rights, support, and participation in society. *Self-advocacy* pertains to an individual's agency, rights, and choices of how to live their life, also extending across the community to increase widespread awareness and acceptance of autism. Autistic self-advocates have adopted the motto, "nothing about us without us," indicating that autistic people must be involved in discussions and efforts to expand autism care, policy, and education to the broader public. These goals have been emphasized by both advocates and researchers, who aim to better align future efforts with a neurodiversity framework and include autistic individuals as team members and partners in the research process (Leadbitter et al., 2021). *More information and resources about autistic advocacy can be found at the Autistic Self Advocacy Network: https://autisticadvocacy.org.*

Closely linked with the approaches of autistic advocacy is the need to utilize a *strength-based approach* in working with individuals with autism. This approach similarly celebrates the ways in which differences of autistic individuals represent unique qualities and strengths, rather than deficits; importantly including, not in spite of, autistic symptomatology. For example, many individuals bond with others and specialize in topics or fields pertaining to their special interests, often incredibly efficient and successful in careers that draw on these passions. These qualities can be leveraged to help autistic people thrive in settings, activities, and careers that fit their skills and preferences. Furthermore, individuals with ASD often prefer sameness and routine. These preferences can be helpful for day-to-day functioning, problem-solving, and job performance due to a person's strengths in carrying out systematic approaches and effectively sticking to routines. Many autistic individuals also have enhanced memory, visual–spatial abilities, and heightened awareness that allows them to process the world differently than others, bringing new and unique insights. Autistic people often have a propensity for logical or hierarchical reasoning, which can be incredibly useful in educational or professional settings. Regardless of functioning level, all people on the spectrum have unique traits and interests, are capable of building relationships, and have the potential to experience independence and fulfillment in ways that may differ from neurotypical individuals. It is important to celebrate this neurodiversity and approach each individual with autism from a strength-based perspective to enhance their independence, success, and overall well-being.

Developmental Course of ASD and Outcomes

While not usually diagnosed until later in toddlerhood, subtle features of autism may be present from early infancy. Eye tracking studies have documented differences as early as the first 2–6 months of life, as infants at risk for ASD may not show typical infant reflexes, such as orienting to faces or voices, and may instead be more attuned to the physical, rather than social, world around them (Jones & Klin, 2013). Within the first year of life, babies with ASD may show delayed or a lack of babbling, reduced receptive language, sleep challenges, eating difficulties, reduced social gaze and social responsivity, preferences to be alone, and may be less likely to demonstrate positive affect while being extremely reactive and difficult to soothe (DiCicco-Bloom et al., 2006; Ozonoff et al., 2010; Zwaigenbaum et al., 2005). These differences may become more pronounced around 12–14 months, at which infants with ASD show delayed or lack of language, decreased pointing or waving, and may exhibit unusual play by lining up, banging, or inspecting objects rather than engaging in functional or symbolic play (DiCicco-Bloom et al., 2006). Some children may develop more typically and then experience a regression or plateau around age 2, in which verbal abilities reduce, alongside limited social interest and engagement. As such, it is recommended to repeat screening for ASD around age 2 (Landa et al., 2007; Zwaigenbaum et al., 2013). During the second year of life, infants may demonstrate more overt signs of autism, such as repetitive behaviors, atypical body movements, and communication atypicalities, such as echolalia or unusual pitch and intonation, alongside reduced language, motor abilities, and slowed acquisition of cognitive skills (Chawarska et al., 2007; Zwaigenbaum et al., 2013). Children may not exhibit normative joint attention or imitation, such as playing peek-a-boo with a parent (Charman et al., 1997). The autism diagnosis holds high stability in the second year of life and throughout childhood (Chawarska et al., 2007).

Most children with ASD are diagnosed by the time they exit preschool (CDC, 2023). During the preschool years, autistic children may not engage with peers as expected, resulting in one-sided interactions or parallel play in which they play near or next to, rather than engaging with, other children. Around 3–4 years of age, difficulties with transitions and new experiences may lead to frequent tantrums, exacerbating social issues already present (Dawson et al., 2004).

In elementary and middle school years, children with ASD often grow in their language and learned social skills with intervention; however, additional issues such as attentional difficulties and challenges with academic instruction and comprehension may emerge. Furthermore, social differences in autism become more pronounced in the school-aged years, during which children may struggle to make and maintain friendships and may experience bullying or teasing as a result of their differences.

As individuals on the spectrum go through puberty and enter adolescence, additional difficulties with bodily changes, hormones, and emotional dysregulation may arise. Comorbid mental health challenges often emerge around this time, causing autistic individuals to feel more isolated from their peers, distressed by high school environments, and struggle navigating romantic interests and sexuality. Transitions from the school system, which individuals in special education "age out" from at age 22, can be challenging, but can be addressed by transition resources and helpful social support. Adolescents on the spectrum often work with their parents and care providers to identify if postsecondary education, employment, or different living arrangements may be ideal options in adulthood. Many adults on the spectrum go on to live fulfilling lives in which they build relationships, communities, or begin families of their own, find hobbies and careers, and establish a lifestyle of meaning for them.

Of note, the current DSM-5-TR includes severity-level specifiers for the autism diagnosis. Since ASD is a spectrum condition, these specifiers are used in lieu of previous labels, such as "high functioning" or "severely impaired." Rather, specifiers range from levels 1 to 3, with Level 1 requiring "Support," Level 2 requiring "Substantial Support," and Level 3 specifying "Requiring Very Substantial Support." These specifiers are primarily utilized to indicate eligibility for insurance coverage of ASD services and for communication amongst care providers. Furthermore, an autistic individual can change in their symptom severity and support needed across the lifespan. For example, many children begin with higher needs, reflected by

Level 2 or 3, and may later drop to Level 1 as adolescents after building skills and no longer requiring as many supports. It is important to remember that these specifiers are intended to enhance specificity and sensitivity for ASD-related care and services, rather than place individuals in definitive categories based on their abilities.

Prevalence and Epidemiology

The prevalence of autism has increased since its initial discovery, as one in 36 children under age 8 in the United States are estimated to have autism (Maenner et al., 2023). Autism's accelerating prevalence is thought to reflect increased awareness, screening, and detection of ASD alongside increased insurance mandates (Mandell et al., 2016). Autism is 3.8 times more prevalent in boys than in girls and has only recently shifted to be less prevalent in non-Hispanic White children than among other racial and ethnic groups (Maenner et al., 2023).

Research has demonstrated that autism stems from multiple causes, and manifests differently across individuals who may share common symptoms. Integrated hypotheses of autism include various genetic and environmental factors associated with ASD.

Genetics

While evidence is mixed surrounding the emphasis of genetic versus environmental factors, autism is thought to be largely genetic in basis as seen through chromosomal variations, abnormalities in white and grey brain matter, and over- and under-connectivity between neural assemblies in the brain (Ratajczak, 2011, Tick et al., 2016). Genetic differences can be seen in the early prenatal period during *neurogenesis*, in which the nervous system tissue grows to formulate new neurons and connections that eventually form the brain and central nervous system. Difficulties in the regulation of this neuronal migration can lead to structural disorganization in the cortex and deficits in brain circuitry. These abnormalities stem from differences in neuronal density and volume, leading to over- or under-connectivity between neural regions and causing challenges in overall brain circuitry and processing (Bhat et al., 2014; Reiner et al., 2016). Susceptibility to ASD has moderate genetic heritability and can occur due to known variations in copy number variants or due to random or sporadic genetic mutations that affect gene structure (Bhat et al., 2014; Hallmayer, 2011; Iossifov et al., 2014; Pan et al., 2019; Reiner et al., 2016). For example, several known genetic or chromosomal conditions, such as fragile X syndrome or tuberous sclerosis, are associated with increased genetic risk for ASD. Overall, ASD recurrence is heightened in siblings of individuals with ASD, estimated at a total of 18.7% (Ozonoff et al., 2011). Twin studies have similarly shown heritability rates from 34% to 76%, higher in monozygotic versus dizygotic twins (Frazier et al., 2014). Family genetic studies have also documented several cognitive dimensions, such as reduced empathy or theory of mind and "tendencies to systemize" that characterize sub-clinical traits of autistic characteristics seen in close relatives of individuals with ASD (Baron-Cohen, 2004). While there is a growing number of known genes implicated in ASD development, known genes or genetic diseases only account for 1–2% of cases of autism, and data from whole-genome research in families with ASD suggest how there are likely many rare mutations that manifest in autism, rather than a singular major genetic cause (DeFrancesco, 2001).

Environmental Factors

As noted, there are mixed findings regarding the strength of genetic versus environmental factors on autism development, with many studies suggesting that genetic and environmental factors may interact (Hallmayer, 2011; Frazier et al., 2014). ASD is more common in children born at pre-term and with low birth weight (Gardener et al., 2009). In addition, children with older parental age at birth have an estimated increased risk of 106% for parents above age 40 versus below age 30 (Gardener et al., 2009; Ratajczak, 2011; Schendel & Bhasin, 2008). Some hypotheses suggest that in addition to genetic alterations, factors, such as parental age and brain inflammation, such as due to viral infections during pregnancy, premature births, environmental toxins, or encephalitis in the child after birth, can pose as contributing risk factors for ASD development due to impacts on the developing brain (Atladóttir et al., 2010; Bhat et al., 2014; Croen et al., 2019; Gardener et al., 2009; Ratajczak, 2011; Schendel & Bhasin, 2008).

Vaccines and Autism
Research demonstrates that there is no causal link between vaccines and autism. However, in 1998, an article published in The Lancet proposed initial allegations for a link between the measles–mumps–rubella (MMR) vaccine and autism, sparking controversy and confusion that has lasted since. This hypothesis has been repeatedly refuted by published evidence, and the study was later retracted by editors of The Lancet due to poor scientific practices. Nevertheless, public confusion regarding impacts of vaccines on development continues to persist, as seen through a decreased rate of vaccinations, especially so in younger siblings of children with ASD (Zerbo et al., 2018). For additional information, see the retraction note from The Editors of the Lancet, (2010) and reports published by the Institute of Medicine of the National Academy of Science's Immunization Safety Review Committee (2004).

Diagnosis and Assessment

Screening

Receiving an autism diagnosis can often be a long process usually occurring in the first few years of life. Autism has strong diagnostic stability and can be accurately diagnosed before age 3, making children eligible for early intervention and ASD-specific services (Chawarska et al., 2007; Ozonoff et al., 2015; Webb & Jones, 2009). However, not all children with initial delays or signs of autism receive a diagnosis at first concerns. Many diagnoses of autism can be fully missed until later childhood or even adulthood, demonstrating the importance of repeated screenings (Eisenhower et al., 2021; Pierce et al., 2016; Zwaigenbaum et al., 2009). The American Academy of Pediatrics recommends screening for autism at 18 and 30 months. Often, pediatricians and primary care providers first utilize *screening tools*, usually geared toward identifying developmental delays more generally (Level 1 screening tools), and then provide more ASD-specific screening for at-risk children with established developmental concerns (Level 2 screening tools).

Diagnosis

Children with ASD are often diagnosed by an *interdisciplinary team*, and diagnoses can be made by psychologists, neuropsychologists, physicians, such as developmental pediatricians or pediatric neurologists, or speech-language pathologists with ASD-specific knowledge and experience. Other key members of a diagnostic team may include behavioral analysts, social workers, genetic counselors, occupational therapists, and educators. The autism diagnostic process often includes a medical examination for differential diagnosis and to rule out other high-risk conditions commonly seen with or confused for autism, such as motor impairments, developmental delays, specific language disorders, hearing or visual challenges, and medical conditions (Yates & Couteur, 2013). Autism itself is diagnosed via a thorough examination of developmental history, gathering data about a child's behavior across different contexts, such as home and school, as well as a standardized in-person assessment. *Caregiver interviews* such as the Autism Diagnostic Interview-Revised (ADI-R; Lord et al., 1994) or the Vineland Adaptive Behavioral Scales (Sparrow & Cicchetti, 1989) can be used to detect delays or variations in developmental trajectories, assessing for a child's use of language, babbling, gestures, and eye contact, as well as adaptive and daily living skills, such as feeding and toileting. To assess the child, the *Autism Diagnostic Observation Schedules* (ADOS-2; Lord et al., 2012) is the gold-standard diagnostic tool comprised of various standardized social situations and activities, which can elicit different behaviors key to understanding autism, such as social responsivity and play skills. Finally, additional *standardized questionnaires* may be included in the diagnostic process to supplement the ADOS-2 and caregiver reports, such as the Childhood Autism Rating Scales, Second Edition (CARS 2, Schopler et al., 2010), Social Communication Questionnaire (SCQ; Rutter et al., 2003), Autism Spectrum Rating Scales (ASRS, Goldstein & Naglieri, 2009), or the Social Responsiveness Scale-Second Edition (SRS-2; Constantino & Gruber, 2012). These types of questionnaire tools can also be utilized as screeners earlier on in the diagnosis process, for example, by primary care providers, to identify concerns for autism and at-risk populations in young children. Oftentimes, home or school observations may be taken to supplement these tools with *behavioral reports* of a child's functioning in their daily environments. Ultimately, none of these methods in isolation can diagnose autism; rather, it is the clinical judgement and expertise of an interdisciplinary team that can most effectively diagnose autism.

Associated Conditions

Autism is associated with various co-occurring neurodevelopmental, medical, and psychiatric conditions. Essential to autism care is a firm understanding of these various *comorbidities*, and an awareness of the dangers of *diagnostic overshadowing*, or the tendency to attribute symptoms to a primary diagnosis, versus considering the potential for differential diagnosis or comorbid diagnoses.

Neurodevelopmental Disorders

Intellectual disability (ID), characterized by an IQ score of 70 or below, is highly common in ASD, with a recent meta-analysis estimating ID rates of 17.7–29.2% in individuals with ASD (Mutluer et al., 2022). Those with ID and ASD have a broad range in cognitive, verbal, and adaptive abilities, with different needs than those with just ID or autism alone. Of note, a more recent classification structured autism severity in terms of verbal and intellectual functioning, defining "profound autism" as an IQ of 50 or below and/or being nonverbal or minimally verbal (Lord et al., 2022). A recent study estimated a 26.7% prevalence of profound autism in 8-year-old autistic children (Hughes et al., 2023). Within this profound autism group, 56.7% of individuals had ID, a majority of whom showed IQ below 50, as compared to a 43.3% of ID in non-profound autism (Hughes et al., 2023). This conceptualization of autism severity in relation to ID is still new, and thus, more research is required to delineate the relationship of IQ severity and DSM-5-TR autism severity levels and support needs. ID disproportionally impacts those with profound autism, associated with lower levels of adaptive skills, increased RRBs, as well as increased social and behavioral challenges persisting across the lifespan (Matson & Shoemaker, 2009). Specifically, this population experiences social withdrawal and isolation, increased stereotypies, sensory differences, and co-occurring medical or mental health challenges (Matson & Shoemaker, 2009; Matson et al., 2008). Autistic individuals who are more impacted by ID are under-represented in research and have some of the lowest reported resources and outcomes of early interventions (Cervantes & Matson, 2015; Matson & Shoemaker, 2009).

Attention-Deficit Hyperactivity Disorder (ADHD) is also highly common in ASD, seen in about a quarter to a third of individuals with autism (Mutluer et al., 2022). Previous research and the DSM-IV suggested that ADHD and ASD could not co-occur, but recent research has supported the distinction between the two disorders, the idea that they increase risk for one another, and the notion that comorbid ASD and ADHD represents a distinct subtype (Antshel & Russo, 2019). Overlapping symptoms between ADHD and ASD includes impulsivity, inattention, and hyperactivity as seen through RRBs or excessive energy. As a result, individuals with both disorders have increased social vulnerabilities, such as susceptibility to bullying, and challenges with learning, inhibitory control, and processing speed (Mannion & Leader, 2013; Mutluer et al., 2022). Overall, it may be difficult to parse apart if these symptoms stem from ASD-related RRBs as well as social and processing differences, core difficulties in inattention and hyperactivity, or both (Mutluer et al., 2022).

Psychiatric Conditions

Psychological disorders in population-based studies of ASD have been found to be lower than in clinical and referenced samples, potentially indicative of the need for more sensitive diagnostic tools that address the nuances of psychopathology in ASD (Mutluer et al., 2022). Mental health conditions occur in higher rates in autistic youth without ID, suggesting the role of verbal abilities and intelligence in how individuals express and comprehend their internalizing and externalizing challenges (Salazar et al., 2015; Selten et al., 2015). Autistic individuals may be more prone to *internalizing disorders* such as anxiety, depression, and obsessive–compulsive disorder (OCD) due to social communication challenges, sensory sensitivity, emotional dysregulation, and inflexibility, increasing vulnerability for these diagnoses (Mutluer et al., 2022). For example, anxiety disorders are estimated to occur in 8.6–14.1% of autistic individuals; however, this may be an underestimate due to challenges accurately expressing emotional states (Mutluer et al., 2022). It is especially important to recognize how anxiety symptoms may present as irritability or overlap with autistic symptoms, and to use tools that accurately capture anxiety, OCD, and depression within the scope of autism. *Externalizing disorders* such as disruptive behavioral disorders have also been seen in autistic populations, again, likely to overlap with symptoms of RRBs, rigidity, and impulsivity. Similar to neurotypical populations, individuals who are autistic may respond well to cognitive behavioral therapy (CBT) techniques as well as social stories, visual supports, and medications in some cases to address psychiatric conditions.

Genetic Conditions

Several genetic conditions are highly associated with or known causes of autism. As such, is it important to distinguish between *syndromic autism*, or autism associated with other clinical signs, versus *idiopathic autism*, or autism with unknown or other causes. Many genetic syndromes have distinct behavioral, social, and communicative symptoms already present, containing autism-like phenomenology and elevating risk for ASD and ID (Moss & Howlin, 2009). Thus, it is important to parse apart the causes of various symptoms to avoid diagnostic overshadowing. For example, *Fragile X Syndrome (FXS)* is the most common inherited cause of ID and highest known cause of ASD, occurring in about 5% of individuals with autism, and with ASD rates ranging from 21% to 50% in males with FXS (Cohen et al., 2005; Moss & Howlin, 2009). FXS is associated with mild to severe ID, learning disabilities, ADHD-like symptoms, anxiety and unstable moods, and ASD-like symptoms such as social difficulties, language challenges, and stereotyped behaviors (Cohen et al., 2005). Strong correlations have been found between the degree of ID, impairments in verbal abilities, and the presence of ASD-like characteristics in FXS (Moss & Howlin, 2009). Other genetic disorders strongly associated with ASD include Angelman Syndrome, Prader–Willi Syndrome, Tuberous Sclerosis, Rett's Syndrome, and Cordelia de Lange Syndrome. Associated symptoms of ID, stereotyped behaviors, and hyperactivity may relate and contribute to syndromic autism presentations in these groups. Nevertheless, there are subtle, qualitative differences in autism presentations across syndromic autism profiles, suggested to be influenced by genetic causes and degrees of ID (Moss & Howlin, 2009).

Medical Conditions

Medical challenges may be interrelated with or cascade from ASD symptoms, as many researchers suggest that alterations in one biological or neurocognitive system increase likelihood for differences across other systems and networks (Tye et al., 2019). Very common in individuals with ID, *epilepsy* has an estimated prevalence of 25–30% in autistic individuals, as compared to 2–3% in the general population (Mannion & Leader, 2013; Tye et al., 2019). Further, multiple reports have documented abnormal EEG findings in ASD not associated with clinical seizure activity (Tye et al., 2019). Epilepsy rates increase with severity of ID, associated with greater social, motor, and behavioral difficulties (Amiet et al., 2008; Mannion & Leader, 2013; Tye et al., 2019). Theorized to be related to cognitive and developmental trajectories, epilepsy, especially when combined with ASD and ID, can lead to greater challenges with daily living skills as well as reduced long-term quality of life outcomes (Mannion & Leader, 2013; Tye et al., 2019). *Sleep challenges* are estimated to occur in 50–80% of individuals who are autistic, including insomnia (difficulties with sleep initiation, duration, and quality), as well as sleep-disordered breathing, and challenges initiating sleep and sleeping independently (Tye et al., 2019). A recent meta-analysis estimated a 19.7% sleep disorder occurrence rate in ASD, reported at much higher levels compared to neurotypical populations (Mutluer et al., 2022; Tye et al., 2019). Sleep challenges may be caused or exacerbated by core ASD symptoms such as reduced sensitivity to environmental cues, perseveration, limited communication, and behavioral challenges, increasing with severity of ASD and ID (Mannion & Leader, 2013; Tye et al., 2019). Sleep difficulties are also associated with and likely have a bidirectional relationship with behavioral dysregulation, aggression, and severity of social and behavioral difficulties, impacting daily functioning and quality of life for autistic individuals and their families (Mannion & Leader, 2013; Tye et al., 2019). *Feeding disorders* and selective eating also occurs frequently in ASD, as individuals with ASD often have a narrower range of foods eaten, exhibiting food selectivity and food refusal (Mannion & Leader, 2013; Sharp et al., 2017). Associated with sensory processing challenges, eating difficulties may also relate to rigidity, behavioral inflexibility, increased GI complaints, and may stem from the texture, taste, color, or presentation of foods (Nadon et al., 2011; Mannion & Leader, 2013). *Gastrointestinal (GI) difficulties* also range from 9% to 71% in the ASD population, with some reports estimating up to 91%, ranging from mild reflux to constipation, abdominal pain, and diarrhea (Mannion & Leader, 2013; McElhanon et al., 2014; Tye et al., 2019). Some research has suggested links between GI symptoms, anxiety, and sensory sensitivity in people who are autistic (Mazurek et al., 2013). Further, it is important to assess for signs of GI challenges in individuals who may be less verbal, such as through vocal, motor, and behavioral signs like sleep disturbance, irritability, and facial grimacing (Buie et al., 2010). GI issues are associated with increased irritability, social withdrawal, and hyperactivity, as well as increased aggression in nonverbal individuals (Tye et al., 2019). Further, individuals on the spectrum often take longer to toilet train, often due to fears of the toilet, communicative difficulties, or rigidity, and experience higher rates of incontinence, posing a barrier to independent functioning and quality of life (Mannion & Leader, 2013).

Interventions for Autism

Current recommended intervention practices typically utilize developmental, behavioral, and naturalistic techniques and can be implemented in homes, schools, and community settings. Of note, while *psychopharmacological medication* is often prescribed for individuals with ASD, these medications primarily treat associated conditions of autism, such as epilepsy, sleep disturbances, or irritability. There are no currently known medications for core symptoms of autism alone. Further, several types of *alternative interventions* such as sensory integration, special diets, animal therapy, and technology-based interventions currently have little evidence for their effectiveness (Sandbank et al., 2020). Overall, service navigation may be challenging and can be best addressed through a multidisciplinary team and coordinated care in which medical, psychological, and educational providers communicate with one another (Rogers & Vismara, 2008). The goal of most interventions aim to build behavioral and communication skills, using age-appropriate learning activities, and implementing goals into naturalistic settings to increase daily functioning and skill generalization. Some interventions include social goals or involve neurotypical peers to increase social interactions, very often including parents and family members to provide additional support and practice (Rogers & Vismara, 2008). Interventions are best delivered when tailored to an individual's needs to address unique strengths and challenges.

Early Intervention

The *Individuals with Disabilities Education Act (IDEA), Part C* was established in 1990 to fund state-coordinated early intervention services for children ages 3 and under with developmental delays. A large majority of children utilize these services prior to receiving ASD diagnoses due to early delays (Monteiro et al., 2016). There is significant variability across states regarding guidelines, policies, and practices of available services, however, most states elect to use an eclectic model in which services are based on child needs and assessed continually (Stahmer & Mandell, 2007). Early intervention services are particularly important because they aim to support children during a critical period for neuroplasticity, social learning, and developmental growth, building foundational skills from which children with ASD can continue to learn and improve. A variety of evidence-based early intervention approaches are available for ASD, shown to improve developmental functioning and reduce maladaptive behaviors and ASD symptom severity (Rogers & Vismara, 2008). Further, research has emphasized that children who receive early interventions show significant gains in developmental skills later in childhood (Rogers & Vismara, 2008). A recent comprehensive review found significant positive effects of behavioral, developmental, and *naturalistic developmental–behavioral intervention (NDBI)* approaches for enhancing developmental outcomes (Sandbank et al., 2020). Recommendations for children under age 3 suggest a combination of these developmental and behavioral intervention approaches, beginning as early as possible, and focusing on core and related features of ASD such as social communication, emotional and behavioral regulation, and adaptive skills. Interventions typically include both clinician and caregiver involvement, with efforts to incorporate learning opportunities into the child's naturalistic environments and daily living routines to boost skill generalization (Sandbank et al., 2020; Zwaigenbaum et al., 2015). Interventions should be mindful and accommodating of family sociocultural beliefs, dynamics, and resources and supports available in their community.

Behavioral Interventions

Behavioral interventions were first established in the 1980s as intensive treatments for autistic children, initially studied in children ages 4 and under by Ivar Lovaas and colleagues. Their model focused on areas for skill growth and improvement, initially falsely promising "recovery" from autism (Lovaas & Smith, 1988). Of note, while interventions can reduce developmental delays and enhance potential for skill growth, there is no evidence that ASD can be "cured" or that children can improve enough to no longer qualify for a diagnosis. Behavioral interventions use functional analysis to understand the causes and functions of a child's current skills or challenging behaviors, utilizing strategies like direct teaching, prompting, repetition, and positive reinforcement to build more positive and functional skills and replace maladaptive behaviors (Vismara & Rogers, 2010). *Applied behavioral analysis (ABA)* emerged out of this work, now one of the most common EI approaches for ASD, used to teach and increase appropriate skills while reducing any maladaptive behaviors. While ABA has a difficult history of more severe and demanding discrete trial approaches, modern-day approaches to ABA emphasize the use of naturalistic techniques and trauma-informed perspectives to respect child preferences and enhance quality of life (Rajaraman et al., 2022; Schuck et al., 2022). ABA practices are compatible with neurodiversity goals when they are strength-based, modeled on naturalistic learning and play opportunities, select targets of importance to autistic individuals, aim to maximize potential and decrease barriers, and are implemented in naturalistic environments (Schuck et al., 2022).

Developmental Intervention Approaches

Developmental intervention approaches similarly focus on core skill challenges seen in ASD, such as social communication, aiming to build a foundation of skills with the idea that gains in one area will facilitate cascading growth in other domains, such as language, learning, and social skills (Sandbank et al., 2020). Many developmental interventions are play-based, centered around joint engagement and relationship building, and implemented via scaffolding and child-directed or naturalistic instruction to increase child engagement, social skills, and more diverse symbolic and functional play (Lang et al., 2009).

Naturalistic Developmental Behavioral Interventions

Naturalistic Developmental Behavioral Interventions (NDBIs) blend these traditional behavioral and developmental approaches, useful for building social communication, language, and play skills. NDBIs use a strength-based model to enhance child motivations and interests in teaching useful and functional skills, improving well-being and optimizing potential (Sandbank et al., 2020; Schreibman et al., 2015). Schreibman and colleagues (2015) described the core components of NDBIs as having learning targets of coordinated skill domains, various contexts of engaging and meaningful activities centered around child preferences and motivation, and naturalistic strategies that enhance learning through developmentally appropriate opportunities. This means that interventionists follow a child's interests and preferences despite their age, and that expectations for growth match a child's developmental ability. In general, NDBI strategies are very commonly used in early intervention programs, however, provider, intervention, and organization-level factors may impact a given NDBI program's focus, structure, and implementation (Pickard et al., 2021).

Social Skills Interventions

As autistic children grow in their language and adaptive skills in middle childhood, social skills training may become useful to increase peer engagement, support interaction, and build friendships. Social skill interventions aim to target several areas, first building social motivation through self-esteem and enjoyment, then working to increase social initiations, responding, and sometimes targeting any interfering behaviors that may affect social relationships, ultimately promoting skill generalization (Williams White et al., 2007). Social skill trainings can also help to enhance related behaviors such as language and play, problem solving, greeting, sharing, and asking for help (McConnell, 2002). Most social skills trainings include modeling and reinforcement techniques that involve social stories, peer-mediated practice opportunities, and structured activities through which children learn and practice social skills (Matson et al., 2007).

Individualized Education Planning

When children with ASD enter school, an *individualized education plan (IEP)* is put in place through a comprehensive assessment of their abilities, strengths, and areas for improvement. IEPs are federally required programs that ensure continual support and monitoring for children with various special needs and conditions throughout their education. Once established, schools follow IEPs to implement a series of accommodations and interventions within the school setting. IEPs under IDEA aim to identify the *least restrictive environment (LRE)* for an individual with autism, meaning that children who require special education services are integrated into settings with as many neurotypical peers as possible. Some children may be placed in a fully special education setting, whereas others may be in integrated classrooms with special needs and neurotypical peers, or even successfully learn in general education classroom settings. IEPs often include education and learning supports within the classroom such as visual supports, 1-on-1 aides, or curriculum modification, in addition to supports for peer engagement and relationship development.

These plans also may include family support to help children stay on-track with educational goals and expectations. Elementary schools often provide services such as speech therapy, occupational therapy, and individual or small group tutoring through pull-out services during school (Kurth & Mastergeorge, 2010). In the teenage years, IEP goals may focus more on daily living, independence, and higher-order academic skills to help adolescents prepare for transitions to other educational or vocational settings (Kurth & Mastergeorge, 2010).

Cognitive Behavioral Therapy

In older children, adolescents, and adults who are autistic, psychotherapy such as *cognitive behavioral therapy (CBT)* can be helpful to support emotional and social adjustment. CBT has mostly been studied in adolescents and adults with ASD who are high functioning, meaning that they have higher intellectual or verbal abilities, demonstrating robust evidence for reducing emotional challenges, such as anxiety symptoms (Ung et al., 2015). CBT provides psychoeducation about emotional, cognitive, and behavioral challenges and how these factors affect one another, helping individuals identify their feelings, change their behaviors, and learn new skills to manage emotions and meet goals, such as building relationships or independence. Individuals on the spectrum may particularly struggle to understand and recognize emotional experiences in themselves and others, benefitting from additional CBT accommodations or modifications such as social stories, visual aids, structured worksheets, parental involvement, and direct coaching (Moree & Davis, 2010; Ung et al., 2015). Both individual and group CBT formats have been found to be useful in ASD populations.

Vocational Skills Training

As autistic adults face a disproportionately high unemployment rate, they may benefit from vocational skills training to support the job application and retention process (Shattuck et al., 2012). Job success is often predicted by vocational "fit," preparedness, and support. As such, vocational training can be essential, including teaching prevocational skills like interviewing, daily job skills such as office conduct and administrative tasks, and even "soft skills" of dressing for work, coworker interaction, and time management (Seaman & Cannella-Malone, 2016). Training in job skills can be essential to ensure preparedness and job stability, helping autistic adults gain additional benefits, such as self-esteem, independence, and enhanced social relationships (Nicholas et al., 2015). Furthermore, many jobs include supported employment structures in which individualized support is provided to employers and employees to ensure proper resources, support, and a nurturing work environment (Nicholas et al., 2015).

Residential Programs

While an estimated 57% of adults on the spectrum remain in their homes with caregivers, some autistic individuals also elect to live in residential or supported living facilities (Graetz, 2010). These living environments range in level of support, with the least amount of assistance being *supported or supervised living,* which provides more independence for adults to rent their own residences and interact with neighbors in a lightly supervised setting. *Group homes* are common community-focused housing arrangements for special needs populations, including all-day staff, provided meals, and assistance with daily living needs, such as shopping, laundry, and cleaning. Other residential programs, such as those on agricultural farms, may include intensive daily schedules, such as including farm or warehouse maintenance, and have more structured expectations in which individuals work in exchange for housing and food (Atsmon et al., 2019). Overall, adults with ASD may choose to transition from the home setting depending on maturity, readiness, and eagerness, as self-motivation is important for transitions into residential living.

Telehealth Assessment and Intervention Approaches

Of note, recent efforts in ASD research and practice have aimed to increase access to care and reduce barriers through use of technology and virtual modes of service delivery (Simacek et al., 2021). For example, recent work has aimed to increase the accessibility of ASD assessment using telehealth approaches. While psychometric data have not yet been published, a series of synchronous diagnostic tools have been developed for clinicians to guide caregivers through a series of activities via telehealth. These assessments can examine spontaneous behaviors and elicit different skills to aid in the autism diagnosis (Berger et al., 2022). While in the early stages of development and adoption, virtual assessments represent a potential solution to speeding up the diagnostic process and more directly reaching under-resourced populations to provide quality diagnoses. Many studies have also demonstrated preliminary efficacy of telehealth intervention modalities, such as virtual parent training, CBT groups, or care coordination (Grynszpan et al., 2014; Solomon & Soares, 2020). However, it is not yet clear how to best engage autistic children remotely, and these methodologies may not be ideal for all families in the autism community (Sutherland et al., 2018). In adolescents and adults able to use technology themselves, research has found technology-supported interventions to be useful in enhancing academic, behavioral, social, and independent living skills (Odom et al., 2015). Additional research on how to effectively implement these telehealth services, and which services are most optimal for remote implementation, is warranted to inform decision-making about virtual service options.

Access to Care for Diverse Autistic Populations

In general, receiving an ASD diagnosis and care can be incredibly difficult due to the need for specialized providers, leading to long waitlists for families in need. Many factors including stigma, mistrust of medical professionals, stress of the diagnostic process, and difficulties with accessing services impact marginalized groups in particular. ASD has been historically under-diagnosed in diverse populations. Research has found autism to be more common in white children and boys in comparison to African American or Hispanic children, with autistic African American children more often diagnosed later, and more likely to have more severe and alternate diagnoses, compared to white children with ASD (Falco, 2014; Mandell et al., 2002). More recent studies have found that children of color and publicly insured children are more likely to screen positive on first-stage developmental screenings, while the ASD diagnosis rate in 8-year-olds who are Hispanic, Asian, or Pacific Islander (AAPI) and Black is for the first time higher than that among white children, potentially representing conscious efforts to improve equitable identification of autism (Eisenhower et al., 2021; Maenner et al., 2023; Shaw et al., 2023). Higher rates of ASD diagnosis and use of services have been associated with parental socioeconomic status (SES) and education, as disparities still emerge between white and Hispanic/Latino families even within higher SES communities (Dickerson et al., 2017; Shattuck et al., 2009). ASD prevalence and care are more often available and utilized in urban areas, with many schools being primary sources for ASD services, leading to challenges for families in rural communities (Antezana et al., 2017). Even within those who successfully receive care, research has found disparities in various aspects of health care quality indicators for children with ASD, indicating exacerbated difficulties in services for Latino and Black children (Magaña et al., 2012). Furthermore, minority racial and ethnic groups, those with lower SES, and families with less English proficiency and ASD-knowledge demonstrate fewer follow-up screenings for autism and fewer therapy hours of services used, potentially exhibiting provider mistrust and reduced willingness to follow up on developmental risks (Eisenhower et al., 2021; Shenouda et al., 2022; Zuckerman et al., 2017). It is recommended that providers take these challenges, in addition to cultural and linguistic strengths, such as family cohesion and religiosity, into account when working with diverse families with ASD (Angell et al., 2018).

Summary

Autism is a lifelong developmental disability characterized by deficits in social cognition, reciprocity, and communication in addition to patterns of restricted, repetitive behaviors, and thinking styles. It is referred to as a spectrum disorder because of the various constellations of symptoms present as well as factors of co-occurring medical and psychological conditions, levels of intellectual and verbal ability, and life stage. Autism may interact with other identities a person may hold, and individuals on the spectrum hold unique relationships with their autism diagnosis. ASD presentations change across developmental stages, first detectable in early nuances in infant behavior through reductions in eye contact, reciprocity with caregivers, and speech and gesture patterns. These symptoms often become more pronounced as a child ages, as seen through challenges in social interactions in which autistic individuals diverge from their peers. Autism is increasing in prevalence and thought to result from a combination of genetic and environmental factors. To accurately diagnose autism, infants and toddlers typically receive early developmental screenings by primary care providers, requiring a full multidisciplinary evaluation including a child assessment, caregiver interviews, and data gathered from other informants. Interventions for ASD typically focus on naturalistic developmental and behavioral approaches to support social learning skills, such as communication. Early intervention is an important component of early autism care, with the potential to boost social communication and learning skills to promote later developmental outcomes. Children with ASD usually receive naturalistic behavioral and/or developmental interventions integrated into home and daily settings, while older children may add on services while in school, such as academic interventions and social skills training. Autistic adolescents and adults often benefit from vocational skills training, residential or supported independent living programs, and cognitive behavioral therapy to support independence and address highly common co-occurring mental health challenges. Recent strides in research and policy aim to address the many barriers to accessing autism care, which stem from systemic marginalization of diverse groups and the need for highly specialized care providers for ASD. Overall, it is essential to provide neurodiversity-affirming care and strengths-based approaches when working with the autism community. Greater efforts to increase public autism awareness and acceptance, address barriers to access for autism care, and maximize outcomes through appropriate intervention can help to support independence, success, and quality of life for the diverse spectrum of individuals with ASD and their families.

Conflicts of Interest We have no known conflicts of interest to disclose.

References

American Psychiatric Association. (1987). *Diagnostic and statistical manual of mental disorders: DSM-III-R*. (3rd ed., rev.).

American Psychiatric Association. (2022). *Diagnostic and statistical manual of mental disorders* (5th ed., text rev.).

Amiet, C., Gourfinkel-An, I., Bouzamondo, A., Tordjman, S., Baulac, M., Lechat, P., Mottron, L., & Cohen, D. (2008). Epilepsy in autism is associated with intellectual disability and gender: Evidence from a meta-analysis. *Biological Psychiatry, 64*(7), 577–582.

Angell, A. M., Empey, A., & Zuckerman, K. E. (2018). Chapter four—A review of diagnosis and service disparities among children with autism from racial and ethnic minority groups in the

United States. In R. M. Hodapp & D. J. Fidler (Eds.), *International review of research in developmental disabilities* (Vol. 55, pp. 145–180). Academic.

Antezana, L., Scarpa, A., Valdespino, A., Albright, J., & Richey, J. A. (2017). Rural trends in diagnosis and services for autism spectrum disorder. *Frontiers in Psychology, 8*, 590.

Antshel, K. M., & Russo, N. (2019). Autism Spectrum disorders and ADHD: Overlapping phenomenology, diagnostic issues, and treatment considerations. *Current Psychiatry Reports, 21*(5), 34.

Arnhart, C., Neale, M., Collins, C., Chesher, T., Coffey, S., Rogers, T. C., ... & Hartwell, M. (2022). The use of person-centered language in scientific research articles focused on autism. *Journal of Developmental & Behavioral Pediatrics*, 43(2), 63–70.

Atladóttir, H. Ó., Thorsen, P., Østergaard, L., Schendel, D. E., Lemcke, S., Abdallah, M., & Parner, E. T. (2010). Maternal infection requiring hospitalization during pregnancy and autism spectrum disorders. *Journal of Autism and Developmental Disorders, 40*, 1423–1430.

Atsmon, T., Yaakobi (Gorga), L., & Lowinger, S. (2019). Adults on the autism spectrum and their families: Residential issues. In S. Lowinger & S. Pearlman-Avnion (Eds.), *Autism in adulthood* (pp. 155–181). Springer International Publishing.

Bacon, A. L., Fein, D., Morris, R., Waterhouse, L., & Allen, D. (1998). The responses of autistic children to the distress of others. *Journal of Autism and Developmental Disorders, 28*, 129–142.

Baranek, G. T. (1999). Autism during infancy: A retrospective video analysis of sensory-motor and social behaviors at 9–12 months of age. *Journal of Autism and Developmental Disorders, 29*, 213–224.

Baron-Cohen, S. (2004). Autism: Research into causes and intervention. *Pediatric Rehabilitation, 7*(2), 73–78.

Baron-Cohen, S., Leslie, A. M., & Frith, U. (1985). Does the autistic child have a "theory of mind" ? *Cognition, 21*(1), 37–46.

Baron-Cohen, S., Campbell, R., Karmiloff-Smith, A., Grant, J., & Walker, J. (1995). Are children with autism blind to the mentalistic significance of the eyes? *British Journal of Developmental Psychology, 13*(4), 379–398.

Berger, N. I., Wainer, A. L., Kuhn, J., Bearss, K., Attar, S., Carter, A. S., Ibanez, L. V., Ingersoll, B. R., Neiderman, H., Scott, S., & Stone, W. L. (2022). Characterizing available tools for synchronous virtual assessment of toddlers with suspected autism Spectrum disorder: A brief report. *Journal of Autism and Developmental Disorders, 52*(1), 423–434.

Bhat, S., Acharya, U. R., Adeli, H., Bairy, G. M., & Adeli, A. (2014). Autism: Cause factors, early diagnosis and therapies. *Reviews in the Neurosciences, 25*(6).

Bodfish, J. W., Symons, F. J., Parker, D. E., & Lewis, M. H. (2000). Varieties of repetitive behavior in autism: Comparisons to mental retardation. *Journal of Autism and Developmental Disorders, 30*, 237–243.

Boucher, J. (2012). Research review: Structural language in autistic spectrum disorder—Characteristics and causes. *Journal of Child Psychology and Psychiatry, 53*(3), 219–233.

Buie, T., Campbell, D. B., Fuchs, G. J., III, Furuta, G. T., Levy, J., VandeWater, J., et al. (2010). Evaluation, diagnosis, and treatment of gastrointestinal disorders in individuals with ASDs: A consensus report. *Pediatrics, 125*(Supplement_1), S1–S18.

CDC. (2023). *ADDM network community report on autism*. Centers for Disease Control and Prevention. https://www.cdc.gov/ncbddd/autism/addm-community-report/index.html

Cervantes, P. E., & Matson, J. L. (2015). Comorbid symptomology in adults with autism spectrum disorder and intellectual disability. *Journal of Autism and Developmental Disorders, 45*(12), 3961–3970.

Chapman, R. (2021). Neurodiversity and the social ecology of mental functions. *Perspectives on Psychological Science, 16*(6), 1360–1372.

Charman, T., Swettenham, J., Baron-Cohen, S., Cox, A., Baird, G., & Drew, A. (1997). Infants with autism: An investigation of empathy, pretend play, joint attention, and imitation. *Developmental Psychology, 33*(5), 781.

Chawarska, K., Klin, A., Paul, R., & Volkmar, F. (2007). Autism spectrum disorder in the second year: Stability and change in syndrome expression. *Journal of Child Psychology and Psychiatry, 48*(2), 128–138.

Cohen, D., Pichard, N., Tordjman, S., Baumann, C., Burglen, L., Excoffier, E., Lazar, G., Mazet, P., Pinquier, C., Verloes, A., & Héron, D. (2005). Specific genetic disorders and autism: Clinical contribution towards their identification. *Journal of Autism and Developmental Disorders, 35*(1), 103–116.

Constantino, J. N., & Gruber, C. P. (2012). *Social responsiveness scale* (2nd ed.). Western Psychological Services.

Croen, L. A., Qian, Y., Ashwood, P., Zerbo, O., Schendel, D., Pinto-Martin, J., Daniele Fallin, M., Levy, S., Schieve, L. A., Yeargin-Allsopp, M., Sabourin, K. R., & Ames, J. L. (2019). Infection and fever in pregnancy and autism spectrum disorders: Findings from the study to explore early development. *Autism Research, 12*(10), 1551–1561.

Dawson, G., Toth, K., Abbott, R., Osterling, J., Munson, J., Estes, A., & Liaw, J. (2004). Early social attention impairments in autism: Social orienting, joint attention, and attention to distress. *Developmental Psychology, 40*(2), 271–283.

Dawson, G., Webb, S. J., & McPartland, J. (2005). Understanding the nature of face processing impairment in autism: Insights from behavioral and electrophysiological studies. *Developmental Neuropsychology, 27*(3), 403–424.

De Vries, A. L., Noens, I. L., Cohen-Kettenis, P. T., van Berckelaer-Onnes, I. A., & Doreleijers, T. A. (2010). Autism spectrum disorders in gender dysphoric children and adolescents. *Journal of Autism and Developmental Disorders, 40*, 930–936.

DeFrancesco, L. (2001). Scientists question rise in autism. *Nature Medicine, 7*(6), 645–646.

DiCicco-Bloom, E., Lord, C., Zwaigenbaum, L., Courchesne, E., Dager, S. R., Schmitz, C., et al. (2006). The developmental neurobiology of autism spectrum disorder. *Journal of Neuroscience, 26*(26), 6897–6906.

Dickerson, A. S., Rahbar, M. H., Pearson, D. A., Kirby, R. S., Bakian, A. V., Bilder, D. A., et al. (2017). Autism spectrum disorder reporting in lower socioeconomic neighborhoods. *Autism, 21*(4), 470–480.

Dwyer, P., Ryan, J. G., Williams, Z. J., & Gassner, D. L. (2022). First do no harm: Suggestions regarding respectful autism language. *Pediatrics, 149*(Supplement 4).

Eisenhower, A., Martinez Pedraza, F., Sheldrick, R. C., Frenette, E., Hoch, N., Brunt, S., & Carter, A. S. (2021). Multi-stage screening in early intervention: A critical strategy for improving ASD identification and addressing disparities. *Journal of Autism and Developmental Disorders, 51*(3), 868–883.

Falco, M. (2014). Autism rates now 1 in 68 US children: CDC. *CNN News*.

Frazier, T. W., Thompson, L., Youngstrom, E. A., Law, P., Hardan, A. Y., Eng, C., & Morris, N. (2014). A twin study of heritable and shared environmental contributions to autism. *Journal of Autism and Developmental Disorders, 44*(8), 2013–2025.

Gardener, H., Spiegelman, D., & Buka, S. L. (2009). Prenatal risk factors for autism: Comprehensive meta-analysis. *British Journal of Psychiatry, 195*(1), 7–14.

Gev, T., Avital, H., Rosenan, R., Oliver Aronson, L., & Golan, O. (2021). Socio emotional competence in young children with ASD during interaction with their typically developing peers. *Research in Autism Spectrum Disorders, 86*, 101818.

Glidden, D., Bouman, W. P., Jones, B. A., & Arcelus, J. (2016). Gender dysphoria and autism spectrum disorder: A systematic review of the literature. *Sexual Medicine Reviews, 4*(1), 3–14.

Goldstein, S., & Naglieri, J. A. (2009). *Autism spectrum rating scales*. Multi-Health Systems, Inc.

Graetz, J. E. (2010). Autism grows up: Opportunities for adults with autism. *Disability & Society, 25*(1), 33–47.

Grynszpan, O., Weiss, P. L. T., Perez-Diaz, F., & Gal, E. (2014). Innovative technology-based interventions for autism spectrum disorders: A meta-analysis. *Autism: The International Journal of Research and Practice, 18*(4), 346–361.

Hallmayer, J. (2011). Genetic heritability and shared environmental factors among twin pairs with autism. *Archives of General Psychiatry, 68*(11), 1095.

Hillier, A., Gallop, N., Mendes, E., Tellez, D., Buckingham, A., Nizami, A., & OToole, D. (2019). LGBTQ+ and autism spectrum disorder: Experiences and challenges. *International Journal of Transgender Health, 21*(1), 98–110.

Hughes, M. M., Shaw, K. A., DiRienzo, M., Durkin, M. S., Esler, A., Hall-Lande, J., Wiggins, L., Zahorodny, W., Singer, A., & Maenner, M. J. (2023). The prevalence and characteristics of children with profound autism, 15 sites, United States, 2000–2016. *Public Health Reports, 138*(6), 971–980.

Iossifov, I., O'Roak, B. J., Sanders, S. J., Ronemus, M., Krumm, N., Levy, D., Stessman, H. A., Witherspoon, K. T., Vives, L., Patterson, K. E., Smith, J. D., Paeper, B., Nickerson, D. A., Dea, J., Dong, S., Gonzalez, L. E., Mandell, J. D., Mane, S. M., Murtha, M. T., Sullivan, C. A., et al. (2014). The contribution of de novo coding mutations to autism spectrum disorder. *Nature, 515*(7526), 216.

Johnson, M. H., Griffin, R., Csibra, G., Halit, H., Farroni, T., De Haan, M., Tucker, L. A., Baron-Cohen, S., & Richards, J. (2005). The emergence of the social brain network: Evidence from typical and atypical development. *Development and Psychopathology, 17*(03), 599.

Jones, W., & Klin, A. (2013). Attention to eyes is present but in decline in 2–6-month-old infants later diagnosed with autism. *Nature, 504*(7480), Article 7480.

Kapp, S. K. (2020). Introduction. In S. K. Kapp (Ed.), *Autistic community and the neurodiversity movement: Stories from the frontline* (pp. 1–19). Springer.

Kasari, C., Freeman, S., & Paparella, T. (2006). Joint attention and symbolic play in young children with autism: A randomized controlled intervention study. *Journal of Child Psychology and Psychiatry, 47*(6), 611–620.

Klin, A., Jones, W., Schultz, R., Volkmar, F., & Cohen, D. (2002). Visual fixation patterns during viewing of naturalistic social situations as predictors of social competence in individuals with autism. *Archives of General Psychiatry, 59*(9), 809.

Klin, A., Danovitch, J. H., Merz, A. B., & Volkmar, F. R. (2007). Circumscribed interests in higher functioning individuals with autism spectrum disorders: An exploratory study. *Research and Practice for Persons with Severe Disabilities, 32*(2), 89–100.

Klin, A., Micheletti, M., Klaiman, C., Shultz, S., Constantino, J. N., & Jones, W. (2020). Affording autism an early brain development re-definition. *Development and Psychopathology, 32*(4), 1175–1189.

Kurth, J., & Mastergeorge, A. M. (2010). Individual education plan goals and Services for Adolescents with Autism: Impact of age and educational setting. *The Journal of Special Education, 44*(3), 146–160.

Landa, R. J., Holman, K. C., & Garrett-Mayer, E. (2007). Social and communication development in toddlers with early and later diagnosis of autism spectrum disorders. *Archives of General Psychiatry, 64*(7), 853.

Lang, R., O'Reilly, M., Rispoli, M., Shogren, K., Machalicek, W., Sigafoos, J., & Regester, A. (2009). Review of interventions to increase functional and symbolic play in children with autism. *Education and Training in Developmental Disabilities*, 481–492.

Leadbitter, K., Buckle, K. L., Ellis, C., & Dekker, M. (2021). Autistic self-advocacy and the neurodiversity movement: Implications for autism early intervention research and practice. *Frontiers in Psychology, 12*.

Leekam, S. R., Prior, M. R., & Uljarevic, M. (2011). Restricted and repetitive behaviors in autism spectrum disorders: a review of research in the last decade. *Psychological Bulletin, 137*(4), 562.

Lord, C., Rutter, M., & Le Couteur, A. (1994). Autism diagnostic interview-revised: A revised version of a diagnostic interview for caregivers of individuals with possible pervasive developmental disorders. *Journal of Autism and Developmental Disorders, 24*(5), 659–685.

Lord, C., Shulman, C., & DiLavore, P. (2004). Regression and word loss in autistic spectrum disorders. *Journal of Child Psychology and Psychiatry, 45*(5), 936–955.

Lord, C., Rutter, M., DiLavore, P. C., Risi, S., Gotham, K., & Bishop, S. (2012). *Autism diagnostic observation schedule* (2nd ed.). Western Psychological Services.

Lord, C., Charman, T., Havdahl, A., Carbone, P., Anagnostou, E., Boyd, B., Carr, T., de Vries, P. J., Dissanayake, C., Divan, G., Freitag, C. M., Gotelli, M. M., Kasari, C., Knapp, M., Mundy, P., Plank, A., Scahill, L., Servili, C., Shattuck, P., et al. (2022). The lancet commission on the future of care and clinical research in autism. *Lancet (London, England), 399*(10321), 271–334.

Lovaas, O. I., & Smith, T. (1988). Intensive behavioral treatment for young autistic children. In *Advances in clinical child psychology* (pp. 285–324). Springer US.

Maenner, M. J., Warren, Z., Williams, A. R., Amoakohene, E., Bakian, A. V., Bilder, D. A., et al. (2023). Prevalence and characteristics of autism spectrum disorder among children aged 8 years—Autism and developmental disabilities monitoring network, 11 sites, United States, 2020. *MMWR Surveillance Summaries, 72*(2), 1.

Magaña, S., Parish, S. L., Rose, R. A., Timberlake, M., & Swaine, J. G. (2012). Racial and ethnic disparities in quality of health care among children with autism and other developmental disabilities. *Intellectual and Developmental Disabilities, 50*(4), 287–299.

Mandell, D. S., Listerud, J., Levy, S. E., & Pinto-Martin, J. A. (2002). Race differences in the age at diagnosis among medicaid-eligible children with autism. *Journal of the American Academy of Child & Adolescent Psychiatry, 41*(12), 1447–1453.

Mandell, D. S., Barry, C. L., Marcus, S. C., Xie, M., Shea, K., Mullan, K., & Epstein, A. J. (2016). Effects of autism Spectrum disorder insurance mandates on the treated prevalence of autism Spectrum disorder. *JAMA Pediatrics, 170*(9), 887–893.

Mannion, A., & Leader, G. (2013). Comorbidity in autism spectrum disorder: A literature review. *Research in Autism Spectrum Disorders, 7*(12), 1595–1616.

Manwaring, S. S., Stevens, A. L., Mowdood, A., & Lackey, M. (2018). A scoping review of deictic gesture use in toddlers with or at-risk for autism spectrum disorder. *Autism & Developmental Language Impairments, 3*, 239694151775189.

Matson, J. L., & Shoemaker, M. (2009). Intellectual disability and its relationship to autism spectrum disorders. *Research in Developmental Disabilities, 30*(6), 1107–1114.

Matson, J. L., Matson, M. L., & Rivet, T. T. (2007). Social-skills treatments for children with autism Spectrum disorders: An overview. *Behavior Modification, 31*(5), 682–707.

Matson, J. L., Wilkins, J., & Ancona, M. (2008). Autism in adults with severe intellectual disability: An empirical study of symptom presentation. *Journal of Intellectual & Developmental Disability, 33*(1), 36–42.

Mazurek, M. O., Vasa, R. A., Kalb, L. G., Kanne, S. M., Rosenberg, D., Keefer, A., et al. (2013). Anxiety, sensory over-responsivity, and gastrointestinal problems in children with autism spectrum disorders. *Journal of Abnormal Child Psychology, 41*, 165–176.

McConnell, S. R. (2002). Interventions to facilitate social interaction for young children with autism: Review of available research and recommendations for educational intervention and future research. *Journal of Autism and Developmental Disorders, 32*(5), 351–372.

McElhanon, B. O., McCracken, C., Karpen, S., & Sharp, W. G. (2014). Gastrointestinal symptoms in autism spectrum disorder: A meta-analysis. *Pediatrics, 133*(5), 872–883.

Mendelson, J. L., Gates, J. A., & Lerner, M. D. (2016). Friendship in school-age boys with autism spectrum disorders: A meta-analytic summary and developmental, process-based model. *Psychological Bulletin, 142*(6), 601.

Mishra, A., Ceballos, V., Himmelwright, K., McCabe, S., & Scott, L. (2021). Gesture production in toddlers with autism spectrum disorder. *Journal of Autism and Developmental Disorders, 51*(5), 1658–1667.

Monteiro, S. A., Dempsey, J., Broton, S., Berry, L., Goin-Kochel, R. P., & Voigt, R. G. (2016). Early intervention before autism diagnosis in children referred to a regional autism clinic. *Journal of Developmental & Behavioral Pediatrics, 37*(1), 15–19.

Moree, B. N., & Davis, T. E. (2010). Cognitive-behavioral therapy for anxiety in children diagnosed with autism spectrum disorders: Modification trends. *Research in Autism Spectrum Disorders, 4*(3), 346–354.

Moss, J., & Howlin, P. (2009). Autism spectrum disorders in genetic syndromes: Implications for diagnosis, intervention and understanding the wider autism spectrum disorder population. *Journal of Intellectual Disability Research, 53*(10), 852–873.

Mutluer, T., Aslan Genç, H., Özcan Morey, A., Yapici Eser, H., Ertinmaz, B., Can, M., & Munir, K. (2022). Population-based psychiatric comorbidity in children and adolescents with autism Spectrum disorder: A meta-analysis. *Frontiers in Psychiatry, 13*, 856208.

Nadon, G., Feldman, D. E., Dunn, W., & Gisel, E. (2011). Mealtime problems in children with autism spectrum disorder and their typically developing siblings: A comparison study. *Autism, 15*(1), 98–113.

Nicholas, D. B., Attridge, M., Zwaigenbaum, L., & Clarke, M. (2015). Vocational support approaches in autism spectrum disorder: A synthesis review of the literature. *Autism, 19*(2), 235–245.

Odom, S. L., Thompson, J. L., Hedges, S., Boyd, B. A., Dykstra, J. R., Duda, M. A., Szidon, K. L., Smith, L. E., & Bord, A. (2015). Technology-aided interventions and instruction for adolescents with autism spectrum disorder. *Journal of Autism and Developmental Disorders, 45*(12), 3805–3819.

Özçalışkan, Ş., Adamson, L. B., & Dimitrova, N. (2016). Early deictic but not other gestures predict later vocabulary in both typical development and autism. *Autism, 20*(6), 754–763.

Ozonoff, S., Iosif, A. M., Baguio, F., Cook, I. C., Hill, M. M., Hutman, T., et al. (2010). A prospective study of the emergence of early behavioral signs of autism. *Journal of the American Academy of Child & Adolescent Psychiatry, 49*(3), 256–266.

Ozonoff, S., Young, G. S., Carter, A., Messinger, D., Yirmiya, N., Zwaigenbaum, L., et al. (2011). Recurrence risk for autism spectrum disorders: A baby siblings research consortium study. *Pediatrics, 128*(3), e488–e495.

Ozonoff, S., Young, G. S., Landa, R. J., Brian, J., Bryson, S., Charman, T., Chawarska, K., Macari, S. L., Messinger, D., Stone, W. L., Zwaigenbaum, L., & Iosif, A.-M. (2015). Diagnostic stability in young children at risk for autism spectrum disorder: A baby siblings research consortium study. *Journal of Child Psychology and Psychiatry, 56*(9), 988–998.

Pan, Y.-H., Wu, N., & Yuan, X.-B. (2019). Toward a better understanding of neuronal migration deficits in autism Spectrum disorders. *Frontiers in Cell and Developmental Biology, 7*, 205.

Patel, S. P., Winston, M., Guilfoyle, J., Nicol, T., Martin, G. E., Nayar, K., Kraus, N., & Losh, M. (2023). Neural processing of speech sounds in ASD and first-degree relatives. *Journal of Autism and Developmental Disorders, 53*(8), 3257–3271.

Pickard, K., Mellman, H., Frost, K., Reaven, J., & Ingersoll, B. (2021). Balancing fidelity and flexibility: Usual care for young children with an increased likelihood of having autism spectrum disorder within an early intervention system. *Journal of Autism and Developmental Disorders*, 1–13.

Pierce, K., Courchesne, E., & Bacon, E. (2016). To screen or not to screen universally for autism is not the question: Why the task force got it wrong. *The Journal of Pediatrics, 176*, 182–194.

Rajaraman, A., Austin, J. L., Gover, H. C., Cammilleri, A. P., Donnelly, D. R., & Hanley, G. P. (2022). Toward trauma-informed applications of behavior analysis. *Journal of Applied Behavior Analysis, 55*(1), 40–61.

Rapin, I., & Dunn, M. (1997). Language disorders in children with autism. *Seminars in Pediatric Neurology, 4*(2), 86–92. WB Saunders.

Ratajczak, H. V. (2011). Theoretical aspects of autism: Causes—A review. *Journal of Immunotoxicology, 8*(1), 68–79.

Reiner, O., Karzbrun, E., Kshirsagar, A., & Kaibuchi, K. (2016). Regulation of neuronal migration, an emerging topic in autism spectrum disorders. *Journal of Neurochemistry, 136*(3), 440–456.

Richler, J., Bishop, S. L., Kleinke, J. R., & Lord, C. (2007). Restricted and repetitive behaviors in young children with autism spectrum disorders. *Journal of Autism and Developmental Disorders, 37*(1), 73–85.

Rogers, S. J., & Ozonoff, S. (2005). Annotation: What do we know about sensory dysfunction in autism? A critical review of the empirical evidence. *Journal of Child Psychology and Psychiatry, and Allied Disciplines, 46*(12), 1255–1268.

Rogers, S. J., & Vismara, L. A. (2008). Evidence-based comprehensive treatments for early autism. *Journal of Clinical Child & Adolescent Psychology, 37*(1), 8–38.

Rutter, M., Bailey, A., & Lord, C. (2003). *The social communication questionnaire*. Western Psychological Services.

Salazar, F., Baird, G., Chandler, S., Tseng, E., O'sullivan, T., Howlin, P., et al. (2015). Co-occurring psychiatric disorders in preschool and elementary school-aged children with autism spectrum disorder. *Journal of Autism and Developmental Disorders, 45*, 2283–2294.

Sandbank, M., Bottema-Beutel, K., Crowley, S., Cassidy, M., Dunham, K., Feldman, J. I., Crank, J., Albarran, S. A., Raj, S., Mahbub, P., & Woynaroski, T. G. (2020). Project AIM: Autism intervention meta-analysis for studies of Young children. *Psychological Bulletin, 146*(1), 1–29.

Scheeren, A. M., Koot, H. M., & Begeer, S. (2012). Social interaction style of children and adolescents with high-functioning autism Spectrum disorder. *Journal of Autism and Developmental Disorders, 42*(10), 2046–2055.

Schendel, D., & Bhasin, T. K. (2008). Birth weight and gestational age characteristics of children with autism, including a comparison with other developmental disabilities. *Pediatrics, 121*(6), 1155–1164.

Schopler, E., Van Bourgondien, M. E., Wellman, G. J., & Love, S. R. (2010). *The childhood autism rating scale* (2nd ed.). Western Psychological Services.

Schreibman, L., Dawson, G., Stahmer, A. C., Landa, R., Rogers, S. J., McGee, G. G., et al. (2015). Naturalistic developmental behavioral interventions: Empirically validated treatments for autism spectrum disorder. *Journal of Autism and Developmental Disorders, 45*, 2411–2428.

Schuck, R. K., Tagavi, D. M., Baiden, K. M. P., Dwyer, P., Williams, Z. J., Osuna, A., Ferguson, E. F., Jimenez Muñoz, M., Poyser, S. K., Johnson, J. F., & Vernon, T. W. (2022). Neurodiversity and autism intervention: Reconciling perspectives through a naturalistic developmental behavioral intervention framework. *Journal of Autism and Developmental Disorders, 52*(10), 4625–4645.

Seaman, R. L., & Cannella-Malone, H. I. (2016). Vocational skills interventions for adults with autism spectrum disorder: A review of the literature. *Journal of Developmental and Physical Disabilities, 28*(3), 479–494.

Selten, J. P., Lundberg, M., Rai, D., & Magnusson, C. (2015). Risks for nonaffective psychotic disorder and bipolar disorder in young people with autism spectrum disorder: A population-based study. *JAMA Psychiatry, 72*(5), 483–489.

Sharp, W. G., Volkert, V. M., Scahill, L., McCracken, C. E., & McElhanon, B. (2017). A systematic review and meta-analysis of intensive multidisciplinary intervention for pediatric feeding disorders: How standard is the standard of care? *The Journal of Pediatrics, 181*, 116–124.e4.

Shattuck, P. T., Durkin, M., Maenner, M., Newschaffer, C., Mandell, D. S., Wiggins, L., et al. (2009). Timing of identification among children with an autism spectrum disorder: Findings from a population-based surveillance study. *Journal of the American Academy of Child & Adolescent Psychiatry, 48*(5), 474–483.

Shattuck, P. T., Narendorf, S. C., Cooper, B., Sterzing, P. R., Wagner, M., & Taylor, J. L. (2012). Postsecondary education and employment among youth with an autism spectrum disorder. *Pediatrics, 129*, 1042–1049. https://doi.org/10.1542/peds.2011-2864

Shaw, K. A., Bilder, D. A., McArthur, D., Williams, A. R., Amoakohene, E., Bakian, A. V., et al. (2023). Early identification of autism Spectrum disorder among children aged 4 years—Autism and developmental disabilities monitoring network, 11 sites, United States, 2020. *MMWR Surveillance Summaries, 72*(1), 1.

Shenouda, J., Barrett, E., Davidow, A. L., Sidwell, K., Halperin, W., Silenzio, V. M., & Zahorodny, W. (2022). Disparities in early intervention program participation by children with autism spectrum disorder in a US metropolitan area, 2006 to 2016. *JAMA Pediatrics, 176*(9), 906–914.

Shultz, S., Klin, A., & Jones, W. (2018). Neonatal transitions in social behavior and their implications for autism. *Trends in Cognitive Sciences, 22*(5), 452–469.

Simacek, J., Elmquist, M., Dimian, A. F., & Reichle, J. (2021). Current trends in telehealth applications to deliver social communication interventions for young children with or at risk for autism spectrum disorder. *Current Developmental Disorders Reports, 8*(1), 15–23.

Solomon, D., & Soares, N. (2020). Telehealth approaches to care coordination in autism spectrum disorder. In M. B. McClain, J. D. Shahidullah, & K. R. Mezher (Eds.), *Interprofessional care coordination for pediatric autism spectrum disorder: Translating research into practice* (pp. 289–306). Springer International Publishing.

Sparrow, S. S., & Cicchetti, D. V. (1989). *The Vineland adaptive behavior scales*. Allyn & Bacon.

Stagg, S. D., & Vincent, J. (2019). Autistic traits in individuals self-defining as transgender or nonbinary. *European Psychiatry, 61*, 17–22.

Stahmer, A. C., & Mandell, D. S. (2007). State infant/toddler program policies for eligibility and services provision for young children with autism. *Administration and Policy in Mental Health and Mental Health Services Research, 34*(1), 29–37.

Stone, W. L., Ousley, O. Y., Yoder, P. J., Hogan, K. L., & Hepburn, S. L. (1997). Nonverbal communication in two-and three-year-old children with autism. *Journal of Autism and Developmental Disorders, 27*, 677–696.

Sutherland, R., Trembath, D., & Roberts, J. (2018). Telehealth and autism: A systematic search and review of the literature. *International Journal of Speech-Language Pathology, 20*(3), 324–336.

Tager-Flusberg, H., Paul, R., & Lord, C. (2005). Language and communication in autism. In F. R. Volkmar, R. Paul, A. Klin, & D. Cohen (Eds.), *Handbook of autism and pervasive developmental disorders* (1st ed., pp. 335–364). Wiley.

Talbott, M. R., Young, G. S., Munson, J., Estes, A., Vismara, L. A., & Rogers, S. J. (2020). The developmental sequence and relations between gesture and spoken language in toddlers with autism spectrum disorder. *Child Development, 91*(3), 743–753.

Tick, B., Bolton, P., Happé, F., Rutter, M., & Rijsdijk, F. (2016). Heritability of autism spectrum disorders: A meta-analysis of twin studies. *Journal of Child Psychology and Psychiatry, 57*(5), 585–595.

Trevisan, D. A., Hoskyn, M., & Birmingham, E. (2018). Facial expression production in autism: A meta-analysis. *Autism Research, 11*(12), 1586–1601.

Tsai, C.-H., Chen, K.-L., Li, H.-J., Chen, K.-H., Hsu, C.-W., Lu, C.-H., Hsieh, K.-Y., & Huang, C.-Y. (2020). The symptoms of autism including social communication deficits and repetitive and restricted behaviors are associated with different emotional and behavioral problems. *Scientific Reports, 10*(1), 20509.

Turner, D., Briken, P., & Schöttle, D. (2017). Autism-spectrum disorders in adolescence and adulthood: Focus on sexuality. *Current Opinion in Psychiatry, 30*(6), 409–416.

Tye, C., Runicles, A. K., Whitehouse, A. J. O., & Alvares, G. A. (2019). Characterizing the interplay between autism spectrum disorder and comorbid medical conditions: An integrative review. *Frontiers in Psychiatry, 9*, 751.

Ung, D., Selles, R., Small, B. J., & Storch, E. A. (2015). A systematic review and meta-analysis of cognitive-behavioral therapy for anxiety in youth with high-functioning autism spectrum disorders. *Child Psychiatry & Human Development, 46*(4), 533–547.

Vismara, L. A., & Rogers, S. J. (2010). Behavioral treatments in autism spectrum disorder: What do we know? *Annual Review of Clinical Psychology, 6*, 447–468.

Webb, S. J., & Jones, E. J. H. (2009). Early identification of autism. *Infants and Young Children, 22*(2), 100–118.

Williams White, S., Keonig, K., & Scahill, L. (2007). Social skills development in children with autism spectrum disorders: A review of the intervention research. *Journal of Autism and Developmental Disorders, 37*(10), 1858–1868.

Yates, K., & Couteur, A. L. (2013). Diagnosing autism. *Paediatrics and Child Health, 23*(1), 5–10.

Zerbo, O., Modaressi, S., Goddard, K., Lewis, E., Fireman, B. H., Daley, M. F., et al. (2018). Vaccination patterns in children after autism spectrum disorder diagnosis and in their younger siblings. *JAMA Pediatrics, 172*(5), 469–475.

Zuckerman, K. E., Lindly, O. J., Reyes, N. M., Chavez, A. E., Macias, K., Smith, K. N., & Reynolds, A. (2017). Disparities in diagnosis and treatment of autism in Latino and non-Latino white families. *Pediatrics, 139*(5).

Zwaigenbaum, L., Bryson, S., Rogers, T., Roberts, W., Brian, J., & Szatmari, P. (2005). Behavioral manifestations of autism in the first year of life. *International Journal of Developmental Neuroscience, 23*(2–3), 143–152.

Zwaigenbaum, L., Bryson, S., Lord, C., Rogers, S., Carter, A., Carver, L., Chawarska, K., Constantino, J., Dawson, G., Dobkins, K., Fein, D., Iverson, J., Klin, A., Landa, R., Messinger, D., Ozonoff, S., Sigman, M., Stone, W., Tager-Flusberg, H., & Yirmiya, N. (2009). Clinical assessment and management of toddlers with suspected autism spectrum disorder: Insights from studies of high-risk infants. *Pediatrics, 123*(5), 1383–1391.

Zwaigenbaum, L., Bryson, S., & Garon, N. (2013). Early identification of autism spectrum disorders. *Behavioural Brain Research, 251*, 133–146.

Zwaigenbaum, L., Bauman, M. L., Choueiri, R., Kasari, C., Carter, A., Granpeesheh, D., et al. (2015). Early intervention for children with autism spectrum disorder under 3 years of age: Recommendations for practice and research. *Pediatrics, 136*(Supplement_1), S60–S81.

Neuroimaging's Role in Intellectual and Developmental Disability Research

Nancy Raitano Lee, Meghan O'Brien, and Hannah E. Grosman

Abstract Neuroimaging uses sophisticated technology to characterize the anatomy and physiology of the living human brain. Some forms of neuroimaging have been in use for close to a century. However, it was not until the end of the twentieth century that neuroimaging studies of individuals with intellectual and developmental disabilities (IDD) emerged in greater frequency in the research literature. This chapter will review four noninvasive technologies that are used with children and adults with IDD—magnetic resonance imaging, electroencephalography, magnetoencephalography, and functional near-infrared spectroscopy. An overview of "how" each technology works as well as "what" it can tell us about "where" and "when" different neural processes occur in the brain is provided. A brief synopsis of what this research has taught us about different forms of IDD is provided along with a discussion of the strengths and weaknesses of the different approaches. Finally, ethical considerations and the limitations of what we can learn from these studies are discussed.

Keywords Brain · Neuroimaging · Magnetic resonance imaging · Electroencephalography · Functional near-infrared spectroscopy

Brain development is believed to be a dynamic process that begins in utero and continues into young adulthood (Stiles & Jernigan, 2010). Prior to the advent of noninvasive neuroimaging methodologies such as magnetic resonance imaging (MRI), our understanding of the anatomy of the developing brain—both within the context of typical and atypical development—was limited to postmortem (i.e., after an individual has died) investigations of the brain or studies of patients with known brain lesions (i.e., abnormalities in brain structure due to acquired injury or illness).

N. R. Lee (✉) · M. O'Brien · H. E. Grosman
Department of Psychological and Brain Sciences, Drexel University, Philadelphia, PA, USA
e-mail: nrl39@drexel.edu

M. G. Valdovinos (ed.), *Intellectual and Developmental Disabilities*, Contemporary Clinical Neuroscience,
https://doi.org/10.1007/978-3-031-66932-3_16

Beginning with the explosion of developmental neuroimaging research in the late twentieth century (e.g., Giedd et al., 1996), our understanding of the structure of the human brain and how it changes over the course of infancy, childhood, and adolescence has never been greater. By closely studying typical brain development, researchers have provided an important benchmark for understanding how brain development among individuals with intellectual and developmental disabilities (IDD) differs from typical expectations. This, in turn, provides an opportunity to identify aspects of brain anatomy or physiology that may underlie atypical cognition and behavior in these groups, thus providing a biological target for interventions that aim to optimize developmental outcomes for individuals with IDD.

Importantly, neuroimaging offers the opportunity to not only study how the brains of individuals with IDD differ from typical peers at one point in time, but it also permits the opportunity to study brain development longitudinally. More specifically, by studying the same individuals over time, longitudinal neuroimaging allows researchers to identify how brain *development* deviates from or adheres to typical expectations at different points across the lifespan. This is crucial for the study of IDD, as unlike traumatic brain injury which can interrupt the brain at a discrete point in time, atypical brain development among individuals with IDD evolves and changes over time. Thus, there may be times in which individuals with IDD are more similar to typical peers and times in which significant deviations from typical expectations emerge. Understanding not only *what* differences are present but *when* those differences emerge is crucial to characterize the neural underpinnings of IDD more effectively and to identify developmentally attuned treatments to support optimal outcomes in these groups.

Overview of Neuroimaging Methodologies

To quantify brain characteristics in vivo (i.e., in the living entity), many different neuroimaging tools are available. Some of these tools focus on anatomy, such as MRI or computed tomography. When used to quantify the anatomical characteristics of the brain, these tools are sometimes referred to as structural imaging methodologies (i.e., they characterize the physical structure of different components of the brain). In contrast, some tools focus on characterizing the electrophysiology of the brain. As reviewed in Kovelman (2011), these include tools that quantify electrical activity, such as electroencephalography (EEG) and its event-based component, event-related potential (ERP), as well as magnetoencephalography (MEG). Complementing these measures are those that measure changes in brain–blood flow secondary to electrical activity changes. These include functional magnetic resonance imaging (fMRI), positron emission tomography (PET), and functional near-infrared spectroscopy (fNIRS). Helpful acronyms are summarized in the pop out box below for reference.

Acronyms:
BOLD: blood oxygen level dependent
CT: computed tomography
EEG: electroencephalography
ERP: event-related potential
fMRI: functional magnetic resonance imaging
fNIRS: functional near-infrared spectroscopy
IDD: intellectual and developmental disabilities
MEG: magnetoencephalography
MRI: magnetic resonance imaging
PET: positron emission tomography

This chapter will provide detailed descriptions of MRI, EEG, MEG, and fNIRS. These technologies were selected because they tend to be used more frequently with children and individuals with IDD, and they are also safe and noninvasive. Although PET has been used in some large-scale studies of IDD, like Down syndrome (Handen et al., 2020), it differs from the other technologies described in this chapter in that it requires exposure to radionuclides (i.e., radiation-containing molecules) via injection or inhalation (Gardner et al., 1992). Thus, there are more risks associated with PET than the other technologies described above, and as a result, researchers must weigh whether the knowledge gained from PET (e.g., a greater understanding of an atypical neurotransmitter system in a particular disorder) outweighs the risks (see Cervenka et al., 2022 for a review).

Prior to reviewing the first neuroimaging technology in greater depth, it is important to note that these different technologies vary not only in whether they describe brain structure or function. They also differ with regard to the precision of the information they provide on two key dimensions discussed in the functional neuroimaging literature—namely, spatial and temporal resolutions. Spatial resolution refers to how precise the technology is at identifying the location of physiological events that are being measured. If this was compared to finding a house on a map, some technologies may be able to tell us the town in which the house is located, others the street on which it is located, and still others may be able to provide information that approximates the house's location on a particular street. The other dimension of interest is temporal resolution, which refers to how precisely the neuroimaging technology captures the timing of neural events. If this was compared to estimating when someone arrives at an event, some technologies may be able to describe the time of the day (AM vs. PM), others the hour, and still others the minute or second of arrival. Other dimensions on which these technologies differ that will be discussed in the chapter include their ability to tolerate participant movement (i.e., the amount the individual is able to move during the acquisition of neuroimaging data and still provide data that are interpretable), the tactile input involved (i.e., whether the technology involves tighter fitting equipment or exposure to different gels), the noise the technology makes, the cost of the technology, and the technology's portability.

MRI: Magnetic Resonance Imaging

Although receiving an MRI scan is commonplace today, MRI technology is relatively 'new' within the context of different tools that have been used to learn about the human body across history. According to the Department of Energy, the first MRI scanner became commercially available less than 50 years ago, in 1980, and just 2 years later, the technology was beginning to be used to measure the brain's oxygen levels (Department of Energy, n.d.; Raichle, 2009). Unlike the other two neuroimaging technologies described in detail below, MRI is unique in that it allows us to measure both brain physiology (via information obtained about blood oxygenation levels) and brain anatomy. In fact, MRI permits the acquisition of exquisite three-dimensional images of the brain (and other body structures), allowing clinicians and researchers to quantify the volume of different brain structures and to describe other physical characteristics of the brain, such as the surface area or thickness of different brain structures. The ability to describe brain structures with a high degree of precision alongside quantifying brain physiology comes at a high cost. MRI machines, particularly ones used in research settings, cost hundreds of thousands of dollars (or more) to purchase and are also very costly to maintain. Moreover, the equipment itself is large (see Fig. 1), and certain safeguards need to be in place, such as the use of non-magnetic materials in the room that houses the MRI scanner. Thus, researchers using MRI as a tool must have adequate funds—often via federal or private grants—to conduct this research. This contrasts with many of the methods described below which are substantially lower in cost (though, not inexpensive, of course).

MRI capitalizes upon how hydrogen ions react to a strong magnetic field (generated via the MRI scanner's superconducting magnet) to create 'images' of human tissue, including brain tissue (MacIver et al., 2021). Different types of brain tissue, such as gray matter (comprised mostly of cell bodies and their synapses), white matter (comprised mostly of myelinated axons or the 'tentacles' that permit long-distance neuronal communication), and cerebral spinal fluid, respond differently to the magnetic field and this results in their differential appearance in brain imaging scans.

When describing brain imaging findings, scientists may quantify the number of voxels in a particular brain region. Voxels are discrete three-dimensional units that often range in size from 1 to 3 ml (though sizes are decreasing with improved technology, allowing for even more detailed descriptions of the brain; Giedd et al., 1996). It is important to note that thousands (or millions) of neurons may 'fit' into one voxel (Hassabis et al., 2009). Thus, when scientists interpret MRI data, they often cannot infer information about the cellular content of each voxel. As a result, it is important to acknowledge that in most cases, MRI cannot tell scientists about neuronal structure or function per se, as MRI does not offer the level of resolution needed to do so in most cases.

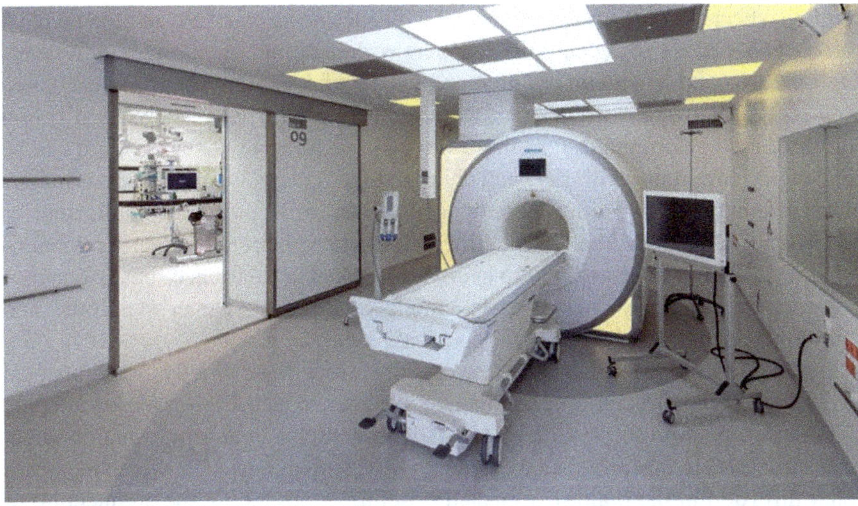

Fig. 1 MRI scanner. (Reprinted from "MRI measurement of the effects of moderate and deep neuromuscular blockade on the abdominal working space during laparoscopic surgery, a clinical study" by P. Krijtenburg, M.H.D. Bruintjes, J.J. Fütterer, G. van de Steeg, F., d'Ancona, G.J. Scheffer, C. Keijzer, & M.C. Warlé, 2023, *BMC Anesthesiology, 23*(1), 238.CC BY 4.0 (http://creativecommons.org/licenses/by/4.0/))

Structural Neuroimaging

When completing structural neuroimaging (i.e., neuroimaging focused on anatomy) with individuals with IDD or other clinical conditions, descriptions of brain structure may be provided at the whole brain level, at the level of the cortical lobes (e.g., frontal lobe volume) or subcortical structures (e.g., basal ganglia volume), or at a much more granular level. In the latter case, subregions of different structures or voxel-level descriptions of different brain regions are investigated, and statistical analyses are completed to evaluate whether there are statistically significant 'differences' between clinical and non-clinical populations. Structural MRI studies have identified differences in brain structure in infants, children, and adults with diverse forms of IDD—from autism spectrum disorder (Wallace et al., 2010) to Down syndrome (Pinter et al., 2001) to Williams syndrome (Thompson et al., 2005) to Fragile X syndrome (Meguid et al., 2012). These studies have been completed using various data collection approaches and have examined participants representing a wide spectrum of ages. For example, MRI investigations of early brain development have been completed while infants are sleeping, a frequent methodology used to study the infant siblings of children with autism spectrum disorder (e.g., Hazlett et al., 2012). They have also been completed with children and young adults who are sedated (e.g., Raznahan et al., 2013). However, this method is not frequently employed because of the risks associated with anesthesia exposure and the medical personnel needed to complete MRI scanning with sedation. More often, MRI investigations are completed with children and young adults without sedation (e.g., Lee et al., 2015). However, the demands of completing an MRI (i.e., lying still in a confined, loud environment for minutes to an hour or more in some cases) can be challenging for groups with IDD and children more generally, limiting descriptions of brain anatomy for groups of individuals with more profound intellectual or behavioral difficulties. Finally, MRI research has been helpful in describing neuroanatomical changes due not only to neuro*developmental* differences but also neuro-*degenerative* differences, as is seen among adults with Down syndrome and comorbid Alzheimer's disease (Pearlson et al., 1998).

Before describing functional neuroimaging, it will be briefly noted that an advanced form of structural neuroimaging, referred to as diffusion tensor imaging or DTI, can be used to provide scientists with complementary information about brain structures to that provided by standard MRI technology. DTI, which measures water diffusion in the brain, is especially useful for describing the architecture of white matter tracts (Ranzenberger et al., 2023). Because white matter is primarily comprised of myelinated axons, which are like the phone lines that allow communication between distant neurons in the brain, it is thought to be central to the development of highly distributed neural networks that underly complex cognition in the brain (Feldman et al., 2010). As such, information obtained about disruptions in white matter architecture from DTI may provide important insights into the hypothesized neural underpinnings of the complex learning challenges faced by people with IDD.

Functional Neuroimaging

Functional neuroimaging, or fMRI, refers to investigations of brain physiology in which the changes in the quantity of oxygen contained in the blood's hemoglobin are measured (Raichle, 1998). The resulting MRI 'signal' that is measured is referred to as the blood-oxygen-level-dependent or BOLD signal. The so-called BOLD signal can increase in certain areas and decrease in others in response to some event or task demand. fMRI investigations can either be "task-based fMRI," meaning that the individual is completing a task while receiving an MRI, or resting-state fMRI, meaning that the individual is resting comfortably while the MRI is completed. In the latter scenario, MRI aims to quantify spontaneous fluctuations in the BOLD signal (Raichle, 2009).

When compared to the other neuroimaging technologies reviewed in this chapter, fMRI offers the highest spatial resolution, but temporal resolution is lower than EEG and MEG in which the brain's electrical activity is being quantified. fMRI can also be greatly impacted by participant motion (Havsteen et al., 2017). Finally, unlike the other three technologies, an MRI scanner is exceptionally loud (as loud as an aircraft; Von Grätz, 2020) and safeguards (e.g., earplugs and/or headphones) must be used to protect the individual's ears.

fMRI investigations have revealed exciting insights into the neural underpinnings of atypical cognition and behavior among individuals with IDD. For example, task-based fMRI investigations have identified brain regions involved with atypical performance on social cognition tasks in autism spectrum disorder (Sugranyes et al., 2011), spatial processing difficulties in Williams syndrome (Meyer-Lindenberg et al., 2004), and hunger in Prader Willi syndrome (Nieuwpoort et al., 2021). These task-based investigations have not only described brain activation during a task at one point in time, but they have also been used to measure changes in brain functioning following psychoeducational (e.g., Barquero et al., 2014) or psychotherapeutic (Schrammen et al., 2022) interventions. Complementing these task-based studies are resting-state fMRI investigations. For example, resting-state fMRI has been used to describe functional connectivity in Down syndrome (Csumitta et al., 2022) and numerous other forms of IDD.

EEG: Electroencephalography

EEG is another noninvasive approach that can measure brain activity. EEG technology was first discovered by the German psychiatrist Hans Berger (1929). In his paper, "Über das Elektrenkephalogramm des Menschen," Berger wrote extensively on the topic of EEG and conducted studies using the technology to assess changes in brain functioning relating to attention and mental effort as well as disorders such as epilepsy and acquired brain injury.

Neurons are the main signaling unit of the brain and fire by propagating action potentials. When an action potential is triggered, the membrane potential of the neuron rapidly rises and falls causing adjacent neurons to fire. An EEG measures the summation of billions of action potentials and records waves of electrical change in the brain. During EEGs, surface electrodes are placed on the scalp to measure these changes in electrical activity. When required for specific medical reasons, such as management of a diagnosis of epilepsy, intracranial electrodes can be placed to measure brain activity in subcortical structures. Each electrode that is placed measures brain activity and produces an output that is referred to as a brain wave. Brain waves vary in frequency (cycles per second) and amplitude (voltage) according to a person's mental state (Teplan, 2002). EEG is often used to diagnose epilepsy, a disorder characterized by abnormal electrical signaling in the brain that results in seizure activity (Smith, 2005).

When EEG is used to measure the brain's response to a specific event, it is referred to as an event-related potential (ERP). While EEG can tell researchers a lot about brain activity at various states of arousal, ERPs help to measure specific brain processes. In an ERP experiment, researchers carefully analyze the brain's response to the onset of a stimulus. The waveform of an ERP contains components that are associated with specific brain processes. For example, the N1 component (a negative potential occurring on average 100 ms after stimulus onset), named to reflect the polarity (positive or negative potential) and timing of the response after stimulus onset, is associated with the visual response. Attenuated (i.e., reduced) N1 response has been shown to be associated with decreased visual attention and this is a well-established finding in the general population (Van Voorhis & Hillyard, 1977).

There is variability in the number of electrodes used in an EEG. Some EEG recordings are completed with just a few electrodes strategically placed on the scalp corresponding to underlying brain regions (e.g., the parietal lobe). An EEG technician places electrodes on the scalp after they are coated in a patient-friendly gel to support the conduction of the electrical signal. Some EEG technologies use electrodes that have a small sponge on the end that is soaked in a saline solution instead of a gel. In some cases, a cap is used to help space electrodes across the head and some of these allow for up to 256 electrodes (though this is quite a lot!). See Fig. 2 for an image of a child wearing an EEG cap.

Studies using EEG and ERP methodologies have helped make significant contributions to the literature on IDD. The technology has been used to characterize brain development (e.g., Marshall et al., 2002), support the identification of diagnostic brain markers of IDD (e.g., Jeste et al., 2015), and characterize atypical brain activity in specific disorders. For example, EEG findings have been used to support the impaired mirror neuron system theory of brain function in autism (Oberman et al., 2005). Mobile EEG has been used with some success to measure brain activity in naturalistic settings with children with IDD (Lau-Zhu et al., 2019).

In some ways, EEG is a more patient-friendly technology than fMRI. Participants are not required to be immobile to complete an EEG and there are no sounds that accompany its use. However, the technology can be very sensitive to movement; therefore, it can be difficult to use with younger or more active participants. As

Fig. 2 Child wearing an EEG cap. (Reprinted from "The big baby experiment" by L. Geddes, 2015, *Nature, 527*(7576), 22–25)

noted, a gel or saline solution is used to help the EEG sensors pick up the electrical signal from the scalp. This can be a challenging sensory experience for those with IDD, and a technology such as fNIRS, which also measures brain activity on the scalp, may be preferred. EEG has other limitations, such as poor spatial resolution (but this is accompanied by high temporal resolution). One group of researchers put it eloquently by stating "we are like blind men trying to understand the workings of a factory by listening outside the walls" (Hassett, 1978).

MEG: Magnetoencephalography

MEG is another noninvasive neuroimaging tool that can offer unique insights into the dynamic interplay of neural activity. Much like EEG and fMRI, MEG has emerged in recent years as a vital method to characterize the neural underpinnings of cognition, perception, and behavior. MEG utilizes a novel technology in this process. Unlike other neuroimaging modalities, MEG detects the magnetic fields produced by naturally occurring electrical activity in the brain (Supek & Aine, 2016). Magnetic fields penetrate biological tissues, including the skull, with minimal distortion, allowing for precise localization of neural activity using highly sensitive detectors, referred to as superconducting quantum interference devices (Hämäläinen et al., 1993).

In order to use MEG technology, the apparatus must operate within a specialized shielded room specifically designed to mitigate environmental interference. Participants are positioned with their heads snugly nestled in a helmet-like array of sensors that hover just around the head (see Fig. 3 for an image of MEG technology). Throughout the experimental session, MEG records the magnetic field fluctuations with remarkable precision, typically at sampling rates exceeding 1000 Hz (Wilson et al., 2016). This high temporal resolution enables the tracking of neural activity to the millisecond, capturing the fleeting nuances of cognitive processes. Leveraging advanced mathematical models, researchers infer the locations and strengths of underlying neuronal sources, providing invaluable insights into the brain regions orchestrating specific cognitive functions (Supek & Aine, 2016).

While MEG may not be as widely used as the previously described methods due to its more recent development, it has been used effectively to study epilepsy, Alzheimer's disease, Parkinsonism, autism spectrum disorder, schizophrenia, and attention-deficit/hyperactivity disorder (Fred et al., 2022; Yen et al., 2023). MEG has been successfully implemented in studies of infants, demonstrating infants' detection and prediction of auditory patterns (Zhao & Kuhl, 2016). One study showed that 11-month-old infants' neural response to non-native speech sounds predicted differences in their spoken grammar skills at 6 years old using MEG (Zhao et al., 2021). This study also found that differences in the prefrontal MEG response at 11 months accurately discriminated between participants with and without speech-language disorders at 6 years, demonstrating the utility of MEG for early identification of developmental differences. This is just one of several investigations

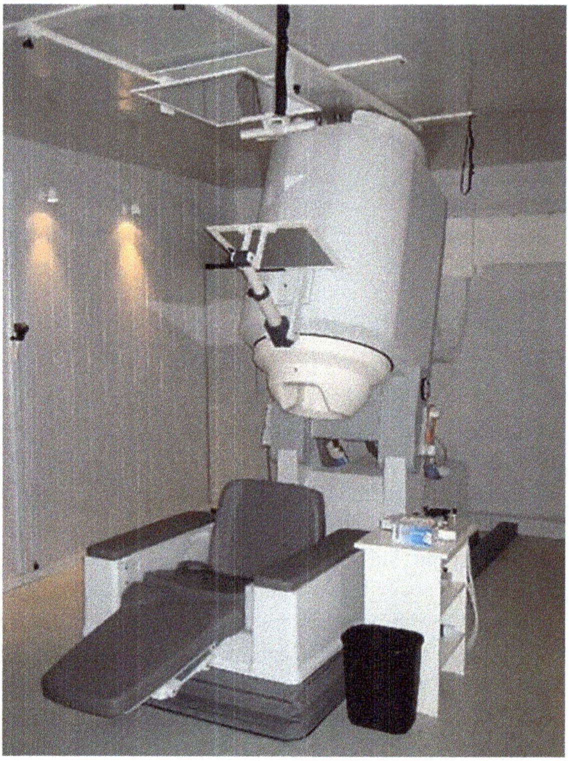

Fig. 3 MEG scanner. (Reprinted from "Magnetoencephalography" by E.S. Schwartz, J.C. Edgar, W.C. Gaetz, T.P.L Roberts, 2010, *Pediatric Radiology, 40*(1), 50–58)

that have utilized MEG to describe neural activity and lateralization among infants and children with language and communication disorders, such as specific language impairment, autism, dyslexia, and stuttering (Liang et al., 2021a, b). In autism, MEG has also been used to demonstrate neural differences that underlie challenges in social cognitive functions, such as orienting to eye gaze (Richard et al., 2013) and processing human faces (Bailey et al., 2005). Similarly, MEG has shown differences in neural coupling in autistic children that has been associated with lower social-communicative functioning (Lajiness-O'Neill et al., 2018). Currently, there are few studies using MEG to investigate other forms of IDD, but recent research has explored the utility of MEG to identify early neural markers of mild cognitive impairment and Alzheimer's disease in adults with Down syndrome (Fernández et al., 2022; García-Alba et al., 2019; Ramírez-Toraño et al., 2021).

Just like the other technologies reviewed thus far, MEG has both strengths and limitations relative to MRI, EEG, and fNIRS. Like EEG, MEG's greatest strength is its high temporal resolution. This is accompanied by poorer spatial resolution. For example, MEG's spatial resolution is not as precise as MRI. Like EEG, MEG is silent; thus, for participants with auditory sensitivities, it may be a preferred modality. Finally, similar to MRI, MEG systems are quite expensive to purchase and house (e.g., they need to be enclosed in a special shielded location) and are not portable. Thus, these are drawbacks to the approach.

fNIRS: Functional Near-Infrared Spectroscopy

fNIRS is a neuroimaging technique that has gained popularity in recent years for its portability, cost effectiveness, and ease of use. Like fMRI, fNIRS measures changes in blood oxygenation to track activation in the brain during different tasks or when participants are at 'rest' (as described in the MRI section). This technology uses near-infrared light, which lies just outside the visible light spectrum in the range of 650–950 nm and easily passes through skin, bone, and brain tissue (Almajidy et al., 2019). Notably, unlike fMRI, fNIRS can measure changes in both oxygenated and deoxygenated hemoglobin, which absorb near-infrared light at different frequencies.

As reviewed in Almajidy et al. (2019), the fNIRS device projects near-infrared light onto the surface of the brain via small, wearable sensors on the scalp. Sensors are organized in pairs of "sources" and "detectors." Source probes emit light, which passes through brain tissue before being received by detectors. As the light passes through the tissue between these sensors, some of the light is lost. The overall loss of light between source and detector probes represents changes in absorption by the brain's hemoglobin. This allows researchers to measure the increase in oxyhemoglobin and decrease in deoxyhemoglobin, called the hemodynamic response, that occurs in active areas of the brain (Almajidy et al., 2019; Lloyd-Fox et al., 2010, 2014; Wilcox & Biondi, 2015). Figure 4 shows how the hemodynamic response is observed using fNIRS.

The discovery that near-infrared light could pass through biological tissue to observe and monitor intact organs came in the 1970s (Jöbsis, 1977), and early uses of near-infrared spectroscopy included monitoring cerebral oxygenation during medical intervention (Scheeren et al., 2012; Wolf et al., 2007). The development of fNIRS to measure the brain's hemodynamic response during cognitive tasks did not come until 1993 (Hoshi & Tamura, 1993). With the refinement of the technology into its current portable design came a new focus on measuring the brain activity of infants and other individuals for whom existing neuroimaging techniques had proved difficult or inaccessible.

Unlike fMRI, fNIRS does not require participants to remain immobile. The sensors can be placed on a participant's head in a matter of minutes, and the fNIRS device can be transported to settings outside of the laboratory, which allow for more ecologically valid investigations and increased accessibility for research participants. However, fNIRS is limited by the depth to which near-infrared light can travel; because of the rapid diffusion of near-infrared light, fNIRS is not able to capture neural activation that occurs deeper than 1–3 cm beneath the cortical (i.e.,

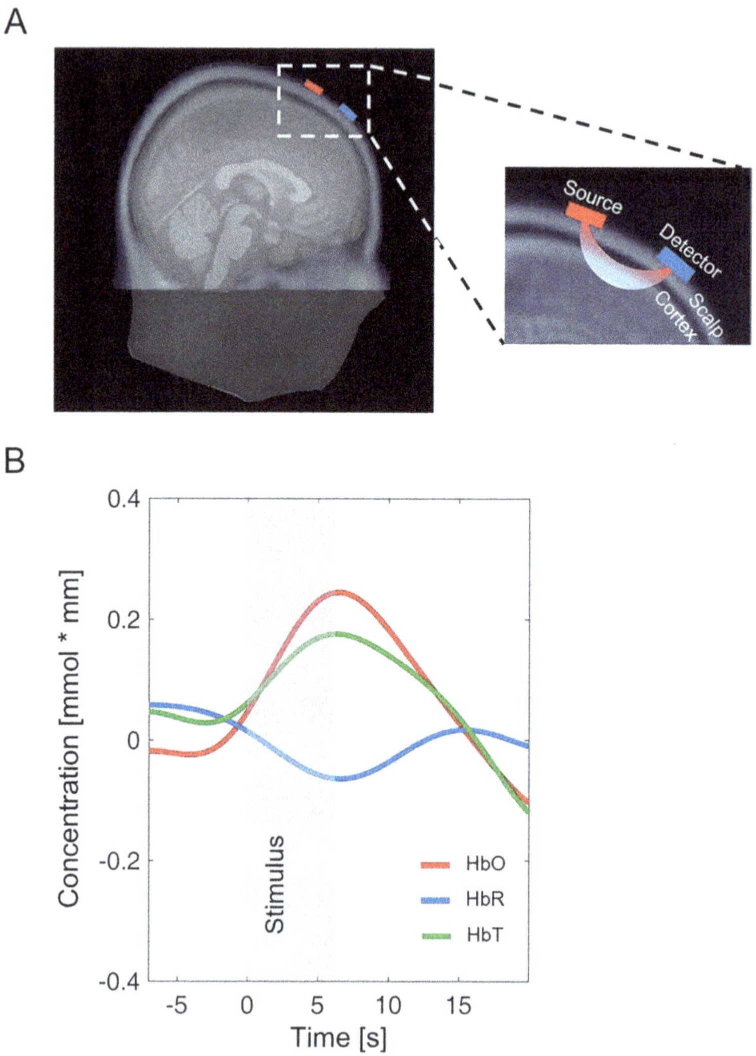

Fig. 4 (**a**) Depiction of the source-detector path followed by near-infrared light in fNIRS and (**b**) the representative fNIRS response. (Reprinted from "Applications of Functional Near-Infrared Spectroscopy in Fatigue, Sleep Deprivation, and Social Cognition" by Y. Pan, G. Borragán, & P. Peigneux, 2019, *Brain Topography, 32*(6), 998–1012)

outer, rind-like layer of the brain) surface (Almajidy et al., 2019). fNIRS may also be limited by the number of sensors placed on the scalp. Designs that prioritize portability may be limited to observations of the frontal lobe (see Fig. 5 for an example of a child wearing such a device), while observations of other areas of the cerebral cortex are possible with caps that contain more sensors but may be more difficult to set up or more uncomfortable to wear.

Fig. 5 Child wearing fNIRS sensor completing a task on an iPad with a researcher

fNIRS has been utilized to observe neural activation during mathematical or language tasks in naturalistic classroom settings (Soltanlou et al., 2018), characterize brain development throughout the lifespan (Liang et al., 2021a, b), measure changes in brain activation during cognitive testing after different types of intervention (Herold et al., 2018), identify objective markers of neural differences that result from risk factors such as malnutrition and poverty (Pinti et al., 2020), and more. Investigations of infant brain activation using fNIRS have been an area of particular focus due to the device's relative comfort and its ability to tolerate movement. These studies have contributed not only to our understanding of typical cognitive and perceptual development in infancy, but also to improvements in the mechanical design of the fNIRS technology to increase both comfort and efficiency. Such adaptations include the incorporation of a cloth headband and modifications to the optic fibers to be more flexible and lightweight (Lloyd-Fox et al., 2010).

The benefits of fNIRS make it particularly useful for investigations involving children with IDD. Because fNIRS is relatively insensitive to movement artifacts and requires less time than other neuroimaging techniques, it can be useful for children who have difficulty sitting still. Studies of children with attention-deficit/hyperactivity disorder using fNIRS have shown differences in the hemodynamic response of the prefrontal cortex during inhibitory control tasks (Ishii et al., 2017; Nagamitsu et al., 2012) and have been useful in evaluating the effects of medication and improving medication adherence (Kawai et al., 2021). The portability of fNIRS lends well to investigations of social communication, which may require a naturalistic setting in which participants can move and speak freely (Bhat et al., 2019; Li et al., 2022; Pirazzoli et al., 2022). fNIRS has the additional benefit of being silent, allowing for the incorporation of auditory stimuli in studies of typical and atypical language development (Gallagher et al., 2023). For participants who are sensitive to loud noises or confined spaces, fNIRS offers a more comfortable and accessible avenue for engaging in research. For these reasons, fNIRS has been used effectively in studies of children with autism spectrum disorder (Li & Yu, 2018; Xu et al., 2021), Down syndrome (Xu et al., 2020), Fragile X syndrome (Li et al., 2022), and psychiatric conditions (Gallagher et al., 2023). It has also been shown to be a viable alternative to the invasive tests of language lateralization that are widely used as part of the pre-surgical evaluation in epilepsy (Gallagher et al., 2007).

Longitudinal and cross-sectional studies of development across age groups have employed fNIRS to look for early diagnostic markers of different forms of IDD (Braukmann et al., 2018; Conti et al., 2022; Lloyd-Fox et al., 2018) and brain-based differences that may predict long-term developmental outcomes (Su et al., 2023). Understanding early indicators of atypical development increases opportunities for early intervention and psychoeducation which can improve outcomes as well as reduce stress in families. Finally, fNIRS has shown utility in extending brain development research to under-resourced areas. Its portability and relatively low cost have allowed researchers in low- and middle-income countries to study the impact of socio-economic and environmental factors on brain development in at-risk populations (Blasi et al., 2019; Pirazzoli et al., 2022). For example, a pilot study in 2014 using fNIRS was the first functional neuroimaging study of infants in Africa and successfully replicated results collected in the UK, emphasizing the practicality of global fNIRS application (Lloyd-Fox et al., 2014).

Comparing MRI, EEG, MEG, and fNIRS

In comparing these methodologies, it is important to consider which neuroimaging technologies will most effectively answer study questions and which technology is best suited to the research participants. For example, if a researcher wants to characterize brain structure, the only appropriate methodology of the ones reviewed here is MRI. In contrast, if the researcher is interested in characterizing brain function, all of the technologies would be appropriate; however, the choice of technology may be dictated by other factors, such as neural processes of interest as well as the temporal and spatial resolution of the information obtained. In addition, the time to set up the technology and acquire the data may be important to consider, particularly for individuals with IDD. For example, fNIRS takes little time to set up and is relatively more comfortable to wear than some other modalities, but it is limited in the amount of neural information it can access (i.e., it cannot readily measure brain function in deep brain structures). While fMRI can produce comprehensive images of the whole brain with high spatial resolution, EEG and MEG offer high temporal resolution, representing the timing of neural responses to the millisecond. This is in part due to differences in how brain activity is measured using each technology. Fig. 6 shows a comparison of how some technologies (e.g., EEG) measure electrophysiology, while others (e.g., fNIRS and fMRI) measure the hemodynamic response.

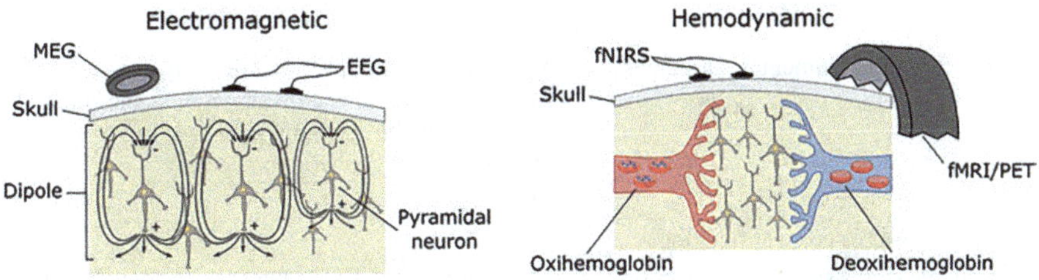

Fig. 6 Measuring the brain through the electromagnetic vs. hemodynamic response. (Reprinted from "Brain Imaging Methods in Social and Affective Neuroscience: A Machine Learning Perspective" by L.R. Trambaiolli, C.E. Biazoli, & J.R. Sato, in P.S. Boggio, T.S.H. Wingenbach, M.L. da Silveira Coêlho, W.E. Comfort, L. Murrins Marques, & M.V.C. Alves (Eds.), *Social and Affective Neuroscience of Everyday Human Interaction: From Theory to Methodology*, Chapter 13, 2022, Springer. CC BY 4.0. (http://creativecommons.org/licenses/by/4.0/))

The choice of neuroimaging to use will also be influenced by what technology is best suited to the population of interest. Many, but not all, individuals with IDD have specific sensory needs and may be unable to tolerate the loud MRI machine or the electrodes in some EEG setups. Unlike MRI, the flexibility of most fNIRS and EEG setups allows for children to sit on a parent's lap or in a comfortable seat, which may be more accessible for individuals who are highly active or anxious. Other individuals, however, may be more comfortable lying down, or may prefer not to wear sensors on the scalp, and, therefore, may be better suited for MRI. When using MEG, participants can be either in a seated position or lying down, and there are no loud noises, unlike with MRI. Finally, decisions about which modality to use may also come down to cost; while all neuroimaging techniques require expensive and difficult-to-replace equipment, EEG is often the most cost-accessible, and may be available for as "low" as $2000 for portable systems.

In summary, when choosing the appropriate technology for a particular research study, scientists must consider the type of information they need (e.g., structural or functional? sub-cortical or cortical activity? high temporal or spatial resolution?), the ability of participants to tolerate the equipment, the physical limitations of the lab space, and the financial implications of purchasing and using the equipment. Table 1 summarizes this information for easy reference. To capitalize on the strengths of each technology and minimize limitations, many studies employ multimodal methods (i.e., using multiple neuroimaging methodologies simultaneously). For example, fNIRS is easily paired with other methodologies including fMRI and EEG to provide a more comprehensive understanding of brain function (Scarapicchia et al., 2017). Prior to concluding this chapter, approaches that have been used by scientists to make neuroimaging more accessible to populations with IDD will be described along with a review of the broader limitations of these technologies, including a discussion of the ethical considerations surrounding neuroimaging research.

Table 1 Comparison of different neuroimaging techniques reviewed in this chapter

Description of equipment & participant positioning	MRI: head placed in coil while body in tube EEG: cap worn on head with gel or saline solution MEG: equipment surrounding head fNIRS: band or cap worn on head
Measures brain structure	MRI only
Directly measures cortical function	All methods
Directly measures subcortical function	MRI; intracranial EEG (available in some hospital settings)
Temporal resolution	EEG and MEG > fNIRS and MRI[a] (> denotes greater precision)
Spatial resolution	MRI > MEG and fNIRS and EEG[b] (> denotes greater precision)
Degree of participant immobility required	MRI > EEG and MEG > fNIRS (> denotes greater immobility)
Portability	fNIRS > MRI and MEG; EEG varies (some portable models) (> denotes more portability)
Auditory input	MRI > EEG and MEG and fNIRS (> denotes louder equipment)
Cost	MRI and MEG > EEG and fNIRS (> denotes greater cost to purchase)

[a]Some sources suggest that EEG has relatively better temporal resolution than MEG, and that fNIRS has relatively better temporal resolution than MRI
[b]Some sources suggest that MEG has relatively better spatial resolution than fNIRS and EEG

Accommodations to Make Neuroimaging More Accessible

Accommodating the unique needs of individuals with IDD in neuroimaging protocols is crucial for ensuring equitable participation and obtaining usable and reliable data to answer research questions. Various accommodations can be implemented, as detailed below, and research suggests that these accommodations are successful at making neuroimaging research more accessible for individuals with significant learning or communication difficulties (Kuschner et al., 2021). For example, simplified task instructions and visual aids are common accommodations that can enhance participant understanding and cooperation (Rao & Gagie, 2006). Additionally, exposure to and practice with the methods in advance of the process can be helpful. Researchers and clinicians alike use mock scanners for familiarization, which, anecdotally, serve to reduce anxiety and, in the research data, improve compliance (Suzuki et al., 2023). Moreover, researchers have decorated actual MRI scanners to be more child friendly (see Fig. 7). To familiarize children with EEG or fNIRS, researchers may provide participants with a tightly fitting cap or band, such as a swim cap or a thick headband, and ask family members to have the individual wear this in the days or even weeks leading up to a session to allow research participants to become accustomed to the sensations of the neuroimaging technology being used. Similarly, researchers using MRI may send a CD or audio file to a family with

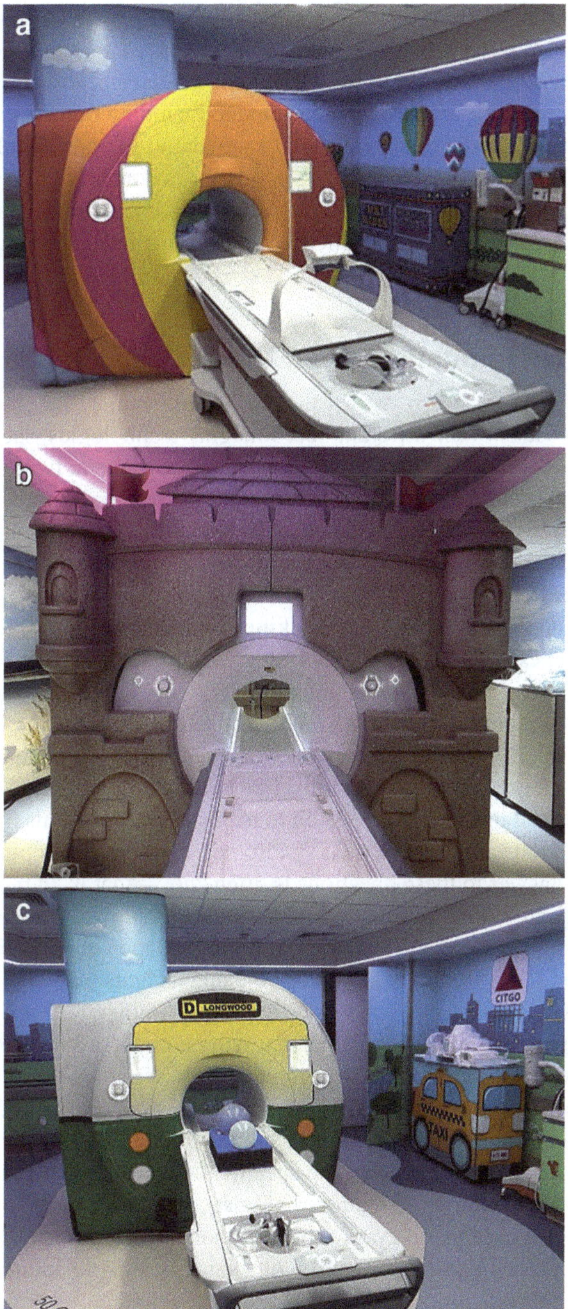

Fig. 7 Decorated MRI scanners for a child-friendly environment. (Reprinted from "Strategies to perform magnetic resonance imaging in infants and young children without sedation" by S.G. Harrington, C. Jaimes, K.M Weagle, M.C. Greer, & M.S. Gee, 2022, *Pediatric Radiology, 52*(2), 374–381)

the ‘sounds’ of the MRI scanner along with ear plugs (that are often worn as a form of protection from the noise of the MRI scanner) and ask that families practice listening to the sounds and wearing the earplugs in advance of the research visit. It is important to note that a supportive environment, including trained staff with expertise in working with individuals with IDD, is essential for successful participation. Ultimately, providing these accommodations helps to contribute to a more inclusive and accurate neuroimaging research base, promoting a better understanding of brain function in individuals with IDD with diverse learning and communication needs.

Neuroimaging Limitations and Ethical Considerations

Although drawbacks of the different neuroimaging technologies, such as their ability to accommodate motion or the costs associated with them, have been described in the preceding sections, there are other limitations to discuss more broadly. In this section, two questions will be discussed: “What can’t neuroimaging methods tell us?” and “What are some ethical considerations surrounding the use of these technologies in research?”

An important consideration when thinking about the utility of different neuroimaging technologies is the distinction between what a neuroimaging technology can tell us about a *group of individuals* (e.g., children with Williams syndrome) versus what it can tell us about *a single individual*. The latter consideration, focused on the individual, is often something families wonder about when deciding whether to enroll their loved one in a research study. For example, families may wonder: “Can this fMRI study tell me about how my child learns best?” “Can this fNIRS study tell me about my adult child with Down syndrome’s risk for developing Alzheimer’s disease?” The answer to these questions is often “no”—that is, in most instances (a few exceptions are noted below), research studies employing neuroimaging technologies cannot provide clinically meaningful information about an individual participant. Rather, these studies can provide valuable insights into what is true about a group of individuals. This is not to say that these studies will not inform future research, and possibly clinical practice, that will be impactful for individuals down the road; however, it is important to recognize that there may be no direct benefits to the individual who participates in a neuroimaging study. A few exceptions to this should be noted. If an MRI study offers a so-called ‘clinical scan’ (i.e., an anatomical scan acquired using the appropriate specifications that would permit the review of the individual’s brain by a neuroradiologist), information about an individual participant, such as whether they have brain tumor, can be provided. Similarly, if an EEG study includes a clinical evaluation of seizure activity in addition to research-related EEG data collection, then this information would be helpful to the individual. However, these benefits are not always offered (or available with the given technology).

Families of individuals with IDD, who may be eager for information that will help their loved ones, may be especially at risk of being taken advantage of by the promise of new, high-tech research methodologies or treatments. As an example, over the years, alternative treatments for IDD have emerged, many of which have been subsequently debunked or discredited by scientific research. One such example is chelation therapy, a medical procedure designed to remove heavy metals from the body. While initially suggested as a potential treatment for autism spectrum disorder based on the unfounded hypothesis that mercury exposure may be linked to the disorder, rigorous scientific studies have consistently refuted any efficacy of chelation therapy in improving autism symptoms (Brent, 2013; Davis et al., 2013). Despite this, some families continue to seek out these alternative treatments; similarly, families may seek answers or alternative explanations through the use of imaging technologies. Ethically, it is important to acknowledge this and inform families and individuals with IDD about the limitations of these technologies. One important way to address this issue is through clear communication during the informed consent process to ensure participants and/or their families understand the benefits and limitations of participating in neuroimaging research.

The explosion of neuroimaging has additionally introduced new ethical questions relating to informed consent, privacy, and data sharing. Research and ethical standards have raced to keep up with the evolving technology, but there remains a lack of consensus on how to handle issues such as properly deidentifying (i.e., removing the individual's name or other personal information) neuroimages, meaningfully explaining to participants how their data will be stored and shared, and the use of artificial intelligence for analysis of brain images (Eke et al., 2021). Some high-resolution MRI scans can generate images that include facial features, meaning that even deidentified images remain potentially identifiable (Kulynych, 2002), and data protection is especially complicated for remote neuroimaging data storage that relies on the use of the cloud (Shen et al., 2020).

Brain imaging research also faces the question of how to deal with incidental findings. What is the researcher's responsibility if a scan reveals a potentially dangerous health concern? It is important that a pathway for handling these findings is established and clearly communicated to the research team and participants before the study is conducted. Further, decisions about the handling of incidental findings should take into consideration the investigator's expertise and professional training, the participant's health and welfare, informed consent, subject autonomy (i.e., the right not to know), and the financial implications of extensive follow-up (Illes et al., 2006). Some recommendations for research teams include involving a physician who is qualified to evaluate brain images, establishing a clear referral pathway to support participants who may need to seek medical attention, and, most importantly, using unambiguous language during informed consent to communicate how incidental findings will be handled (Wolf et al., 2008).

Biomedical research, including neuroimaging studies as well as genetic investigations of the etiology of different forms of IDD, has also been criticized for its implication that different forms of IDD are "diseases" that need to be cured. For example, the Autistic Self Advocacy Network (ASAN) published an official stance on genetic research, which voices concerns related to the risks of emphasizing a "cure" for autism (Autistic Self Advocacy Network, 2022). In their statement, they suggested that "people who want to 'cure' autism think autism is a bad thing." They elaborated upon this, stating that "ASAN thinks these people are wrong. We want them to see autism as just another way of being a person." While ASAN's statement specifically addresses the possible risks of genetic research, neuroimaging research raises similar concerns. Neuroimaging (and other "high-tech") research often captures public attention and is well-funded, but there is little evidence to show that brain imaging research is actually the research avenue that is the most widely beneficial to individuals with IDD (Hinton, 2002). For example, stakeholder research has shown that autistic individuals prioritize applied and service-oriented research over biomedical research (Roche et al., 2021), despite only 9% of funds being dedicated to these areas (Cervantes et al., 2021).

As discussed already, there can be significant challenges implementing imaging technologies with youth with IDD. In general, these challenges are related to difficulties with task comprehension, attention, and/or sensory differences. This has resulted in the systematic exclusion of subgroups of individuals with IDD from research studies, particularly those who are minimally verbal or with intellectual disability (Jack & Pelphrey, 2017). As a result, research utilizing imaging technologies may include only a subset of individuals with a particular form of IDD that can 'tolerate' the neuroimaging procedures, and research findings from these studies may not be generalizable to individuals with the disorder with a different ability level. There is an urgent need to include individuals of all ability levels in neuroimaging studies to obtain larger and more representative samples that permit equity in the improvements in clinical care that may result from research focused on individuals with diverse forms of IDD.

One other way to include individuals who may be excluded from some neuroimaging investigations is to include them in biomedical studies that use alternative technologies, such as eye tracking and heart rate monitoring. For example, eye tracking has been shown to be a valuable tool for assessing gaze patterns and visual attention in individuals with IDD, particularly autism spectrum disorder (Boraston & Blakemore, 2007; Guillon et al., 2014). By analyzing eye movements, researchers gain insights into social cognition and atypical visual processing associated with autism spectrum disorder. Additionally, heart rate monitoring has proven instrumental in studying physiological arousal and stress responses in individuals with IDD, with researchers employing heart rate variability measures to explore autonomic nervous system functioning and emotional regulation (Benevides & Lane, 2015; Van Hecke et al., 2009).

Conclusion

Although there are limitations and ethical considerations associated with neuroimaging research, exciting insights into the biological underpinnings of IDD have been made over the past several decades thanks to advances in neuroimaging technology. As technology continues to advance, researchers may be able to answer questions that have not been thought of yet, and if care is taken to consider the limitations and ethics of biomedical research, our understanding of IDD will only continue to expand. There is hope that with increased understanding, there will be increases in the quantity and quality of the supports that can be provided to individuals with IDD, and that this in turn, will result in increased quality of life for those impacted.

References

Almajidy, R. K., Mankodiya, K., Abtahi, M., & Hofmann, U. G. (2019). A newcomer's guide to functional near infrared spectroscopy experiments. *IEEE Reviews in Biomedical Engineering, 13*, 292–308. https://doi.org/10.1109/RBME.2019.2944351

Autism Self Advocacy Network. (2022). *ASAN Statement on Genetic Research and Autism*. Retrieved December 13, 2023 from https://autisticadvocacy.org/genetic-research/

Bailey, A. J., Braeutigam, S., Jousmäki, V., & Swithenby, S. J. (2005). Abnormal activation of face processing systems at early and intermediate latency in individuals with autism spectrum disorder: A magnetoencephalographic study. *European Journal of Neuroscience, 21*(9), 2575–2585. https://doi.org/10.1111/j.1460-9568.2005.04061.x

Barquero, L. A., Davis, N., & Cutting, L. E. (2014). Neuroimaging of reading intervention. *PLoS One, 9*(1). https://doi.org/10.1371/journal.pone.0083668

Benevides, T. W., & Lane, S. J. (2015). A review of cardiac autonomic measures: Considerations for examination of physiological response in children with autism spectrum disorder. *Journal of Autism and Developmental Disorders, 45*, 560–575. https://10.1007/s10803-013-1971-z

Berger, H. (1929). Über das elektroenkephalogramm des menschen. *Archiv für Psychiatrie und Nervenkrankheiten, 87*(1), 527–570.

Bhat, A., McDonald, N., Eilbott, J., & Pelphrey, K. (2019). Exploring cortical activation and connectivity in infants with and without familial risk for autism during naturalistic social interactions: A preliminary study. *Infant Behavior and Development, 57*, 101337. https://doi.org/10.1016/j.infbeh.2019.101337

Blasi, A., Lloyd-Fox, S., Katus, L., & Elwell, C. E. (2019). fNIRS for tracking brain development in the context of global health projects. *Photonics, 6*, 89. https://doi.org/10.3390/photonics6030089

Boraston, Z., & Blakemore, S. J. (2007). The application of eye-tracking technology in the study of autism. *The Journal of Physiology, 581*(3), 893–898. https://doi.org/10.1113/jphysiol.2007.133587

Braukmann, R., Lloyd-Fox, S., Blasi, A., Johnson, M. H., Bekkering, H., Buitelaar, J. K., & Hunnius, S. (2018). Diminished socially selective neural processing in 5-month-old infants at high familial risk of autism. *European Journal of Neuroscience, 47*(6), 720–728. https://doi.org/10.1111/ejn.13751

Brent, J. (2013). Commentary on the abuse of metal chelation therapy in patients with autism spectrum disorders. *Journal of Medical Toxicology, 9*, 370–372. https://doi.org/10.1007/s13181-013-0345-4

Cervantes, P. E., Matheis, M., Estabillo, J., Seag, D. E., Nelson, K. L., Peth-Pierce, R., Hoagwood, K. E., & Horwitz, S. M. (2021). Trends over a decade in NIH funding for autism spectrum disorder services research. *Journal of Autism and Developmental Disorders, 51*, 2751–2763. https://doi.org/10.1007/s10803-020-04746-3

Cervenka, S., Frick, A., Bodén, R., & Lubberink, M. (2022). Application of positron emission tomography in psychiatry—Methodological developments and future directions. *Translational Psychiatry, 12*(1), 248. https://doi.org/10.1038/s41398-022-01990-2

Conti, E., Scaffei, E., Bosetti, C., Marchi, V., Costanzo, V., Dell'Oste, V., Mazziotti, R., Dell'Osso, L., Carmassi, C., & Muratori, F. (2022). Looking for "fNIRS signature" in autism spectrum: A systematic review starting from preschoolers. *Frontiers in Neuroscience, 16*, 785993. https://doi.org/10.3389/fnins.2022.785993

Csumitta, K. D., Gotts, S. J., Clasen, L. S., Martin, A., & Raitano Lee, N. (2022). Youth with Down syndrome display widespread increased functional connectivity during rest. *Scientific Reports, 12*(1), 9836. https://doi.org/10.1038/s41598-022-13437-1

Davis, T. N., O'Reilly, M., Kang, S., Lang, R., Rispoli, M., Sigafoos, J., Lancioni, G., Copeland, D., Attai, S., & Mulloy, A. (2013). Chelation treatment for autism spectrum disorders: A systematic review. *Research in Autism Spectrum Disorders, 7*(1), 49–55. https://doi.org/10.1016/j.rasd.2012.06.005

Department of Energy. (n.d.). *History of PET and MRI*. Retrieved December 13, 2023 from https://www.doemedicalsciences.org/historypetmri.shtml

Eke, D., Aasebø, I. E., Akintoye, S., Knight, W., Karakasidis, A., Mikulan, E., ... & Zehl, L. (2021). Pseudonymisation of neuroimages and data protection: Increasing access to data while retaining scientific utility. *Neuroimage: Reports, 1*(4), 100053.

Feldman, H. M., Yeatman, J. D., Lee, E. S., Barde, L. H., & Gaman-Bean, S. (2010). Diffusion tensor imaging: A review for pediatric researchers and clinicians. *Journal of Developmental & Behavioral Pediatrics, 31*(4), 346–356. https://doi.org/10.1097/DBP.0b013e3181dcaa8b

Fernández, A., Ramírez-Toraño, F., Bruña, R., Zuluaga, P., Esteba-Castillo, S., Abásolo, D., Moldenhauer, F., Shumbayawonda, E., Maestú, F., & García-Alba, J. (2022). Brain signal complexity in adults with Down syndrome: Potential application in the detection of mild cognitive impairment. *Frontiers in Aging Neuroscience, 14*, 988540. https://doi.org/10.3389/fnagi.2022.988540

Fred, A. L., Kumar, S. N., Kumar Haridhas, A., Ghosh, S., Purushothaman Bhuvana, H., Sim, W. K. J., Vimalan, V., Givo, F. A. S., Jousmäki, V., Padmanabhan, P., & Gulyás, B. (2022). A brief introduction to magnetoencephalography (MEG) and its clinical applications. *Brain Sciences, 12*(6), 788. https://doi.org/10.3390/brainsci12060788

Gallagher, A., Thériault, M., Maclin, E., Low, K., Gratton, G., Fabiani, M., Gagnon, L., Valois, K., Rouleau, I., & Sauerwein, H. C. (2007). Near-infrared spectroscopy as an alternative to the Wada test for language mapping in children, adults and special populations. *Epileptic Disorders, 9*(3), 241–255. https://doi.org/10.1684/epd.2007.0118

Gallagher, A., Wallois, F., & Obrig, H. (2023). Functional near-infrared spectroscopy in pediatric clinical research: Different pathophysiologies and promising clinical applications. *Neurophotonics, 10*(2), 023517–023517. https://doi.org/10.1117/1.NPh.10.2.023517

García-Alba, J., Ramírez-Toraño, F., Esteba-Castillo, S., Bruña, R., Moldenhauer, F., Novell, R., Romero-Medina, V., Maestú, F., & Fernández, A. (2019). Neuropsychological and neurophysiological characterization of mild cognitive impairment and Alzheimer's disease in Down syndrome. *Neurobiology of Aging, 84*, 70–79. https://doi.org/10.1016/j.neurobiolaging.2019.07.017

Gardner, S. F., Green, J. A., Bednarczyk, E. M., Farnett, L., & Miraldi, F. (1992). Principles and clinical applications of positron emission tomography. *American Journal of Hospital Pharmacy, 49*(6), 1499–1506.

Geddes, L. (2015). The big baby experiment. *Nature, 527*(7576), 22–25. https://doi.org/10.1038/527022a

Giedd, J. N., Snell, J. W., Lange, N., Rajapakse, J. C., Casey, B., Kozuch, P. L., Vaituzis, A. C., Vauss, Y. C., Hamburger, S. D., & Kaysen, D. (1996). Quantitative magnetic resonance imaging of human brain development: Ages 4–18. *Cerebral Cortex, 6*(4), 551–559. https://doi.org/10.1093/cercor/6.4.551

Guillon, Q., Hadjikhani, N., Baduel, S., & Rogé, B. (2014). Visual social attention in autism spectrum disorder: Insights from eye tracking studies. *Neuroscience & Biobehavioral Reviews, 42*, 279–297. https://doi.org/10.1016/j.neubiorev.2014.03.013

Hämäläinen, M., Hari, R., Ilmoniemi, R. J., Knuutila, J., & Lounasmaa, O. V. (1993). Magnetoencephalography—Theory, instrumentation, and applications to noninvasive studies of the working human brain. *Reviews of Modern Physics, 65*(2), 413. https://doi.org/10.1103/RevModPhys.65.413

Handen, B. L., Lott, I. T., Christian, B. T., Schupf, N., OBryant, S., Mapstone, M., Fagan, A. M., Lee, J. H., Tudorascu, D., & Wang, M. C. (2020). The Alzheimer's biomarker consortium-Down syndrome: Rationale and methodology. *Alzheimer's & Dementia: Diagnosis, Assessment & Disease Monitoring, 12*(1), e12065. https://doi.org/10.1002/dad2.12065

Harrington, S. G., Jaimes, C., Weagle, K. M., Greer, M.-L. C., & Gee, M. S. (2022). Strategies to perform magnetic resonance imaging in infants and young children without sedation. *Pediatric Radiology, 52*(2), 374–381. https://doi.org/10.1007/s00247-021-05062-3

Hassabis, D., Chu, C., Rees, G., Weiskopf, N., Molyneux, P. D., & Maguire, E. A. (2009). Decoding neuronal ensembles in the human hippocampus. *Current Biology, 19*(7), 546–554. https://doi.org/10.1016/j.cub.2009.02.033

Hassett, J. (1978). *A primer of psychophysiology*. Freeman.

Havsteen, I., Ohlhues, A., Madsen, K. H., Nybing, J. D., Christensen, H., & Christensen, A. (2017). Are movement artifacts in magnetic resonance imaging a real problem? A narrative review. *Frontiers in Neurology, 8*, 232. https://doi.org/10.3389/fneur.2017.00232

Hazlett, H. C., Gu, H., McKinstry, R. C., Shaw, D. W., Botteron, K. N., Dager, S. R., Styner, M., Vachet, C., Gerig, G., & Paterson, S. J. (2012). Brain volume findings in 6-month-old infants at high familial risk for autism. *American Journal of Psychiatry, 169*(6), 601–608. https://doi.org/10.1176/appi.ajp.2012.11091425

Herold, F., Wiegel, P., Scholkmann, F., & Müller, N. G. (2018). Applications of functional near-infrared spectroscopy (fNIRS) neuroimaging in exercise–cognition science: A systematic, methodology-focused review. *Journal of Clinical Medicine, 7*(12), 466. https://doi.org/10.3390/jcm7120466

Hinton, V. J. (2002). Ethics of neuroimaging in pediatric development. *Brain and Cognition, 50*(3), 455–468. https://doi.org/10.1016/S0278-2626(02)00521-3

Hoshi, Y., & Tamura, M. (1993). Detection of dynamic changes in cerebral oxygenation coupled to neuronal function during mental work in man. *Neuroscience Letters, 150*(1), 5–8. https://doi.org/10.1016/0304-3940(93)90094-2

Illes, J., Kirschen, M. P., Edwards, E., Stanford, L. R., Bandettini, P., Cho, M. K., et al. (2006). Incidental findings in brain imaging research. *Science, 311*(5762), 783–784. https://doi.org/10.1126/science.1124665

Ishii, S., Kaga, Y., Tando, T., Aoyagi, K., Sano, F., Kanemura, H., Sugita, K., & Aihara, M. (2017). Disinhibition in children with attention-deficit/hyperactivity disorder: Changes in [oxy-Hb] on near-infrared spectroscopy during "rock, paper, scissors" task. *Brain and Development, 39*(5), 395–402. https://doi.org/10.1016/j.braindev.2016.12.005

Jack, A., & Pelphrey, K. (2017). Annual research review: Understudied populations within the autism spectrum–current trends and future directions in neuroimaging research. *Journal of Child Psychology and Psychiatry, 58*(4), 411–435. https://doi.org/10.1111/jcpp.12687

Jeste, S. S., Frohlich, J., & Loo, S. K. (2015). Electrophysiological biomarkers of diagnosis and outcome in neurodevelopmental disorders. *Current Opinion in Neurology, 28*(2), 110. https://doi.org/10.1097/WCO.0000000000000181

Jöbsis, F. F. (1977). Noninvasive, infrared monitoring of cerebral and myocardial oxygen sufficiency and circulatory parameters. *Science, 198*(4323), 1264–1267. https://doi.org/10.1126/science.929199

Kawai, C., Mori, K., Tanioka, T., Betriana, F., Mori, K., Mori, T., & Ito, H. (2021). Usefulness of near-infrared spectroscopy (NIRS) for evaluating drug effects and improvements in medication adherence in children with attention deficit hyperactivity disorder (ADHD). *The Journal of Medical Investigation, 68*(1.2), 53–58. https://doi.org/10.2152/jmi.68.53

Kovelman, I. (2011). Neuroimaging methods. In E. Hoff (Ed.), *Research methods in child language: A practical guide* (Vol. 4, pp. 43–59). John Wiley & Sons.

Krijtenburg, P., Bruintjes, M. H. D., Fütterer, J. J., van de Steeg, G., d'Ancona, F., Scheffer, G. J., Keijzer, C., & Warlé, M. C. (2023). MRI measurement of the effects of moderate and deep neuromuscular blockade on the abdominal working space during laparoscopic surgery, a clinical study. *BMC Anesthesiology, 23*(1), 238. https://doi.org/10.1186/s12871-023-02201-1

Kulynych, J. (2002). Legal and ethical issues in neuroimaging research: Human subjects protection, medical privacy, and the public communication of research results. *Brain and Cognition, 50*(3), 345–357. https://doi.org/10.1016/S0278-2626(02)00518-3

Kuschner, E. S., Kim, M., Bloy, L., Dipiero, M., Edgar, J. C., & Roberts, T. P. (2021). MEG-PLAN: A clinical and technical protocol for obtaining magnetoencephalography data in minimally verbal or nonverbal children who have autism spectrum disorder. *Journal of Neurodevelopmental Disorders, 13*, 1–17. https://doi.org/10.1186/s11689-020-09350-1

Lajiness-O'Neill, R., Brennan, J. R., Moran, J. E., Richard, A. E., Flores, A.-M., Swick, C., Goodcase, R., Andersen, T., McFarlane, K., Rusiniak, K., Kovelman, I., Wagley, N., Ugolini, M., Albright, J., & Bowyer, S. M. (2018). Patterns of altered neural synchrony in the default mode network in autism spectrum disorder revealed with magnetoencephalography (MEG): Relationship to clinical symptomatology. *Autism Research, 11*(3), 434–449. https://doi.org/10.1002/aur.1908

Lau-Zhu, A., Lau, M. P., & McLoughlin, G. (2019). Mobile EEG in research on neurodevelopmental disorders: Opportunities and challenges. *Developmental Cognitive Neuroscience, 36*, 100635. https://doi.org/10.1016/j.dcn.2019.100635

Lee, N. R., Adeyemi, E. I., Lin, A., Clasen, L. S., Lalonde, F. M., Condon, E., Driver, D. I., Shaw, P., Gogtay, N., & Raznahan, A. (2015). Dissociations in cortical morphometry in youth with Down syndrome: Evidence for reduced surface area but increased thickness. *Cerebral Cortex, 26*(7), 2982–2990. https://doi.org/10.1093/cercor/bhv107

Li, Y., & Yu, D. (2018). Variations of the functional brain network efficiency in a young clinical sample within the autism spectrum: A fNIRS investigation. *Frontiers in Physiology, 9*, 67. https://doi.org/10.3389/fphys.2018.00067

Li, R., Bruno, J. L., Lee, C. H., Bartholomay, K. L., Sundstrom, J., Piccirilli, A., Jordan, T., Miller, J. G., Lightbody, A. A., & Reiss, A. L. (2022). Aberrant brain network and eye gaze patterns during natural social interaction predict multi-domain social-cognitive behaviors in girls with fragile X syndrome. *Molecular Psychiatry, 27*(9), 3768–3776. https://doi.org/10.1038/s41380-022-01626-3

Liang, A., Wu, H., Ma, Y., Liang, X., Jia, Y., Gao, Y., Ning, X., & Fang, J. (2021a). Whole-head magnetoencephalogram and its application in developmental communication disorders research: A review. *IEEE Access, 9*, 42515–42532. https://doi.org/10.1109/ACCESS.2021.3063054

Liang, Z., Tian, H., Yang, H.-C. S., Arimitsu, T., Takahashi, T., Sassaroli, A., Fantini, S., Niu, H., Minagawa, Y., & Tong, Y. (2021b). Tracking brain development from neonates to the elderly by hemoglobin phase measurement using functional near-infrared spectroscopy. *IEEE Journal of Biomedical and Health Informatics, 25*(7), 2497–2509. https://doi.org/10.1109/JBHI.2021.3053900

Lloyd-Fox, S., Blasi, A., & Elwell, C. (2010). Illuminating the developing brain: The past, present and future of functional near infrared spectroscopy. *Neuroscience & Biobehavioral Reviews, 34*(3), 269–284. https://doi.org/10.1016/j.neubiorev.2009.07.008

Lloyd-Fox, S., Papademetriou, M., Darboe, M. K., Everdell, N. L., Wegmuller, R., Prentice, A. M., Moore, S. E., & Elwell, C. E. (2014). Functional near infrared spectroscopy (fNIRS) to assess cognitive function in infants in rural Africa. *Scientific Reports, 4*(1), 4740. https://doi.org/10.1038/srep04740

Lloyd-Fox, S., Blasi, A., Pasco, G., Gliga, T., Jones, E. J., Murphy, D., Elwell, C., Charman, T., Johnson, M. H., & Team, B. (2018). Cortical responses before 6 months of life associate with later autism. *European Journal of Neuroscience, 47*(6), 736–749. https://doi.org/10.1111/ejn.13757

MacIver, C. L., Ebden, S., & Tallantyre, E. C. (2021). MRI: How to understand it. *Practical Neurology, 21*(3), 216–224. https://doi.org/10.1136/practneurol-2020-002905

Marshall, P. J., Bar-Haim, Y., & Fox, N. A. (2002). Development of the EEG from 5 months to 4 years of age. *Clinical Neurophysiology, 113*(8), 1199–1208. https://doi.org/10.1016/s1388-2457(02)00163-3

Meguid, N. A., Fahim, C., Sami, R., Nashaat, N. H., Yoon, U., Anwar, M., El-Dessouky, H. M., Shahine, E. A., Ibrahim, A. S., & Mancini-Marie, A. (2012). Cognition and lobar morphology in full mutation boys with fragile X syndrome. *Brain and Cognition, 78*(1), 74–84. https://doi.org/10.1016/j.bandc.2011.09.005

Meyer-Lindenberg, A., Kohn, P., Mervis, C. B., Kippenhan, J. S., Olsen, R. K., Morris, C. A., & Berman, K. F. (2004). Neural basis of genetically determined visuospatial construction deficit in Williams syndrome. *Neuron, 43*(5), 623–631. https://doi.org/10.1016/j.neuron.2004.08.014

Nagamitsu, S., Yamashita, Y., Tanaka, H., & Matsuishi, T. (2012). Functional near-infrared spectroscopy studies in children. *BioPsychoSocial medicine, 6*, 1–7. https://doi.org/10.1186/1751-0759-6-7

Nieuwpoort, I. C. V., Slagboom, T. N., Jakobsdóttir, S., Deijen, J. B., Veltman, D. J., Curfs, L. M., & Drent, M. L. (2021). Food-related brain activation measured by fMRI in adults with Prader–Willi syndrome. *Journal of Clinical Medicine, 10*(21), 5133. https://10.3390/jcm10215133

Oberman, L. M., Hubbard, E. M., McCleery, J. P., Altschuler, E. L., Ramachandran, V. S., & Pineda, J. A. (2005). EEG evidence for mirror neuron dysfunction in autism spectrum disorders. *Cognitive Brain Research, 24*(2), 190–198. https://doi.org/10.1016/j.cogbrainres.2005.01.014

Pan, Y., Borragán, G., & Peigneux, P. (2019). Applications of functional near-infrared spectroscopy in fatigue, sleep deprivation, and social cognition. *Brain Topography, 32*(6), 998–1012. https://10.1007/s10548-019-00740-w

Pearlson, G. D., Breiter, S. N., PhD, E. H. A., Warren, A. C., Grygorcewicz, M., Frangou, S., PhD, P. E. B. M., & PhD, M. B. P. (1998). MRI brain changes in subjects with down syndrome with and without dementia. *Developmental Medicine & Child Neurology, 40*(5), 326–334.

Pinter, J. D., Eliez, S., Schmitt, J. E., Capone, G. T., & Reiss, A. L. (2001). Neuroanatomy of Down's syndrome: A high-resolution MRI study. *American Journal of Psychiatry, 158*(10), 1659–1665. https://doi.org/10.1176/appi.ajp.158.10.1659

Pinti, P., Tachtsidis, I., Hamilton, A., Hirsch, J., Aichelburg, C., Gilbert, S., & Burgess, P. W. (2020). The present and future use of functional near-infrared spectroscopy (fNIRS) for cognitive neuroscience. *Annals of the New York Academy of Sciences, 1464*(1), 5–29. https://doi.org/10.1111/nyas.13948

Pirazzoli, L., Sullivan, E., Xie, W., Richards, J. E., Bulgarelli, C., Lloyd-Fox, S., Shama, T., Kakon, S. H., Haque, R., & William, A., Jr. (2022). Association of psychosocial adversity and social information processing in children raised in a low-resource setting: An fNIRS study. *Developmental Cognitive Neuroscience, 56*, 101125. https://doi.org/10.1016/j.dcn.2022.101125

Raichle, M. E. (1998). Behind the scenes of functional brain imaging: A historical and physiological perspective. *Proceedings of the National Academy of Sciences, 95*(3), 765–772. https://doi.org/10.1073/pnas.95.3.765

Raichle, M. E. (2009). A brief history of human brain mapping. *Trends in Neurosciences, 32*(2), 118–126. https://doi.org/10.1016/j.tins.2008.11.001

Ramírez-Toraño, F., García-Alba, J., Bruña, R., Esteba-Castillo, S., Vaquero, L., Pereda, E., Maestú, F., & Fernández, A. (2021). Hypersynchronized magnetoencephalography brain networks in patients with mild cognitive impairment and Alzheimer's disease in down syndrome. *Brain Connectivity, 11*(9), 725–733. https://doi.org/10.1089/brain.2020.0897

Ranzenberger, L. R., Das J, M., & Snyder, T. (2023). Diffusion tensor imaging. In *StatPearls*. StatPearls Publishing.

Rao, S. M., & Gagie, B. (2006). Learning through seeing and doing: Visual supports for children with autism. *Teaching Exceptional Children, 38*(6), 26–33. https://doi.org/10.1177/004005990603800 6

Raznahan, A., Lenroot, R., Thurm, A., Gozzi, M., Hanley, A., Spence, S. J., Swedo, S. E., & Giedd, J. N. (2013). Mapping cortical anatomy in preschool aged children with autism using surface-based morphometry. *NeuroImage: Clinical, 2*, 111–119. https://doi.org/10.1016/j.nicl.2012.10.005

Richard, A. E., Lajiness-O'Neill, R. R., & Bowyer, S. M. (2013). Impaired prefrontal gamma band synchrony in autism spectrum disorders during gaze cueing. *Neuroreport, 24*(16), 894. https://doi.org/10.1097/WNR.0000000000000015

Roche, L., Adams, D., & Clark, M. (2021). Research priorities of the autism community: A systematic review of key stakeholder perspectives. *Autism, 25*(2), 336–348. https://doi.org/10.1177/1362361320967790

Scarapicchia, V., Brown, C., Mayo, C., & Gawryluk, J. R. (2017). Functional magnetic resonance imaging and functional near-infrared spectroscopy: Insights from combined recording studies. *Frontiers in Human Neuroscience, 11*, 419. https://doi.org/10.3389/fnhum.2017.00419

Scheeren, T., Schober, P., & Schwarte, L. (2012). Monitoring tissue oxygenation by near infrared spectroscopy (NIRS): Background and current applications. *Journal of Clinical Monitoring and Computing, 26*, 279–287. https://doi.org/10.1007/s10877-012-9348-y

Schrammen, E., Roesmann, K., Rosenbaum, D., Redlich, R., Harenbrock, J., Dannlowski, U., & Leehr, E. J. (2022). Functional neural changes associated with psychotherapy in anxiety disorders–a meta-analysis of longitudinal fMRI studies. *Neuroscience & Biobehavioral Reviews, 104895*. https://doi.org/10.1016/j.neubiorev.2022.104895

Schwartz, E. S., Edgar, J. C., Gaetz, W. C., & Roberts, T. P. L. (2010). Magnetoencephalography. *Pediatric Radiology, 40*(1), 50–58. https://doi.org/10.1007/s00247-009-1451-y

Shen, F. X., Wolf, S. M., Gonzalez, R. G., & Garwood, M. (2020). Ethical issues posed by field research using highly portable and cloud-enabled neuroimaging. *Neuron, 105*(5), 771–775. https://doi.org/10.1016/j.neuron.2020.01.041

Smith, S. J. (2005). EEG in the diagnosis, classification, and management of patients with epilepsy. *Journal of Neurology, Neurosurgery & Psychiatry, 76*(suppl 2), ii2–ii7. https://doi.org/10.1136/jnnp.2005.069245

Soltanlou, M., Sitnikova, M. A., Nuerk, H.-C., & Dresler, T. (2018). Applications of functional near-infrared spectroscopy (fNIRS) in studying cognitive development: The case of mathematics and language. *Frontiers in Psychology, 9*, 277. https://doi.org/10.3389/fpsyg.2018.00277

Stiles, J., & Jernigan, T. L. (2010). The basics of brain development. *Neuropsychology Review, 20*(4), 327–348. https://doi.org/10.1007/s11065-010-9148-4

Su, W.-C., Colacot, R., Ahmed, N., Nguyen, T., George, T., & Gandjbakhche, A. (2023). The use of functional near-infrared spectroscopy in tracking neurodevelopmental trajectories in infants and children with or without developmental disorders: A systematic review. *Frontiers in Psychiatry, 14*, 1210000. https://doi.org/10.3389/fpsyt.2023.1210000

Sugranyes, G., Kyriakopoulos, M., Corrigall, R., Taylor, E., & Frangou, S. (2011). Autism spectrum disorders and schizophrenia: Meta-analysis of the neural correlates of social cognition. *PLoS One, 6*(10), e25322. https://doi.org/10.1371/journal.pone.0025322

Supek, S., & Aine, C. J. (2016). *Magnetoencephalography*. Springer-Verlag.

Suzuki, A., Yamaguchi, R., Kim, L., Kawahara, T., & Ishii-Takahashi, A. (2023). Effectiveness of mock scanners and preparation programs for successful magnetic resonance imaging: A systematic review and meta-analysis. *Pediatric Radiology, 53*(1), 142–158. https://doi.org/10.1007/s00247-022-05394-8

Teplan, M. (2002). Fundamentals of EEG measurement. *Measurement Science Review, 2*(2), 1–11.

Thompson, P. M., Lee, A. D., Dutton, R. A., Geaga, J. A., Hayashi, K. M., Eckert, M. A., Bellugi, U., Galaburda, A. M., Korenberg, J. R., & Mills, D. L. (2005). Abnormal cortical complexity and thickness profiles mapped in Williams syndrome. *Journal of Neuroscience, 25*(16), 4146–4158. https://doi.org/10.1523/JNEUROSCI.0165-05.2005

Trambaiolli, L. R., Biazoli, C. E., Jr, & Sato, J. R. (2022). Brain imaging methods in social and affective neuroscience: A machine learning perspective. In P. S. Boggio et al., *Social and affective neuroscience of everyday human interaction: From theory to methodology*. (pp. 213–230).

Van Hecke, A. V., Lebow, J., Bal, E., Lamb, D., Harden, E., Kramer, A., Denver, J., Bazhenova, O., & Porges, S. W. (2009). Electroencephalogram and heart rate regulation to familiar and unfamiliar people in children with autism spectrum disorders. *Child Development, 80*(4), 1118–1133. https://doi.org/10.1111/j.1467-8624.2009.01320.x

Van Voorhis, S., & Hillyard, S. A. (1977). Visual evoked potentials and selective attention to points in space. *Perception & Psychophysics, 22*(1), 54–62. https://doi.org/10.3758/BF03206080

Von Grätz, P.G. (2020, November 26). *What's that knocking?* Siemens Healthineers. https://www.siemens-healthineers.com/perspectives/mso-whats-that-knocking#:~:text=MRI%20scanners%20certainly%20make%20plenty,easily%20reach%20aircraft%20volume%20levels

Wallace, G. L., Dankner, N., Kenworthy, L., Giedd, J. N., & Martin, A. (2010). Age-related temporal and parietal cortical thinning in autism spectrum disorders. *Brain, 133*(12), 3745–3754. https://doi.org/10.1093/brain/awq279

Wilcox, T., & Biondi, M. (2015). fNIRS in the developmental sciences. *Wiley Interdisciplinary Reviews: Cognitive Science, 6*(3), 263–283. https://doi.org/10.1002/wcs.1343

Wilson, T. W., Heinrichs-Graham, E., Proskovec, A. L., & McDermott, T. J. (2016). Neuroimaging with magnetoencephalography: A dynamic view of brain pathophysiology. *Translational Research, 175*, 17–36.

Wolf, M., Ferrari, M., & Quaresima, V. (2007). Progress of near-infrared spectroscopy and topography for brain and muscle clinical applications. *Journal of Biomedical Optics, 12*(6), 062104-062104-062114. https://doi.org/10.1117/1.2804899

Wolf, S. M., Lawrenz, F. P., Nelson, C. A., Kahn, J. P., Cho, M. K., Clayton, E. W., Fletcher, J. G., Georgieff, M. K., Hammerschmidt, D., & Hudson, K. (2008). Managing incidental findings in human subjects research: Analysis and recommendations. *Journal of Law, Medicine & Ethics, 36*(2), 219–248. https://doi.org/10.1111/j.1748-720X.2008.00266.x

Xu, S.-Y., Lu, F.-M., Wang, M.-Y., Hu, Z.-S., Zhang, J., Chen, Z.-Y., Armada-da-Silva, P. A., & Yuan, Z. (2020). Altered functional connectivity in the motor and prefrontal cortex for children with Down's syndrome: An fNIRS study. *Frontiers in Human Neuroscience, 14*, 6. https://doi.org/10.3389/fnhum.2020.00006

Xu, L., Sun, Z., Xie, J., Yu, J., Li, J., & Wang, J. (2021). Identification of autism spectrum disorder based on short-term spontaneous hemodynamic fluctuations using deep learning in a multi-layer neural network. *Clinical Neurophysiology, 132*(2), 457–468. https://doi.org/10.1016/j.clinph.2020.11.037

Yen, C., Lin, C.-L., & Chiang, M.-C. (2023). Exploring the frontiers of neuroimaging: A review of recent advances in understanding brain functioning and disorders. *Life (Basel), 13*(7), 1472. https://doi.org/10.3390/life13071472

Zhao, T. C., & Kuhl, P. K. (2016). Musical intervention enhances infants' neural processing of temporal structure in music and speech. *Proceedings of the National Academy of Sciences, 113*(19), 5212–5217. https://doi.org/10.1073/pnas.1603984113

Zhao, T. C., Boorom, O., Kuhl, P. K., & Gordon, R. (2021). Infants' neural speech discrimination predicts individual differences in grammar ability at 6 years of age and their risk of developing speech-language disorders. *Developmental Cognitive Neuroscience, 48*, 100949. https://doi.org/10.1016/j.dcn.2021.100949

Behavioral Research Methods in Intellectual and Developmental Disabilities

Yingying Yang, Edward C. Merrill, and Sonia Conde

Abstract Impactful, high-quality, and innovative research in intellectual and developmental disabilities (IDD) is built upon rigorous research methods. Different from those commonly used in the study of a general, neurotypical sample, there are unique issues, challenges, and opportunities in IDD research methods. Here, we briefly reviewed common IDD research methods including ethical considerations, and discussed common challenges to the research process together with strategies to address them. IDD research presents critical opportunities to impact the IDD community directly and indirectly. For instance, participants with IDD and their family/caretakers could be involved as research partners, an approach called participatory research that has been gaining popularity in the field. IDD research results can and should be implemented into practice and policy decision-making to better the IDD community and the broader society. Finally, we considered future directions in IDD research methods.

Keywords Research methods · Research ethics · Matching method · Participatory research · Practice and policies

It is not possible to cover all behavioral research methods relevant to intellectual and developmental disabilities (IDD) in a single chapter. Indeed, entire books are devoted exclusively to detailing research methods used in the study of IDD (e.g., Hott et al., 2021). Therefore, our goal is to highlight the unique issues, challenges, and opportunities in IDD research methods that make them different from those commonly used in the study of a general, neurotypical sample. We will first briefly review common research methods in IDD, including a discussion of research ethics

Y. Yang (✉) · S. Conde
Department of Psychology, Montclair State University, Montclair, NJ, USA
e-mail: yangyi@montclair.edu

E. C. Merrill
Department of Psychology, The University of Alabama, Tuscaloosa, AL, USA

M. G. Valdovinos (ed.), *Intellectual and Developmental Disabilities*, Contemporary Clinical Neuroscience,
https://doi.org/10.1007/978-3-031-66932-3_17

that need to be considered prior to conducting research with people having special needs. Next, we discuss common challenges and corresponding strategies to address those challenges in applying experimental and quasi-experimental designs in the study of IDD. Then, we discuss participatory research and involving parents in IDD research. Finally, we discuss methods for implementing research results into practice and policy decision making. Future directions in IDD research methods are considered at the end.

Brief Overview of Research Methods in IDD

Almost all research methods to study the general population could be used or modified to study people with IDD. This is true for quantitative research focusing on numbers and qualitative research focusing on content and descriptions. Quantitative research methods usually include experimental designs, quasi-experimental designs, correlational designs, survey research, etc. Qualitative research typically employs surveys, interviews, focus groups, case studies, etc. When age or time is treated as a variable, it can be evaluated using cross-sectional designs, examining different age groups at the same time; longitudinal designs, examining the same groups across different times (ages); and pre–post design, examining groups before and after treatments. Data from existing research can be re-examined, involving approaches such as secondary data analysis, meta-analysis, and systematic literature review. There are also multiple ways to measure chosen outcome variables, including cognitive tests, psychological measures, behavioral observations, neurological measures, and parent or teacher/caretaker reports.

Of course, a single study can employ multiple methods, and there are many complex variations of the basic designs. In single-subject designs, participants serve as their own control, whereas in group designs, often another group serves as the control. For instance, May et al. (2022) employed an ABABC reversal design (i.e., baseline–intervention–baseline–intervention–altered intervention) to study whether a token reinforcement system could increase water consumption among 14 adults with IDD in a community living center. In another example, Ringshaw et al. (2022) used both a quasi-experimental design and correlational approach to compare moral decision-making between 38 6- and 12-year-old boys with autism spectrum disorder and 38 neurotypical boys. The researcher manipulated variables such as recipient merit, wealth, and health in an experimental design in the moral decision-making task about third-party resource allocation. They also performed correlation analyses correlating the theory of mind and moral decision-making. In a third example, Will et al. (2019) conducted a prospective longitudinal study where they followed the same participants [i.e., 24 children with autism and fragile X syndrome, 18 children with only fragile X syndrome, and 40 typically developing (TD) children] over a period (i.e., 9, 12, 24, 36, 48, and 60 months of age) and collected real-time data (fine and gross motor skills) through both direct measure and parent report. These examples demonstrate the complexity of research design in IDD and that multiple measures, methods, and tools are commonly utilized to investigate research questions.

Research Ethics in IDD Research

A consideration of research ethics is a necessary first step when conducting research, and it governs the standards of conduct during the research process. Researchers must adhere to ethical standards to protect the dignity, rights, and welfare of research participants. According to the Belmont Report (National Commission for the Protection of Human Subjects of Biomedical and Behavioral Research, 1979), the basic ethical principles of the research process include respect for persons (individuals have the right of self-determination, hence the basis of informed consent), beneficence (researchers must minimize harm and maximize potential benefits), and justice (researchers must ensure fairness in both the burdens and benefits of research). Furthermore, research should provide anonymity (i.e., participants are not identified by name) and confidentiality (i.e., data are securely stored and only used for the purpose explained to the participants). After the data are collected, researchers should not commit scientific fraud, including falsifying or fabricating data, or misrepresenting others' work as their own (i.e., plagiarism).

People with IDD are vulnerable because of their limited capacity to provide informed, voluntary consent for research participation. Research should be carefully designed to consider the rights and perspectives of their participants with IDD and to safeguard them from harm and exploitation. The consent and assent processes could be similar to those of typically developing (TD) children if the participants with IDD are under 18 years. For instance, the parent/guardian reads and signs the informed consent form, and the child participant with IDD completes the assent form. However, if the participants with IDD are over 18 years, researchers might decide to use consent forms for parents/guardians and assent forms for participants with IDD, consent forms for both parents/guardians and participants with IDD, consent forms for only participants with IDD, or other approaches. The specific situations will depend on whether the participants with IDD have the capacity to comprehend study information and make informed, voluntary decisions. It may also require special care when the participants with IDD do not have a legal guardian, and alternative approaches might be sought out (McDonald & Kidney, 2012). However, for specific information regarding how to collect consent/assent and what to include, one would have to consult the institutional review board of their home institute as specific requirements may vary, and best practice standards may change based on refined policies to protect the rights of people with IDD.

Informed consent typically includes (1) study purpose, procedure, and time commitment, (2) study risks to the participants, (3) study benefits, (4) compensation, (5) information confidentiality and participation anonymity, (6) assurance about the voluntary nature of the study, and (7) contact information. When designing consent and assent forms for people with IDD, particular care should be taken to ensure that participants truly understand. For instance, to promote understanding, researchers should consider using simple, jargon-free language, symbols, and visual aids. Researchers may also read the consent form to participants with IDD and repeat information when necessary while allowing participants ample time to decide and to ask questions (McDonald & Kidney, 2012).

People with IDD could be especially vulnerable to being coerced into doing research. Coercion could come in many forms: some are easy to detect (e.g., disproportionately high monetary reward and repeatedly asking participants after refusal), while others may be more covert. For instance, does the participant with IDD feel that they can say "no" when their therapist or counselor is asking for consent? Could the prospect of interacting with new people for those with intellectual disabilities and little social activities serve as an enticement to participation? Researchers should recognize and preemptively avoid the different ways of potentially coercing their participants. Instead, careful measures should be taken to make sure that participants with IDD and their families are completely voluntary and autonomous to make decisions. After obtaining initial consent, researchers should constantly evaluate whether their participants remain willing to participate during the entire research process. Research with participants who are nonverbal and have profound intellectual disabilities may be more challenging. However, this should not preclude them from participating in research. Mietola et al. (2017) have demonstrated some effective approaches in ethnography such as building one-to-one relationships with the participants, adjusting research practices based on sustained and careful observation of participants, and learning about participants from proxies (e.g., family members and caretakers). Hence, it may be necessary for researchers to tailor the consent approaches according to the characteristics of the participants and the nature of the research.

Finally, researchers should recognize that family/caretakers play an important role in the consent process. Researchers should ensure that the consent process with participants with IDD and their families/caretakers is educational, transparent, communicative, and ongoing. Simply getting signatures from the participants with IDD and their family/caretakers without clearly and thoroughly communicating about the study and the consent process may harm the interest of the IDD community, jeopardize the integrity of the research, and make future research endeavors more difficult. Participatory research, as discussed later, which takes into account how people with IDD and their family/caretakers view the research processes, may help promote ethical, inclusive, and strong research that is sensitive to the needs of people with IDD (McDonald & Kidney, 2012).

Unique Challenges in Experimental and Quasi-Experimental Design

When designing experimental and quasi-experimental studies, researchers often face challenges in IDD research that are uncommon in studies of TD children and adults. Below, we discuss issues such as comparison group matching and small sample size, which can restrict the kinds of interpretations appropriate for the research results. We also discuss some common research and statistical methods, such as developmental trajectory analyses and longitudinal study design, which can be used to overcome these research challenges.

Matching

One of the most common ways to study cognition and behavior in people with IDD is to compare them with neurotypical groups. For instance, a researcher may want to establish whether people with Down syndrome could perform at certain levels in working memory, a critical cognitive function for many daily tasks such as remembering what their teachers have just taught and applying it to new problems, asking for directions and then using them, and meeting new people and remembering their names in conversations. However, many questions arise in this process: (1) Which IDD group to study? (2) What neurotypical groups to choose for comparison? (3) How to match the two groups? (4) What if the groups cannot be matched?

Earlier psychological research on intellectual disabilities rarely distinguished etiologies or causes and studied them as a unitary group (e.g., Atwell et al., 2003; Belmont, 1966; Merrill, 1990). However, since approximately the late 1990s, etiology-specific studies started to dominate the field, with researchers focused on one or a few specific etiologies (e.g., Down syndrome, Fragile X syndrome, William syndrome, and fetal alcohol syndrome). Intellectual disabilities may also be a co-occurring condition for certain developmental disabilities/disorders/conditions such as autism spectrum disorder, cerebral palsy, and attention-deficit hyperactivity disorder. Due to great heterogeneity within each etiology of IDD, researchers also examined other individual factors that distinguish different subgroups, based on factors such as age, gender, developmental level, and comorbidity (e.g., Miezah et al., 2021; Will et al., 2019; Van Deusen et al., 2023). It is important to note that it is still of great utility to study people with IDD of unknown etiologies. In fact, for about one-third of people with intellectual disabilities, the cause remains unknown (American Association on Health and Disability, n.d.). Ignoring this population for the sake of "etiology purity" would be a missed opportunity to serve this population.

Once the target population of the research is identified, researchers must decide if a comparison group is necessary and, if it is, how that group should be selected. Typically, the target sample can be compared with TD peers (e.g., children, adolescents, or adults) of similar chronological age (CA) (e.g., Atwell et al., 2003), TD peers of a similar cognitive level [age-equivalent scores or mental age (MA)], peers of similar chronological age (CA) and/or MA but with unknown or mixed etiologies of IDD, or peers of similar CA and/or MA but with another etiology of IDD. Of course, multiple comparison groups could be employed in the same study. The choice of comparison group will depend on the specific research question. For instance, when comparing the target sample with TD children of similar general MA, it helps to answer whether general cognitive ability, rather than life experience accumulated (e.g., CA-related), may explain group differences and whether the strengths or weaknesses in the IDD group are above and beyond what should be expected based on general cognitive ability. Comparing the target sample with peers with IDD of another etiology, mixed etiologies, or unknown etiologies helps to understand whether a particular strength, weakness, or relationship is unique to the target group with IDD, rather than a variable associated with people with IDD in general.

The most common method of comparison is to compare with MA-matched TD peers. However, because of the heterogeneity of cognitive abilities within the target sample, researchers must decide which of several variables (e.g., verbal ability, non-verbal ability, and vocabulary) would provide the best match for the research question and how to match groups. With respect to how, matching can be done at the individual level such that each individual from the IDD group is matched with someone from the comparison group on the matching variable with a few points' tolerance, or at the group level such that the means on the matching variables are matched between two groups. Groups should be as well-matched as possible on control variables that impact the outcome variables. Common ways to match MA could be through verbal MA, non-verbal MA, or general MA. Many times, other non-cognitive variables such as sex, handedness, and family SES may also be considered. For instance, a study of reading skills may match two groups on verbal MA measured by standardized vocabulary tests such as the British Picture Vocabulary Scale, whereas a study of spatial reasoning may match two groups on nonverbal MA measured by standardized tests such as Leiter International Performance Scale or Raven's Matrices Tests (e.g., Davis et al., 2014; Hershkovich et al., 2023; Næss et al., 2012). Researchers may also opt to match groups on more comprehensive, standardized intelligence tests composed of both verbal and nonverbal components, such as the Kaufman Brief Intelligence Test, the Wechsler series of intelligence scales, and the Stanford–Binet test, among others (e.g., Shields et al., 2022). However, as stated earlier, cognitive abilities are not even for people with IDD; they may show strengths in verbal abilities and weaknesses in nonverbal abilities or vice versa (Glenn & Cunningham, 2005). For instance, people with Down syndrome show significant weakness in speech, language, verbal working memory, and motor processes, but relative strengths in sequential spatial memory, social development, and daily living skills (e.g., Abbeduto et al., 2005; Fidler, 2006; Grieco et al., 2015; Silverman, 2007; Yang et al., 2014). In contrast, people with William syndrome show significant weakness in visuospatial, motor, and daily living skills, but relative strengths in some aspects of language such as fluency, comprehension and vocabulary, and social skills (e.g., Karmiloff-Smith, et al., 2003; Martens et al., 2008; Mervis et al., 2000, 2001; Miezah et al., 2021). Therefore, matching on general MA, averaging verbal and nonverbal abilities, may create a comparison group that outperforms the IDD group on certain abilities but underperforms on others (Jarrold & Brock, 2004). In other words, matching on a global measure cannot guarantee that groups are matched on the trait relevant to the variables of interest. All in all, researchers should choose the appropriate matching variables relevant to the outcome variables, guided by previous empirical research and established theoretical frameworks.

There are other issues that researchers should consider when matching groups. First, the researcher should try to avoid floor effects (i.e., participants scoring near the lower limit of the testing instrument) as much as possible, which would render result interpretation difficult or incorrect (Strauss, 2001). Floor effects would indicate that true abilities were not measured appropriately. Unfortunately, it is quite common when measuring the abilities of people with profound intellectual disabilities to find that they perform at floor levels because the test sensitivity is typically low at the extreme ends (Hessl et al., 2009). In this case, caution should be taken when interpreting the results, and alternative, more sensitive measures should be considered. Second, a standardized test can provide multiple measures of ability; what scores should be used for matching? For example, should we use a raw score or mental age score? Age equivalent or mental age scores are the median CA at which a particular raw score was obtained, which means that age equivalent scores are not on an interval scale (Mervis & Klein-Tasman, 2004). A similar problem exists with raw scores in that the performance difference between two adjacent scores might not be equal-distanced due to varying item difficulty. Thus, raw scores may not reflect an interval scale. A good alternative may be to use some conversion scores that make scale corrections of raw scores to accommodate for item difficulty variability, such as the growth scores in the Leiter International Performance Scale (Phillips et al., 2014). Unfortunately, not all standardized tests have such scores available to use. Third, how can one decide statistically that the two groups are indeed matched? Conventional wisdom may suggest that the t test p value comparing two groups should be greater than 0.05, which leads to accepting the null hypothesis of no difference between groups. However, a p value greater than 0.05 actually means a greater than 5% chance that the groups do not differ, which does not necessarily mean that the groups are matched or equivalent (Mervis & Klein-Tasman, 2004). Using a standard of 0.05 may increase the chances of making a type II error, assuming that there is no difference when a difference does exist. Mervis and Klein-Tasman (2004) recommend using a p value of at least 0.5. Researchers may also consider Bayesian analyses, which view probability as a *degree of belief* in an event based on prior knowledge, and this view contrasts with the traditional, frequentist approach, which views probability as the limit of the relative frequency of an event after many trials (Gelman et al., 2013). The Bayes factor, BF01, can be used to indicate the ratio between the likelihood of the null hypothesis (i.e., equivalence) relative to that of the alternative hypothesis (i.e., difference). BF01 values greater than 3, 10, and 30 are considered some evidence, strong evidence, and very strong evidence of equivalence, respectively (e.g., Rouder et al., 2009).

Some scholars have proposed that the matching method may be optimal when the age range of the IDD group is narrow and the outcome variable is also sensitive around the covered age range, which ultimately helps to avoid floor and ceiling (i.e., participants scoring near the upper limit of the testing instrument) effects (Mervis & Klein-Tasman, 2004; Thomas et al., 2009). However, one drawback of the matching method is that it is usually point-based, comparing two or multiple average scores, rather than the full range of performance, and thus might have neglected important variability in the outcome variables. Facon et al. (2011) suggested that matching on the mean is not sufficient and that groups should also be matched on their variances and the distribution shapes of matching variables. However, this practice has largely been ignored in the field as it is common to see studies with variance ratios well above 2.0 for the matching variable. Standardized tests are typically less sensitive at the lower ends of the score continuum. Therefore, extremely low ability scores found in IDD populations could be associated with greater measurement errors. To enhance scientific transparency and replicability, researchers should consider listing group means, standard deviations, skewness and kurtosis, and their standard errors, and visuals such as box plots of the matching or confounding variables to readers (Facon et al., 2011).

Developmental Trajectory and Longitudinal Study Approaches

Because matching methods have largely been about point-based comparisons, they tend to ignore the wide distribution of age and abilities of the target sample and potentially yield floor and ceiling effects, thereby limiting the interpretability and generalizability of results. Against this backdrop, developmental trajectory analyses, rooted in growth curve modeling (a statistical approach that analyzes changes over time), have been increasingly applied in the field of IDD research (Annaz et al., 2008; Karmiloff-Smith, 1997; Thomas et al., 2009; Yang et al., 2021). It is more advantageous when the age, age equivalence measure, and outcome measure have a wide range. Furthermore, the experimental task still needs to be sensitive within the range (i.e., avoid floor and ceiling effects).

Developmental trajectory analyses place development "at the heart of explanations of development deficits" (Thomas et al., 2009, p. 339) because developmental differences, as seen in IDD, may not be present at birth but emerge later, either gradually or abruptly, at a critical developmental period. The primary approach is to construct a function linking task performance with a criterion (e.g., chronological age, mental age) for each participant group in each experimental task and then compare this function between different groups or tasks (Thomas et al., 2009). Different from point-based comparisons between groups or tasks, developmental trajectory analyses are able to demonstrate differences in several behavioral indicators such as the start point (i.e., intercept), the shape of development (e.g., linear, non-linear, and curvilinear), and the rate of development (i.e., slope). For instance, through developmental trajectory analyses, Bailey et al. (2019) found that children with IDD showed significantly more externalizing behaviors at a young age (i.e., 3 years) than TD children and less substantial reduction in externalizing behaviors than TD children over time.

Developmental trajectory analyses are usually based on a cross-sectional design, which does not analyze behaviors over a period of time and cannot establish causal effects. Therefore, researchers also resort to longitudinal designs, which could complement and validate developmental trajectory analyses. A longitudinal study design focuses on the sensitive age where changes in performance are more likely to happen. The specific design details, such as the frequency of assessment and duration, should be tailored to answer specific research questions. For instance, Baghdadli et al. (2012) tracked the adaptive behaviors over 10 years in children with autism spectrum disorder when they were 5, 8, and 15 years of age. In another example, Conners et al. (2018) assessed phonological memory and language in adolescents and young adults with Down syndrome (10–21 years) twice, 2 years apart. In a third example, for fine and motor skills, Will et al. (2019) tracked their participants with autism and Fragile X syndrome between 9, 12, 24, 36, 48, and 60 months of age.

Longitudinal studies can be particularly insightful in IDD research as they help to understand the developmental trajectory and potentially explore causal inference, which is usually hard to achieve with cross-sectional, quasi-experimental, and correlational designs. Although, longitudinal studies may be more difficult to conduct, especially considering the high time commitment from the researchers and participants and the associated cost. Some specific strategies for recruiting and retaining participants in longitudinal studies (which also apply to other study designs) include removing barriers to participation (e.g., providing transport, study reminders, minimizing burden during data collection processes, and consistency in research staff), incentives, emphasizing study benefits, and developing a project community and study bond among participants (McCarron et al., 2022).

Small Sample Size

It is common to see a relatively small sample size in IDD research. For instance, a literature review by Martens et al. (2008) found that, except for some parental report studies, studies measuring cognition and behavior in people with William syndrome typically have a median sample size of 6–17, which greatly limits the generalizability of the findings. In an example of empirical research, a pilot study examined the effect of a 12-week exercise program on everyday executive function in people with Down syndrome, including 9 participants in the experimental group and 11 in the control group (Shields et al., 2022).

Maes et al. (2021) listed five reasons why it is difficult to recruit participants with IDD, especially those with profound intellectual disabilities. First, specific forms of IDD, such as Down syndrome, are of low prevalence, occurring in about 1 in every 700 babies (Centers for Disease Control and Prevention, 2023). Furthermore, for many other forms of IDD, children might not have been diagnosed until after 2 or 3 years of age, which makes it difficult to identify these potential participants early in life. Second, significant physical (e.g., mobility) and mental health issues (e.g., social anxiety) may hinder participation. Third, the recruitment channels are limited. There are many national organizations for IDD in the USA such as The Arc of the United States, the Special Olympics, the National Down Syndrome Society, the National Autism Association, and the National Fragile X Foundation. They usually also have their local chapters. However, because there are relatively few organizations and service groups for people with IDD, many of them can be easily over-subscribed and overwhelmed by the large number of requests to recruit participants. Fourth, families may be hard to reach and not interested in research due to being concerned about time constraints and the emotional burden associated with potential research activities. Finally, parents and families may underestimate the abilities of people with IDD and not be convinced that they could complete certain tasks or benefit from certain interventions. Despite these challenges, multiple efforts could be employed to increase research participation, such as using social media effectively and adopting participatory research (as described later). Furthermore, researchers can establish regional, national, and international collaborations between different research centers and between researchers and service providers/ advocacy organizations, which could increase participation from the IDD community.

All the drawbacks associated with small sample size studies in general apply to IDD research. Small sample sizes may reduce the power to detect group differences and limit the ability to generalize the results to a broader population. More specifically, when the sample size is small, it may be challenging to validate the data distribution, and many other assumptions for parametric tests (i.e., tests that make assumptions about the population distribution based on the sample) are usually not met. In this case, non-parametric tests (i.e., tests that generally do not make assumptions about the population distribution) may be a more appropriate way to analyze the data. However, non-parametric tests have less power to detect significant results relative to parametric tests. Furthermore, when the sample size is small, the heterogeneity within the group is less well accommodated and it may be impossible to further distinguish sub-groups within an already small sample size. This can also limit the generalizability of the results to the broader population.

From a design perspective, there are certain ways to overcome the obstacles of small sample size. For instance, when appropriate, the researchers may consider single-case experimental designs (Maes et al., 2021), which involve researchers manipulating an independent variable and repeatedly measuring the dependent variable before and after manipulating the independent variable. In this case, more data on the dependent variables can be collected despite the small number of participants to start with. For instance, Nijs et al. (2018) employed a single-case experimental design to examine whether training staff could promote peer-directed behaviors among people with severe IDD. Four staff members first received training on peer-directed behaviors in people with IDD and then created an activity plan to promote peer-directed behaviors based on the training. Four dyads of people with IDD participated in the activity with the staff members over 10 weeks. Researchers video-recorded the activities at baseline, week 3, week 10, and follow-up (16 weeks). After every activity, the staff member or a support worker also completed questionnaires about themselves and the participants with IDD during the activity. Results showed an increase in staff members' social scaffolding behaviors and peer-directed behaviors among participants with IDD after training. Nevertheless, these two measures dropped at the 16-week follow-up. This study showed the utility of intensive data collection and that single-case experimental design can serve as an important first step or pilot project for larger-scale assessment and intervention studies in the future. Other ways to compensate for small sample sizes include case studies, longitudinal data collection, and mixed methods combining qualitative and quantitative approaches. Of course, the specific sample size to use will depend on the specific research questions, prior research findings, theoretical framework for the study, and, most importantly, power analysis (Faul et al., 2009).

Despite small sample sizes being prevalent in IDD research, it has also become more likely in recent years to see relatively larger sample sizes as the number of multi-site, large-scale, grant-funded collaborative projects increases. For instance, in a project examining cognitive flexibility, Van Deusen et al. (2023) recruited and tested 72 children with Down syndrome between 2 and 8 years old across the USA, including the West, Mountain West, and Midwest. In addition, to examine cross-sectional and longitudinal associations between obesity and physical fitness with cardiovascular disease risk factors, researchers from Spain, Izquierdo-Gómez et al. (2023), recruited and tested 90 adolescents with Down syndrome. Finally, based on the Autism Speaks–Autism Treatment Network clinical registry study, Shui et al. (2018) used multivariate modeling to identify predictors of sleep problems among 1045 children with autism spectrum disorder enrolled at 17 sites from 2008 to 2016.

Other Methodological Approaches in IDD Research

Involving Parents in IDD Research

Many of the empirical research examples addressed above have studied the participants with IDD exclusively. However, there is also a large body of literature devoted to parents and other caretakers/stakeholders of people with IDD (teachers, therapists, siblings, etc.). Here, we briefly discuss research on parents of people with IDD. In many parts of the world, individuals with IDD live with a family caregiver, and parents often provide lifelong support to their children with IDD. There are several important reasons why researchers engage with and are interested in parents of children with IDD. First, the IDD literature has long recognized the value of using multiple informants and gathering information via both self and caregiver reports (e.g., Fisher et al., 2014; Halvorsen et al., 2023; Miller et al., 2017). For people with profound intellectual disabilities who have more limited cognitive, communicative, and reflective abilities, they may not be able to provide meaningful self-reports, especially regarding subjective experiences such as emotional, personal, and preferential experiences and behaviors. In such cases, parents can provide insightful proxy reports for their children with IDD. Nevertheless, researchers should also be aware of the social desirability bias associated with proxy reports and that proxy reports often need to be cross-validated using other measures such as direct observations of people with IDD (Maes et al., 2021). Second, parents of children with IDD are worth studying in their own right. They often report higher rates of mental and physical health symptoms compared with parents of TD children (Lee et al., 2017; Scherer et al., 2019). Knowledge acquired from parents about their unique caregiving needs can be used to guide future interventions and support services. Third, engaging caregivers in the research process, especially in participatory research (as detailed later), may help ensure that the research goals better align with the needs of children and families and can inform future research practices (Fletcher-Watson et al., 2019).

How popular are studies involving parents of people with IDD? Grove et al. (2022) screened 7845 published papers in 7 high-impact IDD-focused journals between 2010 and 2019 and found that 19.37% of all screened papers included at least one parental measure. They also coded the measurement tools and found that questionnaires and interviews were the predominant measurement tools for gathering parent information, accounting for 88.0% of all the measurement tools. In contrast, other measures such as direct observation, focus groups, biological/medical records, and others accounted for less than 5% each. This review also touched upon geographical diversity, an interesting topic that has been gaining popularity in other fields of developmental psychology (Draper et al., 2022) but has been largely neglected in IDD research so far. Nevertheless, the review found that the majority of studies involving parents of people with IDD, i.e., 76%, came from North America and Europe, despite these places accounting for less than 20% of the world population. In contrast, other continents, such as Africa and South America, represented less than 1% of all studies. Overall, this review showed the popularity of parental measures in IDD research and the limited diversity in research methods and geographical locations.

Participatory Research

Participatory research allows participants to be more involved in the study by looking into the experiences of the participants involved (Gilbert, 2004). It has deep roots in disability activism of "Nothing about us without us" (McDonald et al., 2013) and is particularly important in IDD research. It emphasizes the collaboration between researchers and the IDD community to address community concerns. Rather than mere research participants, people with IDD, their family members, and other stakeholders (e.g., therapists, teachers) are co-researchers, partners, and collaborators. Participatory research has become increasingly popular since the early 2000s. In a literature review, Stack and McDonald (2014) identified 21 participatory research projects with people with IDD in peer-reviewed journal articles published in English. The review also showed that all studies but one had strong qualitative components and usually focused on social issues important to people with IDD (e.g., quality of life, self-advocacy, education and employment, social relationships, and health).

For instance, while studying suicidal thoughts and behaviors in siblings of people with IDD, Wolff et al. (2023) recruited seven siblings (10–24 years old) to serve on the advisory board. They advised on the study design such as using sensitive and appropriate language and also piloted the study. In another study, Kramer et al. (2019) formed the Intellectual and Developmental Disability Mental Health Research Partnership to learn about the lived experiences of individuals with IDD and co-occurring mental health conditions. The partnership was comprised of six young adults with IDD, five professionals working with people with IDD, a research lab, and the representations from the Arc of the United States and Self Advocates Becoming Empowered, who worked collaboratively for 1 year. The partnership designed and conducted a national survey and several national storytelling sessions with people with IDD and co-occurring mental health conditions and their parents. Their study helped to identify disparities, priority areas of concern, and recommendations about access to mental health services for this population.

While engaging in participatory research, it is important for researchers to engage in reflexivity, which involves disclosing, reflecting, and examining the researchers' own beliefs, experience, background, judgment, and practice during the study design, data collection, and results interpretation process (Kerr et al., 2023; Rinaldi, 2013). A neurotypical researcher with little or no direct life experience of IDD in their own or family members may have a more limited understanding of certain daily, functional issues pertinent to the IDD community. It is especially important because of the unbalanced power dynamic between researchers and participants and the institutional and structural oppression of people with IDD. Engaging in reflexivity helps understand how researchers themselves may impact the research process and potentially enhance the overall quality of the research (Dorozenko et al., 2016).

Participatory research benefits the researchers by allowing them to gain different insights about the issues of interest, gaining greater interest and engagement from participants, and potentially generating more impactful research. It may also benefit the research partners with IDD directly by gaining new experience, learning new skills about scientific thinking, forming new friendships, and feeling respected, valued, and more confident (García Iriarte et al., 2014). Furthermore, if the focus of the participatory research is to address community concerns and has been driven by community needs, it could benefit the community by addressing those needs and providing research results that are more accessible, executable, and meaningful in improving everyday lives.

However, participatory research also has unique challenges, such as competing obligations, power imbalance between the researchers and the partners, and communicating research ideas (e.g., statistics, design, and confounds). Morgan et al. (2014) related one occasion when a research partner with IDD received a question during a joint conference presentation (p. 1310). The gist of the question was, "How and what have you (the research partner) taught the lead researcher?" As noted by the researchers, this question was complex and multifaceted and might have required additional prompts for the research partner to understand. Their dilemma was whether or not the lead researcher should answer the question for the research partner. If they did, it might erode the sense of trust, highlight the disability of the research partner, and unbalance the power dynamic. But if they did not, it could leave the research partner struggling with words, which might make one feel inadequate and awkward. This example shows that implementing participatory research is a continuous learning process. Nevertheless, in the end, the reward is immense for conducting participatory research.

How IDD Research Is Used: Practices and Policies

There have been ample examples where behavioral science research can be and has been implemented to impact policy (Gopalan & Pirog, 2017). For instance, behavioral research has shown that people are more likely to participate in beneficial programs when hassles are minimized and the tasks are easy. Accordingly, automating enrollment is one of the most effective tools to promote program participation. A good example of using this principle is that many employers automatically enroll their employees in retirement plans unless the employees explicitly opt-out. As a result, automatic enrollment increases employee savings for the future and the number of employee participation (IRS, 2023). Similarly, the US government (The National Science and Technology Council, 2016) launched pilot programs in several states that allowed states to automatically enroll students into the National School Lunch Program based on Medicaid Data. These projects helped reduce the need for parents from low-income households to file applications, hence streamlining access and increasing the integrity of the program. Behavioral research in IDD can also be implemented to positively influence the lives of people with IDD and their families and communities. However, the process from basic research to application is not always straightforward. Below, we discuss evidence-based practice, implementation science, and some tips to translate research into broader policy changes for the betterment of the IDD community.

Evidence-based practice refers to practices and programs with positive and meaningful effects on outcomes backed by high-quality empirical research (Cook & Odom, 2013). This is commonly seen in education, where the emphasis has been on teaching students via evidence-based practices. For instance, Spooner et al. (2019) reviewed evidence-based practices in teaching mathematics to students with moderate and severe developmental disabilities. Between 2005 and 2016, 36 studies were identified to have empirically examined teaching mathematics to students with IDD, including 33 single-case studies and 3 group-experimental studies. Several evidence-based practices have been used in teaching mathematics to people with IDD, including systematic instruction, technology-aided instruction, graphic organizers, manipulatives, and explicit instructions.

Implementation science is “the study of methods to promote the adoption and integration of evidence-based practices, interventions, and policies into routine health care and public health settings to improve the impact on population health.” (U.S. Department of Health & Human Services, 2022; also see Eccles & Mittman, 2006). It starts from evidence-based practice, usually on a small scale, and ends in routine practice, usually on a larger scale. For instance, two decades of research on mindfulness have tested and shown its effectiveness on behavioral, cognitive, and emotional functioning (e.g., Chiesa et al., 2011; Hofmann et al., 2010). This also applies to the IDD community as research has shown that mindfulness-based programs positively impact the caregivers, such as reducing aggression in people with IDD and stress in caregivers, which in turn cascade to positively affect people with IDD in their care (e.g., Singh, 2020). Furthermore, people with mild or borderline IDD can learn from mindful-based programs to improve their self-regulation in stressful situations, such as reducing aggression and self-injury behaviors (Hwang & Singh, 2016). Based on these evidence-based practices, Singh (2020) proposed a three-tiered system integrating schools, families, and agencies (e.g., vocational care facilitates) to implement mindfulness-based programs and enhance the quality of life for people with IDD and their caretakers.

Ultimately, implementation science may change not only practice but also policies. Research centers and funding agencies such as the National Science Foundation and National Institute of Health (NIH) in the USA and the National Institute for Health and Care Research in the UK and other similar agencies across the world have been steadily increasing the recognition and emphasis on the importance of "broader impact": How does research benefit society? Behavioral research in IDD has the great potential to influence policy relevant to people with IDD. Unfortunately, there is a gap between research and policy communities. Scientists are usually not trained to communicate with policymakers and might not be keen to engage the general public, who elect and influence lawmakers. However, existing science in IDD can shape policy and some already did. For instance, extensive research has documented that restraint and seclusion in school settings are ineffective and harmful and that these techniques have been disproportionately used on students with IDD (e.g., Gage et al., 2022; Westling et al., 2010). Together with strong advocacy, this line of research has propelled legislation to prohibit schools from restraining and secluding students except to ensure safety and to better train school personnel about evidence-based intervention strategies (Freeman & Sugai, 2013; Worthington, 2018).

IDD researchers could impact policy through some generalizable strategies (Horne, 2022). For instance, researchers may need to disseminate the results in a form that is relevant and accessible to policymakers and provide detailed recommendations for applying research findings. Policymakers may not be interested in the specific research methods and detailed statistics but may be more focused on results and conclusions, such as what works and what does not in the real-world context. The research conclusions and specific recommendations should also be practical, feasible, affordable, equitable, and scalable. Researchers should also put their work "out there" through social media and network channels and connect with policymakers and government offices. Furthermore, groups of researchers from diverse perspectives should consider working together to deliver policy messages on particular policy topics.

Future Directions

In response to the call to update NIH research plans on Down syndrome in 2020, major Down syndrome organizations worked with the scientific community to craft recommendations for a research agenda for the NIH for the next decade (Hendrix et al., 2020). Overall priorities have been given to clinical and genetic phenotyping, cellular and animal models, longitudinal studies, randomized controlled trials, centralized biorepository, open access centralized data, research training for clinicians and scientists, and research inclusion. Although the agenda promotes new directions for Down syndrome research, many research priorities may also apply to IDD in general.

Relative to research on TD individuals, there are much fewer studies on IDD using neurocognitive tools such as MRI (magnetic resonance imaging), fMRI (functional magnetic resonance imaging), EEG (Electroencephalography), TMS (transcranial magnetic stimulation), and fNIRS (functional near-infrared spectroscopy; Hendrix et al., 2020). However, these studies could shed light on the link between cognitive performance and underlying brain structure and have important clinical implications for pharmacological development. Fortunately, we are seeing more and more studies combining human behavioral research methods with neurological, biological, and genetic measures (e.g., Lanillos et al., 2020; McCullagh et al., 2020; Uddin, 2021; Vacca et al., 2019) and expect this trend to continue.

We are also witnessing many exciting new research possibilities through advanced technology and innovation. In particular, machine learning is an emerging but powerful form of artificial intelligence. It leverages big data to extract patterns without relying on determined equations and self-improves model efficiency, and precision as the number of samples increases. For instance, in order to rapidly detect autism risk and reduce diagnosis time, Duda et al. (2014) employed machine learning to identify a small subset of items (i.e., 28%) from the Autism Diagnostic Observation Schedule (ADOS) score sheets, which were then derived into a machine learning classifier. This classifier is very promising and showed 97% sensitivity (detecting autism correctly) and 77% specificity (detecting non-autism correctly) compared with the full ADOS-2 and ADOS-G(generic) algorithms. Liu et al. (2021) reviewed 47 papers published between 2011 and 2021 that have applied machine learning to analyze fMRI data in autistic individuals and their neurotypical controls, which has significant implications for identifying biomarkers of autism spectrum disorder (also see Hyde et al., 2019). We expect technology innovations related to machine learning and artificial intelligence will continue to transform the field of IDD research.

The Open Science movement refers to making all parts of scientific research (e.g., data, materials, publications, etc.) publicly accessible to everyone, which promotes scientific transparency and reproducibility (Vicente-Saez & Martinez-Fuentes, 2018). It has gained considerable momentum in recent years. Researchers of IDD should consider pre-registering their study before collecting data and making their data, materials, and analysis publicly available. Websites such as the Open

Science Foundation and many open-source repositories (e.g., Digital Commons) hosted by universities support such endeavors. Open science in IDD research may help increase the transparency and replicability of research and enable data synthesis across different studies, making it possible to answer new research questions in IDD (van Dijk et al., 2021).

There are still many challenges in behavioral research in IDD, such as reliable and valid clinical outcome measures, producing etiology-specific "developmental norms," and describing, explaining, and predicting the wide heterogeneity across individuals and groups of the same etiology. Furthermore, IDD research has generally suffered from a lack of diversity in terms of participant characteristics. Greater efforts should be made in the future for a wider inclusion of participants from diverse social, cultural, economic, and geographic backgrounds. We are not able to cover all these topics, and interested readers should further explore them. We are optimistic that challenges come with opportunities and solutions, and progress will be made continuously to solve these issues in the twenty-first century.

Concluding Remarks

We hope that future researchers aspiring to study IDD using behavioral research methods feel interested or even inspired and motivated about empirical research of people with IDD and their community. There are still so many important research topics in IDD waiting to be discovered, investigated, and reformed for basic and applied science, improving the life quality of people with IDD, and the betterment of their community and society in general. The journey of innovative, rigorous, impactful IDD research has just begun, and we invite more people, whether undergraduate students, graduate students, clinicians/educators/researchers, or community members, to join this incredibly rewarding journey.

References

Abbeduto, L., Chapman, R. S., Fletcher, P., & Miller, J. (2005). Language development in down syndrome and fragile X syndrome. *Trends in Language Acquisition Research, 4*, 53–72. https://doi.org/10.1075/tilar.4.05abb

American Association on Health and Disability. (n.d.). *Population specific fact sheet—Intellectual disability*. https://nationaldisabilitynavigator.org/ndnrc-materials/fact-sheets/population-specific-fact-sheet-intellectual-disability/#:~:text=There%20are%20over%207%20million,condition%2C%20the%20cause%20remains%20unknown

Annaz, D., Karmiloff-Smith, A., & Thomas, M. S. C. (2008). The importance of tracing developmental trajectories for clinical child neuropsychology. In J. Reed & J. Warner-Rogers (Eds.), *Child neuropsychology: Concepts, theory, and practice* (pp. 7–18). Wiley-Blackwell.

Atwell, J. A., Conners, F. A., & Merrill, E. C. (2003). Implicit and explicit learning in young adults with mental retardation. *American Journal on Mental Retardation, 108*(1), 56–68. https://doi.org/10.1352/0895-8017(2003)108<0056:IAELIY>2.0.CO;2

Baghdadli, A., Assouline, B., Sonié, S., Pernon, E., Darrou, C., Michelon, C., Picot, M. C., Aussilloux, C., & Pry, R. (2012). Developmental trajectories of adaptive behaviors from early childhood to adolescence in a cohort of 152 children with autism spectrum disorders. *Journal of Autism and Developmental Disorders, 42*, 1314–1325. https://doi.org/10.1007/s10803-011-1357-z

Bailey, T., Totsika, V., Hastings, R. P., Hatton, C., & Emerson, E. (2019). Developmental trajectories of behaviour problems and prosocial behaviours of children with intellectual disabilities in a population-based cohort. *Journal of Child Psychology and Psychiatry, 60*(11), 1210–1218. https://doi.org/10.1111/jcpp.13080

Belmont, J. M. (1966). Long-term memory in mental retardation. *International Review of Research in Mental Retardation, 1*, 219–255. https://doi.org/10.1016/S0074-7750(08)60107-X

Centers for Disease Control and Prevention. (2023, January 28). *Facts about Down Syndrome.* https://www.cdc.gov/ncbddd/birthdefects/downsyndrome.html#print

Chiesa, A., Calati, R., & Serretti, A. (2011). Does mindfulness training improve cognitive abilities? A systematic review of neuropsychological findings. *Clinical Psychology Review, 31*(3), 449–464. https://doi.org/10.1016/j.cpr.2010.11.003

Conners, F. A., Tungate, A. S., Abbeduto, L., Merrill, E. C., & Faught, G. G. (2018). Growth and decline in language and phonological memory over two years among adolescents with down syndrome. *American Journal on Intellectual and Developmental Disabilities, 123*(2), 103–118. https://doi.org/10.1352/1944-7558-123.2.103

Cook, B. G., & Odom, S. L. (2013). Evidence-based practices and implementation science inspecial education. *Exceptional Children, 79*(3), 135–144. https://doi.org/10.1177/001440291307900201

Davis, M., Merrill, E. C., Conners, F. A., & Roskos, B. (2014). Patterns of differences in wayfinding performance and correlations among abilities between persons with and without Down syndrome and typically developing children. *Frontiers in Psychology, 5*, 1446. https://doi.org/10.3389/fpsyg.2014.01446

Dorozenko, K. P., Bishop, B. J., & Roberts, L. D. (2016). Fumblings and faux pas: Reflections on attempting to engage in participatory research with people with an intellectual disability. *Journal of Intellectual & Developmental Disability, 41*(3), 197–208. https://doi.org/10.3109/13668250.2016.1175551

Draper, C. E., Barnett, L. M., Cook, C. J., Cuartas, J. A., Howard, S. J., McCoy, D. C., Merkley, R., Molano, A., Maldonado-Carreño, C., Obradović, J., Scerif, G., Valentini, N. C., Venetsanou, F., & Yousafzai, A. K. (2022). Publishing child development research from around the world: An unfair playing field resulting in most of the world's child population underrepresented in research. *Infant and Child Development, e2375*. https://doi.org/10.1002/icd.2375

Duda, M., Kosmicki, J. A., & Wall, D. P. (2014). Testing the accuracy of an observation-based classifier for rapid detection of autism risk. *Translational Psychiatry, 4*(8), e424. https://doi.org/10.1038/tp.2014.65

Eccles, M. P., & Mittman, B. S. (2006). Welcome to *implementation science. Implementation Science, 1*(1), 1–3. https://doi.org/10.1186/1748-5908-1-1

Facon, B., Magis, D., & Belmont, J. M. (2011). Beyond matching on the mean in developmental disabilities research. *Research in Developmental Disabilities, 32*(6), 2134–2147. https://doi.org/10.1016/j.ridd.2011.07.029

Faul, F., Erdfelder, E., Buchner, A., & Lang, A. G. (2009). Statistical power analyses using G*Power 3.1: Tests for correlation and regression analyses. *Behavior Research Methods, 41*, 1149–1160. https://doi.org/10.3758/BRM.41.4.1149

Fidler, D. J. (2006). The emergence of a syndrome-specific personality profile in young children with down syndrome. *Down Syndrome, Research and Practice, 10*(2), 53–60. https://doi.org/10.3104/reprints.305

Fisher, M. H., Mello, M. P., & Dykens, E. M. (2014). Who reports it best? A comparison between parent-report, self-report, and the real life social behaviors of adults with Williams syndrome. *Research in Developmental Disabilities, 35*(12), 3276–3284. https://doi.org/10.1016/j.ridd.2014.08.011

Fletcher-Watson, S., Larsen, K., Salomone, E., & COST ESSEA Working Groups. (2019). What do parents of children with autism expect from participation in research? A community survey about early autism studies. *Autism, 23*(1), 175–186. https://doi.org/10.1177/1362361317728436

Freeman, J., & Sugai, G. (2013). Recent changes in state policies and legislation regarding restraint or seclusion. *Exceptional Children, 79*(4), 427–438. https://doi.org/10.1177/001440291307900403

Gage, N. A., Pico, D. L., & Evanovich, L. (2022). National trends and school-level predictors of restraint and seclusion for students with disabilities. *Exceptionality, 30*(1), 1–13. https://doi.org/10.1080/09362835.2020.1727327

García Iriarte, E., O'Brien, P., & Chadwick, D. (2014). Involving people with intellectual disabilities within research teams: Lessons learned from an Irish experience. *Journal of Policy and Practice in Intellectual Disabilities, 11*(2), 149–157. https://doi.org/10.1111/jppi.12081

Gelman, A., Carlin, J. B., Stern, H. S., Dunson, D. B., Vehtari, A., & Rubin, D. B. (2013). *Bayesian data analysis* (3rd ed.). Chapman and Hall/CRC.

Gilbert, T. (2004). Involving people with learning disabilities in research: Issues and possibilities. *Health & Social Care in the Community, 12*(4), 298–308. https://doi.org/10.1111/j.1365-2524.2004.00499.x

Glenn, S., & Cunningham, C. (2005). Performance of young people with Down syndrome on the Leiter-R and British picture vocabulary scales. *Journal of Intellectual Disability Research, 49*(4), 239–244. https://doi.org/10.1111/j.1365-2788.2005.00643.x

Gopalan, M., & Pirog, M. A. (2017). Applying behavioral insights in policy analysis: Recent trends in the United States. *Policy Studies Journal, 45*(S1), S82–S114. https://doi.org/10.1111/psj.12202

Grieco, J., Pulsifer, M., Seligsohn, K., Skotko, B., & Schwartz, A. (2015). Down syndrome: Cognitive and behavioral functioning across the lifespan. *American Journal of Medical Genetics Part C: Seminars in Medical Genetics, 169*(2), 135–149. https://doi.org/10.1002/ajmg.c.31439

Grove, L., Yang, Y., Diggs, D., & Hershkovich, A. (2022). Parental participation in intellectual and developmental disability research: A Review of diversity. *Disabilities, 2*(2), 219–234. https://doi.org/10.3390/disabilities2020016

Halvorsen, M. B., Helverschou, S. B., Axelsdottir, B., Brøndbo, P. H., & Martinussen, M. (2023). General measurement tools for assessing mental health problems among children and adolescents with an intellectual disability: A systematic review. *Journal of Autism and Developmental Disorders, 53*(1), 132–204. https://doi.org/10.1007/s10803-021-05419-5

Hendrix, J. A., Amon, A., Abbeduto, L., Agiovlasitis, S., Alsaied, T., Anderson, H. A., Bain, L. J., Baumer, N., Bhattacharyya, A., Bogunovic, D., Botteron, K. N., Capone, G., Chandan, P., Chase, I., Chicoine, B., Cieuta-Walti, C., DeRuisseau, L. R., Durand, S., Esbensen, A., et al. (2020). Opportunities, barriers, and recommendations in down syndrome research. *Translational Science of Rare Diseases, 5*(3–4), 99–129. https://doi.org/10.3233/trd-200090

Hershkovich, A., Lasc, D., Grove, L., Sullivan, D., & Yang, Y. (2023). Visuospatial perspective taking in people with Down syndrome. *Research in Developmental Disabilities, 139*, 1–13. https://doi.org/10.1016/j.ridd.2023.104565

Hessl, D., Nguyen, D. V., Green, C., Chavez, A., Tassone, F., Hagerman, R. J., Senturk, D., Schneider, A., Lightbody, A., Reiss, A. L., & Hall, S. (2009). A solution to limitations of cognitive testing in children with intellectual disabilities: The case of fragile X syndrome. *Journal of Neurodevelopmental Disorders, 1*, 33–45. https://doi.org/10.1007/s11689-008-9001-8

Hofmann, S. G., Sawyer, A. T., Witt, A. A., & Oh, D. (2010). The effect of mindfulness-based therapy on anxiety and depression: A meta-analytic review. *Journal of Consulting and Clinical Psychology, 78*(2), 169–183. https://doi.org/10.1037/a0018555

Horne, R. (2022, October 11). Five tips for translating research findings into smart public policy. *SpectrumNews*. https://doi.org/10.53053/BGFG9555

Hott, B. L., Brigham, F. J., & Peltier, C. (2021). *Research methods in special education*. SLACK Incorporated.

Hwang, Y. S., & Singh, N. N. (2016). Mindfulness. In N. N. Singh (Ed.), *Handbook of evidence-based practices in intellectual and developmental disabilities* (pp. 311–346). Springer.

Hyde, K. K., Novack, M. N., LaHaye, N., Parlett-Pelleriti, C., Anden, R., Dixon, D. R., & Linstead, E. (2019). Applications of supervised machine learning in autism spectrum disorder research: A review. *Review Journal of Autism and Developmental Disorders, 6*, 128–146. https://doi.org/10.1007/s40489-019-00158-x

IRS. (2023, May 5). *Retirement topics – Automatic enrollment*. https://www.irs.gov/retirement-plans/plan-participant-employee/retirement-topics-automatic-enrollment

Izquierdo-Gómez, R., Esteban-Cornejo, I., Cabanas-Sánchez, V., Marcos, A., Gómez-Martínez, S., Castro-Piñero, J., Veiga, Ó. L., & UP&DOWN Study Group. (2023). Are obesity and physical fitness associated with cardiovascular disease risk in adolescents with Down syndrome? The longitudinal UP&DOWN study. *Journal of Intellectual & Developmental Disability, 48*(3), 247–259. https://doi.org/10.3109/13668250.2023.2188876

Jarrold, C., & Brock, J. (2004). To match or not to match? Methodological issues in autism-related research. *Journal of Autism and Developmental Disorders, 34*(1), 81–86. https://doi.org/10.1023/b:jadd.0000018078.82542.ab

Karmiloff-Smith, A. (1997). Crucial differences between developmental cognitive neuroscience and adult neuropsychology. *Developmental Neuropsychology, 13*(4), 513–524. https://doi.org/10.1080/87565649709540693

Karmiloff-Smith, A., Brown, J. H., Grice, S., & Paterson, S. (2003). Dethroning the myth: Cognitive dissociations and innate modularity in williams syndrome. *Developmental Neuropsychology, 23*(1–2), 227–242. https://doi.org/10.1080/87565641.2003.9651893

Kerr, J., Sharry, J., & Wilson, C. (2023). Parents' experiences of raising adolescents with intellectual or developmental disabilities. *Journal of Intellectual & Developmental Disability, 48*(2), 206–214. https://doi.org/10.3109/13668250.2022.2057843

Kramer, J. M., Schwartz, A. E., Watkins, D., Peace, M., Luterman, S., Barnhart, B., Bouma-Sims, J., Riley, J., Shouse, J., Maharaj, R., Rosenberg, C. R., Harvey, K., Huereña, J., Schmid, K., & Alexander, J. S. (2019). Improving research and practice: Priorities for young adults with intellectual/developmental disabilities and mental health needs. *Journal of Mental Health Research in Intellectual Disabilities, 12*(3–4), 97–125. https://doi.org/10.1080/19315864.2019.1636910

Lanillos, P., Oliva, D., Philippsen, A., Yamashita, Y., Nagai, Y., & Cheng, G. (2020). A review on neural network models of schizophrenia and autism spectrum disorder. *Neural Networks, 122*, 338–363. https://doi.org/10.1016/j.neunet.2019.10.014

Lee, M. H., Park, C., Matthews, A. K., & Hsieh, K. (2017). Differences in physical health, and health behaviors between family caregivers of children with and without disabilities. *Disability and Health Journal, 10*(4), 565–570. https://doi.org/10.1016/j.dhjo.2017.03.007

Liu, M., Li, B., & Hu, D. (2021). Autism spectrum disorder studies using fMRI data and machine learning: A review. *Frontiers in Neuroscience, 15*, 338–363. https://doi.org/10.3389/fnins.2021.697870

Maes, B., Nijs, S., Vandesande, S., Van Keer, I., Arthur-Kelly, M., Dind, J., Goldbart, J., Petitpierre, G., & Van der Putten, A. (2021). Looking back, looking forward: Methodological challenges and future directions in research on persons with profound intellectual and multiple disabilities. *Journal of Applied Research in Intellectual Disabilities, 34*(1), 250–262. https://doi.org/10.1111/jar.12803

Martens, M. A., Wilson, S. J., & Reutens, D. C. (2008). Research review: Williams syndrome: A critical review of the cognitive, behavioral, and neuroanatomical phenotype. *Journal of Child Psychology and Psychiatry, and Allied Disciplines, 49*(6), 576–608. https://doi.org/10.1111/j.1469-7610.2008.01887.x

May, B. K., Britwum, K., & Nicol, G. E. (2022). Improving beverage choice in adults with developmental disabilities: Implementation of a token reinforcement system in a community residential setting. *Journal of Autism and Developmental Disorders, 52*(4), 1523–1535. https://doi.org/10.1007/s10803-021-05051-3

McCarron, M., McCausland, D., McGlinchey, E., Bowman, S., Foley, M., Haigh, M., Burke, E., & McCallion, P. (2022). Recruitment and retention in longitudinal studies of people with intellectual disability: A case study of the Intellectual Disability Supplement to the Irish Longitudinal Study on Ageing (IDS-TILDA). *Research in Developmental Disabilities, 124*, 1–10. https://doi.org/10.1016/j.ridd.2022.104197

McCullagh, E. A., Rotschafer, S. E., Auerbach, B. D., Klug, A., Kaczmarek, L. K., Cramer, K. S., Kulesza, R. J., Jr., Razak, K. A., Lovelace, J. W., Lu, Y., Koch, U., & Wang, Y. (2020). Mechanisms underlying auditory processing deficits in fragile x syndrome. *The FASEB Journal, 34*(3), 3501–3518. https://doi.org/10.1096/fj.201902435R

McDonald, K. E., & Kidney, C. A. (2012). What is right? Ethics in intellectual disabilities research. *Journal of Policy and Practice in Intellectual Disabilities, 9*(1), 27–39. https://doi.org/10.1111/j.1741-1130.2011.00319.x

McDonald, K. E., Kidney, C. A., & Patka, M. (2013). 'You need to let your voice be heard': research participants' views on research. *Journal of Intellectual Disability Research, 57*(3), 216–225. https://doi.org/10.1111/j.1365-2788.2011.01527.x

Merrill, E. C. (1990). Attentional resource allocation and mental retardation. *International Review of Research in Mental Retardation, 16*, 51–88. https://doi.org/10.1016/S0074-7750(08)60092-0

Mervis, C. B., & Klein-Tasman, B. P. (2004). Methodological issues in group-matching designs: α levels for control variable comparisons and measurement characteristics of control and target variables. *Journal of Autism and Developmental Disorders, 34*(1), 7–17. https://doi.org/10.1023/B:JADD.0000018069.69562.b8

Mervis, C. B., Robinson, B. F., Bertrand, J., Morris, C. A., Klein-Tasman, B. P., & Armstrong, S. C. (2000). The Williams syndrome cognitive profile. *Brain and Cognition, 44*(3), 604–628. https://doi.org/10.1006/brcg.2000.1232

Mervis, C. B., Klein-Tasman, B. P., & Mastin, M. E. (2001). Adaptive behavior of 4- through 8-year-old children with Williams syndrome. *American Journal of Mental Retardation, 106*(1), 82–93. https://doi.org/10.1352/0895-8017(2001)106<0082:ABOTYO>2.0.CO;2

Mietola, R., Miettinen, S., & Vehmas, S. (2017). Voiceless subjects? Research ethics and persons with profound intellectual disabilities. *International Journal of Social Research Methodology, 20*(3), 263–274. https://doi.org/10.1080/13645579.2017.1287872

Miezah, D., Porter, M., Rossi, A., Kazzi, C., Batchelor, J., & Reeve, J. (2021). Cognitive profile of young children with Williams syndrome. *Journal of Intellectual Disability Research, 65*(8), 784–794. https://doi.org/10.1111/jir.12860

Miller, L. E., Perkins, K. A., Dai, Y. G., & Fein, D. A. (2017). Comparison of parent report and direct assessment of child skills in toddlers. *Research in Autism Spectrum Disorders, 41-42*, 57–65. https://doi.org/10.1016/j.rasd.2017.08.002

Morgan, M. F., Cuskelly, M., & Moni, K. B. (2014). Unanticipated ethical issues in a participatory research project with individuals with intellectual disability. *Disability & Society, 29*(8), 1305–1318. https://doi.org/10.1080/09687599.2014.934440

Næss, K. A., Melby-Lervåg, M., Hulme, C., & Lyster, S. A. H. (2012). Reading skills in children with down syndrome: A meta-analytic review. *Research in Developmental Disabilities, 33*(2), 737–747. https://doi.org/10.1016/j.ridd.2011.09.019

National Commission for the Protection of Human Subjects of Biomedical and Behavioral Research. (1979). *The Belmont report: Ethical principles and guidelines for the protection of human subjects of research.* U.S. Department of Health and Human Services. https://www.hhs.gov/ohrp/regulations-and-policy/belmont-report/read-the-belmont-report/index.html

Nijs, S., Vlaskamp, C., & Maes, B. (2018). Promoting social scaffolding behaviors in staff members and peer-directed behaviors among persons with profound intellectual and multiple disabilities: An intervention study. *Journal of Policy and Practice in Intellectual Disabilities, 15*(2), 124–135. https://doi.org/10.1111/jppi.12251

Phillips, B. A., Loveall, S. J., Channell, M. M., & Conners, F. A. (2014). Matching variables for research involving youth with down syndrome: Leiter-R versus PPVT-4. *Research in Developmental Disabilities, 35*(2), 429–438. https://doi.org/10.1016/j.ridd.2013.11.016

Rinaldi, J. (2013). Reflexivity in research: Disability between the lines. *Disability Studies Quarterly, 33*(2), 10.18061/dsq.v33i2.3711.

Ringshaw, J. E., Hamilton, K., & Malcolm-Smith, S. (2022). Theory of mind and moral decision-making in the context of autism spectrum disorder. *Journal of Autism and Developmental Disorders, 52*(4), 1693–1711. https://doi.org/10.1007/s10803-021-05055-z

Rouder, J. N., Speckman, P. L., Sun, D., Morey, R. D., & Iverson, G. (2009). Bayesian t tests for accepting and rejecting the null hypothesis. *Psychonomic Bulletin & Review, 16*(2), 225–237. https://doi.org/10.3758/PBR.16.2.225

Scherer, N., Verhey, I., & Kuper, H. (2019). Depression and anxiety in parents of children with intellectual and developmental disabilities: A systematic review and meta-analysis. *PLoS One, 14*(7), e0219888. https://doi.org/10.1371/journal.pone.0219888

Shields, N., Mizzi, N., Buhlert-Smith, K., Strydom, A., Prendergast, L., & Hocking, D. R. (2022). A 12-week exercise programme has a positive effect on everyday executive function in young people with down syndrome: A pilot non-randomised controlled trial. *Journal of Intellectual Disability Research, 66*(12), 924–938. https://doi.org/10.1111/jir.12979

Shui, A. M., Katz, T., Malow, B. A., & Mazurek, M. O. (2018). Predicting sleep problems in children with autism spectrum disorders. *Research in Developmental Disabilities, 83*, 270–279. https://doi.org/10.1016/j.ridd.2018.10.002

Silverman, W. (2007). Down syndrome: Cognitive phenotype. *Mental Retardation and Developmental Disabilities Research Reviews, 13*(3), 228–236. https://doi.org/10.1002/mrdd.20156

Singh, N. N. (2020). Implementation science of mindfulness in intellectual and developmental disabilities. *American Journal on Intellectual and Developmental Disabilities, 125*(5), 345–348. https://doi.org/10.1352/1944-7558-125.5.345

Spooner, F., Root, J. R., Saunders, A. F., & Browder, D. M. (2019). An updated evidence-based practice review on teaching mathematics to students with moderate and severe developmental disabilities. *Remedial and Special Education, 40*(3), 150–165. https://doi.org/10.1177/0741932517751055

Stack, E., & McDonald, K. E. (2014). Nothing about us without us: Does action research in developmental disabilities research measure up? *Journal of Policy and Practice in Intellectual Disabilities, 11*(2), 83–91. https://doi.org/10.1111/jppi.12074

Strauss, M. E. (2001). Demonstrating specific cognitive deficits: A psychometric perspective. *Journal of Abnormal Psychology, 110*(1), 6–14. https://doi.org/10.1037/0021-843X.110.1.6

The National Science and Technology Council. (2016). *Social and behavioral sciences team, 2016* [Annual report]. Social and Behavioral Sciences Team. https://sbst.gov/download/2016%20SBST%20Annual%20Report.pdf

Thomas, M. S., Annaz, D., Ansari, D., Scerif, G., Jarrold, C., & Karmiloff-Smith, A. (2009). Using developmental trajectories to understand developmental disorders. *Journal of Speech, Language, and Hearing Research, 52*(2), 336–358. https://doi.org/10.1044/1092-4388(2009/07-0144)

U.S. Department of Health & Human Services. (2022). *About implementation science.* National Cancer Institute: Division of Cancer Control & Population Sciences. https://cancercontrol.cancer.gov/is/about

Uddin, L. Q. (2021). Brain mechanisms supporting flexible cognition and behavior in adolescents with autism spectrum disorder. *Biological Psychiatry, 89*(2), 172–183. https://doi.org/10.1016/j.biopsych.2020.05.010

Vacca, R. A., Bawari, S., Valenti, D., Tewari, D., Nabavi, S. F., Shirooie, S., Sah, A. N., Volpicella, M., Braidy, N., & Nabavi, S. M. (2019). Down syndrome: Neurobiological alterations and therapeutic targets. *Neuroscience and Biobehavioral Reviews, 98*, 234–255. https://doi.org/10.1016/j.neubiorev.2019.01.001

Van Deusen, K., Prince, M. A., Thurman, A. J., Esbensen, A. J., Patel, L. R., Abbeduto, L., Walsh, M. M., Daunhauer, L. A., Feigles, R. T., & Fidler, D. J. (2023). Evaluating an adapted reverse categorisation task to assess cognitive flexibility in young children with down syndrome. *Journal of Intellectual Disability Research, 67*(8), 734–745. https://doi.org/10.1111/jir.13040

van Dijk, W., Schatschneider, C., & Hart, S. A. (2021). Open science in education sciences. *Journal of Learning Disabilities, 54*(2), 139–152. https://doi.org/10.1177/0022219420945267

Vicente-Saez, R., & Martinez-Fuentes, C. (2018). Open Science now: A systematic literature review for an integrated definition. *Journal of Business Research, 88*, 428–436.

Westling, D. L., Trader, B. R., Smith, C. A., & Marshall, D. S. (2010). Use of restraints, seclusion, and aversive procedures on students with disabilities. *Research and Practice for Persons with Severe Disabilities, 35*(3–4), 116–127. https://doi.org/10.2511/rpsd.35.3-4.116

Will, E. A., Bishop, S. L., & Roberts, J. E. (2019). Developmental divergence: Motor trajectories in children with fragile x syndrome with and without co-occurring autism. *Journal of Neurodevelopmental Disorders, 11*(1), 1–10. https://doi.org/10.1186/s11689-019-9281-1

Wolff, B., Franco, V. R., Magiati, I., Pestell, C. F., & Glasson, E. J. (2023). Psychosocial and neurocognitive correlates of suicidal thoughts and behaviours amongst siblings of persons with and without neurodevelopmental conditions. *Research in Developmental Disabilities, 139*, 1–18. https://doi.org/10.1016/j.ridd.2023.104566

Worthington, J. (2018). *Restraint and Seclusion Guidance for Students with Disabilities*. State of New Jersey Department of Education (NJ DOE). https://www.nj.gov/education/specialed/home/docs/071018Restraint.pdf

Yang, Y., Conners, F. A., & Merrill, E. C. (2014). Visuo-spatial ability in individuals with down syndrome: Is it really a strength? *Research in Developmental Disabilities, 35*(7), 1473–1500. https://doi.org/10.1016/j.ridd.2014.04.002

Yang, Y., Himmelberger, Z. M., Robinson, T., Davis, M., Conners, F., & Merrill, E. (2021). Everyday memory in people with down syndrome. *Brain Sciences, 11*(5), 1–15. https://doi.org/10.3390/brainsci11050551

Index

M. G. Valdovinos (ed.), *Intellectual and Developmental Disabilities*, Contemporary Clinical Neuroscience,
https://doi.org/10.1007/978-3-031-66932-3

www.ingramcontent.com/pod-product-compliance
Lightning Source LLC
Chambersburg PA
CBHW062117170326
41400CB00003B/2521
9783031669316